Pathogenic *Escherichia coli*

Molecular and Cellular Microbiology

Edited by

Stefano Morabito

EU Reference Laboratory for *E. coli*
Istituto Superiore di Sanità
Rome
Italy

Caister Academic Press

Copyright © 2014

Caister Academic Press
Norfolk, UK

www.caister.com

British Library Cataloguing-in-Publication Data
A catalogue record for this book is available from the British Library

ISBN: 978-1-908230-37-9 (hardback)
ISBN: 978-1-908230-99-7 (ebook)

Cover design adapted from Figures 10.1 and 12.2.

Printed and bound in Great Britain

Contents

Contributors

Nadia Boisen
Department of Pediatrics
University of Virginia School of Medicine
Charlottesville, VA
USA

nb9f@eservices.virginia.edu

Eric Cox
Laboratory of Immunology
Faculty of Veterinary Medicine
Ghent University
Merelbeke
Belgium

Eric.Cox@ugent.be

Felipe Del Canto
Programa de Microbiología y Micología
Instituto de Ciencias Biomédicas
Facultad de Medicina
Universidad de Chile
Santiago
Chile

felipedelcanto@med.uchile.cl

Bert Devriendt
Laboratory of Immunology
Faculty of Veterinary Medicine
Ghent University
Merelbeke
Belgium

b.devrient@ugent.be

Ulrich Dobrindt
Institute of Hygiene
University of Münster
Münster
Germany

dobrindt@uni-muenster.de

John M. Fairbrother
GREMIP
Faculty of Veterinary Medicine
University of Montreal
Saint Hyacinthe, QC
Canada

john.morris.fairbrother@montreal.ca

John P. Geibel
Yale University School of Medicine
BML 265
New Haven, CT
USA

john.geibel@yale.edu

Bruno M. Goddeeris
Department of Biosystems
Faculty of Bioscience Engineering
KULeuven
Leuven
Belgium

bruno.goddeeris@biw.kuleuven.be

Julian A. Guttman
Department of Biological Sciences
Shrum Science Centre
Simon Fraser University
Burnaby, BC
Canada

jguttman@sfu.ca

Samuel Juillot
Spemann Graduate School of Biology and Medicine
Centre for Biological Signalling Studies
Institute of Biology II
Albert Ludwigs University Freiburg
Freiburg
Germany

samuel.juillot@sgbm.uni-freiburg.de

Helge Karch
Institute of Hygiene
University of Münster
Münster
Germany

hkarch@uni-muenster.de

Sascha Kopic
Yale School of Medicine
Yale University
New Haven, CT
USA

sascha.kopic@gmail.com

Karen A. Krogfelt
Department of Microbiology and Control
Statens Serum Institute
Copenhagen
Denmark

kak@ssi.dk

H.T. Law
Department of Biological Sciences
Shrum Science Centre
Simon Fraser University
Burnaby, BC
Canada

hlaw@sfu.ca

Shana R. Leopold
Institute of Hygiene
University of Münster
Münster
Germany

sleopold@uni-muenster.de

Adam J. Lewis
Division of Microbiology and Immunology
Pathology Department
University of Utah School of Medicine,
Salt Lake City, UT
USA

adam.lewis@path.utah.edu

Jacques G. Mainil
Bacteriology
Department of Infectious Diseases
Faculty of Veterinary Medicine
University of Liege
Liege
Belgium

jg.mainil@ulg.ac.be

Vesna Melkebeek
Laboratory of Immunology
Faculty of Veterinary Medicine
Ghent University
Merelbeke
Belgium

vesna.melkebeek@ugent.be

Alexander Mellmann
Institute of Hygiene
University of Münster
Münster
Germany

mellmann@uni-muenster.de

Valeria Michelacci
European Union Reference Laboratory for *Escherichia coli*
Istituto Superiore di Sanità
Dipartimento di Sanità Pubblica Veterinaria e Sicurezza Alimentare
Rome
Italy

valeria.michelacci@iss.it

Stefano Morabito
European Union Reference Laboratory for *Escherichia coli*
Istituto Superiore di Sanità
Dipartimento di Sanità Pubblica Veterinaria e Sicurezza Alimentare
Rome
Italy

stefano.morabito@iss.it

Matthew A. Mulvey
Division of Microbiology and Immunology
Pathology Department
University of Utah School of Medicine
Salt Lake City, UT
USA

mulvey@path.utah.edu

Maite Muniesa
Department of Microbiology
University of Barcelona
Barcelona
Spain

mmuniesa@ub.edu

James P. Nataro
Department of Pediatrics
University of Virginia School of Medicine
Charlottesville, VA
USA

JPN2R@hscmail.mcc.virginia.edu

Jan Oscarsson
Department of Odontology
Umeå University
Umeå
Sweden

jan.oscarsson@odont.umu.se

Elizabeth M. Ott
Division of Microbiology and Immunology
Pathology Department
University of Utah School of Medicine
Salt Lake City, UT
USA

betsy.ott@path.utah.edu

Adrienne W. Paton
Research Centre for Infectious Diseases
School of Molecular and Biomedical Science
University of Adelaide
Adelaide
Australia

adrienne.paton@adelaide.edu.au

James C. Paton
Research Centre for Infectious Diseases
School of Molecular and Biomedical Science
University of Adelaide
Adelaide
Australia

james.paton@adelaide.edu.au

Winfried Römer
Centre for Biological Signalling Studies
Institute of Biology II
Albert Ludwigs University Freiburg
Freiburg
Germany

winfried.roemer@bioss.uni-freiburg.de

Ahmad Saleh
Yale School of Medicine
Yale University
New Haven, CT
USA

ahmad.saleh@yale.edu

Flemming Scheutz
WHO Collaborating Centre for Reference and Research on *Escherichia* and *Klebsiella*
Department of Microbiology and Infection Control
Statens Serum Institute
Copenhagen
Denmark

fsc@ssi.dk

Herbert Schmidt
Institute of Food Science and Biotechnology
Department of Food Microbiology
University of Hohenheim
Stuttgart
Germany

herbert.schmidt@uni-hohenheim.de

Domonkos Sváb
Institute for Veterinary Medical Research
Centre for Agricultural Research
Hungarian Academy of Sciences
Budapest
Hungary

svab@vmri.hu

Alfredo G. Torres
Department of Microbiology and Immunology
Department of Pathology and Sealy Center for Vaccine
    Development
University of Texas Medical Branch
Galveston, TX
USA

altorres@utmb.edu

István Tóth
Institute for Veterinary Medical Research
Centre for Agricultural Research
Hungarian Academy of Sciences
Budapest
Hungary

tothi@vmri.hu

Rosangela Tozzoli
European Reference Laboratory for *E. coli*
Dipartimento di Sanità Pubblica Veterinaria e
    Sicurezza Alimentare
Istituto Superiore di Sanità
Rome
Italy

rosangela.tozzoli@iss.it

Bernt E. Uhlin
Department of Molecular Biology
Laboratory for Molecular Infection Medicine Sweden
    (MIMS)
Umeå University
Umeå
Sweden

bernt.eric.uhlin@mims.umu.se

Daisy Vanrompay
Department of Molecular Biotechnology
Faculty of Bioscience Engineering
Ghent University
Ghent
Belgium

daisy.vanrompay@ugent.be

Sun N. Wai
Department of Molecular Biology
Laboratory for Molecular Infection Medicine Sweden
    (MIMS)
Umeå University
Umeå
Sweden

sun.nyunt.wai@molbiol.umu.se

Hui Wang
Research Centre for Infectious Diseases
School of Molecular and Biomedical Science
University of Adelaide
Adelaide
Australia

hui.wang@adelaide.edu.au

Travis J. Wiles
Division of Microbiology and Immunology
Pathology Department
University of Utah School of Medicine
Salt Lake City, UT
USA

twiles@uoregon.edu

Weiping Zhang
Diagnostic Medicine/Pathobiology
Kansas State University College of Veterinary
    Medicine
Manhattan, KS
USA

wpzhang@vet.k-state.edu

Foreword

Our greatest responsibility is to be good ancestors.

(Jonas Salk 1914–1995)

Theodor Escherich described, at the end of the nineteenth century, a bacterium isolated from stool samples of healthy infants that he named *Bacterium coli commune*. This happened a decade after Robert Koch proved the association between microbes and disease. At that time *Bacterium coli commune*, later named *Escherichia coli* after the man who discovered it, was considered basically a commensal bacterial species and it took other 50 years for it to be associated with a human infection.

Today *E. coli* is known as one of the bacterial species with the widest adaptability to an amazing variety of niches either within organisms or outside in the environment. As a matter of fact it colonizes the gastrointestinal tract of humans and warm-blooded animals during the first phases of the life establishing mutual beneficial relationships with the host and playing an important role in maintaining the equilibrium between the numerous bacterial species constituting the gut microflora. At the same time, *E. coli* is one of the most diffuse bacterial species in the environment, being present in almost all the niches including water and soil.

Further to its *de jure* role as a natural member of intestinal flora, *E. coli* has also developed the ability to cause a number of diverse diseases to humans and animals. Two main groups of pathogenic *E. coli*, defined on the basis of which part of the organism they colonize, are recognized: the diarrhoeagenic *E. coli* (DEC), which alter the functions of the gastrointestinal tract, and the extraintestinal pathogenic *E. coli* (ExPEC), infecting body compartments other than the enteric environment. DEC cause gastroenteritis in humans and animals and are responsible for a range of diarrhoeal diseases and syndromes. They include several pathotypes marked by peculiar colonization mechanisms or specific toxin production: verocytotoxin-producing *E. coli* (VTEC), also termed Shiga toxin-producing *E. coli* (STEC), enteropathogenic *E. coli*, (EPEC), enteroinvasive *E. coli* (EIEC), enterotoxigenic *E. coli* (ETEC), enteroaggregative *E. coli* (EAEC), diffusely adherent *E. coli* (DAEC). On the other hand, ExPEC include the uropathogenic *E. coli* (UPEC), which cause 70–95% of community acquired urinary tract infections and about half of all cases occurring during hospitalization, the neonatal meningitis-associated *E. coli* as well as strains causing septicaemia.

The success of *E. coli* as a pathogen is witnessed by numbers: urinary tract infection is one of the most common bacterial infections in Europe and North America with an estimated 150 million cases occurring each year, while DEC, collectively, are the most common bacterial pathogens worldwide. Some of the DEC pathotypes represent a leading cause of illness and death in low-income countries, such as the EPEC, ETEC, EIEC and EAEC while others constitute a challenge for the public health systems in the industrialized areas of the globe. The latter case is well illustrated by the STEC O104:H4 that caused one of the largest and most severe outbreaks of food-borne infection ever occurred in the European Union. The episode, occurred in May–June 2011, counted about 4000 cases with more than 800 hospitalizations and the heavy toll of around 50 deaths.

The epidemiology of the infections caused by pathogenic *E. coli* is as diversified as the ecology of the different types is. Most DEC have a human reservoir and an inter-human circulation. The infections caused by these strains are thus more common in the developing area of the world where hygienic conditions are generally poor and re-infection cycles via the oral-faecal route account for the persistence of these pathogens in the environment. On the other hand, other pathogenic *E. coli*, such as STEC, are zoonotic. They have an animal reservoir and are mainly transmitted to humans by consumption of contaminated food. The latter pathogenic groups, on the contrary, are more diffused in the industrialized countries where intensive farming and large food distribution foster their circulation and the spreading of the infections.

The success of *E. coli* in colonizing such a wide range of hosts and environments is basically due to a noticeable ductility in exploiting the available resources. In turn, what makes its physiology so adaptable is an extraordinary genomic plasticity, the capability to exchange genetic material with other bacteria often belonging to other species, through horizontal gene transfer (HGT). This mechanism is mediated by the action of mobile genetic elements (MGE), DNA molecules capable of integrating themselves in the bacterial chromosome, such as bacteriophages, transposons and pathogenicity islands, or autonomously replicating in the bacterial cell cytoplasm such as the plasmids. These MGE are by definition capable of self-mobilization and can convey part of the genetic information from a bacterial cell to a new one by following cycles of integration–excision–integration. This process is the engine generating new bacterial variants and drives the evolution of the species.

Some 80 years have passed since *E. coli* was first recognized as a human pathogen and the emergence of new pathotypes seems not declining though. Studies on the virulence of this bacterial species has been conducted since their initial association with human disease but it was only in the last dozen years, with the advent of the genomic era and the high-throughput technologies, that the real mosaic structure of the genome of pathogenic *E. coli* became apparent.

In 2001 the journal *Nature* published the whole genome sequence of STEC O157 (Nature 2001, *409*:529–533). The sequence was completely annotated and unveiled how much of the STEC O157 genome, more than one million base pairs out of a total of 5.2 millions, had been acquired by HGT from different bacterial species. This was indeed the spur for the many comparative genomic studies that populated the scientific literature in the following twelve years and up to this day and age. The many efforts the scientific community has put into this endeavour basically shaped the bricks used to build our knowledge of the molecular mechanisms underlying the pathogenesis of the diseases caused by *E. coli* and led to the deployment, at least in the industrialized countries, of measures to counteract the diffusion of these infections with the aim of improving the citizens' health protection and reducing the economical losses caused by the *E. coli* diseases affecting animal husbandry.

Much has still to be done in the low-income countries, where the hygienic conditions must be raised to higher standards in order to limit the high morbidity and mortality plaguing especially infants and children. Such an improvement would also be an important step in limiting the recombination events occurring between different *E. coli* pathotypes and resulting in the emergence of mosaic pathogenic forms, often displaying a devastating virulence, as the STEC O104:H4 that caused the German outbreak in 2011. Such a perspective would trigger a mutual benefit for the people living in these regions and the global food market that increasingly looks at these countries as sources of exotic and low-cost food commodities or raw materials entering the food chains in the industrialized areas of the globe.

This publication portrays the molecular bases of *E. coli* pathogenetic mechanisms as well as the interactions with the host at the cellular level. It aims at being a wide-angle lens to scientists, students or simply to enquiring minds for exploring what is known on the subject and at being a stimulus and a starting point for those investigators who want to follow the pathway towards the discovery of this amazing organism.

Stefano Morabito

Diarrhoeagenic *Escherichia coli* Infections in Humans

Rosangela Tozzoli and Flemming Scheutz

Abstract

Escherichia coli is the predominant component of the mammals' gastrointestinal tract microbiota. It is usually a harmless commensal. Nevertheless, some strains have evolved the capability to cause disease in humans and are subdivided in groups depending on which part of the body they affect and of their particular pathogenic mechanism. Clinical syndromes resulting from infection with pathogenic *E. coli* strains include (i) urinary tract infection, (ii) sepsis/meningitis, and (iii) enteric/diarrhoeal disease. Strains causing the first two clinical syndromes are altogether termed extraintestinal pathogenic *E. coli* (ExPEC), whereas the strains inducing gastroenteric disease are known as diarrhoeagenic *E. coli* (DEC). DEC are subdivided in different pathotypes based on their adhesion/colonization mechanism and the toxins produced, which are enteropathogenic *E. coli* (EPEC), enterotoxigenic *E. coli* (ETEC), enteroinvasive *E. coli* (EIEC), enteroaggregative *E. coli* (EAggEC), diffusely adherent *E. coli* (DAEC) and Shiga toxin-producing *E. coli* (STEC), which are also referred to as verocytotoxin-producing *E. coli* (VTEC).

This chapter focuses on the human infections caused by DEC, providing general knowledge on the virulence traits and pathogenic mechanisms, the symptoms induced, known incidence and routes of transmission of DEC belonging to the different pathotypes.

Introduction

Escherichia coli colonize the gastrointestinal tract of mammals during the first phases of life, being commensal bacteria exerting an advantageous effect to the host. Bacteria belonging to this species establish mutual beneficial relationships with the host and play a pivotal role in maintaining the equilibrium between the numerous other bacterial species constituting the gut microflora. Distinct lineages have developed the capability to cause disease in specific hosts and these strains are classified into pathogenic groups or pathogroups based on the presence of a large variety of virulence factors, the kind of clinical disease they cause and the organs they affect.

There are two main groups of pathogenic *E. coli* – the extraintestinal pathogenic *E. coli* (ExPEC) and the diarrhoeagenic *E. coli* (DEC) – based on whether they cause disease outside or inside the gastrointestinal tract respectively (Nataro and Kaper, 1998). These two *E. coli* pathogroups are then subdivided into more homogeneous sub-populations identified by the mechanisms of pathogenesis or the presence of virulence factors and termed pathotypes.

ExPEC include uropathogenic *E. coli* (UPEC), which are the aetiological agents of urinary tract infections, and the neonatal meningitis-associated *E. coli* (NMEC). ExPEC also include strains causing bacteraemia and thereof systemic infections in both humans and animals.

DEC comprise strains able to cause gastroenteritis in humans and animals and are responsible for a range of diarrhoeal diseases and syndromes. The currently recognized DEC categories, based on their specific virulence factors and phenotypic traits, include enteropathogenic *E. coli* (EPEC), enterotoxigenic *E. coli* (ETEC), enteroinvasive *E. coli* (EIEC), enteroaggregative *E. coli* (EAggEC), diffusely adherent *E. coli* (DAEC), and Shiga toxin-producing *E. coli* (STEC), which are also referred to as verocytotoxin-producing *E. coli* (VTEC).

This chapter will focus on DEC and the different pathotypes will be described below with emphasis on their pathogenesis and epidemiology of human infections.

Enteropathogenic *E. coli* (EPEC)

Enteropathogenic *Escherichia coli* (EPEC) is a leading cause of infantile diarrhoea in developing countries. As a consequence of the improvement of general standards of hygiene, the frequency of the infections caused by these organisms has decreased in industrialized countries, during the years, but still they continue to be an important cause of diarrhoea (Nataro and Kaper, 1998).

In the past, enteropathogenic *E. coli* was the general denomination of bacteria belonging to this species and causing enteric disorders. Traditionally, EPEC were defined by their O group and as associated with infantile diarrhoea. These organisms were identified in routine diagnosis by determining their O groups, corresponding to the bacterial lipopolysaccharide surface antigen and later by the flagellar H antigen. For a long time the O:H serotype was the unique characteristic used to discriminate pathogenic from commensal strains. Even though the different DEC pathotypes were not known at that time, it soon became clear that serogroup/serotype determination resulted in an over-representation of EPEC and was not an efficacious method for classification (Ochoa and Contreras, 2011). Indeed, there was no agreement on which *E. coli* O-antigens should be considered as flags for the true EPEC strains. During a meeting organized by the World Health Organization in 1987 to address this matter a consensus on the definition of the following EPEC serogroups was obtained: O26, O55, O86, O111, O114, O119, O125, O126, O127, O128, O142 and O158 (WHO, 1987). However, in the following years, while studies on *E. coli* biology and genetics progressed, it was recognized that the O group determination was not sufficient to precisely identify EPEC and

that the same serotype could refer to more than one *E. coli* pathotype.

In 1995, a definition of EPEC was adopted at the Second International Symposium on EPEC, São Paulo, Brazil, which identified the most important characteristics of EPEC as its ability to cause attaching and effacing (A/E) lesion (see Chapter 10) and its inability to produce Shiga toxins (Chapter 5) (Kaper, 1996). It was also agreed that EPEC belonged to specific O:H serotypes but a consensus on which types was not agreed, acknowledging that the EPEC group is not homogeneous, displaying different genetic and ecological features. The EPEC pathotype was divided into two groups termed typical and atypical EPEC (tEPEC and aEPEC). The tEPEC were defined as EPEC harbouring a large virulence plasmid with a marker termed EPEC Adherence Factor (EAF), and which was later shown to harbour genes encoding the bundle-forming pili (BFP) and genes involved in the regulation of the adhesion to the enterocyte. tEPEC infections are characterized by the inter-human transmission *via* the oral-faecal route and are very common in developing countries. aEPEC lack the EAF plasmid and share additional virulence genes of another *E. coli* pathotype, the Shiga toxin-producing *E. coli* (STEC; see below). Such genes include those vehiculated by a large virulence plasmid carried by STEC and encoding, among other virulence factors, the enterohaemolysin (*E-hly*). Similarly to STEC, aEPEC may have an animal reservoir. They have been isolated from a wide range of diarrhoeic and healthy animal species (Aidar *et al.*, 2000; Aktan *et al.*, 2004; Blanco *et al.*, 2005; Ishii *et al.*, 2007; Krause *et al.*, 2005; Leomil *et al.*, 2005; Stephan *et al.*, 2004; Wani *et al.*, 2007; Yuste *et al.*, 2006) and may therefore be transmitted to humans by contaminated food of animal origin.

Today the O:H serotype determination continues to be a tool to identify EPEC, particularly when other means to detect specific genetic traits are not available. Efforts have been made to compile a list of *E. coli* serogroups associated with EPEC (Table 1.1). Although not exhaustive, it can be used as reference for the first assignment to this pathotype of strains isolated from cases of diarrhoea. Of the EPEC O:H serotypes shown in Table 1.1, only 13 of the 182 recognized O groups

Table 1.1 O:H serotypes regarded as classical and newly recognized EPEC O:H serotypes[a]

O group	H antigen[b]	Comments
O26	H–; H11	O26:H– and O26:H11 may also be STEC/VTEC (Bitzan *et al.*, 1991; Levine *et al.*, 1987; Scotland *et al.*, 1990)
O39	H–	New type (Hedberg *et al.*, 1997)
O55	H–; H6; H7; H34	O55:H7, H10 and H– may also be STEC/VTEC (Dorn *et al.*, 1989)
O86	H–; H 8; H34	O86:H– may also be EAggEC (Albert *et al.*, 1993; Schmidt *et al.*, 1995; Smith *et al.*, 1997;Tsukamoto and Takeda, 1993) H8 is a new type
O88	H–; H25	New type (Gomes *et al.*, 1989a; Pedroso *et al.*, 1993; Tsukamoto *et al.*, 1992)
O103	H2	New type
O111	H–; H2; H7; H8; H9; H25	O111:H– may also be STEC/VTEC (Allerberger *et al.*, 1996; Bitzan *et al.*, 1991; Cameron *et al.*, 1995; Caprioli *et al.*, 1994; Dorn *et al.*, 1989) or EAggEC (Chan *et al.*, 1994; Monteiro-Neto *et al.*, 1997; Morabito *et al.*, 1998; Schmidt *et al.*, 1995; Scotland *et al.* 1991, 1994; Tsukamoto and Takeda 1993)
O114	H–; H2	
O119	H–; H2; H6	
O125ac	H–; H6; H21	O125 may also be EAggEC (Do Valle *et al.*, 1997; Smith *et al.*, 1997; Tsukamoto and Takeda 1993)
O126	H–; H2; H12; H21; H27	O126:H12: (Barlow *et al.*, 1999)
O127	H–; H4; H6; H9; H21; H40	O127:H4: (Smith *et al.*, 1996)
O128ab	H–; H2; H7; H12	O128:H2 may also be STEC/VTEC (Beutin *et al.*, 1993; Kudo *et al.*, 1994)
O142	H–; H6; H34	
O145	H–; H45	New type (Gomes *et al.*, 1989a; Pedroso *et al.*, 1993)
O157	H–; H8; H16; H45	New types (Makino *et al.*, 1999; Scotland *et al.*, 1992)
O158	H–; H23	

[a]O18:H–; H7; H14, O26:H34 and O44:H34 have also been listed but only in Knutton *et al.*, 1991. O18 strains are not EPEC (Cimolai *et al.*, 1997; Knutton *et al.*, 1989; Orskov and Orskov, 1992). O44:[H18] is now considered to belong to the group of enteroaggregative *E. coli* (Smith *et al.*, 1994).
[b]Non-motile strains of *E. coli* are regarded as descendants of motile strains that have lost their motility by mutation(s). Their original H antigen was often deduced from comparison of biochemical reactions (Kauffman and Dupont, 1950; Staley *et al.*, 1969).

are represented and only a few H antigens, e.g. H2, H6, H7, H12 and H21 occur among several of these 13 O groups. Recently, other serotypes possessing EPEC-associated virulence markers have been added to the list. Some of these include O39:H–, O88:H25, O103:H2, O127:H40, O142:H34, O145:H45 and O157:H45.

EPEC-induced disease consists primarily of persistent diarrhoea, and many cases develop protracted or chronic EPEC diarrhoea (Donnenberg, 1995; Levine and Edelman, 1984). The infection can often be quite severe, and many clinical reports emphasize the severity of the disease (Bower *et al.*, 1989; Rothbaum *et al.*, 1982). In the mid-20th century EPEC caused outbreaks in the USA and the UK, where mortality rates of 25–50% were reported (Donnenberg, 1995; Levine and Edelman, 1984). While outbreaks are a rare event in developed countries, they have been reported in the past in developing countries with 30% mortality (Senerwa *et al.*, 1989). In addition to profuse watery diarrhoea, vomiting and low-grade fever are common symptoms of EPEC infection. Faecal leucocytes are seen only occasionally, but more sensitive tests for inflammatory diarrhoea such as an anti-lactoferrin latex bead agglutination test are frequently positive with EPEC infection (Miller *et al.*, 1994). Proximal small intestinal mucosal biopsy specimens often, but not always, show intimately adherent bacteria and the classic A/E histopathology (Fagundes-Neto, 1996; Rothbaum *et al.*, 1982; Sherman *et al.*, 1989). The presence of the A/E lesion is associated with disarrangement of the digestive-absorptive enzyme system, leading to malabsorption of nutrients (Fagundes-Neto 1996; Hill *et al.*, 1991; Rothbaum *et al.*, 1982). Multiple steps are involved in producing the characteristic A/E lesion. A proposed three-stage model of EPEC pathogenesis consists of (i) localized adherence, (ii) signal transduction, and (iii) intimate adherence (Donnenberg and Kaper, 1992). The first stage is characterized by initial interaction of bacteria with the enterocyte layer, and the initial attachment is mediated by the bundle-forming pilus, encoded on the EAF plasmid. In the second stage, *eae* and other genes are activated, causing dissolution of the normal microvillar of the intestinal epithelial cells. In the third stage, the bacterium binds closely to the epithelial membrane via the *eae*-encoded protein intimin. The disruption of the cytoskeleton producing the microvilli effacement, its rearrangement as a consequence of the phosphorylation of cellular proteins mediated by other bacterial gene products, lead to the A/E histopathology (see Chapter 10).

The public health significance of EPEC in developing countries has declined in published literature of the last decades (Okeke, 2009; Ochoa *et al.*, 2008), but could be due to underdiagnosis in the primary diagnostic clinics, which may have abolished the traditional slide agglutination techniques and not replaced them with modern molecular diagnostic detection methods such as PCR for the *eae* gene. This was exemplified in a Danish study where classical EPEC was only isolated from children below 2 years of age and was the second most common bacterial pathogen in this age group, indicating that EPEC continues to be an important cause of sporadic diarrhoeal illness in infants (Olesen, 2005). The decline could also be the effect of active interventions, e.g. as a consequence of the UNICEF/WHO 0–6 month exclusive breastfeeding campaign (Unicef, 1992, Gendrel *et al.*, 1984; Cravioto, 1991), or it could be the effect of the overestimation caused by earlier studies that based diagnosis on O– or O:H-typing compared to more recent ones in which EPEC identification was based on molecular methods.

Despite these uncertainties EPEC remain an important cause of diarrhoeal disease (Okeke, 2009). EPEC infections are most common, and most serious, in children under one year of age, and continue to be associated with disease in this age group in developing countries (Mandomando *et al.*, 2007; Creek *et al.*, 2009). Additionally, EPEC has emerged as an important cause of diarrhoea in HIV-infected individuals.

A systematic review of paediatric diarrhoea aetiology was made using 266 studies published between 1990 and 2002. EPEC were still identified among the most important pathogens, with a median prevalence of 8.8% [inter-quartile range (IQR) 6.6–13.2] in the community setting, 9.1% (IQR 4.5–19.4) in the outpatient setting and 15.6% (IQR 8.3–27.5) in the inpatient setting (Lanata *et al.*, 2002). In the inpatient setting EPEC was the second most common cause of diarrhoea after rotavirus (25.4%), although showing important regional and temporal variations. In a recent study of hospitalized diarrhoeal patients in India, EPEC was responsible for 3.2% of 648 diarrhoea samples in children younger than 5 year of age (Nair *et al.*, 2010).

In a study aiming at defining the diarrhoeagenic *E. coli* prevalence in Peru, 8000 *E. coli* strains previously isolated from eight different studies in children mainly younger than 36 months of age have been subjected to molecular analysis for the determination of the pathotype (Ochoa *et al.*, 2011). Overall, the average EPEC prevalence in diarrhoeal samples ($n = 4243$) was 8.5% (95% CI 7.6–9.3%) and increased with age. EPEC was found in 3% of samples from children less than 6 months of age, in 11% of children in the age range 6–12 months, and in 16% of children belonging to the 13–24 months age range. It has to be noted, that in this setting exclusive breastfeeding is very prevalent and therefore it may explain the lower prevalence in children less than three months. Small infants may indeed be protected from both asymptomatic and symptomatic EPEC infection when being exclusively breast-fed rather than being supplemented or non-breast-fed (Quiroga *et al.*, 2000).

Interestingly, among control samples collected from asymptomatic children from the above-mentioned Peruvian study sites ($n = 3760$), EPEC was the most prevalent pathotype with an average prevalence of 10.9% (95% CI 9.4–11.4%). This is quite paradigmatic of the circulation of enteric pathogens in the low-income countries. Many studies have found a significant association of EPEC with infant diarrhoea compared to control samples (Alikani *et al.*, 2006; Scaletsky *et al.*, 2002; Araujo *et al.*, 2007); however, other studies have described EPEC carriage with similar frequencies among diarrhoeal and control samples (Ochoa and Contreras, 2011, Knutton *et al.*, 2001; Afset *et al.*, 2003). In the Danish study this was also found for other *E. coli* causing the attaching and effacing lesion but not belonging to the classical EPEC serotypes (Olesen *et al.*, 2005). These discrepancies could be explained by the fact that there is no clear definition of EPEC types that are pathogenic. Healthy carriage of enteric pathogens is indeed very common in developing and developed countries, where the choice between colonization and illness may reside in the interplay of multiple factors including host susceptibility, related to child's age, breastfeeding, nutritional and immunological status, presence of specific bacterial virulence factors and environmental dynamics linked to poor hygiene and high faecal contamination.

Enterotoxigenic *E. coli* (ETEC)

ETEC causes around 840 million gastrointestinal infections (Wenneras and Erling, 2004) and about 380,000 deaths worldwide each year (Anon, 1990). ETEC infections are a major cause of diarrhoea among infants in developing countries where these strains are the most commonly isolated enteric pathogenic bacteria in children below 5 years of age. ETEC are also one of the leading causes of travellers' diarrhoea that affects individuals from industrialized countries travelling to these regions causing an estimated 10 million cases worldwide each year (Black, 1990; Northey *et al.*, 2007). Some authors have proposed that individuals with the genetic predisposition to produce high levels of interleukin 10 (IL-10) are more likely to experience symptomatic ETEC travellers' diarrhoea (Flores *et al.*, 2008).

ETEC colonizes the gastro intestinal tract of the host by attaching to specific receptors on the enterocytes in the intestinal lumen by the action of hair-like fimbriae. More than 20 types of fimbrial antigens, called *E. coli* surface antigens (CSs) or colonization factor antigens (CFAs) have been described, which define strain-specific antigenicity (Isidean *et al.*, 2011). CFAs and their mechanism of action are described in detail in Chapter 11.

A typical feature of ETEC is the ability to express one or more heat-stable (ST) or heat-labile (LT) enterotoxins (Chapter 8) encoded by genes vehiculated by plasmids. One of these toxins, LTI, resembles the toxin produced by *Vibrio cholerae* strains. Another heat-labile toxin (LTII) is chromosomally encoded and is similar to LTI both in mode of action and in structure. The plasmid-encoded heat-stable enterotoxins belong to two groups, STa/STI and STb/STII (Isidean *et al.*, 2011).

The infectious dose of ETEC is estimated to range from 10^6 to 10^{10} organisms (Levine *et al.*, 1979). However, vulnerable populations such as children and the elderly may be susceptible to infection at lower doses.

The high infectious dose of ETEC makes direct person-to-person spread rare. Instead, poor food-handling practices and infected food-handlers likely contribute to ETEC outbreaks, which are primarily transmitted through contaminated food, mainly raw vegetables and soft cheeses, and water. Poor drinking water quality, lack of a sewage system and feeding of supplementary foods are risk factors for ETEC infection, as described in several studies (Brussow *et al.*, 1992; Quadri *et al.*, 2005). The epidemiology of ETEC seems to be related to poor socioeconomic conditions and in fact, countries with the highest incidence of ETEC infection tend to have rapid population growth, semi-urban areas with poor infrastructure combined with rapid urbanization, poor water sanitization facilities causing the drinking water to become sewage-contaminated. Heavy rainfall and flooding related to climate change can also lead to contamination by ETEC of drinking water supplies or crops not intended for heat treatment.

Disease induced by ETEC follows ingestion of contaminated food or water and is characterized by profuse watery diarrhoea lasting for several days with little or no fever, a condition that often leads to dehydration, sometimes life-threatening, and malnutrition in young children. Extraintestinal complications, such as bacteraemia, do not generally occurr during ETEC infections.

Either LT or ST, both of which stimulate chloride secretion and inhibit chloride absorption in small intestine epithelial cells, can cause secretory watery diarrhoea.

ETEC ingestion is followed by an incubation period of between 10 h and 3 days, while the illnesses typically last for 3–5 days, although either shorter or longer duration have been described (Dalton *et al.*, 1999).

ETEC colonizes both animal and human intestines. However, there are differences in the range of the virulence factors, such as CFs and LT and ST subtypes (Table 1.2), indicating species specificity (Porat *et al.*, 1998; Steinsland *et al.*, 2002). Anyway, in some cases, animal strains may cause illness in humans (Nishikawa *et al.*, 1998).

ETEC often belong to a limited number of O: K: H serotypes and O groups and can possess LT and ST, which can be found alone or together. Table 1.3 reports the most common ETEC serotypes isolated from human disease.

Enteroinvasive *E. coli* (EIEC)

EIEC was first described in 1944 as a bacterium called 'paracolon bacillus', but later on, a prototype EIEC was identified as an *E. coli* strain belonging to serogroup O124. In the 1950s, the capability of these bacteria to invade the host cells was identified by carrying out the Serény-test, an assay based on the ability of invasive strains to cause keratoconjuntivitis in guinea pig. By using this approach, a group of invasive *E. coli* was identified and initially classified under the species *Shigella* as *Shigella manolovi*, *S. sofia*, *Shigella* strain 13, and *S. metadysenteriae*. Later on these strains were placed in a new *E. coli* subgroup, EIEC, and named *E. coli* O164 (Bando *et al.*, 1998; Monolov, 1959; Rowe *et al.*, 1977).

EIEC and *Shigella* spp. share remarkably similar phenotypic and genetic traits and many of these similarities may be attributed to the invasive lifestyle. As a matter of fact both these organisms spend much of their lifetime within eukaryotic cells and therefore have evolved to use the nutrients coming from the cellular environment, an ecosystem much more homogeneous and not subjected to the variations present in the niches characterizing the external environment (Lan and Reeves, 2002).

The majority of EIEC are non-motile, are not able to decarboxylate lysine and approximately 70% of EIEC strains do not ferment lactose. These features are shared with most of the strains belonging to the genus *Shigella* although some strains of *S. sonnei* are able to slowly ferment lactose, but are considered atypical characteristics when compared to those present in the majority of the other *E. coli* strains. Moreover, they seem to belong to a very limited number of serotypes, as reported in Table 1.4.

The clinical symptoms caused by EIEC infection are almost identical to those occurring during shigellosis, being characterized by diarrhoea, abdominal cramps, vomiting, fever, chills,

Table 1.2 Variants of enterotoxins found in ETEC strains. Suffixes commonly used are: h for human variant and p for porcine variant

LT: Heat-labile enterotoxins

LT-I: *eltI*	LTp (LTp-I): *eltIp*	LTh (LTh-I): *eltIh*	Associated with disease in both humans and animals
LT-II: *eltII*	LT-IIa: *eltIIa*	LT-IIb: *eltIIb*	No specific association with disease. Rare in human isolates

ST: Heat-stable enterotoxins

STa (STI) *estA*	STp (STIa) *estAp*	STh (STIb) *estAh*	Produced by ETEC and several other Gram-negative bacteria. In ETEC, ST is often found together with the genes (astA) for EAST1.
STb (STII): *estB*			Induces histological damage in the intestinal epithelium. Most often found in porcine ETEC

Table 1.3 O:(K:)H serotypes of human ETEC

O group	(K):H antigen	
O6	H–	K15:H16
O7	H–	H18
O8	K47:H–	K25:H9; K40:H9; H10; K87:H19
O9	H–	K9; K84:H2
O11		H27
O15		H11; H15; H45
O17		K23:H45; H18
O20	H–	H30
O21		H21
O25	H–	K7:H42; H16
O27		H7; H20; H27
O29		H?
O48		H26
O55		H7
O56	H–	
O63		H12; H30
O64	H–	
O65		H12
O71		H36
O73		H45
O77		H45
O78	H–	K2; H11; H12
O85		H7
O86		H2
O88		H25
O105		Hru
O114	H–	H21
O115	H–	H21; H40; H51
O119		H6
O126	H–	H9; H12
O128ac		H7; H12; H19; H21
O133		H16
O138		K81
O139		H28
O141	H–	H4
O147		H?
O148		H28
O149		H4; H10; H19
O153		H10
O159	H–	H4; H5; H12; H20; H21; H27; H34; H37
O166		H27
O167		H5
O?		H2; H10; H28; K39:H32

Table 1.4 O:H serotypes of EIEC and the O antigen identity to *Shigella* O antigens

O group	H antigen	*Shigella* O antigen
O28ac	H–	*S. boydii* 13
O29	H–	
O112ac	H–	*S. dysenteriae* 2 O112ab ~ *S. boydii* 15
O115	H–	
O121	H–	*S. dysenteriae* 7
O124	H–; H7; H30; H32	*S. dysenteriae* 3
O135	H–	
O136	H–	
O143	H–	*S. boydii* 8
O144	H–; H25	*S. dysenteriae* 10
O152	H–	*S. dysenteriae* 12
O159	H2	
O164	H–	
O167	H–; H4; H5	*S. boydii* 3
O173	H–	

and a generalized malaise. Like *Shigella* they are capable of invading and multiplying in the distal large bowel host cells in the human gut. Dysentery caused by this organism is generally self-limiting with no known complications. Common symptoms include watery diarrhoea that may precede the production of stools containing blood and mucus. Ulceration of the bowel can also occur in severe cases. The symptoms appear 12–72 h after the ingestion of the pathogen depending on the dose assumed.

The ability of EIEC to invade and destroy colonic tissue is associated with the presence of a high molecular weight (140 MDa) plasmid (pINV) which is also present in *Shigella* and harbours genes encoding virulence factors such as the invasion antigens named as IpaA to IpaH (Lan *et al.*, 2004). The large plasmid also harbours the *mxi-spa* genes encoding a type three secretion system, which allows the injection of the bacterial proteins necessary for the pathogenesis (Marteyn *et al.*, 2012) and the *icsA* gene coding for the protein essential for the migration to the adjacent cells.

EIEC are human pathogens and infected humans appear to be the primary source of infection as there are no known animal reservoirs. Infection occurs via the faecal–oral route by person-to-person transmission. Sometimes, food and water may be implicated as vehicles of infection, but they are generally contaminated by a human source.

Enteroaggregative *E. coli* (EAggEC)

At the end of the 1980s, the studies on the interactions between *E. coli* strains and cultured Hep-2 cells monolayers introduced the characterization of enteropathogenic *E. coli* on the basis of their pattern of adhesion. One group of *E. coli* strains produced, in this system, a peculiar diffuse adhesion pattern resembling the position of the bricks in a wall and was called 'aggregative adherence' (AA). These strains were thus identified as entero-aggregative *E. coli* (Nataro *et al.*, 1987). Despite their late identification, EAggEC are supposed to be agents of diarrhoeal disease at least since 1920 based on the retrospective analysis of the symptoms typically associated to the infections.

The clinical features of EAggEC infections include watery, mucoid, secretory diarrhoea with low-grade fever and little or no vomiting (Bhan Raj *et al.*, 1989; Paul *et al.*, 1994), with blood in the stools of about 30% of patients (Cravioto *et al.*, 1991a). The duration of EAggEC diarrhoea is its most striking feature, being longer than that associated with any other enteric pathogen lasting up to 17 days (Bahn Raj *et al.*, 1989). The incubation period of EAggEC diarrhoeal illness ranges from 8 to 18 h. Although observations exist suggesting that EAggEC infection may be accompanied by mucosal inflammation, any clinical evidence is lacking in most patients. Pathogenesis of EAggEC infections includes the colonization of the gut by a strong adherence where the bacteria are embedded in a mucus-containing biofilm, whose formation may contribute to its ability to cause long-lasting colonization.

EAggEC strains produce a number of virulence factors that are involved in disease. These include the plasmid-encoded aggregative adhesion fimbriae (AAF/I to AAF/V), the main determinant intervening in the AA pattern of adhesion (also termed 'stacked-brick'), whose coding genes are controlled by the transcriptional regulator AggR, a key factor in EAggEC pathogenesis.

About half of EAggEC strains elaborate toxins, comprising the enteroaggregative heat-stable toxin-1 (EAST-1), the plasmid encoded Pet toxin and the *Shigella* enterotoxin-1 (ShET1) (Nataro and Kaper, 1998).

During the initial years after the discovery of EAggEC there were a lot of doubts about the pathogenicity of this category of diarrhoeagenic *E. coli* (DEC) (Echeverria *et al.*, 1992; Gomes *et al.*, 1989b). However, Nataro *et al.* (1995) showed that a reference strain of EAggEC could cause diarrhoea in a volunteer study. EAggEC are now considered a leading cause of gastroenteritis in developing countries, but EAggEC infections are reported also in industrialized countries as causing sporadic cases and outbreaks, sometimes involving a large number of cases.

A growing number of studies have supported the association of EAggEC with diarrhoea in developing countries, most prominently in association with persistent diarrhoea (Bhan *et al.*, 1989a–c; Fang *et al.*, 1995; Lima *et al.*, 1992; Bhatnagar *et al.*, 1993; Bouzari *et al.*, 1994; Gonzalez *et al.*, 1997; Nataro *et al.*, 1987). Many of the epidemiological surveys that identified EAggEC as a diarrhoeal pathogen were done in developing countries. However, EAggEC has been found to be associated with diarrhoea in developed countries as well. In a Scandinavian case–control study the prevalence and the association of EAggEC with diarrhoea was greater than for EPEC (Bhatnagar *et al.*, 1993). Another study conducted in east London showed that EAggEC could be recovered from children with acute and persistent diarrhoea (Chan *et al.*, 1994). A clear association of EAggEC with diarrhoea in children in Germany was shown by Huppertz *et al.* (1997), who recovered EAggEC from 16 of 798 children with diarrhoea but none from 580 healthy controls. Other European studies in children (Knutton *et al.*, 2001; Presterl *et al.*, 1999) also indicate that EAggEC may be a leading cause of diarrhoeal disease in developed as well as developing countries.

A large outbreak of EAggEC diarrhoea occurred in Gifu Prefecture, Japan, in 1993, when 2697 children in 16 schools became ill after consuming suspected contaminated school lunches. A single EAggEC strain was implicated, but the organism was not found in any of the foods served

Table 1.5 O:H serotypes of enteroaggregative *E. coli* (EAggEC)

O group	H antigen	References listing these serotypes as EAggEC
O3	H2	Albert *et al*. (1993)
O15	H18	Albert *et al*. (1993), Scotland *et al*. (1994), Tsukamoto and Takeda (1993), Vial *et al*. (1988)
O44	H–, H18	Schmidt *et al*. (1995), Scotland *et al*. (1991 (1996), Smith *et al*. (1994), Tsukamoto and Takeda (1993)
O86	H–	Albert *et al*. (1993), Schmidt *et al*. (1995), Smith *et al*. (1997), Tsukamoto and Takeda (1993)
O111	H12, H21	Chan *et al*. (1994), Monteiro-Neto *et al*. (1997), Morabito *et al*. (1998), Schmidt *et al*. (1995), Scotland *et al*. (1991, 1994, 1996), Tsukamoto and Takeda (1993)
O125	H9, H21	doValle *et al*. (1997), Smith *et al*. (1997), Tsukamoto and Takeda (1993)

in the implicated lunch (Itoh *et al*., 1997). Four outbreaks of EAggEC-related diarrhoea occurred in the UK in 1994 (Smith *et al*., 1997). In two of the outbreaks a single EAggEC strain was implicated. Also, in this case no single vehicles could be identified, although each of these outbreaks was epidemiologically linked to the consumption of a meal.

An outbreak episode possibly associated to contaminated cheese occurred in Italy in 2008. An EAggEC strain of serotype O92:H33 was isolated from six participants to a banquet as well as from one member of the restaurant's staff. A retrospective cohort study indicated a pecorino cheese made with unpasteurized sheep milk as the possible source. However, although samples of the suspected cheese had *E. coli* counts higher than 10^6 CFU/g, the outbreak strain could not be isolated (Scavia *et al*., 2008).

Other studies have shown that this pathogen is an important cause of travellers' diarrhoea (Adachi *et al*., 2001; Gascón *et al*., 1998; Schultsz *et al*., 2000). However, documented reports are less common because EAggEC is not sought in many studies. Travellers to all developing countries are at risk and it has been speculated that EAggEC infections may account for the over 25% of cases for which no pathogen is recovered (Adachi *et al*., 2001).

EAggEC infections are usually self-limiting and responsive to oral rehydration therapy (Huang *et al*., 2004). Antimicrobial therapy for travellers' diarrhoea and paediatric diarrhoea should be based on an individual basis, and remains largely an empirical treatment (DuPont and Ericsson, 1993). Antimicrobial susceptibility patterns of EAggEC strains vary by geographic region. Some

studies have reported EAggEC to have moderate-to high-level resistance to ampicillin, tetracycline, trimethoprim, sulfamethoxazole and chloramphenicol (Sobieszczanska *et al*., 2003). Table 1.5 reports the most common serotypes of EaggEC isolated from human disease.

Diffusely adherent *E. coli* (DAEC)

Diffusely adherent *E. coli* (DAEC) are a genetically non-homogeneous group of DEC characterized, as suggested by their name, by a diffuse pattern of adhesion to epithelial cells. A subset of DAEC strains produces Daa adhesins from the Afa/Dr family. Other diffuse adhesins include fimbrial and non-fimbrial adhesins AIDA-1, Sfa and P-pili (Le Bouguenec and Servin, 2006), which are also detected in uropathogenic *E. coli* (UPEC) as well as in commensals *E. coli* strains. Some DAEC strains possess the *aap* gene coding for the dispersin, which is commonly found in EAggEC (Jafari *et al*., 2012). The role of DAEC in diarrhoeal disease, if any, is still unclear along with their classification and epidemiology. Nonetheless, *daaD*-positive DAEC were recently associated with disease in adults in Ghana (Opintan *et al*., 2010).

Shiga toxin-producing *E. coli* (STEC) or verocytotoxin-producing *E. coli* (VTEC)

Among diarrhoeagenic *E. coli*, VTEC are the most virulent to date. Their main feature is the capability to elaborate potent cytotoxins termed verocytotoxins (VTs) (Chapter 5), which cause cell death by blocking the protein synthesis. VTs are also termed

Shiga toxins (Stxs) because of their structural similarity to the Shiga-toxin produced by *Shigella dysenteriae* type I (Melton Chelsea and O'Brien, 1998); therefore, *E. coli* strains that produce them are also known as Shiga toxin-producing *E. coli* (STEC). VTs family comprises two main antigenically distinct types, VT1 and VT2 (Scotland *et al.*, 1985) and numerous subtypes and variants based on differences in the DNA sequence of the coding genes have been described so far. In particular, three subtypes of VT1 (VT1a, VT1c and VT1d) and seven subtypes of VT2 (VT2a, VT2b, VT2c, VT2d, VT2e, VT2f and VT2g) have been identified to date (Persson *et al.*, 2007; Scheutz *et al.*, 2012). VTEC can produce either VT1 or VT2 alone or both in different combination type/subtype. The VT-coding genes (*vtx*) are vehiculated by temperate lambdoid bacteriophages integrated in the bacterial chromosome. The *vtx*-converting phages can be induced to excision by treatments posing a risk for VTEC strain survival, such as UV light or the antimicrobial mytomicin, and are able to form new phage particles capable to infect an indicator *E. coli* strain.

VTEC are zoonotic pathogens, and ruminants, particularly cattle, are considered their natural reservoirs. During the past thirty years, knowledge about the routes of transmission and the origin of human infections has improved and it seems evident that VTEC may be transmitted from animal reservoirs to humans not only via the ingestion of contaminated foods, but also by contact with VTEC-positive animals or the environment. Moreover, new routes of transmission including water, either drinking or recreational, and other sources such as person-to-person contacts are increasingly reported (Caprioli *et al.*, 2005).

In order to characterize the routes of transmission of VTEC O157, a review of 90 outbreaks confirmed microbiologically occurred in the UK, Ireland, Denmark, Norway, Finland, USA, Canada, and Japan, between 1982, and 2006, has been recently conducted (Pennington, 2010). The study showed that in more than 50% of the outbreaks the source of infection was food. The food vehicles included dairy products, ground beef hamburgers, steak tenderized by injection, steak tartare, kebabs, ready-to-eat cold meats including poultry, pork, and beef products, salami and other fermented meat products, venison jerky, cheese, milk, butter, yoghurt, ice cream, apple juice, grapes, coleslaw, lettuce, spinach, radishes, alfalfa sprouts, and melons. More food items are continuously added. As an example, raw cookie dough was associated with a multistate outbreak in the USA in 2009, with 72 cases of infections and ten HUS (CDC, 2009). Though studies on non-O157 VTEC have started in a second time and less data are available, a picture can be drawn concerning the vehicles of infections, similar to that for VTEC O157.

VTEC are a pathotype of *E. coli* capable to cause a spectrum of diseases upon infection. They constitute a heterogeneous group including strains consistently associated to the most severe forms of infections, considered as highly pathogenic strains and defined as enterohaemorrhagic *E. coli*, but strains isolated from mild disease or possessing features never observed in isolates from diseased subjects are also found and therefore assumed to be non-pathogenic. The disease spectrum covers symptoms from mild diarrhoea to more severe forms such as haemorrhagic colitis (HC) and the life-threatening haemolytic–uraemic syndrome (HUS). The latter represents the most severe form of VTEC infection and usually occurs in children, the elders and immune-compromised patients. HUS is clinically characterized by low platelet count, acute renal failure and non-immune haemolytic anaemia (Banatvala *et al.*, 2001). HUS can be severe with the majority of patients requiring red blood cell transfusions. Other complications include renal failure needing dialysis due to the action of the VTs on the endothelium of glomerula and in a number of cases, neurological impairment due to ischaemic lesions caused by the action of the VT on the endothelium of the brain vessels may also occur. Despite improvements in intensive care and dialysis facilities in developed countries, there is still a 3–5% case-fatality rate for patients affected with HUS (Karch *et al.*, 2005).

Administration of antibiotics for the treatment of VTEC infections is not recommended, since their use seems to favour the progression towards the most severe forms of the disease. In fact, antibiotics can induce the vtx-converting phage, provoking an enhancement in the production of VTs (Kimmirth *et al.*, 2000). Therefore, the management of infection is mainly supportive including rehydration and dialysis for the treatment of HUS.

The presence of the *vtx*-converting bacteriophages seems to be not sufficient for VTEC to cause the disease, at least when the most severe forms are considered: VTEC isolated from bloody diarrhoea or HUS generally produce additional virulence factors that are involved in colonization and, generally, in the pathogenic mechanism. Most of these strains cause a typical histopathological lesion to the enterocyte termed 'attaching and effacing' (A/E) (Chapter 10), a capability shared with EPEC strains, which is genetically governed by genes located in the LEE pathogenicity island.

The majority of VTEC isolated from severe human disease [also termed enterohaemorrhagic *E. coli* (EHEC)] represent a more homogeneous sub-group sharing common genomic and pathogenetic features but exceptions occur and can be of major public health significance as it was seen in the outbreak occurred in Germany in May–June 2011 and caused by an EHEC O104:H4. Moreover, data from HUS surveillance activities have shown that EHEC usually belong to a dozen of serogroups and in particular in 2010, strains belonging to serogroups O157, O26, O111, O145, O121 accounted for most of the HUS cases occurring in the European countries (*EFSA Journal* 2012). In that year, the total number of confirmed VTEC cases in the EU was 4,000, and the most commonly reported serotype was O157:H7 followed by O157: H– and O103:H2.

VTEC serotype O157:H7 is considered as the prototype of EHEC and was the first VTEC ever isolated in 1982 in the US during a large outbreak of haemorrhagic colitis (Riley *et al.*, 1983). Over the years, the range of VTEC causing disease in humans has been extended to more than 400 different serotypes (Table 1.6), raising questions on the opportunity to develop new schemes for their definition, which are not linked to the serotypes. As a matter of fact, the pathogenicity of VTEC completely resides on the presence of virulence genes conveyed by mobile genetic elements, which, by definition, could be virtually acquired by any *E. coli* strain regardless the serogroup it belongs to.

As an example, during summer 2011, Germany experienced an unprecedented outbreak of bloody diarrhoea and HUS caused by an unusual VTEC strain, belonging to serotype O104:H4. Travellers from several countries were affected and a smaller outbreak caused by the same strain occurred in France in July 2011. More than 3800 cases (including 53 deaths) have been reported in Germany, 855 of which evolved to HUS. Unexpectedly, most of the patients with HUS were adults (89%; median age 43 years), and women were overrepresented (68%). The epidemic strain harboured genes typically found in two types of pathogenic *E. coli*, the enteroaggregative *E. coli* (EAggEC) and VTEC, and it specifically produced the verocytotoxin 2a. This could explain the unexpectedly high level of virulence of such a strain negative for the attaching/effacing pathogenicity island. It is indeed conceivable that the enteroaggregative adherence phenotype could have allowed these *E. coli* O104 strains to colonize the intestinal mucosa of the affected patients as efficiently as the typical *eae*-positive STEC/VTEC strains (Scheutz *et al.*, 2011). The different mechanism of adhesion might also account of the reason why this strain was able to cause severe disease in adults rather than in children: adults and children might differ in their susceptibility to the adherence and/or colonization properties of this type of EAggEC strain. Moreover, the enhanced adherence of this strain to intestinal epithelial cells might facilitate systemic absorption of Shiga toxin and could explain the high frequency of progression to HUS (Bielaszewska *et al.*, 2011).

This peculiar combination of virulence characteristics had already been described before, during the investigations of a small outbreak occurred in France in late 1990s (Morabito *et al.*, 1998). In that outbreak the causative strain was an EAggEC O111, which had acquired a VT2-converting phage, which enabled such strain to cause the HUS severe disease.

The possibility for emerging VTEC clones poses a major challenge for public health, which shall be addressed by the implementation of diagnostic procedures able to identify not only classical VTEC, but accounting also for the newly emerging strains widening the epidemiological picture of VTEC infections.

Table 1.6 Serotypes of non-O157 STEC/VTEC isolated from humans (Modified from Scheutz and Strockbine, 2005; Mellmann *et al.*, 2008; Scheutz *et al.*, 2011; EFSA Panel on Biological Hazards (BIOHAZ) 2013)

O1:H–	O20:H19	O69:H11	O103:H18	O121:H–	O153:H25
O1:H1	O21:H5	O70:H8	O103:H21	O121:H2	O153:30
O1:H2	O21:H8	O70:H11	O103:H25	O121:H8	O153:H33
O1:H7	**O21:H?**	O71:H–	O103:HNT	O121:H11	O154:H–
O1:H20	O22:H–	O73:H18	**O104:H–**	O121:H19	**O154:H4**
O1:H42	O22:H1	O73:H34	O104:H2	O123:H2	O154:H19/20
O1:HNT	O22:H5	O74	**O104:H4**	O123:H19	O156:H–
O2:H–	O22:H8	O75:H–	O104:H7	O123:H49	O156:H4
O2:H1	O22:H16	O75:H1	O104:H16	O124:H–	O156:H7
O2:K1:H2	O22:H40	O75:H5	O104:H21	O125:H–	O156:H25
O2:H5	O23:H7	O75:H8	O105ac:H18	O125:H8	O156:H27
O2:H6	O23:H16	O76:H7	O105:H19	O125:H?	O156:HNT
O2:H7	O23:H21	O76:H19	O105:H20	O126:H–	O160:H?
O2:H11	O25:H–	O77:H–	O106	O126:H2	O161:H–
O2:H27	O25:K2:H2	O77:H4	O107:H27	O126:H8	O162:H4
O2:H29	O25:H14	O77:H7	O109:H2	O126:H11	O163:H–
O2:H44	O26:H–	O77:H18	O109:H16	O126:H20	O163:H19
O3:H10	O26:H2	O77:H41	O110:H–	O126:H21	O163:H25
O4:H–	O26:H8	O78:H–	O110:H19	O126:H27	O165:H–
O4:H5	O26:H11	O79:H7	O110:H28	O127	O165:H10
O4:H10	O26:H12	O79:H14	**O111:H–**	O128:H–	**O165:H19**
O4:H40	O26:H32	O79:H23	O111:H2	O128ab:H2	O165:H21
O5:H–	O26:H46	O80:H–	**O111:H7**	O128:H7	**O165:H25**
O5:H16	O27:H–	**O80:H2**	**O111:H8**	O128:H8	O166:H12
O6:H–	O27:H30	O81:H?	O111:H10	O128:H10	O166:H15
O6:H1	O28ab:H–	O82:H–	O111:H11	O128:H12	O166:H28
O6:H2	O28:H25	O82:H5	O111:H21	O128:H25	O168:H–
O6:H4	O28:H35	O82:H8	O111:H30	O128:H31	O169:H–
O6:H12	O30:H2	O83:H–	O111:H34	O128:H45	O171:H–
O6:H28	O30:H21	O83:H1	O111:H40	O129:H–	O171:H2
O6:H29	O30:H23	O84:H–	O111:H49	**O130:H11**	O172:H–
O6:H31	O37:H41	O84:H2	O111:H?	O131:H4	**O172:H?**
O6:H34	O38:H21	O84:H20	O112:H–	O132:H–	O173:H2
O6:H49	O38:H26	O85:H–	O112ab:H2	O133:H–	O174:H–
O7:H4	O39:H4	**O85:H10**	O112:H19	O133:H53	O174:H2
O7:H6	O39:H8	**O85:H23**	O112:H21	**O134:H25**	O174:H8
O7:H8	O39:H28	O86:H–	O113:H2	**O136:Hnt**	**O174:H21**
O8:H–	O40:H2	O86:H10	O113:H4	**O137:H6**	O175:H16
O8:H2	O40:H8	**O86:H27**	O113:H5	**O137:H41**	O176:H–
O8:H9	O41:H2	O86:H40	O113:H7	O138:H2	**O177:H–**
O8:H11	O41:H26	O87:H16	O113:H21	O141:H–	O177:H11
O8:H14	O44	O88:H–	O113:H32	O141:H2	O178:H7
O8:H19	O45:H–	O88:H25	O113:H53	O141:H8	O179:H8

O8:H21	O45:H2	O89:H–	O114:H4	O142	O181:H15
O8:H25	O45:H7	O90:H–	O114:H48	O143:H–	O181:H49
O9ab:H–	O46:H2	O91:H–	**O114:H?**	O144:H–	ONT:H–
O9:H7	O46:H31	O91:H4	**O115:H10**	**O145:H–**	**ONT:H2**
O9:H21	**O46:H38**	O91:H10	O115:H18	O145:H4	ONT:H8
O11:H–	O48:H21	O91:H14	O116:H–	O145:H8	ONT:H18
O11:H2	O49:H–	O91:H15	O116:H4	O145:H16	ONT:H19
O11:H8	**O49:H10**	**O91:H21**	O116:H10	**O145:H25**	ONT:H21
O11:H49	O50:H–	O91:H40	O116:H19	O145:H26	**ONT:H25**
O12:H–	O50:H7	O91:HNT	O117:H–	**O145:H28**	ONT:H41
O14:H–	O51:H49	O92:H3	**O117:H4**	O145:H46	ONT:H47
O15:H–	O52:H19	O92:H11	O117:H7	O145:HNT	**ONT:K39:H48**
O15:H2	O52:H23	O95:H–	O117:K1:H7	O146:H–	Orough:H–
O15:H8	O52:H25	**O96:H10**	O117:H8	O146:H8	Orough:H2
O15:H27	O54:H21	O98:H–	O117:H19	O146:H11	**Orough:H5**
O16:H–	O55:H–	O98:H8	O117:H28	O146:H14	Orough:K1:H6
O16:H6	**O55:H6**	**O100:H25**	O118:H–	O146:H21	Orough:K1:H7
O16:H21	**O55:H7**	O100:H32	O118:H2	O146:H28	**Orough:H11**
O17:H18	O55:H9	**O101:H–**	O118:H12	O148:H28	**Orough:H16**
O17:H41	**O55:H10**	O101:H9	**O118:H16**	O150:H–	Orough:H18
O18:H–	O55:H19	O102:H6	O118:H30	O150:H8	Orough:H20
O18:H7	**O55:H?**	**O103:H–**	O119:H–	O150:H10	Orough:H21
O18:H12	O60:H–	**O103:H2**	**O119:H2**	O152:H4	Orough:H28
O18:H15	O64:H25	O103:H4	**O119:H5**	O153:H2	
O18:H?	O65:H16	O103:H6	O119:H6	O153:H11	
O20:H–	O68:H–	O103:H7	O119:H25	O153:H12	
O20:H7	**O69:H–**	O103:H11	O120:H19	O153:H21	

Serotypes in bold represent strains isolated from patients with HUS. The list of serotypes is compiled, as new serotypes are isolated from human cases of disease.

References

Adachi, J.A., Jiang, Z.D., Mathewson, J.J., Verenkar, M.P., Thompson, S., Martinez-Sandoval, F., Steffen, R., Ericsson, C.D., and DuPont, H.L. (2001). Enteroaggregative *Escherichia coli* as a major etiologic agent in traveler's diarrhoea in 3 regions of the world. Clin. Infect. Dis. 32, 1706–1709.

Afset, J.E., Bergh, K., and Bevanger, L. (2003). High prevalence of atypical enteropathogenic *Escherichia coli* (EPEC) in Norwegian children with diarrhoea. J. Med. Microbiol. 52, 1015–1019.

Aidar, L., Penteado, A.S., Trabulsi, L.R., Blanco, J.E., Blanco, M., Blanco, J., and Pestana de Castro, A.F. (2000). Subtypes of intimin among non-toxigenic *Escherichia coli* from diarrheic calves in Brazil. Can. J. Vet. Res. 64, 15–20.

Aktan, I., Sprigings, K.A., La Ragione, R.M., Faulkner, L.M., Paiba, G.A., and Woodward, M.J. (2004). Characterisation of attaching-effacing *Escherichia coli* isolated from animals at slaughter in England and Wales. Vet. Microbiol. 102, 43–53.

Albert, M.J., Qadri, F., Haque, A., and Bhuiyan, N.A. (1993). Bacterial clump formation at the surface of liquid culture as a rapid test for identification of enteroaggregative *Escherichia coli*. J. Clin. Microbiol. 31, 1397–1399.

Allerberger, F.J., Rossboth, D., Dierich, M.P., Aleksic, S., Schmidt, H., and Karch, H. (1996). Prevalence and clinical manifestations of shiga toxin-producing *Escherichia coli* infections in Austrian children. Eur. J. Clin. Microbiol. Infect. Dis. 15, 545–550.

Alikhani, M.Y., Mirsalehian, A., and Aslani, M.M. (2006). Detection of typical and atypical enteropathogenic

Escherichia coli (EPEC) in Iranian children with and without diarrhoea. J. Med. Microbiol. *55*, 1159–1163.

Anon. (1990). Development of vaccines against cholera and diarrhoea due to enterotoxigenic *Escherichia coli*: memorandum from a WHO meeting. Bull. W.H.O. *68*, 303–312.

Araujo, J.M., Tabarelli, G.F., Aranda, K.R.S., Fabbricotti, S.H., Fagundes-Neto, U., Mendes, C.M.F., and Scaletsky, I.C. (2007). Typical Enteroaggregative and atypical Enteropathogenic types of *Escherichia coli* are the most prevalent diarrhoea-associated pathotypes among Brazilian children. J. Clin. Microbiol. *45*, 3396–3399.

Banatvala, N., Griffin, P.M., Greene, K.D., Barrett, T.J., Bibb, W.F., Green, J.H., Wells, J.G., and Hemolytic Uremic Syndrome Study Collaborators. (2001). The United States National Prospective Hemolytic Uremic Syndrome Study: microbiologic, serologic, clinical, and epidemiologic findings. J. Infect. Dis. *183*, 1063–1070.

Bando, S.Y., doValle, G.R.F., Martinez, M.B., Trabulsi, L.R., and Moreira-Filho, C.A. (1998). Characterization of enteroinvasive *Escherichia coli* and *Shigella* strains by RAPD analysis. FEMS Microbiol. Lett. *165*, 159–165.

Barlow, R.S., Hirst, R.G., Norton, R.E., Ashhurst-Smith, C., and Bettelheim, K.A. (1999). A novel serotype of enteropathogenic *Escherichia coli* (EPEC) as a major pathogen in an outbreak of infantile diarrhoea. J. Med. Microbiol. *48*, 1123–1125.

Beutin, L., Geier, D., Steinrück, H., Zimmermann, S., and Scheutz, F. (1993). Prevalence and some properties of verotoxin (Shiga-like toxin)-producing *Escherichia coli* in seven different species of healthy domestic animals. J. Clin. Microbiol. *31*, 2483–2488.

Bhan, M.K., Bhandari, N., Sazawal, S., Clemens, J., Raj, P., Levine, M.M., and Kaper, J.B. (1989a). Descriptive epidemiology of persistent diarrhoea among young children in rural northern India. Bull. W.H.O. *67*, 281–288.

Bhan, M.K., Raj, P., Levine, M.M., Kaper, J.B., Bhandari, N., and Srivastava, R. (1989b). Enteroaggregative *Escherichia coli* associated with persistent diarrhoea in a cohort of rural children in India. J. Infect. Dis. *159*, 1061–1064.

Bhan, M.K., Sazawal, S., Raj, P., Bhandari, N., Kumar, R., Bhardwaj, Y., Shrivastava, R., and Bhatnagar, S. (1989c). Aggregative *Escherichia coli*, *Salmonella*, and *Shigella* are associated with increasing duration of diarrhoea. Indian J. Pediatr. *56*, 81–86.

Bhatnagar, S., Bhan, M.K., Sommerfelt, H., Sazawal, S., Kumar, R., and Saini, S. (1993). Enteroaggregative *Escherichia coli* may be a new pathogen causing acute and persistent diarrhoea. Scand. J. Infect. Dis. *25*, 579–583.

Bielaszewska, M., Mellmann, A., Zhang, W., Köck, R., Fruth, A., Bauwens, A., Peters, G., and Karch, H. (2011). Characterisation of the *Escherichia coli* strain associated with an outbreak of haemolytic uraemic syndrome in Germany, 2011: a microbiological study. Lancet Infect. Dis. *11*, 671–676.

Bitzan, M., Karch, H., Maas, M.G., Meyer, T., Rüssmann, H., Aleksic, S., and Bockemühl, J. (1991). Clinical and genetic aspects of Shiga-like toxin production in traditional enteropathogenic *Escherichia coli*. Int. J. Med. Microbiol. *274*, 496–506.

Black, R.E. (1990). Epidemiology of travelers' diarrhoea and relative importance of various pathogens. Rev. Infect. Dis. *12*, S73–79.

Blanco, M., Schumacher, S., Tasara, T., Zweifel, C., Blanco, J.E., Dahbi, G., Blanco, J., and Stephan, R. (2005). Serotypes, intimin variants and other virulence factors of *eae* positive *Escherichia coli* strains isolated from healthy cattle in Switzerland. Identification of a new intimin variant gene (*eae*-eta2). BMC Microbiol. *5*, 23.

Bouzari, S., Jafari, A., Farhoudi-Moghaddam, A.A., Shokouhi, F., and Parsi, M. (1994). Adherence of non-enteropathogenic *Escherichia coli* to HeLa cells. J. Med. Microbiol. *40*, 95–97.

Bower, J.R., Congeni, B.L., Cleary, T.G., Stone, R.T., Wanger, A., Murray, B.E., Mathewson, J.J., and Pickering, L.K. (1989). *Escherichia coli* O114:nonmotile as a pathogen in an outbreak of severe diarrhoea associated with a day care center. J. Infect. Dis. *160*, 243–247.

Brussow, H., Rahim, H., and Freire, W. (1992). Epidemiological analysis of serologically determined rotavirus and enterotoxigenic *Escherichia coli* infections in Ecuadorian children. J. Clin. Microbiol. *30*, 1585–1587.

Cameron, S., Walker, C., Beers, M., Rose, N., and Anear, E. (1995). Enterohaemorrhagic *Escherichia coli* outbreak in South Australia associated with the consumption of mettwurst. Comm. Dis. Intell. *19*, 70–71.

Caprioli, A., Luzzi, I., Rosmini, F., Resti, C., Edefonti, A., Perfumo, F., Farina, C., Goglio, A., Gianviti, A., and Rizzoni, G. (1994). Communitywide outbreak of hemolytic- uremic syndrome associated with non-O157 verocytotoxin-producing *Escherichia coli*. J. Infect. Dis. *169*, 208–211.

Caprioli, A., Morabito, S., Brugère, H., and Oswald, E. (2005). Enterohaemorrhagic *Escherichia coli*: emerging issues on virulence and modes of transmission. Vet. Res. *36*, 289–311.

Centers for Disease Control and Prevention. (2009). Multistate outbreak of *E. coli* O157:H7 infections linked to eating raw refrigerated, prepackaged cookie dough. Available at: http://www.cdc.gov/ecoli/2009/0619.html

Chan, K.N., Phillips, A.D., Knutton, S., Smith, H.R., and Walker-Smith, J.A. (1994). Enteroaggregative *Escherichia coli*: another cause of acute and chronic diarrhoea in England? J. Pediatr. Gastroenterol. Nutr. *18*, 87–91.

Cravioto, A., Tello, A., Navarro, A., Ruiz, J., Villafan, H., Uribe, F., and Eslava, C. (1991a). Association of *Escherichia coli* HEp-2 adherence patterns with type and duration of diarrhoea. Lancet *337*, 262–264.

Cravioto, A., Tello, A., Villafán, H., Ruiz, J., del Vedovo, S., and Neeser, J.R. (1991b). Inhibition of localized adhesion of enteropathogenic *Escherichia coli* to HEp-2 cells by immunoglobulin and oligosaccharide fractions of human colostrum and breast milk. J. Infect. Dis. *163*, 1247–1255.

Creek, T.L., Kim, A., Lu, L., Bowen, A., Masunge, J., Arvelo, W., Smit, M., Mach, O., Legwaila, K.,

Motswere, C., *et al.* (2010). Hospitalization and mortality among primarily non-breastfed children during a large outbreak of diarrhoea and malnutrition in Botswana, 2006. J. Acquir. Immune Defic. Syndr. *53*, 14–19.

Dalton, C.B., Mintz, E.D., Wells, J.G., Bopp, C.A., and Tauxe, R.V. (1999). Outbreaks of enterotoxigenic *Escherichia coli* infection in American adults: a clinical and epidemiologic profile. Epidemiol. Infect. *123*, 9–16.

Donnenberg, M.S. (1995). Enteropathogenic *Escherichia coli* infections of the gastrointestinal tract. (Raven Press, Ltd, New York), pp. 709–726.

Donnenberg, M.S., and Kaper, J.B. (1992). Enteropathogenic *Escherichia coli*. Infect. Immun. *60*, 3953–3961.

Dorn, C.R., Scotland, S.M., Smith, H.R., Willshaw, G.A., and Rowe, B. (1989). Properties of Vero cytotoxin-producing *Escherichia coli* of human and animal origin belonging to serotypes other than O157:H7. Epidemiol. Infect. *103*, 83–95.

do Valle, G.R.F., Gomes, T.A.T., Irino, K., and Trabulsi, L.R. (1997). The traditional enteropathogenic *Escherichia coli* (EPEC) serogroup O125 comprises serotypes which are mainly associated with the category of enteroaggregative *E. coli*. FEMS Microbiol. Lett. *152*, 95–100.

DuPont, H.L., and Ericsson, C.D. (1993). Prevention and treatment of traveler's diarrhoea. N. Engl. J. Med. *328*, 1821–1827.

Echeverria, P., Serichantalerg, O., Changchawalit, S., Baudry, B., Levine, M.M., Orskov, F., and Orskov, I. (1992). Tissue culture-adherent *Escherichia coli* in infantile diarrhoea. J. Infect. Dis. *165*, 141–143.

European Food Safety Authority, European Centre for Disease Prevention and Control. (2012). The European Union Summary Report on Trends and Sources of Zoonoses, Zoonotic Agents and Foodborne Outbreaks in 2010. EFSA *10*(2597), 442. Available at: www.efsa.europa.eu/efsajournal

EFSA Panel on Biological Hazards (BIOHAZ) (2013). Scientific Opinion on VTEC-seropathotype and scientific criteria regarding pathogenicity assessment. EFSA Journal *11*(4), 3138. Available at: http://www.efsa.europa.eu/en/efsajournal/doc/3138.pdf

Fagundes-Neto, U. (1996). Enteropathogenic *Escherichia coli* infection in infants: clinical aspects and small bowel morphological alterations. Rev. Microbiol. Sao Paulo *27* (Suppl. 1), 117–119.

Fang, G.D., Lima, A.A., Martins, C.V., Nataro, J.P., and Guerrant, R.L. (1995). Etiology and epidemiology of persistent diarrhoea in northeastern Brazil: a hospital-based, prospective, case–control study. J. Pediatr. Gastroenterol. Nutr. *21*, 137–144.

Flores, J., DuPont, H.L., Lee, S.A., Belkind-Gerson, J., Paredes, M., Mohamed, J.A., Armitige, L.Y., Guo, D.C., and Okhuysen, P.C. (2008). Influence of host interleukin-10 polymorphisms on development of traveler's diarrhoea due to heat-labile enterotoxin-producing *Escherichia coli* in travelers from the United States who are visiting Mexico. Clin. Vaccine Immunol. *15*, 1194–1198.

Gasco'n, J., Vargas, M., Quinto, L., Corachan, M., Jimenez de Anta, M.T., and Vila, J. (1998). Enteroaggregative *Escherichia coli* strains as a cause of traveler's diarrhoea: a case–control study. J. Infect. Dis. *177*, 1409–1412.

Gendrel, D., Akaga, R., Ivanoff, B., Okouoyo, E., and Nguemby-Mbina, C. (1984). Acute gastroenteritis and breast feeding in Gabon. Preliminary results. Med. Trop. *44*, 323–325.

Gomes, T.A.T., Vieira, M.A.M., Wachsmuth, I.K., Blake, P.A., and Trabulsi, L.R. (1989a). Serotype-Specific Prevalence of *Escherichia coli* strains with EPEC Adherence Factor genes in infants with and without diarrhoea in Sao Paulo, Brazil. J. Infect. Dis. *160*, 131–135.

Gomes, T.A., Blake, P.A., and Trabulsi, L.R. (1989b). Prevalence of *Escherichia coli* strains with localized, diffuse, and aggregative adherence to HeLa cells in infants with diarrhoea and matched controls. J. Clin. Microbiol. *27*, 266–269.

Gonzalez, R., Diaz, C., Marino, M., Cloralt, R., Pequeneze, M., and Perez-Schael, I. (1997). Age-specific prevalence of *Escherichia coli* with localized and aggregative adherence in Venezuelan infants with acute diarrhoea. J. Clin. Microbiol. *35*, 1103–1107.

Hedberg, C.W., Savarino, S.J., Besser, J.M., Paulus, C.J., Thelen, V.M., Myers, L.J., Cameron, D.N., Barrett, T.J., Kaper, J.B., Osterholm, M.T., *et al.* (1997). An outbreak of foodborne illness caused by *Escherichia coli* O39:NM, an agent not fitting into the existing scheme for classifying diarrheogenic *E. coli*. J. Infect. Dis. *176*, 1625–1628.

Hill, S.M., Phillips, A.D., and Walker-Smith, J.A. (1991). Enteropathogenic *Escherichia coli* and life threatening chronic diarrhoea. Gut *32*, 154–158.

Huang, D.B., Okhuysen, P.C., Jiang, Z.D., and DuPont, H.L. (2004). Enteroaggregative *Escherichia coli*: an emerging enteric pathogen. Am. J. Gastroenterol. *99*, 383–389.

Huppertz, H.I., Rutkowski, S., Aleksic, S., and Karch, H. (1997). Acute and chronic diarrhoea and abdominal colic associated with enteroaggregative *Escherichia coli* in young children living in western Europe. Lancet *349*, 1660–1662.

Ishii, S., Meyer, K.P., and Sadowsky, M.J. (2007). Relationship between phylogenetic groups, genotypic clusters, and virulence gene profiles of *Escherichia coli* strains from diverse human and animal sources. Appl. Environ. Microbiol. *73*, 5703–5710.

Isidean, S.D., Riddle, M.S., Savarino, S.J., and Porter, C.K. (2011). A systematic review of ETEC epidemiology focusing on colonization factor and toxin expression. Vaccine *29*, 6167–6178.

Itoh, Y., Nagano, I., Kunishima, M., and Ezaki, T. (1997). Laboratory investigation of enteroaggregative *Escherichia coli* O untypeable:H10 associated with a massive outbreak of gastrointestinal illness. J. Clin. Microbiol. *35*, 2546–2550.

Jafari, A., Aslani, M.M., and Bouzari, S. (2012). *Escherichia coli*: a brief review of diarrhoeagenic pathotypes and their role in diarrheal diseases in Iran. Iran J. Microbiol. *4*, 102–117.

Kaper, J.B. (1996). Defining EPEC. Rev. Microbiol. Sao Paulo 27, 130–133.

Karch, H., Tarr, P.I., and Bielaszewska, M. (2005). Enterohaemorrhagic *Escherichia coli* in human medicine. Int. J. Med. Microbiol. 295, 405–418.

Kauffmann, F., and Dupont, A. (1950). *Escherichia* strains from infantile epidemic gastroenteritis. Acta Pathol. Microbiol. Immunol. Scand. 27, 552–564.

Kimmitt, P.T., Harwood, C.R., and Barer, M.R. (2000). Toxin gene expression by Shiga toxin-producing *Escherichia coli*: the role of antibiotics and the bacterial SOS response. Emerg. Infect. Dis. 6, 458–465.

Knutton, S., Baldwin, T., Williams, P.H., and McNeish, A.S. (1989). Actin accumulation at sites of bacterial adhesion to tissue culture cells: basis of new diagnostic test for enteropathogenic and enterohemorrhagic *Escherichia coli*. Infect. Immun. 57, 1290–1298.

Knutton, S., Phillips, A.D., Smith, H.R., Gross, R.J., Shaw, R., Watson, P., and Price, E. (1991). Screening for enteropathogenic *Escherichia coli* in infants with diarrhoea by the fluorescent-actin staining test. Infect. Immun. 59, 365–371.

Knutton, S., Shaw, R., Phillips, A.D., Smith, H.R., Willshaw, G.A., Watson, P., and Price, E. (2001). Phenotypic and genetic analysis of diarrhoea-associated *Escherichia coli* isolated from children in the United Kingdom. J. Pediatr. Gastroenterol. Nutr. 33, 32–40.

Krause, G., Zimmermann, S., and Beutin, L. (2005). Investigation of domestic animals and pets as a reservoir for intimin- (*eae*) gene positive *Escherichia coli* types. Vet. Microbiol. 106, 87–95.

Kudoh, Y.A., Kai, A., Obata, H., Kusunoki, J., Monma, C., Shingaki, M., Yanagawa, Y., Yamada, S., Matsushita, S., Ito, T., and Ohta, K. (1994). Epidemiologic surveys on verocytotoxin-producing *Escherichia coli* infections in Japan. In Recent advances in verocytotoxin-producing *Escherichia coli* infections. (Elsevier Science, Amsterdam), pp. 53–56.

Lan, R., and Reeves, P.R. (2002). *Eschericheria coli* in disguise: molecular origins of *Shigella*. Microbes Infect. 4, 1125–1132.

Lan, R., Alles, M.C., Donohoe, K., Martinez, M.B., and Reeves, P.R. (2004). Molecular evolutionary relationships of enteroinvasive *Escherichia coli* and *Shigella* spp. Infect. Immun. 72, 5080–5088.

Lanata, C.F., Mendoza, W., and Black, R.E. (2002). Improving diarrhoea estimates. WHO. Available at: http://www.who.int/child_adolescent_health/documents/pdfs/improving_diarrhoea_estimates.pdf.

Le Bouguenec, C., and Servin, A.L. (2006). Diffusely adherent *Escherichia coli* strains expressing Afa/Dr adhesins (Afa/Dr DAEC): hitherto unrecognized pathogens. FEMS Microbiol. Lett. 256, 185–194.

Leomil, L., Pestana de Castro, A.F., Krause, G., Schmidt, H., and Beutin, L. (2005). Characterization of two major groups of diarrhoeagenic *Escherichia coli* O26 strains which are globally spread in human patients and domestic animals of different species. FEMS Microbiol. Lett. 249, 335–342.

Levine, M.M., Nalin, D.R., Hoover, D.L., Bergquist, E.J., Hornick, R.B., and Young, C.R. (1979). Immunity to enterotoxigenic *Escherichia coli*. Infect. Immun. 23, 729–736.

Levine, M.M., and Edelman, R. (1984). Enteropathogenic *Escherichia coli* of classic serotypes associated with infant diarrhoea: epidemiology and pathogenesis. Epidemiol. Rev. 6, 31–51.

Levine, M.M., Xu, J.G., Kaper, J.B., Lior, H., Prado, V., Nataro, J.P., Karch, H., and Wachsmuth, I.K. (1987). A DNA probe to identify enterohemorrhagic *Escherichia coli* of O157:H7 and other serotypes that cause hemorrhagic colitis and hemolytic uremic syndrome. J. Infect. Dis. 156, 175–182.

Lima, A.A., Fang, G., Schorling, J.B., de Albuquerque, L., McAuliffe, J.F., Mota, S., Leite, R., and Guerrant, R.L. (1992). Persistent diarrhoea in northeast Brazil: etiologies and interactions with malnutrition. Acta Paediatr. Suppl. 381, 39–44.

Makino, S., Asakura, H., Shirahata, T., Ikeda, T., Takeshi, K., Arai, K., Nagasawa, M., Abe, T., and Sadamoto, T. (1999). Molecular epidemiological study of a mass outbreak caused by enteropathogenic *Escherichia coli* O157:H45. Microbiol. Immunol. 43, 381–384.

Mandomando, I.M., Macete, E.V., Ruiz, J., Sanz, S., Abacassamo, F., Vallès, X., Sacarlal, J., Navia, M.M., Vila, J., Alonso, P.L., and Gascon, J. (2007). Etiology of diarrhoea in children younger than 5 years of age admitted in a rural hospital of southern Mozambique. Am. J. Trop. Med. Hyg. 76, 522–527.

Marteyn, B., Gazi, A., and Sansonetti, P. (2012). *Shigella*: a model of virulence regulation *in vivo*. Gut Microbes 3, 104–120.

Mellmann, A., Bielaszewská, M., Kock, R., Friedrich, A.W., Fruth, A., Middendorf, B., Harmsen, D., Schmidt, M.A., and Karch, H. (2008). Analysis of collection of hemolytic uremic syndrome-associated enterohemorrhagic *Escherichia coli*. Emerg. Infect. Dis. 14, 1287–1290.

Melton-Celsa, A.R., and O'Brien A. (1998). Structure, biology, and relative toxicity of Shiga toxin family members for cells and animals. In *Escherichia coli* O157:H7 and other Shiga toxinproducing *E. coli* strains, Kaper, J.B., and O'Brien, A.D., eds. (American Society for Microbiology, Washington, DC), pp. 121–128.

Miller, J.R., Barrett, L., Kotloff, K., and Guerrant, R.L. (1994). A rapid diagnostic test for infectious and inflammatory enteritis. Arch. Intern. Med. 154, 2660–2664.

Monolov, D.G. (1959). A new type of the genus *Shigella*— 'Shigella 13'. J. Hyg. Epidemiol. Microbiol. Immunol. 3, 184–190.

Monteiro-Neto, V., Campos, L.C., Ferreira, A.J.P., Gomes, T.A.T., and Trabulsi, L.R. (1997). Virulence properties of *Escherichia coli* O111:H12 strains. FEMS Microbiol. Lett. 146, 123–128.

Morabito, S., Karch, H., Mariani Kurkdjian, P., Schmidt, H., Minelli, F., Bingen, E., and Caprioli, A. (1998). Enteroaggregative, shiga toxin-producing *Escherichia coli* O111:H2 associated with an outbreak of hemolytic-uremic syndrome. J. Clin. Microbiol. 36, 840–842.

Nair, G.B., Ramamurthy, T., Bhattacharya, M.K., Krishnan, T., Ganguly, S., Saha, D.R., Rajendran, K., Manna, B., Ghosh, M., Okamoto, K., and Takeda, Y. (2010). Emerging trends in the etiology of enteric pathogens as evidenced from an active surveillance of hospitalized diarrhoeal patients in Kolkata, India. Gut Pathog. 2, 4.

Nataro, J.P., and Kaper, J.B. (1998). Diarrheagenic *Escherichia coli*. Clin. Microbiol. Rev. 1, 142–201.

Nataro, J.P., Kaper, J.B., Robins-Browne, R.M., Prado, V., Vial, P., and Levine, M.M. (1987). Patterns of adherence of diarrhoeagenic *Escherichia coli* to HEp-2 cells. Pediatr. Infect. Dis. J. 6, 829–831.

Nataro, J.P., Deng, Y., Cookson, S., Cravioto, A., Savarino, S.J., Guers, L.D., Levine, M.M., and Tacket, C.O. (1995). Heterogeneity of enteroaggregative *Escherichia coli* virulence demonstrated in volunteers. J. Infect. Dis. 171, 465–468.

Nishikawa, Y., Helander, A., Ogasawara, J., Moyer, N.P., Hanaoka, M., Hase, A., and Yasukawa, A. (1998). Epidemiology and properties of heat-stable enterotoxin-producing *Escherichia coli* serotype O169:H41. Epidemiol. Infect. 121, 31–42.

Northey, G., Evans, M.R., Sarvotham, T.S., Thomas, D.R., and Howard, T.J. (2007). Sentinel surveillance for travellers' diarrhoea in primary care. BMC Infect. Dis. 7, 126.

Okeke, I.N. (2009). Diarrheagenic *Escherichia coli* in sub-Saharan Africa: status, uncertainties and necessities. J. Infect. Dev. Ctries. 3, 817–842.

Ochoa, T.J., and Contreras, C.A. (2011). Enteropathogenic *Escherichia coli* infection in children. Curr. Opin. Infect. Dis. 24, 478–483.

Ochoa, T.J., Barletta, F., Contreras, C., and Mercado, E. (2008). New insights into the epidemiology of enteropathogenic *Escherichia coli* infection. Trans. R. Soc. Trop. Med. Hyg. 102, 852–856.

Ochoa, T.J., Mercado, E.H., Durand, D., Rivera, F.P., Mosquito, S., Contreras, C., Riveros, M., Lluque, A., Barletta, F., Prada, A., and Ruiz, J. (2011). Frequency and pathotypes of diarrhoeagenic *Escherichia coli* in peruvian children with and without diarrhoea. Rev. Peru Med. Exp. Salud Publica 28, 13–20.

Olesen, B., Neimann, J., Böttiger, B., Ethelberg, S., Schiellerup, P., Jensen, C., Helms, M., Scheutz, F., Olsen, K.E.P., Krogfelt, K.A., *et al.* (2005). Etiology of diarrhea in young children in Denmark: a case-control study. J. Clin. Microbiol. 43, 3636–3641.

Opintan, J.A., Bishar, R.A., Newman, M.J., and Okeke, I.N. (2010). Carriage of Diarrhoeagenic *Escherichia coli* by older children and adults in Accra, Ghana. Trans. R. Soc. Trop. Med. Hyg. 104, 504–506.

Ørskov, F., and Ørskov, I. (1992). *Escherichia coli* serotyping and disease in man and animals. Can. J. Microbiol. 38, 699–704.

Paul, M., Tsukamoto, T., Ghosh, A.R., Bhattacharya, S.K., Manna, B., and Chakrabarti, S. (1994). The significance of enteroaggregative *Escherichia coli* in the etiology of hospitalized diarrhoea in Calcutta, India and the demonstration of a new honey-combed pattern of aggregative adherence. FEMS Microbiol. Lett. 117, 319–326.

Pedroso, M.Z., Freymuller, E., Trabulsi, L.R., and Gomes, T.A.T. (1993). Attaching- Effacing lesions and intracellular penetration in HeLa cells and human duodenal mucosa by 2 *Escherichia coli* strains not belonging to the classical Enteropathogenic *E. coli* serogroups. Infect. Immun. 61, 1152–1156.

Pennington, H. (2010). *Escherichia coli* O157. Lancet 376, 1428–1435.

Persson, S., Olsen, K.E., Ethelberg, S., and Scheutz, F. (2007). Subtyping method for *Escherichia coli* shiga toxin (verocytotoxin) 2 variants and correlations to clinical manifestations. J. Clin. Microbiol. 45, 2020–2024.

Porat, N., Levy, A., Fraser, D., Deckelbaum, R.J., and Dagan, R. (1998). Prevalence of intestinal infections caused by diarrhoeagenic *Escherichia coli* in Bedouin infants and young children in Southern Israel. Pediatr. Infect. Dis. J. 17, 482–488.

Presterl, E., Nadrchal, R., Wolf, D., Rotter, M., and Hirschl, A.M. (1999). Enteroaggregative and enterotoxigenic *Escherichia coli* among isolates from patients with diarrhoea in Austria. Eur. J. Clin. Microbiol. Infect. Dis. 18, 209–212.

Qadri, F., Khan, A.I., Faruque, A.S.G., Begum, Y.A., Chowdhury, F., Nair, G.B., Salam, M.A., Sack, D.A., and Svennerholm, A. (2005). Enterotoxigenic *Escherichia coli* and *Vibrio cholerae* diarrhoea, Bangladesh, 2004. Emerg. Infect. Dis. 11, 1266.

Quiroga, M., Oviedo, P., Chinen, I., Pegels, E., Husulak, E., Binztein, N., Rivas, M., Schiavoni, L., and Vergara, M. (2000). Asymptomatic infections by diarrhoeagenic *Escherichia coli* in children from Misiones, Argentina, during the first twenty months of their lives. Rev. Inst. Med. Trop. Sao Paulo 42, 9–15.

Rajendran, P., Ajjampur, S.S., Chidambaram, D., Chandrabose, G., Thangaraj, B., Sarkar, R., Samuel, P., Rajan, D.P., and Kang, G. (2010). Pathotypes of diarrhoeagenic *Escherichia coli* in children attending a tertiary care hospital in South India. Diagn. Microbiol. Infect. Dis. 68, 117–122.

Rothbaum, R., McAdams, A.J., Giannella, R., and Partin, J.C. (1982). A clinicopathological study of enterocyte-adherent *Escherichia coli*: a cause of protracted diarrhoea in infants. Gastroenterology 83, 441–454.

Rowe, B., Gross, R.J., and Woodroof, D.P. (1977). Proposal to recognise serovar 145/146 (synonyms: 147, *Shigella* 13, *Shigella sofia*, and *Shigella manolovii*) as a new *Escherichia coli* O group, O164. Int. J. Syst. Bacteriol. 27, 15–18.

Scaletsky, I.C., Fabbricotti, S.H., Silva, S.O., Morais, M.B., and Fagundes-Neto U. (2002). HEp-2-adherent *Escherichia coli* strains associated with acute infantile diarrhoea, Sao Paulo, Brazil. Emerg. Infect. Dis. 8, 855–858.

Scavia, G., Staffolani, M., Fisichella, S., Striano, G., Colletta, S., Ferri, G., Escher, M., Minelli, F., and Caprioli, A. (2008). Enteroaggregative *Escherichia coli* associated with a foodborne outbreak of gastroenteritis. J. Med. Microbiol. 57, 1141–1146.

Scheutz, F., Nielsen, E.M., Frimodt-Møller, J., Boisen, N., Morabito, S., Tozzoli, R., Nataro, J.P., and Caprioli, A. (2011). Characteristics of the enteroaggregative Shiga toxin/verotoxin-producing *Escherichia coli* O104:H4 strain causing the outbreak of haemolytic uraemic syndrome in Germany, May to June 2011. Euro Surveill. *16*.

Scheutz, F., and Strockbine, N.A. (2005). *Escherichia*. In Bergey's Manual of Systematic Bacteriology, Vol2, PartB The Gammaproteobacteria edition, Garrity, G.M., Brenner, D.J., Krieg, N.R., and Staley, J.T., eds. (Springer, New York, USA), pp. 607–624.

Schmidt, H., Knop, C., Franke, S., Aleksic, S., Heesemann, J., and Karch, H. (1995). Development of PCR for screening of enteroaggregative *Escherichia coli*. J. Clin. Microbiol. *33*, 701–705.

Schultsz, C., van den Ende, J., Cobelens, F., Vervoort, T., van Gompel, A., Wetsteyn, J.C., and Dankert, J. (2000). Diarrheagenic *Escherichia coli* and acute and persistent diarrhoea in returned travelers. J. Clin. Microbiol. *38*, 3550–3554.

Scotland, S.M., Smith, H.R., and Rowe, B. (1985). Two distinct toxins active on Vero cells from *Escherichia coli* O157. Lancet *2*, 885–886.

Scotland, S.M., Willshaw, G.A., Smith, H.R., and Rowe, B. (1990). Properties of strains of *Escherichia coli* O26:H11 in relation to their enteropathogenic or enterohemorrhagic classification. J. Infect. Dis. *162*, 1069–1074.

Scotland, S.M., Smith, H.R., Said, B., Willshaw, G.A., Cheasty, T., and Rowe, B. (1991). Identification of enteropathogenic *Escherichia coli* isolated in Britain as enteroaggregative or as members of a subclass of attaching-and-effacing *E. coli* not hybridising with the EPEC adherence-factor probe. J. Med. Microbiol. *35*, 278–283.

Scotland, S.M., Willshaw, G.A., Cheasty, T., and Rowe, B. (1992). Strains of *Escherichia coli* O157:H8 from human diarrhoea belong to attaching and effacing class of *E. coli*. J. Clin. Pathol. *45*, 1075–1078.

Scotland, S.M., Willshaw, G.A., Cheasty, T., Rowe, B., and Hassall, J.E. (1994). Association of enteroaggregative *Escherichia coli* with travellers' diarrhoea. J. Infect. *29*, 115–116.

Scotland, S.M., Smith, H.R., Cheasty, T., Said, B., Willshaw, G.A., Stokes, N., and Rowe, B. (1996). Use of gene probes and adhesion tests to characterise *Escherichia coli* belonging to enteropathogenic serogroups isolated in the United Kingdom. J. Med. Microbiol. *44*, 438–443.

Senerwa, D., Olsvik, O., Mutanda, L.N., Lindqvist, K.J., Gathuma, J.M., Fossum, K., and Wachsmuth, K. (1989). Enteropathogenic *Escherichia coli* serotype O111:HNT isolated from preterm neonates in Nairobi, Kenya. J. Clin. Microbiol. *27*, 1307–1311.

Sherman, P., Drumm, B., Karmali, M., and Cutz, E. (1989). Adherence of bacteria to the intestine in sporadic cases of enteropathogenic *Escherichia coli*-associated diarrhoea in infants and young children: a prospective study. Gastroenterology *96*, 86–94.

Smith, H.R., Scotland, S.M., Willshaw, G.A., Rowe, B., Cravioto, A., and Eslava, C. (1994). Isolates of *Escherichia coli* O44:H18 of diverse origin are enteroaggregative. J. Infect. Dis. *170*, 1610–1613.

Smith, H.R., Scotland, S.M., Cheasty, T., Willshaw, G.A., and Rowe, B. (1996). Enteropathogenic *Escherichia coli* infections in the United Kingdom. Rev. Microbiol. São Paulo *27*, 45–49.

Smith, H.R., Cheasty, T., and Rowe, B. (1997). Enteroaggregative *Escherichia coli* and outbreaks of gastroenteritis in UK. Lancet *350*, 814–815.

Sobieszczanska, B., Kowalska-Krochmal, B., Mowszet, K., and Pytrus, T. (2003). Susceptibility to antimicrobial agents of enteroaggregative *Escherichia coli* strains isolated from children with diarrhoea. Przegl Epidemiol. *57*, 499–503.

Staley, T.E., Jones, E.W., and Corley, L.D. (1969). Attachment and penetration of *Escherichia coli* into intestinal epithelium of the ileum in newborn pigs. Am. J. Pathol. *56*, 371–392.

Steinsland, H., Valentiner-Branth, P., Perch, M., Dias, F., Fischer, T.K., Aaby, P., Mølbak, K., and Sommerfelt, H. (2002). Enterotoxigenic *Escherichia coli* infections and diarrhoea in a cohort of young children in Guinea-Bissau. J. Infect. Dis. *186*, 1740–1747.

Stephan, R., Borel, N., Zweifel, C., Blanco, M., and Blanco, J.E. (2004). First isolation and further characterization of enteropathogenic *Escherichia coli* (EPEC) O157:H45 strains from cattle. BMC Microbiol. *4*, 10.

Tsukamoto, T., Kimoto, T., Magalhaes, M., and Takeda, Y. (1992). Enteroadherent *Escherichia coli* exhibiting localized pattern of adherence among infants with diarrhoea in Brazil – incidence and prevalence of serotypes. Kansenshogaku Zasshi *66*, 1538–1542.

Tsukamoto, T., and Takeda, Y. (1993). Incidence and prevalence of serotypes of enteroaggregative *Escherichia coli* from diarrhoeal patients in Brazil, Myanmar and Japan. Kansenshogaku Zasshi *67*, 289–294.

UNICEF (1992). Take the baby friendly initiative! New York: United Nations Children's Fund.

Vial, P.A., Robins-Browne, R.M., Lior, H., Prado, V., Kaper, J.B., Nataro, J.P., Maneval, D., Elsayed, A., and Levine, M.M. (1988). Characterization of enteroadherent- aggregative *Escherichia coli*, a putative agent of diarrhoeal disease. J. Infect. Dis. *158*, 70–79.

Wani, S.A., Hussain, I., Nabi, A., Fayaz, I., and Nishikawa, Y. (2007). Variants of *eae* and *stx* genes of atypical enteropathogenic *Escherichia coli* and non-O157 Shiga toxin-producing *Escherichia coli* from calves. Lett. Appl. Microbiol. *45*, 610–615.

Wenneras, C., and Erling, V. (2004). Prevalence of enterotoxigenic *Escherichia coli*-associated diarrhoea and carrier state in the developing world. Journal of Health, Population and Nutrition *22*, 370–382.

WHO 1987. Programme for control of diarrhoeal diseases (CDD/83.3 Rev1). In Manual for Laboratory Investigations of Acute Enteric Infections, (Geneva: World Health Organization) p. 27.

Yuste, M., De La Fuente, R., Ruiz-Santa-Quiteria, J.A., Cid, D., and Orden, J.A. (2006). Detection of the *astA* (EAST1) gene in attaching and effacing *Escherichia coli* from ruminants. J. Vet. Med. B. Infect. Dis. Vet. Public Health *53*, 75–77.

Pathogenic *Escherichia coli* in Domestic Mammals and Birds

2

Jacques G. Mainil and John M. Fairbrother

Abstract

Escherichia coli is an important cause of disease worldwide and occurs in most mammalian species, including humans, and in birds. *E. coli* was first described in 1885 by a German paediatrician, Theodor Escherich, in the faeces of a child suffering from diarrhoea. In 1893, a Danish veterinarian postulated that the *E. coli* species comprises different strains, some being pathogens, others not. Today, pathogenic *E. coli* are classified into categories or pathotypes based on the production of virulence factors and on the clinical manifestations that they cause. The most important categories in animals are those colonizing the intestine and causing diarrhoea in calves, pigs, and most other animal species, those colonizing the intestine and causing a toxaemia, or oedema disease, in pigs, and those residing in the intestine of healthy animals but capable of invading the host in certain conditions and causing septicaemia in young animals of most species, localized or systemic infections in poultry, or urinary tract infections, especially in dogs. The purpose of this chapter is to give an overview of the most relevant pathotypes causing infections in domestic mammals and birds, with emphasis on their history, virulence-associated properties, economic importance, diagnostic procedures, public health hazard, and vaccine potential.

Classification of pathogenic *Escherichia coli*

While most strains of the bacterial species *Escherichia coli* occur as commensal members of the microbiota in the intestinal tract of animals and humans, some are, however, important pathogens that cause a wide spectrum of diseases, ranging from self-limiting to life-threatening intestinal and extraintestinal illnesses (Nataro *et al.*, 2011; Quinn *et al.*, 2011). This was a puzzling observation in the years following its description (Escherich, 1885), because according to the Koch's postulates (Koch, 1884), a bacterial species was either pathogenic or not. In 1893, a Danish veterinarian postulated that the *E. coli* species actually comprises different strains, some being pathogens and others not (Jensen, 1893). Not only did his assertion prove true but also *E. coli* is today subdivided into many pathogenic strains causing different intestinal, urinary tract or invasive infections and pathologies, in all animal species and in humans.

The criteria for the classification and denomination of pathogenic *E. coli* have evolved with time, originally being the target hosts and associated clinical syndromes and now referring to the pathotypes, i.e. the specific combinations of virulence properties, such as adherence, invasion and/or toxicity (Table 2.1). To add to the nomenclature difficulty, the virulence properties were not described simultaneously but rather gradually since the 1960s, many pathogenic strains being therefore assigned to different pathotypes in the course of time.

The purpose of this chapter is to give an overview of the most relevant pathotypes causing enteric, enterotoxaemic, invasive, and urinary tract infections (Table 2.1) in domestic mammals and birds, with emphasis on their economic and public health relevance. On the other hand, the carrier state of several verotoxigenic (shigatoxigenic) *E. coli* in ruminants (Table 2.1) will not be

Table 2.1 Classes of pathogenic *Escherichia coli* in animals and humans (adapted from Mainil, 2003a, 2013; Piérard *et al.*, 2012; http://www.ecl-lab.com/en/ecoli/classification.asp)

Name (acronym)	Primary hosts (secondary hosts)	Clinical signs and diseases	Most relevant virulence properties	Comments/remarks
Diarrhoeagenic *E. coli* (DEC)				
Enterotoxigenic (ETEC)	Ruminants, pigs, dogs, humans	Watery diarrhoea in newborn ruminants, piglets, pups, babies; milder diarrhoea in weaned piglets; travellers' diarrhoea	Fimbrial adhesins (F2 to F6, F17a, F18, F41); afimbrial adhesin (AIDA); heat-stable (STa, STb) and heat-labile (LT1, LT2) enterotoxins; α haemolysin (α-Hly)	
Enteropathogenic (EPEC)	All mammals, including humans	Undifferentiated diarrhoea	Attaching and effacing (AE) lesion; type 4 BFP fimbriae of human typical (t)EPEC; AF/R/Ral and Paa fimbriae of rabbit and porcine atypical (a)EPEC respectively	Localized adherence (LA) of tEPEC and LA-like of aEPEC patterns on cells in culture
Attaching/effacing verotoxigenic (AE-VTEC)	Young ruminants, especially calves	Undifferentiated acute or chronic diarrhoea	AE lesion; Verotoxins (VTx) play no role in ruminants	Limited to a few serotypes (O5, O26, O111, O118)
Necrotoxigenic (NTEC)	NTEC1: all mammals, including humans; NTEC2: ruminants	Diarrhoea and extra-intestinal infections (urinary tract infections, septicaemia)	Cytotoxic necrotizing factors (CNF) 1/2; cytolethal distending toxin III (CDT-III); α-Hly; fimbrial (Pap/Prs, Sfa/F1C and F17b/c) and afimbrial adhesins (AFA family)	
Diffusely adherent (DAEC)	Animals, humans	Diarrhoea and extra-intestinal infections (urinary tract infections, septicaemia)	Afimbrial adhesins: AFA family and AIDA	Diffuse adherence (DA) pattern on cells in culture
Enteroinvasive (EIEC)	Humans, primates	Dysentery	Invasion of and multiplication in the enterocytes	Similar to *Shigella* sp.
Enteroaggregative (EAggEC)	Humans	Diarrhoea	Fimbrial adhesins (AAF/Hda); toxins (Pet, EAST, ShET1); transcriptional activator gene (*aggR*)	Aggregative 'stacked brick' adherence (AggA) pattern on cells in culture
Enterotoxaemic *E. coli*				
Attaching/effacing verotoxigenic (AE-VTEC)	Humans (Dogs)	Diarrhoea, haemorrhagic colitis (HC), haemolytic uraemic syndrome (HUS)	AE lesion; Verotoxins (Vtx) 1/2	Scores of serotypes; ruminants can be healthy carriers (=reservoir hosts)
Verotoxigenic (VTEC)	Pigs, humans	Oedema disease (ED) in piglets; HUS in humans	Vtx 1/2; fimbrial (F18 of porcine VTEC) and afimbrial (Saa of human or AIDA of porcine VTEC) adhesins	Ruminants can be healthy carriers of human VTEC (=reservoir hosts)

Name (acronym)	Primary hosts (secondary hosts)	Clinical signs and diseases	Most relevant virulence properties	Comments/remarks
Extraintestinal pathogenic *E. coli* (ExPEC)				
Septicaemic (SePEC)	Ruminants (other mammals)	Septicaemia, bacteraemia, multisystemic infection, endotoxaemia	Pap/Prs, Sfa/F1C, F17b/c, AFA-VIII, CS31A adhesins; CNF 1/2, CDT-III/ IV and α-Hly toxins; endotoxin; blood vessel barrier crossing (IbeA); antiphagocytosis (O and K antigens); resistance to complement (Iss, TraT); iron-scavenging systems (aerobactin, yersiniabactin, salmochelin);	Of faecal origin
Avian pathogenic (APEC)	Chickens, turkeys, ducks (other poultry)	Septicaemia, multisystemic infection		
Uropathogenic (UPEC)	Humans, dogs, cats (other mammals)	Cystitis, pyelonephritis (bacteraemia, septicaemia)		

reviewed in this chapter, since they are primarily human pathogens.

Diarrhoeagenic *Escherichia coli*

Enterotoxigenic, enteropathogenic, and attaching-effacing verotoxigenic *E. coli* are responsible for diarrhoea in different animal species, and have been collectively designated as diarrhoeagenic *E. coli* or DEC, whereas at least two other pathotypes, necrotoxigenic and diffusely adherent *E. coli*, are also associated with diarrhoea (Table 2.1). Conversely, DEC have not been described in birds, apart from a very few sporadic reports with no evidence of diarrhoea.

Enterotoxigenic *Escherichia coli*

Enterotoxigenic *E. coli* (ETEC) have been described in several domestic mammalian species, more especially in cattle and pigs and, to a lesser extent, in sheep, goats and dogs (Beutin, 1999; Nagy and Fekete, 1999, 2005; Radostits *et al.*, 2007; Gyles and Fairbrother, 2010). Their history actually begins in 1967 with the description of the production of toxins by bovine and porcine diarrhoeagenic *E. coli*, causing fluid accumulation in the intestinal ligated loop assay; hence the names 'enterotoxins' and 'enterotoxigenic *E. coli*' (Smith and Halls, 1967a,b). In the following years the

ETEC enterotoxins were subdivided on the basis of physical, chemical, and biological properties into heat-labile (LT) and heat-stable (STa and STb) toxins and additional variants were later described (LT2a, LT2b, LT2c, EAST) (Table 2.2) (Holland, 1990; Nagy and Fekete, 1999, 2005; Nawar *et al.*, 2010; Gyles and Fairbrother, 2010; Piérard *et al.*, 2012; Mainil, 2013). The heat-labile (LT/LT2a/LT2b/LT2c) enterotoxins are A = B subunit proteins whereas the heat-stable (STa/ STb/EAST) enterotoxins are oligopeptides. LT, STa and STb are the most frequently observed enterotoxins, being found in various combinations in ETEC from calves, lambs, piglets, dogs, and humans. The variants LT2a, LT2b and LT2c are mostly found in ETEC from humans, water buffaloes, ostriches, and from food, and less frequently from cattle and pigs. EAST is also frequently found in *E. coli* pathotypes other than ETEC, from humans and animals with diarrhoea and from food (Ménard and Dubreuil, 2002; Veilleux and Dubreuil, 2006). Following activation of adenyl cyclase (LT, LT2), guanylyl cyclase (STa, EAST), or a still unidentified enzyme (STb), efflux pump activity is increased, leading to increased anion (Cl^-, HCO_3^-) excretion along with water molecules and reduced cation (Na^+, K^+) absorption (Dubreuil, 2008; Gyles and Fairbrother, 2010; Mainil, 2013).

The host-specificity of ETEC is due to the

Table 2.2 Toxins, adhesins and serotypes of ETEC according to the animal host species

Target host species	Adhesins	Enterotoxins	Most relevant combinations of virulence properties	Most relevant serogroups
Ruminants (newborn calves, lambs and kids)	F5, F17a, F41	STa	F5/F41/STa; F5/STa; F5/F17a/STa	O: 8, 9, 20, 64, 101 K: 25, 28, 30, 35
Pigs (newborn piglets)	F4, F5, F6, F41	STa, STb, LT	F4/STb/LT; F5/STa; F5/F41/STa; F6/STa/STb; F6/STa	O: 8, 9, 20, 45, 64, 101, 141, 147, 149, 157
Pigs (weaned piglets)	F4, F18	STa, STb, LT	F4/STb/LT; F4/STa/STb/LT; F18/STb/LT; F18/STa/STb	O: 8, 138, 139, 141, 147, 149, 157
Dogs (suckling pups)	Unknown	STa, STb	STa; STa/STb; STb	O: 4, 6, 8, 42, 70, 105
Cats	No ETEC described			
Rabbits	No ETEC described			
Birds (poultry and cage birds)	No ETEC described			

presence of adhesins (Table 2.2) mediating the colonization of the small intestine that were described as early as the mid-1960s. ETEC adhesins may be fimbriae (thread, fringe or fibre in Latin), fibrillae (thin and small fibres in Latin) (F4, F5, F6, F17, F18 and/or F41), or outer membrane proteins (the autotransporter AIDA protein, after Adhesin Involved in Diffuse Adherence). The F4, F17 and F18 fimbriae occur as three (ab, ac, ad), four (a, b, c, d) and two (ab, ac) antigenic variants respectively, all but the F17b, F17c and F17d being found in ETEC (Tables 2.1 and 2.2) (Debroy et al., 2009; Gyles and Fairbrother, 2010; Mainil, 2013). At the time of their description, the F4 and F5 fimbriae were mistaken as capsular antigens on the basis of their physical properties and received the names of 'K88 and K99 common antigens' respectively, whereas most other fimbriae and fibrillae were initially named according to the number of the strain of origin: 987P for F6, Att25 or FY for F17a (Girardeau, 1985), F107 or 2134P or 8813 for F18, and F41 was so named because it was described on the bovine ETEC B41 (Holland, 1990; Rippinger et al., 1995; Mainil, 2013).

The natural disease associated with ETEC is a watery diarrhoea in newborn calves, lambs, kids and piglets, and though less dramatically, in pups and in weaned piglets, with progressive dehydration, acidaemia, loss of skin elasticity, sinking of the eyes, weakness, recumbency, hypothermia, coma and death within a few hours for the most acute cases. At necropsy, the lesions are scarce and unspecific: the intestine is empty and congestive,

but not inflamed (Radostits et al., 2007; Foster and Smith, 2009; Gyles and Fairbrother, 2010). Experimentally, watery diarrhoea has been reproduced with different ETEC in newborn calves, lambs and piglets. Moreover, the role of different adhesins and enterotoxins in gut colonization and development of diarrhoea respectively was demonstrated by comparing wild-type ETEC and their plasmid-free mutants (Smith and Linggood, 1971, 1972; Smith and Huggins, 1978; Tzipori et al., 1983; Contrepois and Girardeau, 1985; Runnels et al., 1987; Sarmiento et al., 1988; Jensen et al., 2006). Ruminant dams and sows are healthy carriers of ETEC and represent the source of contamination of their offspring. ETEC diarrhoea is, therefore, very important economically in suckling young, particularly in beef cattle, sheep and pigs, but much less in dairy cattle (Radostits et al., 2007). ETEC were described in dogs for the first time in 1980 although they are not very frequent in this species. The source of contamination of pups is probably the dam as in other species, although data on the carrier state in adults are lacking. ETEC do not seem to be present in cats, though some sporadic reports exist in the literature (Josse et al., 1980; Beutin, 1999; Mainil, 2002).

Depending on the target species, ETEC belong to a restricted and host-associated range of serotypes and combinations of virulence properties (Table 2.2). Genes coding for the F4, F5, F6, F18 and AIDA adhesins and for the STa, STb, EAST, and LT enterotoxins are located on plasmids in different combinations, whereas genes coding for the F17a and F41 fimbriae and for the LT2 enterotoxins are chromosomal (Nagy and Fekete, 1999, 2005; Gyles and Fairbrother, 2010; Nawar et al., 2010; Mainil, 2013).

The specific receptors for the adhesins are glycoproteins/glycolipids present on the enterocytes, usually at the level of the small intestinal segments and at certain ages: for example, the receptors for the F5 fimbriae are present in the distal segments of the small intestine of very young calves, lambs and/or piglets and disappear after a few days of life, whereas the receptors for the F18 fimbriae are absent in newborn piglets and increase progressively from three weeks of age onwards. On the other hand, the receptors for the F4 adhesins are present along the different segments of the small intestine in both newborn and weaned piglets, though decreasingly with age. Moreover, piglets can be genetically resistant to colonization by F4 and F18 fimbriae, since their respective receptors are encoded by dominant genes. A recessive homozygote does not synthesize any receptor and is therefore resistant to colonization and develops no disease (Teneberg et al., 1990, 1994; Nagy et al., 1992; Francis et al., 1999; Nagy and Fekete, 1999, 2005; Radostitis et al., 2007; Gyles and Fairbrother, 2010). Piglets also develop age-related resistance to disease, but not to colonization, by F6+ ETEC due to the release of the specific receptors into the mucus layer and intestinal lumen, therefore facilitating bacterial clearance (Dean et al., 1989).

With the exception of the AIDA adhesin that is produced by some porcine ETEC (Benz and Schmidt, 1989), none of the adhesins of animal ETEC are produced by human ETEC, the latter producing their own host-specific adhesins. Moreover, animal ETEC, even those producing the AIDA adhesin belong to host-associated serotypes. Finally, to date there is no report or epidemiological evidence of cross-contamination of humans by animal ETEC (Beutin, 1999; Nagy and Fekete, 1999, 2005; Kaper et al., 2004; Radostits et al., 2007; Gyles and Fairbrother, 2010).

Clinical diagnosis of ETEC infections is relatively straightforward in newborn calves, lambs and piglets due to the characteristic clinical manifestations, but less so in pups and weaned piglets. Inoculation on blood agar of faecal samples or intestinal content from newborn and weaned piglets is helpful since several porcine ETEC are β-haemolytic due to the production of α-haemolysin (α-Hly) whose encoding genes are also plasmid-located (Nagy and Fekete, 1999; Mainil, 2013). Laboratory diagnosis can be confirmed by isolation of E. coli after growth on selective medium such as MacConkey's agar and subsequent identification of the virulence properties of several E. coli colonies using multiplex PCR for detection of the genes coding for the different adhesins and enterotoxins (e.g. Franck et al., 1998; Weiner and Osek, 2004). Genetic tests are preferred to immunological assays, like ELISA, slide agglutination, or latex bead agglutination, as in vitro production of toxins and adhesins can be very much dependent on the growth media composition and temperature (Holland, 1990; Nagy and Fekete, 1999, 2005). Nevertheless, commercial immunological tests (latex agglutination) exist for direct detection of F5+ ETEC, together with rotavirus, coronavirus and/or crytosporidiae.

ETEC are the only pathogenic E. coli for which commercial vaccines are available on the veterinary market for newborn ruminants and piglets (see Chapter 14). The vaccines combine different semi-purified fimbrial/fibrillar adhesins (F4, F5, F6, F17a, and/or F41; but not F18) and the detoxified LT enterotoxin (but not the LT2 enterotoxins) in vaccines for piglets, all being very good immunogens (Contrepois and Girardeau, 1985; Runnels et al., 1987; Nagy and Fekete, 1999, 2005; Radostits et al., 2007). Conversely, the STa, STb, and EAST enterotoxins are poorly immunogenic, being small peptides, and are therefore not included in vaccines, although various attempts have been made to increase the production of antibodies to the STa enterotoxin (review in Ruth et al., 2005). Vaccines are administered to pregnant dams and sows to protect the newborns via colostral immunity (antibodies do not cross the placenta in domestic mammals). These vaccines are very efficient and have been widely used in cattle and pig production. The protection of weaned piglets against post-weaning diarrhoea via colostral immunity is less convincing. A live nonenterotoxigenic F4+ E. coli vaccine for oral administration to pigs immediately following weaning is commercially available in certain countries and effectively protects weaned pigs against post-weaning diarrhoea due to F4[+] ETEC (Nagy and Fekete, 1999, 2005; Radostits *et al.*, 2007; Fairbrother and Gyles, 2011).

Enteropathogenic and attaching–effacing-verotoxigenic *Escherichia coli*

Before the 1980s, the pathogenic mechanisms of several DEC in humans were still totally unknown and their only property was to belong to a limited number of serotypes (Levine *et al.*, 1978; WHO, 1980). Currently, these DEC comprise not only several pathotypes that were identified in the course of time in humans [enteropathogenic, verotoxigenic, enteroadherent, diffusely adherent and necrotoxigenic *E. coli* to cite only the most relevant (Nataro and Kaper, 1998; Kaper *et al.*, 2004)], but also some that have been identified in animals (Quinn *et al.*, 2011), and which will be described in the next two sections.

Among these groups, enteropathogenic (EPEC) and a subset of verotoxigenic (VTEC) *E. coli* have in common the production of a histological and ultramicroscopic lesion characterized by the effacement of the enterocyte microvilli and by intimate (< 10 nm gap) attachment of the bacteria to the exposed cytoplasmic membrane of the enterocytes. This lesion was thus referred to as 'attaching and effacing' (AE) (see Chapter 10) and the strains involved were originally named, 'attaching/effacing *E. coli*' (AEEC) (Moon *et al.*, 1983; Moxley and Smith, 2010; Mainil, 2013). The same lesion can also be produced by strains of *Escherichia albertii* (ex-*Hafnia alvei*) and of *Citrobacter rodentium* (ex-*C. freundii*) (Luperchio *et al.*, 2000; Ooka *et al.*, 2013). The effacement of the microvilli is caused by a reorganization of the actin filament network of the eukaryotic cell as a consequence of the translocation of bacterial effectors into the eukaryotic target cells via a type III secretion system that is entirely encoded by genes located on a pathogenicity island (Pai), the locus for enterocyte effacement (LEE). The intimate attachment is mediated by interaction between a bacterial outer membrane protein, the intimin adhesin, and its receptor, named the 'Translocated Intimin Receptor' or Tir, being a type III-translocated bacterial protein that is incorporated into the eukaryotic cell cytoplasmic membrane. Both intimin and Tir are coded by LEE-located genes.

In addition to the production of verotoxins (VTx), also named Shiga toxins (Stx), EPEC and attaching–effacing (AE)-VTEC differ by several properties, such as the production of different primary colonization factors (review in Bardiau *et al.*, 2010), when identified. Also, EPEC are subdivided in the literature into 'typical (t) EPEC' producing the type 4 fimbriae named 'bundle-forming pili' (BFP) and isolated primarily from humans and 'atypical (a) EPEC' not producing the BFP and isolated from animals and humans (Trabulsi *et al.*, 2002; Hernandes *et al.*, 2009; Gomes *et al.*, 2011; Mainil, 2013). Diarrhoea is considered to be the result of the production of the AE lesion and of the ensuing inflammatory response after the secretion of pro-inflammatory cytokines by the damaged enterocytes. In VTEC infections, VTx are responsible for the intestinal haemorrhages (haemorrhagic colitis or HC) and, after crossing the intestinal wall by transcytosis and resorption into the bloodstream, for internal organ lesions like the renal lesions in the haemolytic uraemic syndrome (HUS) in humans or the oedema disease in piglets (see next section), as a consequence of their cytotoxicity on endothelial cells of blood vessels.

EPEC cause acute and chronic diarrhoea in domestic mammals of various species and age groups (Table 2.1), most dramatically in newborn (up to 1 week of age) and weaned rabbits. Indeed, EPEC is one of the most frequently observed and lethal diarrhoeagenic infectious agents in this species, especially in newborn/suckling rabbits in which a case fatality rate of up to 100% may occur. They are also responsible for high economic losses in weaned rabbits (up to 50% case fatality rate), depending on the serotype of the EPEC. They cause watery yellowish, usually non-haemorrhagic, diarrhoea with dehydration and death. The clinical condition is easily reproduced experimentally in 4-week-old rabbits. At necropsy, the small intestine is congested and inflamed and the caecal contents are yellowish/brownish, liquid, and foul smelling, rarely containing blood (Peeters, 1994; Milon *et al.*, 1999; Wales *et al.*, 2005; Gyles and Fairbrother, 2010). Manifestation of EPEC infections occurs less frequently in other non-ruminant animal species (suckling and weaned piglets, pups and kittens, and also sometimes older dogs and cats), consisting of persistent diarrhoea, although intervention of additional infectious and non-infectious factors influences the severity of the clinical signs. Nevertheless, porcine EPEC may also cause economic losses of some importance in pig farms. AE lesions and clinical signs can be experimentally reproduced in colostrum-deprived newborn calves and piglets and in young pups. In piglets, pups and kittens, the lesions at necropsy are milder than in rabbits: inflammation is not always visible macroscopically and the contents may be liquid, but usually not foul-smelling (Helie et al., 1991; Beutin, 1999; Mainil et al., 2000; Sancak et al., 2004; Wales et al., 2005; Malik et al., 2006; Morato et al., 2009; Gyles and Fairbrother, 2010; Gouveia et al., 2011).

The majority of rabbit and porcine EPEC produce intestinal colonization factors in addition to intimin: the AF/R 1/2 ('adhesive factor/rabbit 1 and 2'), Ral ('REPEC adherence locus') and LPF ('long polar fimbriae') adhesins by rabbit EPEC; the Paa ('porcine attaching-effacing associated') adhesin by porcine EPEC; and the genes coding for LPF are also present in porcine EPEC (Berendson et al., 1983; An et al., 1999; Milon et al., 1999; Batisson et al., 2003; Dow et al., 2005; Newton et al., 2004; Bruant et al., 2009; Bardiau et al., 2010; Gyles and Fairbrother, 2010). Conversely, only a very few canine and feline EPEC harbouring the genes coding for known colonization factors have been reported: genes coding for the Paa adhesin of porcine EPEC or for the BFP type 4 fimbriae of human tEPEC (An et al., 1999; Goffaux et al., 2000; Nakazato et al., 2004; Almeida et al., 2012).

Chronic and acute, sometimes haemorrhagic diarrhoea has also been associated with intestinal colonization and faecal excretion of AE-VTEC in dogs, although evidence of their actual role in the development of the enteritis is somewhat limited. Though other aetiologies are possible, in some instances dogs can also suffer from HUS due to the action of the E. coli VTx, more especially after an episode of bloody diarrhoea, as in humans (Beutin, 1999; Staats et al., 2003; Sancak et al., 2004; Wang et al., 2006; Gyles and Fairbrother, 2010).

EPEC and AE-VTEC may also cause diarrhoea in ruminants, especially in young calves and to a lesser extent in lambs, up to 3 months of age. Although AE-VTEC of a large number of serotypes are associated with a healthy carrier state in ruminants, only a few are naturally and experimentally responsible for enteritis and diarrhoea in calves, more especially O5:H–, O26:H11, O118:H16, and O111:H–. EPEC associated with enteritis and diarrhoea in calves mostly belong to the same serotypes as AE-VTEC, also raising the questions of their actual identity and putative zoonotic role (see infra). The diarrhoea naturally observed or experimentally produced is often quite mild, though sometimes haemorrhagic, but also persistent and therefore is associated with economic losses through reduced growth rate and mortality in 2–3% of cases. Nevertheless, additional infectious (viral infection) and non-infectious (general hygiene level and climate) factors can highly influence the severity of the clinical signs and of the lesions (Mainil, 1999; Mainil and Daube, 2005; Naylor et al., 2005; Wales et al., 2005; Gyles, 2007; Wieler et al., 2007; Gyles and Fairbrother, 2010; Moxley and Smith, 2010; Bolton, 2011). The role of the VTx in pathogenicity in young ruminants, if any, is still uncertain. Indeed, VTx cause no blood vessel damage being unable to cross the intestinal wall in contrast to what happens in humans and piglets, and endothelial cells of ruminants lack receptors for VTx (Pruimboom-Brees et al., 2000; Hoey et al., 2002). Therefore, clinical manifestations of HC and HUS, as in humans or ED, as in piglets (see next section), have never been observed.

Rabbit EPEC are considered to be host-specific, as they belong to specific serotypes and produce specific colonization factors (AF/R1, AF/R2 or Ral) (Milon et al., 1999; Dow et al., 2005). On the other hand, this question remains unresolved for porcine, canine and feline EPEC. Some of these can indeed belong to the same serotypes as the human aEPEC and/or harbour genes coding for similar colonization factors, such as the Paa adhesin or the BFP type 4 fimbriae of human tEPEC and for other general and virulence-associated properties. It could also be argued that these dogs and cats were contaminated by humans in the first place, initiating a circle of mutual cross-contaminations (Trabulsi et al., 2002; Rodrigues *et al.*, 2004; Bentancor, 2006; Malik *et al.*, 2006; Hernandes *et al.*, 2009; Morato *et al.*, 2009; Moura *et al.*, 2009; Almeida *et al.*, 2012).

Similarly, the zoonotic potential of EPEC and AE-VTEC from diarrhoeic young ruminants is the centre of many discussions, because most of them belong to the same serotypes as human aEPEC and AE-VTEC and can harbour similar virulence-associated genes (Trabulsi *et al.*, 2002; Mainil and Daube, 2005; Radostitis *et al.*, 2007; Bardiau *et al.*, 2009; Hernandes *et al.*, 2009; Gyles and Fairbrother, 2010; Moxley and Smith, 2010; Horcajo *et al.*, 2012). Human infection by AE-VTEC via foodstuffs contaminated by the faeces of healthy ruminant carriers is now proven beyond any doubt (Nataro and Kaper, 1998; Kaper *et al.*, 2004; Bolton, 2011). On the other hand, EPEC and AE-VTEC causing diarrhoea in young ruminants have not been well characterized with respect to their virulence or epidemiology. For example, their colonization factors have not yet been identified, although several candidates have been reported (Bardiau *et al.*, 2010). The most interesting one is the Lda ('locus for diffuse adherence') adhesin identified first in bovine O26:H11 EPEC (Szalo *et al.*, 2002; Scaletsky *et al.*, 2005; Bardiau *et al.*, 2009). However, Lda-encoding genes are also present in human aEPEC and (AE-)VTEC belonging to various serogroups (Scaletsky *et al.*, 2005, 2010). More systematic inclusion of EPEC and AE-VTEC from diarrhoeic young calves in comparative studies of isolates from healthy adult cattle and humans would help to determine their host and age specificity, if any (Wieler *et al.*, 1992, 2000; Mainil *et al.*, 2011).

In contrast to the situation in newborn and weaned rabbits, the clinical diagnosis of infections due to EPEC in young suckling or just weaned ruminants and piglets and, sometimes, in adult dogs and cats and to AE-VTEC in calves cannot go further than the observation of a mild to relatively severe and persistent diarrhoea. The aetiology must be confirmed following isolation of *E. coli* colonies by classical diagnostic procedures. Bio- and serotyping are strongly indicative of the rabbit EPEC (Peeters, 1994; Milon *et al.*, 1999) and VTx production may also be detected by ELISA assays (Ball *et al.*, 1994; Paton and Paton, 1998; Willford *et al.*, 2009; Parma *et al.*, 2012). However, these diagnostic tests have now been replaced by PCR targeting appropriate genes, in particular, those coding for the intimin adhesin (*eae* gene) and for the VTx (*vtx* genes) (e.g. China *et al.*, 1996; Franck *et al.*, 1998; Osek, 2003; Rajkhowa *et al.*, 2010; Valadez *et al.*, 2011; Botkin *et al.*, 2012).

Nevertheless it is often difficult to confirm the actual role of EPEC and AE-VTEC in animal infections, as additional infectious and non-infectious factors influence the severity of the clinical signs and the course of the disease (cfr supra). Another critical problem is the loss of genes encoding virulence properties, especially those located on phages and coding for the VTx, during subculture and/or storage, or even *in vivo*, leading to misdiagnosis (Karch *et al.*, 1992; Bielaszewska *et al.*, 2007). Recently, PCR targeting additional genes helped to distinguish 'true' EPEC from VTx-negative derivatives of AE-VTEC and may improve the diagnostic procedure (Bugarel *et al.*, 2011). Finally, it is important to note that no vaccine against EPEC and AE-VTEC is commercially available at this time.

Other diarrhoeagenic *Escherichia coli*

Two other pathotypes of DEC are also associated with diarrhoea in domestic mammals: 'necrotoxigenic' and 'diffusely adherent' *E. coli* (NTEC and DAEC) (Table 2.1). Two additional pathotypes of DEC are restricted to humans and primates: the 'enteroinvasive' and 'enteroaggregative' *E. coli* (EIEC and EAggEC). The latter are characterized by a stacked brick-like or aggregative adherence pattern on cells in culture (Nataro and Kaper, 1998; Kaper *et al.*, 2004) (see Chapter 12). *E. coli* with an aggregative adherence pattern have been recently isolated from diarrheic animals in Brazil, but were negative for the specific adhesins, toxins and expression regulator of human EAggEC (Uber *et al.*, 2006; Weintraub, 2007; Piérard *et al.*, 2012). Thus, to the authors' knowledge, there exists no report about the presence of human-like EAggEC in animals.

Necrotoxigenic *Escherichia coli*

NTEC are defined on the basis of the production of cytotoxic necrotizing factors (CNFs) 1 to 3 (Table 2.1; see Chapter 7). CNF1 is produced by *E. coli* (NTEC1) isolated from humans and all domestic mammals; CNF2 is restricted to *E. coli* (NTEC2) isolated from cattle and sheep; and CNF3 is restricted to *E. coli* (NTEC3) isolated from sheep and goats. NTEC1 are responsible for intestinal (DEC) and different extra-intestinal (extra-intestinal pathogenic *E. coli* or ExPEC) infections, whereas NTEC2 may cause diarrhoea (DEC) and septicaemia (septicaemic *E. coli* or SePEC) and NTEC3 have not been associated with disease to date (De Rycke *et al.*, 1999; Knust and Schmidt, 2010). Nevertheless, since NTEC2 and to a lesser extent NTEC1, can be isolated from the faecal or intestinal content of up to 50% of healthy animals, their diarrhoeagenic role may be questioned. Diarrhoea has effectively been reproduced in a colostrum-deprived newborn calf model following challenge with an NTEC2 isolate, but not with its ΔCNF2 mutant, underlining the diarrhoeagenic role of CNF2 toxin (Van Bost *et al.*, 2003b), and in a newborn pig model following challenge with an NTEC1 isolate, although a diarrhoeagenic role for CNF1 has not been demonstrated (Fournout *et al.*, 2000). In addition to CNF toxins, NTEC from animals and humans can produce various fimbrial and afimbrial adhesins, other cytotoxins and iron-sequestering systems, and resist the bactericidal action of complement, all properties typical of ExPEC (De Rycke *et al.*, 1999; Mainil *et al.*, 1999, 2001; Kaper *et al.*, 2004; Mainil and Van Bost, 2004; Gyles and Fairbrother, 2010). Thus, NTEC will be described in greater detail in the sections on ExPEC infections.

Diffusely adherent *Escherichia coli*

DAEC is characterized by a diffuse adherence pattern on cells in culture (Table 2.1) (Nataro *et al.*, 1987). They have been more especially described in intestinal and extra-intestinal infections in humans, but may also be present in animals. Diffuse adherence (DA) may be mediated by the AIDA adhesin (Benz and Schmidt, 1989) of porcine ETEC (see previous section) and VTEC causing ED (see next section) that has been demonstrated to play a role in the colonization of the pig intestine in certain circumstances (Ngeleka *et al.*, 2003; Ravi *et al.*, 2007; Gyles and Fairbrother, 2010). DA is also mediated by the family of afimbrial adhesins (Afa) (Nowicki *et al.*, 1990;

Le Bouguénec and Servin, 2006). Recently it has been demonstrated that some *E. coli* from intestinal and extraintestinal infections in domestic mammals and birds harbour the genes coding for Afa-VIII adhesin, sometimes also being positive for the CNF1 or CNF2 toxins (Mainil *et al.*, 1997, 1999; Lalioui *et al.*, 1999; Gérardin *et al.*, 2000). Since many DAEC from animals can also produce other fimbrial and afimbrial adhesins, different cytotoxins including CNF and iron-sequestering systems, and resist the bactericidal action of complement, all properties typical of ExPEC (Kaper *et al.*, 2004; Le Bouguénec and Servin, 2006) they will also be further described in the sections on ExPEC infections.

Enterotoxaemic *Escherichia coli*

Another class of verotoxigenic *E. coli* (VTEC) is responsible for the 'oedema disease' (ED) syndrome in recently weaned piglets (Imberechts *et al.*, 1992; Mainil, 1999; Mainil and Daube, 2005; Gyles and Fairbrother, 2010). ED was first described in 1938 in Great Britain and Ireland and was characterized by sudden deaths, nervous signs, fluid infiltration of tissues, and the presence of high numbers of β haemolytic *E. coli* in the bowel (Shanks, 1938). About 20 years later 'the conclusion drawn from clinical and experimental observations was that the naturally occurring disease was caused by the absorption into the system of a specific toxic factor formed in the bowel' (Timoney, 1957). It was also rapidly observed that the β-haemolytic *E. coli* belong to a limited number of serotypes. Historically, the most frequently encountered serotypes were O138:K81, O139:K82, O141:K85 (Sojka *et al.*, 1960). Though the serotypes O139:K82 and O141:K85 remain the most prevalent, with geographic variations, several others have been described in the course of time (Gyles and Fairbrother, 2010).

Again about 20 years later, the 'specific toxic factor' was characterized as a toxin active on Vero cells, and was successively named '*E. coli* neurotoxin', 'oedema disease principle', and finally 'porcine *E. coli* Verotoxin 2e or Shiga toxin 2e' (VTx2e or Stx2e). In contrast to the other *E. coli* VTx and certain VTx2e of human origin, VTx2e associated with ED may not be encoded by genes located on bacteriophages as no VTx-converting phages could be isolated from such strains (Konowalchuk *et al.*, 1977; Dobrescu, 1983; Weinstein *et al.*, 1998; Mainil and Daube, 2005; Gyles, 2007). Enterocytes are not the primary target cells of VTx2e *in vivo* (Waddell *et al.*, 1996). Indeed, VTx2e crosses the porcine intestinal epithelium by transcytosis after binding to its specific Gb4 receptors, most probably travels in the bloodstream (= enterotoxaemia) in association with leucocytes (like in humans), and attaches to Gb4 receptors on the endothelial cells of the blood vessels throughout the body. The inhibition of protein synthesis causes endothelial cell death and damage to the blood vessel walls leading to extravasation (= 'oedema disease') (Gyles and Fairbrother, 2010).

ED occurs in fattening piglets as one of the diseases of the post-weaning period. The clinical picture is characterized by nervous signs and sudden deaths, as a consequence of the cerebral oedema. At necropsy, oedema can be observed, especially in subcutaneous tissues (for example, of the eyelids and face), in the stomach and intestine walls, and in the peritoneum. Typical ED is characterized by the absence of diarrhoea, as VTx2e is not enterotoxigenic. When diarrhoea is observed, the aetiology can be dietary or infectious, with different enteric viruses or bacteria, including *E. coli*. The latter combine the properties of VTEC and of porcine ETEC responsible for post-weaning diarrhoea, especially the production of F4 fimbriae and of STb and/or LT enterotoxins. They are most commonly of the O138:K81, O141:K85, and O149:K– serotypes (Nagy and Fekete, 1999, 2005; Radostits *et al.*, 2007; Gyles and Fairbrother, 2010). The incidence of ED greatly varies between farms and countries. As adult sows are the source of VTEC contamination of their offspring, in the absence of healthy carriers on the farm, ED will never occur. Also, the general management during the post-weaning period can differ from farm to farm. Indeed, the lower the level of stress of the piglets at weaning, the lower the incidence of ED, and of other infectious diseases (Radostits *et al.*, 2007).

The first step in the pathogenesis of ED is the colonization of the small intestine. Porcine VTEC do not produce any type 3 secretion system nor do they induce the AE lesion. However, they produce F18 fimbrial adhesins (Table 2.1), either the F18ab variant that was described in the early 1990s, or also, as demonstrated recently, the F18ac variant, more specifically associated with porcine ETEC (Imberechts *et al.*, 1992; Debroy *et al.*, 2009; Gyles and Fairbrother, 2010). The name F18-VTEC has been proposed for ED-associated porcine VTEC (Piérard *et al.*, 2012). It was later shown that most porcine F18-VTEC are also positive for the AIDA adhesin (Table 2.1). The genes coding for F18 fimbriae are plasmid-located along with the genes coding for the AIDA adhesin (Niewerth *et al.*, 2001; Fekete *et al.*, 2002; Mainil *et al.*, 2002). As observed for F18+ ETEC, recessive homozygote piglets are genetically resistant to colonization by F18-VTEC as no receptor is synthesized, and therefore never develop ED (Radostits *et al.*, 2007; Gyles and Fairbrother, 2010). It has also been suggested, but not yet proven, that α-Hly gives an advantage to porcine VTEC in the colonization of and subsequent multiplication in the porcine intestine (Gyles and Fairbrother, 2010).

ED may be highly suspected based on the age of the affected piglets and the clinical signs observed. Laboratory confirmation can be obtained by growth of (pure) culture of β-haemolytic *E. coli* from the small intestinal content and by characterization of the isolates by serotyping and genetic (PCR) and immunological (ELISA) assays to detect the genes coding for VTx2e and F18 or their production, respectively (Radostits *et al.*, 2007; Barth *et al.*, 2011). The ELISA assay can also be directly applied on faeces or on the intestinal content.

Production of protective antibodies against the VTx2e is possible following parenteral or oral administration of chemically or genetically detoxified toxin, but no commercial vaccine has been marketed to date (MacLeod and Gyles, 1991; Bosworth *et al.*, 1996; Johansen *et al.*, 1997; Makino *et al.*, 2001).

VTx2e-positive *E. coli* have also been isolated, though rather rarely, from humans suffering diarrhoea from the early 1990s, raising the question of the zoonotic potential of ED-associated porcine F18-VTEC (Piérard *et al.*, 1991; Muniesa et al., 2000). However, since human and porcine VTx2e-producing *E. coli* differ with respect to their serotype and virulent properties, human strains being F18 negative (Sonntag *et al.*, 2005), it is unlikely that the latter represent a public health hazard.

Extraintestinal *Escherichia coli*

In contrast to DEC, extraintestinal pathogenic *E. coli* (ExPEC) have been subdivided on the basis of the clinical syndromes they provoke and of their target species and not on the basis of their combinations of particular virulence properties. Overall, animal ExPEC comprise *E. coli* causing invasive infections of the blood (septicaemia) and internal organs in mammals and birds, or local infections of the urinary tract (cystitis and pyelonephritis), genital tract (prostatitis, metritis and pyometra) and mammary gland (mastitis) in mammals, and of the subcutaneous tissue (cellulitis) in birds, as well as those strains associated with serositis, abscesses, wound infections, etc. One important epidemiological feature of all these clinical syndromes is that the causative pathogenic *E. coli* strains are directly or indirectly of faecal origin. Hence, a high standard of general hygiene of the farms, kennels, and immediate environment of the animals is of the upmost importance as prophylactic measure.

The following sections will focus on *E. coli* causing invasive infections in mammals (named 'septicaemic *E. coli*' or SePEC') or birds (named 'avian pathogenic *E. coli*' or APEC') and urinary tract infections (UTI) in mammals (named 'uropathogenic *E. coli*' or UPEC'), since these ExPEC are quite well described (Table 2.1). On the other hand, the clinical syndromes associated with *E. coli* infections of the genital tract, mammary gland and (sub)cutaneous tissue, will not be discussed because much less is known about the associated *E. coli*. In the latter group, host factors are indeed likely to play the most important role. Let us nevertheless mention recent publications describing the intracellular survival adaptation of *E. coli* causing chronic mammary gland infections (White *et al.*, 2010; Dogan *et al.*, 2012). Not only do such strains certainly deserve more investigation in the future to fully understand

their adaptation to this intracellular niche in the mammary gland, but, also, it will be important to carry out surveys looking for strains with similar intracellular survival potential and responsible for chronic infections of other tissues and organs (see also the section on UPEC).

Invasive *Escherichia coli*

Invasive *E. coli* must be able not only to colonize an epithelium surface but also to cross the mucosal layers of different organs (intestine, kidneys, lungs, air sacs, mammary gland, uterus) and to traverse the blood vessel walls in both directions. They must also have the capacity to survive and multiply in the bloodstream and internal organs. Consequently, invasive *E. coli* possess a wide range of virulence factors and properties (Table 2.1): production of colonization factors, crossing of the mucosal and blood vessel barriers, resistance to phagocytosis and to bactericidal activity of the complement, production of iron uptake systems, and/or production of toxins. One exception is the subset of the invasive *E. coli* strains entering the bloodstream via the navel in newborn mammals that do not need to colonize any host mucosal surfaces. In contrast to humans, specific neonatal meningitis-associated *E. coli* (NMEC) have not been described in animals.

Septicaemic *Escherichia coli* of mammals

E. coli can cause septicaemia and infections of internal organs in all domestic mammals, but the most frequent and dramatic cases occur in newborn ruminants. In piglets, foals, pups and kittens, invasive *E. coli* strains are less frequently encountered, being only one of the multiple bacterial species causing neonatal septicaemia. The most important ports of entry are the small intestine and the navel in newborn animals and the tonsils in older animals. Intestinal colonization of newborns is facilitated by the absence of microbiota at birth and the survival and multiplication of invasive bacteria in the bloodstream and internal organs, by the absence of systemic maternal antibodies since these cannot cross the placental barrier during pregnancy. The first antibodies to reach the bloodstream in newborns have traversed the epithelial intestinal barrier, having originated from the colostrum. However, this crossing is limited to only 24–36 h after birth in ruminants and to a few days in other animal species. In adults, *E. coli* can also cause septicaemia but quite infrequently. The ports of entry differ from those in the newborn: pyelic cavity of the kidneys (especially in dogs and cats; see section on UPEC), mammary gland (especially in ruminants) and uterus (especially in dogs in the case of pyometrum), and directly via injuries (Broes, 1993; Beutin, 1999; Radostitis *et al.*, 2007; Fecteau *et al.*, 2009; Gyles and Fairbrother, 2010; Quinn *et al.*, 2011).

In newborn calves, lambs and kids, neonatal colisepticaemia can occur from birth up to only a few days of age (usually < 1 week) whereas it can occur later in other animal species: the younger the animal, the more severe the disease manifestations, although these can also vary depending on the virulence profile of the invasive *E. coli*. It was not unusual in the past, in Western Europe, to see cattle farms with a mortality rate of over 50% in calves in the first week after calving. Since SePEC are excreted in the faeces of healthy adults, newborns are contaminated via the oral route after contact with and licking of faeces-contaminated environment. In peracute cases, no clinical sign is observed, the animals being found dead. In acute cases, the animals manifest non-specific general clinical signs, such as depression, apathy, fever, lack of suckling ability, etc., and usually die within 24–48 h. In these cases, the lesions are non-specific, and include congestion of internal organs and fluid accumulation in the peritoneal and pleural cavities. The bacteria can be recovered in high numbers and in pure culture from the blood and some internal organs. In less acute and subacute cases, involving more resistant animals and/or less pathogenic *E. coli*, more specific clinical signs and lesions in different organs, such as diarrhoea and enteritis, respiratory distress and bronchopneumonia, nervous signs and meningo-encephalitis, lameness and arthritis, small abscesses on internal organs, and petechial haemorrhages on the serous membranes, may be observed. In surviving animals, persistent sequelae, such as arthritis, pneumonia, and/or meningitis may be observed. If the portal of entry is the navel, omphalitis and omphalo-phlebitis are present. During the course of the disease, the causative *E. coli* may be excreted in faeces, nasal secretions and urine (Radostitis *et al.*, 2007; Fecteau *et al.*, 2009).

As expected SePEC possess a wide range of virulence-associated properties and factors (Table 2.1) (MacLaren, 1997; Wray and Woodward, 1997; Beutin, 1999; De Rycke *et al.*, 1999; Mainil, 2003a,b, 2013; Van Bost and Mainil, 2003; Kaper *et al.*, 2004; Mainil and Van Bost, 2004; Radostits *et al.*, 2007; Smith *et al.*, 2007; Fecteau *et al.*, 2009; Gyles and Fairbrother, 2010).

Bovine and porcine SePEC belong to a wide range of O serogroups, such as O15, O45, O78 and O115. O antigens of SePEC from foals, pups and kittens are less characterized. Except for O serogrouping, SePEC from mammals have, in general, not been well characterized. Application of appropriate prophylactic measures, such as strict hygiene to avoid contact with adult faeces, and correct administration of colostrum, is usually sufficient to ensure a dramatic drop in the incidence of colisepticaemia and in the case fatality rate on the affected farm.

The colonization of the small intestine by SePEC can be mediated by one or more of several fimbrial and non-fimbrial adhesins, including P (F165-1 or F11-like/PapGIII), S (Sfa, F165-2 or F1C-like) and F17 (b or c) fimbrial, Afa-VIII afimbrial, and/or CS31A non-fimbrial adhesins. The latter, CS31A, is a plasmid-encoded, capsule-like antigen produced by certain porcine and bovine SePEC, that not only mediates adherence to intestinal cells but also resistance to phagocytosis. However, their precise role has not been proven to date in *in vivo* models. Curli adhesins are also usually produced by SePEC and may intervene in the adherence to the extra-cellular matrix after the crossing of the intestinal epithelium.

SePEC can also produce one or more of three types of cytotoxins: CNFs (1 or 2), CDTs (III or sometimes IV) and α-Hly, whose precise role(s) and *in vivo* target cell(s) remain nevertheless largely undefined. As already mentioned (see section on DEC), NTEC1 are associated with septicaemia in all mammals whereas NTEC2 are restricted to ruminants. Targets of the CNFs may be the epithelial and endothelial cells, which could acquire phagocytic capacity allowing the crossing of the mucosal (e.g. intestinal) and blood vessel barriers by the *E. coli*. In addition, CDTs could also contribute to the weakening of the epithelial and endothelial barriers by their DNase-like activity in actively replicating cells leading to G2/M mitosis arrest, cellular distension and cell death. However, a role for CNF1, CNF2, or CDT-III in the development of septicaemia has not been demonstrated in *in vivo* experimental models in the piglet and calf (Fournout *et al.*, 2000; Van Bost *et al.*, 2003b). CNF1 is produced in combination with P fimbriae and α-Hly and their encoding genes are located on the same pathogenicity island. NTEC1 can also harbour genes coding for S fimbriae alone or in combination with P fimbriae, and though infrequently, for CDT-IV. Conversely, CNF2 is most often associated with F17b or F17c fimbriae or with Afa-VIII adhesin and with CDT-III, whose encoding genes can be located on the same plasmid.

The mechanisms of traversing the intestinal mucosa and the blood vessel walls of the *E. coli* have not been more precisely identified, to date. Various outer membrane proteins seem to be required for mucosal and 'in and out' blood vessel wall crossing by human NMEC (Table 2.1). One of them, the IbeA protein, has been more extensively studied and indeed contributes to *in vitro* and *in vivo* invasion of brain microvascular epithelial cells (BMEC). However, the *ibeA* gene present in NMEC has not been identified in SePEC, as far as the authors know.

In addition to cytotoxins, the lipid A moiety of the LPS, or endotoxin, which is liberated in the bloodstream and internal organs after bacterial lysis, has multiple toxic activities that can lead to a septic shock during Gram-negative infections in mammals. The liberated endotoxin indeed initiates an inflammatory response with fever, vasodilatation, hypotension, activation of the complement cascade, disseminated intravascular coagulation, and shock. Nevertheless, not all host species are equally sensitive to the *E. coli* endotoxins, humans, foals, piglets and young ruminants being the most sensitive.

To successfully invade the bloodstream and internal organs, SePEC must also resist phagocytosis and complement activities, and possess different systems to acquire iron. The first specific iron acquisition system to be identified in SePEC is the plasmid-encoded aerobactin siderophore, but others were later identified, such as the salmochelin siderophore system and the ferric yersiniabactin uptake system. Aerobactin, salmochelin and yersiniabactin contribute to virulence *in vivo*, but the role of the others still needs confirmation.

A few O antigens can prevent the deposit at the bacterial cell surface either of the C3b component and therefore opsonization and phagocytosis, or of the membrane attack complex (MAC) and therefore the MAC-mediated lysis. The actual mechanism of this anti-complement activity is not known but could reside in spatial obstruction by long and ramified O antigens (like the O78 antigen) or in the formation of a capsule-like structure by the O antigens (like the K'V165' capsule by the O115 antigen). Conversely, the K1 antigen of human NMEC is not produced by animal SePEC, as far as the authors know. Surprisingly, some fimbriae, such as the P fimbriae, may also contribute to the resistance to phagocytosis of SePEC in the bloodstream and internal organs. The F165-1 P fimbriae of porcine SePEC, for example, would inhibit the oxidative response by polymorphonuclear leucocytes.

In addition, different outer membrane proteins (OMPs) can play a similar preventive role against complement-mediated activities, amongst others, the Iss and the TraT proteins. The Iss protein is believed to prevent the depositing of the MAC and the TraT protein, the depositing of the C3b component preventing later stages of complement activity, although the mechanisms are not understood. Both Iss and TraT are plasmid-encoded, the *traT* gene actually being present within the gene cluster for plasmid conjugation, whereas the *iss* gene might be part of a large pathogenicity island present only in SePEC and APEC. Genes coding for CNF2 and CDT-III cytotoxins, for F17b, F17c or Afa-VIII adhesins and/or for the aerobactin siderophore can be located on the same plasmid as the *iss* genes.

Routine laboratory diagnosis of SePEC is more straightforward than that of DEC. Indeed, no bacteria are usually present in the blood and internal organs of healthy animals. Therefore, isolation of a 'pure and abundant growth of *E. coli*' from the bloodstream, internal organs, blood marrow and/or joints (especially in animals having survived for several days) is, in theory, sufficient for a presumptive diagnosis of SePEC infection. Nevertheless, confirmation of the pathotype is important since any *E. coli* can invade the cadaver post mortem, especially when necropsy has been delayed (Radostits *et al.*, 2007; Fecteau *et al.*, 2009). Pathotype identification can be performed using multiplex PCR for detection of appropriate virulence-associated genes (e.g. Van Bost *et al.*, 2003a).

Vaccines were available in the past in industrialized countries and are still available in some countries for immunization of the mother before calving in order to produce enough specific colostral antibodies. However, there are some problems impeding the efficacy of such vaccines: (i) the strains present in the vaccine may not reflect the actual SePEC present on the farm or their virulence factors; (ii) the vaccination will not stop the spread of the SePEC on farms where poor hygienic conditions allow frequent contact of newborns with adult faeces (Radostits *et al.*, 2007; Fecteau *et al.*, 2009; Gyles and Fairbrother, 2010).

Since animal SePEC may share certain virulence properties, like O antigens, CNF1 toxin, P and/or S fimbriae, Afa-VIII afimbrial adhesin, etc., with human SePEC or UPEC, the question of their zoonotic potential has been raised (Girardeau *et al.*, 2003; Smith *et al.*, 2007). However, epidemiological data linking animal and human infections are still lacking. In the event that isolates are transmitted to human beings (which is still to be proven) it would certainly not be via foodstuffs, as farm animal SePEC target the newborn, and not slaughter-aged animals.

Avian pathogenic *Escherichia coli*

E. coli causing any localized or systemic infections in poultry, chickens, turkeys, ducks, etc. is termed 'avian pathogenic *E. coli*' (APEC). This section, however, will focus on APEC causing septicaemia and systemic infections of the internal organs, a condition causing severe economic losses with up to 50% case fatality rate in industrialized broiler chicken, turkey and duck farms worldwide. Nevertheless, the severity of the disease and associated lesions in affected birds vary depending on the combinations of virulence factors of APEC (Table 2.1). In the case of severe per-acute

septicaemia, as in mammals, no clinical signs are observed prior to death and no specific lesion is visible at necropsy. In most other cases, loss of appetite, prostration, high fever, and respiratory distress may be observed. At necropsy, multiple lesions may be present in various combinations: fibrineous and caseous aerosacculitis, perihepatitis, pericarditis and/or peritonitis, hepatitis and nephritis with presence of necrotic foci, or salpingitis and yolk egg infection. In subacute cases, lesions of osteomyelitis, arthritis, and cellulitis, may also be observed (Gross, 1994; Dho-Moulin and Fairbrother, 1999; Gyles and Fairbrother, 2010; Mellata, 2013).

Like SePEC, APEC are present in poultry as members of the intestinal microbiota but, unlike SePEC, they contaminate the birds by the respiratory route via inhalation of *E. coli*-carrying dust from dried faeces. Classical APEC belong to limited numbers of O and K serogroups and possess different combinations of virulence factors and properties (Dho-Moulin and Fairbrother, 1999; Gérardin *et al.*, 2000; Stordeur *et al.*, 2002, 2004; Vandemaele *et al.*, 2003; Schouler *et al.*, 2004; Li *et al.*, 2005; Rodriguez-Siek *et al.*, 2005b; Dziva and Stevens, 2008; Gyles and Fairbrother, 2010; Kariyawasam and Nolan, 2011; Wang *et al.*, 2011a,b; Gao *et al.*, 2012; Mellata *et al.*, 2012; Mellata, 2013). The most cited O antigens are O1, O2 and O78 although they represent less than half of the APEC and dozens of other O antigens may be identified with regional variation. The best known K antigens are K1, most frequently associated with the O1 and O2 somatic antigens, K5, K15 and K80.

The colonization of the lungs and air sacs is mediated by fimbrial and non-fimbrial adhesins. For example, F1 fimbriae and P (F11/PapGII) fimbriae are statistically associated with APEC compared to non-pathogenic *E. coli* of the intestinal tract. F1 fimbriae are expressed by more than 70% of APEC during colonization of the trachea, lungs and air sacs and may therefore mediate adherence to the corresponding epithelium layers. However, mutants in the F1-encoding gene cluster are still able to efficiently colonize the respiratory tract in an experimental model in germ-free chickens. Clearly, the role of F1 fimbriae as a colonization factor does not seem primordial and can be taken over by other adhesins. P (F11/PapGII) fimbriae are also expressed in the lungs and air sacs (although not in the trachea), but by only 20–25% of APEC. It is therefore no surprise that the search for other adhesins by different means has identified several candidates whose precise role and importance are nevertheless still a matter for debate: these include F17 fimbriae, Afa-VIII afimbrial adhesin, Tsh haemagglutinin, AC/I S fimbriae, Stg fimbriae, AatA autotransporter protein, and curli, and others are most probably still awaiting recognition.

Within a few hours after initial infection, the APEC enter the bloodstream by crossing the blood vessel walls in the lungs and of air sacs. The mechanism by which APEC reach the bloodstream is still poorly understood. The CNF and CDT toxins have not been described in APEC and the *ibeA* gene has only been statistically associated with APEC of the O1 and O2 serogroups, leaving room for discovery of additional specific factors in the other APEC serogroups.

As with SePEC, APEC can resist phagocytosis and complement-mediated activities. Indeed, certain O and K antigens of APEC can prevent the deposition at the bacterial cell surface either of the C3b component or of the MAC. For example, isogenic mutants of the O78 and the K1 antigens are impaired in their resistance to complement and their potential for invasion. In addition significantly more APEC than non-pathogenic avian *E. coli* possess the plasmid-located *iss* and *traT* genes (see section 'Septicaemic *Escherichia coli* of mammals') and P (F11/PapGII) fimbriae expressed during the internal tissue invasion process can also contribute to the resistance to phagocytosis. Genes coding for several iron acquisition systems are also significantly associated with APEC: these include aerobactin, salmochelin, *Salmonella* iron transport (Sit) system, iron-repressible protein (Irp) and yersiniabactin. The roles of aerobactin, salmochelin, Sit system and yersiniabactin in enhancing the virulence of APEC have been proven *in vivo*, although antibody specific for aerobactin is not protective. In addition to these systems, which contribute to direct acquisition of iron from host siderophores, a haem utilization/transport protein has been found in an O2 APEC, by signature tagged mutagenesis.

Several toxin-encoding genes have also been identified in APEC, but in very limited numbers of isolates, and their expression *in vivo* has not been studied. Thus, their precise roles in pathogenesis remain obscure.

As is the case for SePEC, the initial step in the diagnosis of invasive colibacillosis in poultry is the isolation, preferably in pure and abundant culture, of *E. coli* from the blood and internal organs. This is usually followed by the identification of the predominant somatic antigens, O1, O2 and O78. However, as already mentioned, these three somatic antigens usually represent less than half of the APEC. Identification of APEC is performed using multiplex PCR targeting genes coding for adhesins, IbeA, capsular antigens, siderophores and/or toxins (e.g. Skyberg *et al.*, 2003; Ewers *et al.*, 2005; Johnson *et al.*, 2008). Recently, a confirmatory diagnostic strategy was proposed based on four patterns of virulence genes allowing the confirmation of 70% of APEC (Schouler *et al.*, 2012). This strategy, however, cannot include genes with low prevalence and does not incorporate the most recently described genes whose addition should improve the percentage of identification.

Although the virulence gene profiles of APEC can now be more precisely identified and several vaccines have been commercialized for active or passive immunization of broilers, there is still no fully efficient vaccine against APEC. The reasons are essentially similar to those mentioned for vaccines against SePEC in mammals, but an additional factor may be that poultry colibacillosis is secondary to viral or mycoplasma infections in field conditions. In general, prophylatic measures are based on disinfection of the eggs, control of the sources of infection, and prevention of primary mycoplasmal and viral infections.

The zoonotic potential of APEC has been the centre of much discussion for several years now. Indeed, certain APEC carry genes coding for similar if not identical properties identified in human NMEC and UPEC. For example, many APEC and human UPEC shared common serogroups, phylogenetic groups, and virulence gene profiles. The most striking evidence is that of common virulence gene profiles, including the presence of the *ibeA* gene, in APEC and NMEC of serotype

O18:K1:H7, although the proportion of APEC belonging to this serotype is quite low. Nevertheless, avian and human O18:K1:H7 *E. coli* present very similar, but not identical, PFGE profiles (Rodriguez-Siek *et al.*, 2005a; Moulin-Schouleur *et al.*, 2006; Smith *et al.*, 2007; Ewers *et al.*, 2009; Tivendale *et al.*, 2010; Mellata, 2013). Therefore, appropriate epidemiological studies should be performed to check whether APEC can indeed be transmitted to humans and cause either meningitis or UTI. It would also be most interesting to perform similar epidemiological studies to explore the possibility of cross-contamination of other mammalian species by APEC.

Uropathogenic *Escherichia coli*

E. coli is the most common bacterial species responsible for cystitis and pyelonephritis in all domestic animals, females being more susceptible to UTI than males, for anatomical reasons. UPEC infections occur most frequently in dogs, followed by cats and horses. Pigs and ruminants are considered to be less frequently infected by UPEC, although this may reflect a lack of interest for economical reasons. As a consequence, more data have been published on UPEC in dogs and cats than in any other animal species. UPEC are of faecal origin and colonize the urinary tract usually following interference with micturition due to anatomical abnormalities, long-term catheterization, presence of calculi or injuries, colics and abdominal pain, because they are not completely flushed out and can therefore multiply in the bladder (Warren, 1996; Quinn *et al.*, 1997, 2011; Starcic *et al.*, 2002; Gyles and Fairbrother, 2010).

UTI are usually ascending and begin with the colonization of the urethra and of the bladder, sometimes followed by the colonization of the urethra and of the pyelic cavity of the kidney. Usual clinical signs of acute cystitis are dysuria, pollakiuria, stranguria, and abdominal pain, with occasional presence of blood (haematuria), pus (pyuria) and/or proteins (proteinuria). Local clinical signs of pyelonephritis are similar but more pronounced with frequent and relatively abundant presence of blood, pus and/or necrotic tissue; general clinical signs are also present, such as fever, depression and anorexia. Later clinical signs linked to elevated uraemia can develop:

vomiting, polydipsia, polyuria, and halithiosis. Sublumbar pain is easy to detect by palpating the back between the fingers at the level of the kidneys. As in humans, the kidneys can be the port of entry into the bloodstream of UPEC that possess additional properties enabling them to cause septicaemia and systemic infections. UPEC can also cause metritis and pyometra in females and prostatitis in males, more especially in dogs (Warren, 1996; Quinn *et al.*, 1997, 2011; Gyles and Fairbrother, 2010).

In addition to being adapted to grow in the nutriment-restricted environment found in urine (except in individuals with diabetes mellitus where glucose is present and strongly favours bacterial growth in urine), UPEC possess specific virulence-associated properties and factors not present in other faecal *E. coli* isolates (Warren, 1996; Johnson, 1997; Yuri *et al.*, 1998; Beutin, 1999; Rippere-Lampe *et al.*, 2001; Starcic *et al.*, 2002; Wiles *et al.*, 2008; Gyles and Fairbrother, 2010; Mainil, 2013). For example, UPEC possess a restricted range of O and K antigens and are motile (H antigens). The most prevalent somatic O antigens of dog and cat UPEC are O4 (H–, H4, H5) and O6 (H1, H4, H5, H7, H31), but several others have been identified. The most prevalent K capsular antigens are K1 (associated with somatic antigens O2, O7, O18, O21, O83), K5 (associated with somatic antigens O6, O21, O75), and K53 (associated with somatic antigen O6).

Besides the F1 fimbriae that help to colonize the bladder, but not the pyelic cavity of the kidneys, canine and feline UPEC also produce different types of fimbrial and afimbrial adhesins, the most frequently encountered being F12/PapGIII and F13/PapGIII P fimbriae. Their encoding genes are located on pathogenicity islands (Pai) along with the genes coding for CNF1 and/or α-Hly. S fimbriae include the Sfa and F1C types and their encoding genes are also located on Pai, sometimes together with the genes coding for α-Hly. In addition, some UPEC can produce the F17c (or G) fimbrial adhesin or the Afa-VIII (or M haemagglutinin) afimbrial adhesin. Together, these adhesins enable UPEC to adhere to the uroepithelial cells of the bladder and, especially for the P fimbriae, of the pyelic cavity. This interaction provokes the secretion of pro-inflammatory cytokines that attract neutrophils that can eliminate the bacteria, but also cause some damage to the epithelial and mucosal layers. However, similarly to SePEC, several UPEC, such as those associated with pyelonephritis, can resist phagocytosis and complement activity and also produce similar iron-acquisition systems. Resistance to phagocytosis and complement activity is mediated by different K antigens like K1 and K5, OMP like Iss and TraT proteins, and/or adhesins like P fimbriae.

Additional tissue damage may be induced by CNF1 and α-Hly. All (or almost all) CNF1-positive UPEC produce α-Hly, but not all α-Hly-positive UPEC produce CNF1. As for SePEC, CNF1 may help UPEC to cross the mucosal layers of the kidneys and the walls of the blood vessels to cause septicaemia. Conversely, the prevalence of the *ibeA* gene in UPEC has not been studied yet, to the authors' knowledge.

In addition to causing acute UTI, some UPEC are associated with recurrent infections and/or asymptomatic bacteriuria, as observed in humans. Like the persistent *E. coli* strains in the mammary gland in cattle, persistent UPEC have developed a strategy to enter, survive, and form a biofilm-like intracellular structure inside the uro-epithelial cells (Reigstad *et al.*, 2007; Wiles *et al.*, 2008; Ejrnaes *et al.*, 2011). More research is needed not only to fully understand their survival mechanisms and clinical relevance in humans, but also to study their relevance in animal UTI that may serve as a model for humans.

Diagnosis of UPEC infection follows the same approach as for any bacterial UTI. Urine is preferably sampled via catheter, especially in females, to minimize contamination by vaginal, perineal and faecal microbiota, and a volume of 50 to 100 µl is inoculated onto agar plates. Presumptive *E. coli* colonies are confirmed by biochemical tests and an antibiotic sensitivity profile is performed. Case history and number of colony forming units (CFU) are important elements in the appreciation of the presence of an *E. coli* UTI infection. In dogs, number of CFU must be over 10^3–10^5 per ml and, in cats, over 10^2–10^4 depending on the urine sampling method. Confirmation of the isolation of a UPEC by PCR targeting genes coding for most important virulence properties is rarely performed in routine diagnosis. Prophylaxis of UPEC-associated UTI is based initially on prevention of faecal contamination and of any predisposing cause. No vaccine is currently present on the market (Yamamoto *et al.*, 1995; Quinn *et al.*, 1997, 2011; Starcic *et al.*, 2002).

Since canine, feline and human UPEC share common properties (serotypes, adhesins, toxins, iron acquisition systems), the questions of whether there is cross-contamination and what is the zoonotic potential of animal UPEC have been asked for many years now (Beutin, 1999; Johnson *et al.*, 2000; Kaper *et al.*, 2004; Smith *et al.*, 2007; Gyles and Fairbrother, 2010). Nevertheless, the following points should be stressed: (i) although canine, feline, and human UPEC share similar properties, they are not necessarily identical as the variants of, for example, the major and adhesin subunits of fimbriae can differ (Stroemberg *et al.*, 1990); and (ii) epidemiological evidence is still lacking. Therefore, at this stage we clearly need more field and laboratory studies before providing definite answers to these two questions.

Concluding remarks

E. coli disease is still of great economic importance throughout the world, especially in intensive production animals, in spite of remarkable technological advances allowing rapid molecular characterization of the pathogenic strains of *E. coli*. Much is now known about the molecular basis of *E. coli* pathogenicity and interaction with the host. However, we still have a poor understanding of how certain pathogenic *E. coli* are able to persist and be transmitted in the environment both in the intestine and outside the host. A greater understanding of these areas will certainly allow us to conceive more efficacious control and prevention measures for *E. coli* diseases in animals.

One question, which preoccupies many researchers today, is the 'possible zoonotic potential' of the pathogenic *E. coli* present in animals. Often, conclusions are drawn from incomplete comparative studies. Because genes coding for virulence factors belonging to the same family are identified by PCR in pathogenic strains from animals and humans, the former are considered as representing a 'possible public health hazard'.

However, as already mentioned, we still lack epidemiological evidence of cross-contamination for most of these. Also, we certainly must not forget that if some animal pathogenic *E. coli* can be transmitted by foodstuffs to humans, other food borne pathogenic *E. coli* are specific human pathogens, as observed for the recent Agg-VTEC outbreak in Germany (Piérard *et al.*, 2012). Nevertheless, although the progress in the molecular area is helpful to better understand the fundamental mechanisms and the evolution of the pathogenic bacteria, medical bacteriologists still need to go back to the basics in bacterial pathogenicity and to *in vivo* reproduction of the diseases in animal models to identify their true virulence factors and study their range of target hosts, fulfilling the molecular Koch's postulates (Falkow, 1988).

The history of the pathogenic strains of *E. coli* is fascinating and raises many questions: Where do they originate from?; When did they acquire the genes coding for virulence-associated properties?; How do they evolve today?; and of course, Why do they suddenly rise to the attention of the medical fields? Retrospective studies indeed confirm that most, if not all, pathotypes existed prior to their actual recognition (Mainil, 1999; Van Bost *et al.*, 2001; Zhang *et al.*, 2006). With the huge technical development in the molecular field, whole genome sequencing of bacteria is a matter of hours today and phylogenetic studies of pathogenic *E. coli* can really begin (Zhang *et al.*, 2006; Leopold *et al.*, 2009; Ogura *et al.*, 2009; Kyle *et al.*, 2012). However, pathogenic *E. coli* are still evolving rapidly, selection being determined by conditions in their immediate environment. Continued basic front line monitoring of the virulence gene profile of pathogenic *E. coli* both in clinical cases and in healthy animals is still essential, to permit the rapid detection of the emergence of new variants responsible for the changing disease patterns being observed in animals and to respond with appropriate control and prevention measures.

References

Almeida, P.M., Arais, L.R., Andrade, J.R., Prado, E.H., Irino, K., and de Cerqueira, A.M. (2012). Characterization of atypical enteropathogenic *Escherichia coli* (aEPEC) isolated from dogs. Vet. Microbiol. *158*, 420–424.

An, H., Fairbrother, J.M., Desautels, C., and Harel, J. (1999). Distribution of a novel locus called Paa (porcine attaching and effacing associated) among enteric *Escherichia coli*. Adv. Exp. Med. Biol. *473*, 179–184.

Ball, H.J., Finlay, D., Burns, L., and Mackie, D.P. (1994). Application of monoclonal antibody-based sandwich ELISAs to detect verotoxins in cattle faeces. Res. Vet. Sci. *57*, 225–232.

Bardiau, M., Labrozzo, S., and Mainil, J.G. (2009). Putative adhesins of enteropathogenic and enterohemorrhagic *Escherichia coli* of serogroup O26 isolated from humans and cattle. J. Clin. Microbiol. *47*, 2090–2096.

Bardiau, M., Szalo, M., and Mainil, J.G. (2010). Initial adherence of EPEC, EHEC and VTEC to host cells. Vet. Res. *41*, 57–72.

Barth, S., Schwanitz, A., and Bauerfeind, R. (2011). Polymerase chain reaction-based method for the typing of F18 fimbriae and distribution of F18 fimbrial subtypes among porcine Shiga toxin-encoding *Escherichia coli* in Germany. J. Vet. Diagn. Invest. *23*, 454–464.

Batisson, I., Guimond, M.P., Girard, F., An, H., Zhu, C., Oswald, E., Fairbrother, J.M., Jacques, M., and Harel, J. (2003). Characterization of the novel factor *paa* involved in the early steps of the adhesion mechanism of attaching and effacing *Escherichia coli*. Infect. Immun. *71*, 4516–4525.

Bentancor, A. (2006). Epidemiological role of pets in urban transmisision cycle of STEC. Medicina (B Aires) *66*, 37–41.

Benz, I., and Schmidt, M.A. (1989). Cloning and expression of an adhesin (AIDA-I) involved in diffuse adherence of enteropathogenic *Escherichia coli*. Infect. Immun. *57*, 1506–1511.

Berendson, R., Cheney, C.P., Schad, P.A., and Boedeker, E.C. (1983). Species-specific binding of purified pili (AF/R1) from the *Escherichia coli* RDEC-1 to rabbit intestinal mucosa. Gastroenterol. *85*, 837–845.

Beutin, L. (1999). *Escherichia coli* as a pathogen in dogs and cats. Vet. Res. *30*, 285–298.

Bielaszewska, M., Prager, R., Köck, R., Mellmann, A., Zhang, W., Tschäpe, H., Tarr, P.I., and Karch, H. (2007). Shiga toxin gene loss and transfer *in vitro* and *in vivo* during enterohemorrhagic *Escherichia coli* O26 infection in humans. Appl. Environ. Microbiol. *73*, 3144–3150.

Bolton, D.J. (2011). Verocytotoxigenic (Shiga toxin-producing) *Escherichia coli*: virulence factors and pathogenicity in the farm to fork paradigm. Foodborne Pathog. Dis. *8*, 357–365.

Bosworth, B.T., Samuel, J.E., Moon, H.W., O'Brien, A.D., Gordon, V.M., and Whipp, S.C. (1996). Vaccination with genetically modified Shiga-like toxin IIe prevents edema disease in swine. Infect. Immun. *64*, 55–60.

Botkin, D.J., Galli, L., Sankarapani, V., Soler, M., Rivas, M., and Torres, A.G. (2012). Development of a multiplex PCR assay for detection of Shiga toxin-producing *Escherichia coli*, enterohaemorrhagic *Escherichia coli*, and enteropathogenic *E. coli* strains. Front. Cell. Infect. Microbiol. *2*(8), 1–10.

Broes, A. (1993). Les *Escherichia coli* pathogènes du chien et du chat. Ann. Méd. Vét. *137*, 377–384.

Bruant, G., Zhang, Y., Garneau, P., Wong, J., Laing, C., Fairbrother, J.M., Gannon, V.P., and Harel, J. (2009). Two distinct groups of porcine enteropathogenic *Escherichia coli* strains of serogroup O45 are revealed by comparative genomic hybridization and virulence gene microarray. BMC Genomics *10*, 402.

Bugarel, M., Beutin, L., Scheutz, F., Loukiadis, E., and Fach, P. (2011). Identification of genetic markers for differentiation of Shiga toxin-producing, enteropathogenic and avirulent strains of *Escherichia coli* O26. Appl. Environ. Microbiol. *77*, 2275–2281.

China, B., Pirson, V., and Mainil, J. (1996). Typing of bovine attaching and effacing *Escherichia coli* by multiplex *in vitro* amplification of virulence-associated genes. Appl. Environ. Microbiol. *62*, 3462–3465.

Contrepois, M.G., and Girardeau, J.P. (1985). Additive protective effects of colostral antipili antibodies in calves experimentally infected with enterotoxigenic *Escherichia coli*. Infect. Immun. *56*, 947–949.

De Rycke, J., Milon, A., and Oswald, E. (1999). Necrotoxic *Escherichia coli* (NTEC): two emerging categories of human and animal pathogens. Vet. Res. *30*, 221–234.

Dean, E.A., Whipp, S.C., and Moon, H.W. (1989). Age-specific colonization of porcine intestinal epithelium by 987P-piliated enterotoxigenic *Escherichia coli*. Infect. Immun, *57*, 82–87.

Debroy, C., Roberts, E., Scheuchenzuber, W., Kariyawasam, S., and Jayarao, B.M. (2009). Comparison of genotypes of *Escherichia coli* strains carrying F18ab and F18ac fimbriae from pigs. J. Vet. Diagn. Invest. *21*, 359–364.

Dho-Moulin, M., and Fairbrother, J.M. (1999). Avian pathogenic *Escherichia coli* (APEC). Vet. Res. *30*, 299–316.

Dobrescu, L. (1983). New biological effect of edema disease principle (*Escherichia coli* neurotoxin) and its use as an *in vitro* assay for this toxin. Am. J. Vet. Res. *44*, 31–34.

Dogan, B., Rishniw, M., Bruant, G., Harel, J., Schukken, Y.H., and Simpson, K.W. (2012). Phylogroup and *lpfA* influence epithelial invasion by mastitis-associated *Escherichia coli*. Vet. Microbiol. *159*, 163–170.

Dow M.A., Toth, I., Alexa, P., Davies, M., Malik, A., Oswald, E., and Nagy, B. (2005). Predominance of *afr2* and *ral* fimbrial genes related to those encoding the K88 and CS31A fimbrial adhesins in enteropathogenic *Escherichia coli* isolates from rabbits with postweaning diarrhoea in central Europe. J. Clin. Microbiol. *43*, 1366–1371.

Dubreuil, J.D. (2008). *Escherichia coli* STb toxin and colibacillosis: knowing is half of the battle. FEMS Microbiol. Lett. *278*, 137–145.

Dziva, F., and Stevens, M.P. (2008). Colibacillosis in poultry: unravelling the molecular basis of virulence of avian pathogenic *Escherichia coli* in their natural hosts. Avian Pathol. *37*, 355–366.

Ejrnaes, K., Stegger, M., Reisner, A., Ferry, S., Monsen, T., Holm, S.E., Lundgren, B., and Frimodt-Möller, N. (2011). Characteristics of *Escherichia coli* causing persistence or relapse of urinary tract infections. Phylogenetic groups, virulence factors and biofilm formation. Virulence *2*, 1–10.

Escherich, T. (1885). Die Darmbacterien des Neugeborenen und Saglings. Fortschr. Med. *3*, 515–522.

Ewers, C., Janssen, T., Kiessling, S., Philip, H.C., and Wieler, L.H. (2005). Rapid detection of virulence-associated genes in avian pathogenic *Escherichia coli* by multiplex polymerase chain reaction. Avian Dis. *49*, 269–273.

Ewers, C., Antao, E.M., Diehl, I., Philip, H.C., and Wieler, L.H. (2009). Intestine and environment of the chicken as reservoirs for extra-intestinal pathogenic *Escherichia coli* strains with zoonotic potential. Appl. Environ. Microbiol. *75*, 184–192.

Fairbrother, J.M., and Gyles, C.L. (2011). *Escherichia coli*. In Diseases of Swine (10th Edition), Straw, B.E., Zimmerman, J.J., D'Allaire, S., and Taylor, D.J., eds. (Iowa State University Press, Ames, IA, USA), pp. 723–749.

Falkow, S. (1988). Molecular Koch's postulates applied to microbial pathogenicity. Rev. Infect. Dis. *10*, S274-S276.

Fecteau, G., Smith, B.P., and George, L.W. (2009). Septicemia and meningitis in the newborn calf. Vet. Clin. North Am. Food Anim. Pract. *25*, 195–208.

Fekete, P.Z., Gérardin, J., Jacquemin, E., Mainil, J.G., and Nagy, B. (2002). Replicon typing of the F18 fimbriae encoding plasmids of enterotoxigenic and verotoxigenic *Escherichia coli* strains from porcine postweaning diarrhoea and oedema disease. Vet. Microbiol. *85*, 275–284.

Foster, D.M., and Smith, G.W. (2009). Pathophysiology of diarrhoea in calves. Vet. Clin. North Am. Food Anim. Pract. *25*, 13–36.

Fournout, S., Dozois, C.M., Odin, M., Desautels, C., Pérès, S., Hérault, F., Daigle, F., Segafredo, C., Lafitte, J., Oswald, E., Fairbrother, J.M., and Oswald, I.P. (2000). Lack of a role of cytotoxic necrotizing factor 1 toxin from *Escherichia coli* in bacterial pathogenicity and host cytokine response in infected germfree piglets. Infect. Immun. *68*, 839–847.

Francis, D.H., Erickson, A.K., and Grange, P.A. (1999). K88 adhesins of enterotoxigenic *Escherichia coli* and their porcine enterocyte receptors. Adv. Exp. Med. Biol. *473*, 147–154.

Franck, S.M., Bosworth, B.T., and Moon, H.W. (1998). Multiplex PCR for enterotoxigenic, attaching and effacing, and Shiga toxin-producing *Escherichia coli* strains from calves. J. Clin. Microbiol., *36*, 1795–1797.

Gao, Q., Wang, X., Xu, H., Ling, J., Zhang, D., Gao, S., and Liu, X. (2012). Roles of iron acquisition systems in virulence of extra-intestinal pathogenic *Escherichia coli*: salmochelin and aerobactin contribute more to virulence than heme in a chicken infection model. BMC Microbiol. *12*, 143.

Gérardin, J., Lalioui, L., Jacquemin, E., Le Bouguénec, C., and Mainil, J.G. (2000). The *afa*-related gene cluster in necrotoxigenic and other *Escherichia coli* from animals belongs to the *afa-8* variant. Vet. Microbiol. *76*, 175–184.

Girardeau, J.P., Lalioui, L., Ou Said, A.M., De Champs, C., and Le Bouguénec, C. (2003). Extended virulence genotytpe of pathogenic *Escherichia coli* isolates carrying the *afa-8* operon: evidence of similarities between isolates from humans and animals with extra-intestinal infections. J. Clin. Microbiol. *41*, 218–226.

Goffaux, F., China, B., Janssen, L., and Mainil, J. (2000). Genotypic characterization of enteropathogenic *Escherichia coli* (EPEC) isolated in Belgium from dogs and cats. Res. Microbiol. *151*, 865–871.

Gomes, T.A., Hernandes, R.T., Torres, A.G., Salvador, F.A., Guth, B.E., Vaz, T.M., Irino, K., Silva, R.M., and Vieira, M.A. (2011). Adhesin-encoding genes from Shiga toxin-producing *Escherichia coli* are more prevalent in atypical than in typical enteropathogenic *E. coli*. J. Clin. Microbiol. *49*, 3334–3337.

Gouveia, E.M., Silva, I.S., Nakazato, G., de Araujo, F.R., and Chang, M.R. (2011). Experimental infection with enteropathogenic *Escherichia coli* identified by PCR using enteric-coated capsules in boxer pups. Acta Cir. Bras. *26*, 144–148.

Gross, W.G. (1994). Diseases due to *Escherichia coli* in poultry. In *Escherichia coli* in Domestic Animals and Humans, Gyles, C.L., ed. (Wallingford, Oxon, UK: CAB International), pp. 237–260.

Gyles, C.L. (2007). Shiga-toxin-producing *Escherichia coli*: an overview. J. Anim. Sci. *85(Suppl. 13)*, E45–62.

Gyles, C.L., and Fairbrother, J.M. (2010). *Escherichia coli*. In Pathogenesis of Bacterial Infections in Animals, Gyles, C.L., Prescott, J.F., Songer, J.G., and Thoen, C.O., eds. (Ames, IA: Wiley-Blackwell), pp. 267–308.

Helie, P., Morin, M., Jacques, M., and Fairbrother, J.M. (1991). Experimental infection of newborn pigs with an attaching and effacing *Escherichia coli* O45:K'E65' strain. Infect. Immun. *59*, 814–821.

Hernandes, R.T., Elias, W.P., Vieira, M.A.M., and Gomes, T.A.T. (2009). An overview of atypical enteropathogenic *Escherichia coli*. FEMS Microbiol. Lett. *297*, 137–149.

Hoey, D.E., Currie, C., Else, R.W., Nutikka, A., Lingwood, C.A., Gally, D.L., and Smith, D.G. (2002). Expression of receptors for Verotoxin 1 from *Escherichia coli* O157 on bovine intestinal epithelium. J. Med. Microbiol. *51*, 143–149.

Holland, R.E. (1990). Some infectious causes of diarrhea in young farm animals. Clin. Rev. Microbiol. *3*, 345–375.

Horcajo, P., Dominguez-Bernal, G., de la Fuente, R., Ruiz-Santa-Quiteria, J.A., Blanco, J.E., Blanco, M., Mora, A., Dahbi, G., Lopez, C., Puentes, B., *et al.* (2012). Comparison of ruminant and human attaching and effacing *Escherichia coli* (AEEC) strains. Vet. Microbiol. *155*, 341–348.

Imberechts, H., De Greve, H., and Lintermans, P. (1992). The pathogenesis of edema disease in pigs – A review. Vet. Microbiol. *31*, 221–233.

Jensen, C.O. (1893). Ueber die Kälberruhr und deren Aetiologie. Mh. Tierhelk. *4*, 97–124.

Jensen, G.M., Frydendahl, K., Svendsen, O., Jorgensen, C.B., Cirera, S., Fredholm, M., Nielsen, J.P., and Moller, K. (2006). Experimental infection with *Escherichia coli* O149:F4ac in weaned piglets. Vet. Microbiol. *115*, 243–249.

Johansen, M., Andresen, L.O., Jorsal, S.E., Thomsen, L.K., Waddell, T.E., and Gyles, C.L. (1997). Prevention of edema disease in pigs by vaccination with Verotoxin 2e toxoid. Can. J. Vet. Res. *61*, 280–285.

Johnson, J.R. (1997). Urinary tract infections. In *Escherichia coli*: Mechanisms of Virulence, Sussman, M., ed. (Cambridge University Press, Cambridge, UK), pp. 495–549.

Johnson, J.R., O'Bryan, T.T., Low, D.A., Ling, G., Delavari, P., Fasching, C., Russo, T.A., Carlino, U., and Stell, A.L. (2000). Evidence for commonality between canine and human extra-intestinal pathogenic *Escherichia coli* strains that express *papG* allele III. Infect. Immun. *68*, 3327–3336.

Johnson, T.J., Wannemuehler, Y., Doetkott, C., Johnson, S.J., Rosenberger, S.C., and Nolan, L.K. (2008). Identification of minimal predictors of avian pathogenic *Escherichia coli* virulence for use as a rapid diagnostic tool. J. Clin. Microbiol. *46*, 3987–3996.

Josse, M., Jacquemin, E., and Kaeckenbeeck, A. (1980). Présence chez le chien d'*Escherichia coli* productrices d'une entérotoxine thermostable (STa). Ann. Méd. Vét. *124*, 211–214.

Kaper, J.B., Nataro, J.P., and Mobley, H.L. (2004). Pathogenic *Escherichia coli*. Nat. Rev. Microbiol. *2*, 123–140.

Karch, H., Meyer, T., Rüssman, H., and Heesemann, J. (1992). Frequent loss of Shiga-like toxin genes in clinical isolates of *Escherichia coli* upon subcultivation. Infect. Immun. *60*, 3464–3467.

Kariyawasam, S., and Nolan, L.K. (2011). *papA* gene of avian pathogenic *Escherichia coli*. Avian Dis. *55*, 532–538.

Knust, Z., and Schmidt, G. (2010). Cytotoxic Necrotizing Factors (CNFs)–A growing toxin family. Toxins *2*, 116–127.

Koch, R. (1884). Die Aetiologie der Tuberkulose. Mitt. Kaiser Gesundh. *2*, 1–88.

Konowalchuk, J., Speirs, J.I., and Stavric, S. (1977). Vero response to a cytotoxin of *Escherichia coli*. Infect. Immun. *18*, 775–779.

Kyle, J.L., Cummings, C.A., Parker, C.T., Quiñones, B., Vatta, P., Newton, E., Huynh, S., Swimley, M., Degoricija, L., Barker, M., *et al.* (2012). *Escherichia coli* serotype O55:H7 diversity supports parallel acquisition of bacteriophage at Shiga toxin phage insertion sites during evolution of the O157:H7 lineage. J. Bacteriol. *194*, 1885–1896.

Lalioui, L., Jouve, M., Gounon, P., and Le Bouguénec, C. (1999). Molecular cloning and characterization of the *afa-7* and *afa-8* gene clusters encoding afimbrial adhesins in *Escherichia coli* strains associated with diarrhoea or septicemia in calves. Infect. Immun. *67*, 5048–5059.

Le Bouguénec, C., and Servin, A.L. (2006). Diffusely adherent *Escherichia coli* strains expressing Afa/Dr adhesins (Afa/Dr DAEC): hitherto unrecognized pathogens. FEMS Microbiol. Lett. *256*, 185–194.

Leopold, S.R., Magrini, V., Holt, N.J., Shaikh, N., Mardis, E.R., Cagno, J., Ogura, Y., Iguchi, A., Hayashi, T., Mellmann, A., *et al.* (2009). A precise reconstruction of the emergence and constrained radiations of

Escherichia coli O157 portrayed by backbone concatenomic analysis. Proc. Natl. Acad. Sci. U.S.A. *106*, 8713–8718.

Levine, M.M., Bergquist, E.J., Nalin, D.R., Waterman, D.H., Hornick, R.B., Young, C.R., and Sotman, S. (1978). *Escherichia coli* strains that cause diarrhoea but do not produce heat-labile or heat-stable enterotoxins and are non-invasive. Lancet *1(8074)*, 1119–1122.

Li, G., Laturnus, C., Ewers, C., and Wieler, L.H. (2005). Identification of genes required for avian *Escherichia coli* septicemia by signature-tagged mutagenesis. Infect. Immun. *73*, 2818–2827.

Luperchio, S.A., Newman, J.V., Dangler, C.A., Schrenzel, M.D., Brenner, D.J., Steigerwalt, A.G., and Schauer, D.B. (2000). *Citrobacter rodentium*, the causative agent of transmissible murine colonic hyperplasia, exhibits clonality: synonymy of *C. rodentium* and mouse-pathogenic *Escherichia coli*. J. Clin. Microbiol. *38*, 4343–4350.

MacLaren, D.M. (1997). Soft tissue infection and septicaemia. In *Escherichia coli* – Mechanisms of Virulence, Sussman, M., ed. (University Press, Cambridge, UK), pp. 469–494.

MacLeod, D.L., and Gyles, C.L. (1991). Immunization of pigs with a purified Shiga-like toxin II variant toxoid. Vet. Microbiol. *29*, 309–318.

Mainil, J. (1999). Shiga/Verocytoxins and Shiga/Verotoxigenic *Escherichia coli* in animals. Vet. Res. *30*, 235–257.

Mainil, J. (2002). Les souches pathogènes d'*Escherichia coli* chez les chiens et chats: IV) Discussion générale. Ann. Méd. Vét. *146*, 219–224.

Mainil, J. (2003a). Facteurs de virulence et propriétés spécifiques des souches invasives d'*Escherichia coli*: (I) Les adhésines et facteurs de colonisation. Ann. Méd. Vét. *147*, 105–126.

Mainil, J. (2003b). Facteurs de virulence et propriétés spécifiques des souches invasives d'*Escherichia coli*: (II) Franchissement des muqueuses et propriétés invasives. Ann. Méd. Vét. *147*, 159–171.

Mainil, J. (2013). *Escherichia coli* virulence factors. Vet. Immunol. Immunopathol. *152*, 2–12.

Mainil, J., and Daube, G. (2005). Verotoxigenic *Escherichia coli* from animals, humans and foods: who's who? J. Appl. Microbiol. *98*, 1332–1344.

Mainil, J., and Van Bost, S. (2004). Facteurs de virulence et propriétés spécifiques des souches invasives d'*Escherichia coli*: (IV) Les souches nécrotoxinogènes. Ann. Méd. Vét. *148*, 121–132.

Mainil, J., Jacquemin, E., Hérault, F., and Oswald, E. (1997). Presence of *pap*-, *sfa*- and *afa*-related sequences in necrotoxigenic *Escherichia coli* isolates from cattle: evidence for new variants of the AFA family. Can. J. Vet. Res. *61*, 193–199.

Mainil, J., Jacquemin, E., Pohl, P., Fairbrother, J.M., Ansuini, A., Le Bouguénec C., Ball, H.J., De Rycke, J., and Oswald, E. (1999). Comparison of necrotoxigenic *Escherichia coli* isolates from farm animals and from humans. Vet. Microbiol. *70*, 123–135.

Mainil, J., Janssen, L., Charlier, G., Jacquemin, E., China, B., and Goffaux, F. (2000). Les souches pathogènes d'*Escherichia coli* chez les chiens et les chats: II)

Données cliniques et bactériologiques sur les souches entéropathogènes. Ann. Méd. Vét. *144*, 335–343.

Mainil, J., Wilbaux, M., Jacquemin, E., Oswald, E., Imberechts, H., and Van Bost, S. (2001). Les souches pathogènes d'*Escherichia coli* chez les chiens et les chats: III) Données bactériologiques et cliniques sur les souches nécrotoxinogènes et sur celles positives pour des adhésines. Ann. Méd. Vét. *145*, 343–354.

Mainil, J., Jacquemin, E., Pohl, P., Kaeckenbeeck, A., and Benz, I. (2002). DNA sequences coding for the F18 fimbriae and AIDA adhesin are localized on the same plasmid in *Escherichia coli* isolates from piglets. Vet. Microbiol. *86*, 303–311.

Mainil, J., Bardiau, M., Ooka, T., Ogura, Y., Murase, K., Etoh, Y., Ichihara, S., Horikawa, K., Buvens, G., Piérard, D., Itoh, T., and Hayashi, T. (2011). IS*621*-based multiplex PCR printing method of O26 enterohaemorrhagic and enteropathogenic *Escherichia coli* isolated from humans and cattle. J. Appl. Microbiol. *111*, 773–786.

Makino, S., Watarai, M., Tabuchi, H., Shirahata, T., Furuoka, H., Kobayashi, Y., and Takeda, Y. (2001). Genetically modified Shiga toxin 2e (Stx2e) producing *Escherichia coli* is a vaccine candidate for porcine edema disease. Microb. Pathog. *31*, 1–8.

Malik, A., Toth, I., Beutin, L., Schmidt, H., Taminiau, B., Dow, M.A., Morabito, S., Oswald, E., Mainil, J., and Nagy, B. (2006). Serotypes and intimin types of intestinal and faecal strains of *eae*⁺ *Escherichia coli* from weaned piglets. Vet. Microbiol. *114*, 82–93.

Mellata, M. (2013). Human and avian extraintestinal pathogenic *Escherichia coli*: infections, zoonotic risks and antibiotic resistance trends. Foodborne Pathog. Dis. *10*, 916–932.

Mellata, M., Maddux, J.T., Nam, T., Thomson, N., Hauser, H., Stevens, M.P., Mukhopadhyay, S., Sarker, S., Crabbé, A., Nickerson, C.A., et al. (2012). New insights into the bacterial fitness-associated mechanisms revealed by the characterization of large plasmids of an avian pathogenic *Escherichia coli*. PLoS ONE *7(1)*, e29481.

Ménard, L.P., and Dubreuil, J.D. (2002). Enteroaggregative *Escherichia coli* heat-stable enterotoxin I (EAST1): a new toxin with an old twist. Crit. Rev. Microbiol. *28*, 43–60.

Milon, A., Oswald, E., and De Rycke, J. (1999). Rabbit EPEC: a model for the study of enteropathogenic *Escherichia coli*. Vet. Res. *30*, 203–219.

Moon, H.W., Whipp, S.C., Argenzio, R.A., Levine, M.M., and Giannella, R.A. (1983). Attaching and effacing activities of rabbit and human enteropathogenic *Escherichia coli* in pig and rabbit intestines. Infect. Immun. *41*, 1340–1351.

Morato, E.P., Leomil, E.M., Belisie, B.W., and Holmes, R.K. (2009). Domestic cats constitue a natural reservoir of human enteropathogenic *Escherichia coli* types. Zoon. Pub. Health *56*, 229–237.

Moulin-Schouleur, M., Schouler, C., Tailliez, P., Kao, M.R., Brée, A., Germon, P., Oswald, E., Mainil, J., Blanco, M., and Blanco, J. (2006). Common virulence factors and genetic relationship between O18:K1:H7 *Escherichia coli* isolates of human and avian origin. J. Clin. Microbiol. *44*, 3484–3492.

Moura, R.A., Sircili, M.P., Leomil, L., Matte, M.H., Trabulsi, L.R., Elias, W.P., Irino, K., and de Castro A.F.P. (2009). Clonal relationship among atypical enteropathogenic *Escherichia coli* strains isolated from different animal species and humans. Appl. Environ. Microbiol. *75*, 7399–7408.

Moxley, R.A., and Smith, D.R. (2010). Attaching-effacing *Escherichia coli* infections in cattle. Vet. Clin. North Am. Food Anim. Pract. *26*, 29–56.

Muniesa, M., Recktenwald, J., Bielaszewska, M., Karch, H., and Schmidt, H. (2000). Characterization of a Shiga toxin 2e-converting bacteriophage from an *Escherichia coli* strain of human origin. Infect. Immun. *68*, 4850–4855.

Nagy, B., and Fekete, P.Z. (1999). Enterotoxigenic *Escherichia coli* (ETEC) in farm animals. Vet. Res. *30*, 259–284.

Nagy, B., and Fekete, P.Z. (2005). Enterotoxigenic *Escherichia coli* in veterinary medicine. Int. J. Med. Microbiol. *295*, 443–454.

Nagy, B., Casey, T.A., Whipp, S.C., and Moon, H.W. (1992). Susceptibility of porcine intestine to pilus-mediated adhesion by some isolates of piliated enterotoxigenic *Escherichia coli* increases with age. Infect. Immun. *60*, 1285–1294.

Nakazato, G., Gyles, C.L., Ziebell, K., Keller, R., Trabulsi, L.R., Gomes, T.A., Irino, K., Da Silveira, W.D., and de Castro, A.F.P. (2004). Attaching and effacing *Escherichia coli* isolated from dogs in Brazil: characteristics and serotypic relationship to human enteropathogenic *E. coli* (EPEC). Vet. Microbiol. *101*, 269–277.

Nataro, J.P., and Kaper, J.B. (1998). Diarrheagenic *Escherichia coli*. Clin. Microbiol. Rev. *11*, 142–201.

Nataro, J.P., Kaper, J.B., Robins-Browne, R., Prado, V., Vial, P., and Levine, M.M. (1987). Patterns of adherence of diarrhoeagenic *Escherichia coli* to HEp-2 cells. Pediatr. Infect. Dis. J. *6*, 829–831.

Nataro, J.P., Bopp, C.A., Fields, P.I., Kaper, J.B., and Strockbine, N.A. (2011). *Escherichia*, *Shigella*, and *Salmonella*. In Manual of Clinical Microbiology (10th edition), Versalovic, J., Caroll, K.C., Funke, G., Jorgensen, J.H., Landry, M.L., and Warnock, D.W., eds. (ASM Press, Washington, DC), pp. 603–626.

Nawar, H.F., King-Lyons, N.D., Hu, J.C., Pasek, R.C., and Connell, T.D. (2010). LT-IIc, a new member of the type II heat-labile enterotoxin family encoded by an *Escherichia coli* strain obtained from a non-mammalian host. Infect. Immun. *78*, 4705–4713.

Naylor, S.W., Gally, D.L., and Low, J.C. (2005). Enterohemorrhagic *Escherichia coli* in veterinary medicine. Int. J. Med. Microbiol. *295*, 419–441.

Newton, H.J., Sloan, J., Bennett-Wood, V., Adams, L.M., Robins-Browne, R.M., and Hartland, E.L. (2004). Contribution of long polar fimbriae to the virulence of rabbit-specific enteropathogenic *Escherichia coli*. Infect. Immun. *72*, 1230–1239.

Ngeleka, M., Pritchard, J., Appleyard, G., Middleton, D.M., and Fairbrother, J.M. (2003). Isolation and association of *E. coli* AIDA/STb, rather than EAST1 pathotype with diarrhoea in piglets and antibiotic sensitivity of isolates. J. Vet. Diagn. Invest. *15*, 242–252.

Niewerth, U., Frey, A., Voos, T., Le Bouguénec C., Baljer, G., Franke, S., and Schmidt, M.A. (2001). The AIDA autotransporter system is associated with F18 and Stx2e in *Escherichia coli* isolates from pigs diagnosed with edema disease and postweaning diarrhoea. Clin. Diagn. Lab. Immunol. *8*, 143–149.

Nowicki, B., Labigne, A., Moseley, S., Hull, R., Hull, S., and Moulds, J. (1990). The Dr hemagglutinin, afimbrial adhesins AFA-I and AFA-III, and F1845 fimbriae of uropathogenic and diarrhoea-associated *Escherichia coli* belong to a family of hemagglutinins with Dr receptor recognition. Infect. Immun. *58*, 279–281.

Ogura, Y., Ooka, T., Iguchi, A., Toh, H., Asadulghani, M.D., Oshima, K., Kodama, T., Abe, H., Nakayama, K., Kurokawa, K., et al. (2009). Comparative genomics reveal the mechanism of the parallel evolution of O157 and non-O157 enterohemorrhagic *Escherichia coli*. Proc. Natl. Acad. Sci. U.S.A. *106*, 17939–17944.

Ooka, T., Tokuoka, E., Furukawa, M., Nagamura, T., Ogura, Y., Arisawa, K., Harada, S., and Hayashi, T. (2013). Human gastroenteritis outbreak associated with *Escherichia albertii*, Japan. Emerg. Infect. Dis. *19*, 144–146.

Osek, J. (2003). Development of a multiplex PCR approach for the identification of Shiga toxin-producing *Escherichia coli* strains and their major virulence factor genes. J. Appl. Microbiol. *95*, 1217–1225.

Parma, Y.R., Chacana, P.A., Lucchesi, P.M., Rogé, A., Granobles Velandia, C.V., Krüger, A., Parma, A.E., and Fernandez-Miyakawa, M.E. (2012). Detection of Shiga toxin-producing *Escherichia coli* by sandwich enzyme-linked immunosorbent assay using chicken egg yolk IgY antibodies. Front. Cell. Infect. Microbiol. *2*, 84.

Paton, J.C., and Paton, A.W. (1998). Pathogenesis and diagnosis of Shiga toxin-producing *Escherichia coli* infections. Clin. Microbiol. Rev. *11*, 450–479.

Peeters, J.E. (1994). *Escherichia coli* infections in rabbits, cats, dogs, goats and horses. In *Escherichia coli* in Domestic Animals and Humans, Gyles, C.L., ed. (Wallingford, Oxon, UK: CAB International), pp. 261–283.

Piérard, D., De Greve, H., Haesebrouck, F., and Mainil, J.G. (2012). O157:H7 and O104:H4 Vero/Shiga toxin-producing *Escherichia coli*: respective role of cattle and humans. Vet. Res. *43*, 13.

Piérard, D., Huyghens, L., Lauwers, S., and Lior, H. (1991). Diarrhoea associated with *Escherichia coli* producing porcine edema disease verotoxin. Lancet *338(8769)*, 762.

Pruimboom-Brees, I.M., Morgan, T.W., Ackermann, M.R., Nyström, E.D., Samuel, J.E., Cornick, N.A., and Moon, H.W. (2000). Cattle lack vascular receptors for *Escherichia coli* O157:H7 Shiga toxins. Proc. Natl. Acad. Sci. U.S.A. *97*, 10325–10329.

Quinn, P.J., Donnelly, W.J.C., Carter, M.E., Markey, B.K.J., Torgerson, P.R., and Breathnach, R.M.S. (1997). Microbial and Parasitic Diseases of the Dog and Cat (W.B. Saunders Company, London, UK).

Quinn, P.J., Markey, B.K., Leonard, F.C., FitzPatrick, E.S., Fanning, S., and Hartigan, P.J. (2011). *Enterobacteriaceae*. In Veterinary microbiology and microbial diseases, 2nd edn. (Chichester, West Sussex, UK: Wiley and Blackwell), pp. 263–286.

Radostits, O., Gay, C.C., Hinchcliff, K., and Constable, P.D. (2007). Diseases associated with *Escherichia coli*. In Veterinary medicine – A textbook of the diseases of cattle, horses, sheep, pigs, and goats, 10th edn. (Saunders Elsevier, Philadelphia, PA), pp. 847–896.

Rajkhowa, S., Das, R., Bora, S., Rajkhowa, C., Rahman, H., and Bujarbaruah, K.M. (2010). Detection of Shiga toxin-producing *Escherichia coli* and enteropathogenic *Escherichia coli* in faecal samples of healthy mithun (*Bos frontalis*) by multiplex polymerase chain reaction. Zoon. Publ. Hlth. *57*, 397–401.

Ravi, M., Ngeleka, M., Kim, S.H., Gyles, C., Berthiaume, F., Mourez, M., Middleton, D., and Silko, E. (2007). Contribution of AIDA-I to the pathogenicity of a porcine diarrhoeagenic *Escherichia coli* and to intestinal colonization through biofilm formation in pigs. Vet. Microbiol. *120*, 308–319.

Reigstad, C.S., Hultgren, S.J., and Gordon, J.I. (2007). Functional genomic studies of uropathogenic *Escherichia coli* and host urothelial cells when intracellular bacterial communities are assembled. J. Biol. Chem. *282*, 21259–21267.

Rippere-Lampe, K.E., O'Brien, A.D., Corion, R., and Lockman, H.A. (2001). Mutation of the gene encoding cytotoxic necrotizing factor type 1 (*cnf1*) attenuates the virulence of uropathogenic *Escherichia coli*. Infect. Immun. *69*, 3954–3964.

Rippinger, P., Bertschinger, H.U., Imberechts, H., Nagy, B., Sorg, I., Stamm, M., Wild, P., and Wittig, W. (1995). Designations F18ab and F18ac for the related fimbrial types F107, 2134P and 8813 of *Escherichia coli* isolated from porcine postweaning diarrhoea and from oedema disease. Vet. Microbiol. *45*, 281–295.

Rodrigues, J., Thomazini, C.M., Lopes, C.A., and Danats, L.O. (2004). Concurrent infection in a dog and colonization in a child with human enteropathogenic *Escherichia coli* clone. J. Clin. Microbiol. *42*, 1388–1389.

Rodriguez-Siek, K.E., Giddings, C.W., Doetkott, C., Johnson, T.J., Fakhr, M.K., and Nolan, L.K. (2005a). Comparison of *Escherichia coli* isolates implicated in human urinary tract infection and avian colibacillosis. Microbiology *151*, 2097–2110.

Rodriguez-Siek, K.E., Giddings, C.W., Doetkott, C., Johnson, T.J., and Nolan, L.K. (2005b). Characterizing the APEC pathotype. Vet. Res. *36*, 241–256.

Runnels, P.L., Moseley, S.L., and Moon, H.W. (1987). F41 pili as protective antigens of enterotoxigenic *Escherichia coli* that produce F41, K99, or both pilus antigens. Infect. Immun. *55*, 555–558.

Ruth, N., Mainil, J., Roupie, V., Frère, J.M., Galleni, M., and Huygen, K. (2005). DNA vaccination for the priming of neutralizing antibody against non-immunogenic STa enterotoxin from enterotoxigenic *Escherichia coli*. Vaccine *23*, 3618–3627.

Sancak, A.A., Rutgers, H.C., Hart, C.A., and Batt, R.M. (2004). Prevalence of enteropathic *Escherichia coli* in dogs with acute and chronic diarrhoea. Vet. Rec. *154*, 101–106.

Sarmiento, J.I., Casey, T.A., and Moon, H.W. (1988). Postweaning diarrhoea in swine: experimental model of enterotoxigenic *Escherichia coli* infection. Am. J. Vet. Res. *49*, 1154–1159.

Scaletsky, I.C., Silva, M.L., and Trabulsi, L.R. (1984). Distinctive patterns of adherence of enteropathogenic *Escherichia coli* to HeLa cells. Infect. Immun. *45*, 534–536.

Scaletsky, I.C., Michalski, J., Torres, A.G., Dulguer, M.V., and Kaper, J.B. (2005). Identification and characterization of the locus for diffuse adherence, which encodes a novel afimbrial adhesin found in atypical enteropathogenic *Escherichia coli*. Infect. Immun. *73*, 4753–4765.

Scaletsky, I.C., Aranda, K.R.S., Souza, T.B., and Silva, N.P. (2010). Adherence factors in atypical enteropathogenic *Escherichia coli* strains expressing the localized adherence-like pattern in HEp-2 cells. J. Clin. Microbiol. *48*, 302–306.

Schouler, C., Koffmann, F., Amory, C., Leroy-Sétrin, S., and Moulin-Schouleur, M. (2004). Genomic subtraction for the identification of putative new virulence factors of an avian pathogenic *Escherichia coli* strain of O2 serogroup. Microbiology *150*, 2973–2984.

Schouler, C., Schaeffer, B., Brée, A., Mora, A., Dahbi, G., Biet, F., Oswald, E., Mainil, J., Blanco, J., and Moulin-Schouleur, M. (2012). Diagnostic strategy for identifying avian pathogenic *Escherichia coli* based on four patterns of virulence genes. J. Clin. Microbiol. *50*, 1673–1678.

Shanks, P.L. (1938). An unusual condition affecting the digestive organs of the pig. Vet. Rec. *50*, 356–358.

Skyberg, J.A., Horne, S.M., Giddings, C.W., Wooley, R.E., Gibbs, P.S., and Nolan, L.K. (2003). Characterizing avian *Escherichia coli* isolates with multiplex polymerase chain reaction. Avian Dis. *47*, 1441–1447.

Smith, H.W., and Halls, S. (1967a). Observation by the ligated intestinal segment and oral inoculation methods on *Escherichia coli* infections in pigs, calves, lambs, and rabbits. J. Pathol. Bacteriol. *93*, 499–529.

Smith, H.W., and Halls, S. (1967b). Studies on *Escherichia coli* enterotoxin. J. Pathol. Bacteriol. *93*, 531–543.

Smith, H.W., and Huggins, M.B. (1978). The influence of plasmid-determined and other characteristics of enteropathogenic *Escherichia coli* on their ability to proliferate in the alimentary tract of piglets, calves and lambs. J. Med. Microbiol. *11*, 471–492.

Smith, H.W., and Linggood, M.A. (1971). Observations on the pathogenic properties of the K88, HLY and Ent plasmids of *Escherichia coli* with particular reference to porcine diarrhoea. J. Med. Microbiol. *4*, 467–485.

Smith, H.W., and Linggood, M.A. (1972). Further observations on *Escherichia coli* enterotoxins with particular regard to those produced by atypical piglet strains and by calf and lamb strains: the transmissible nature of these enterotoxins and of a K antigen possessed by calf and lamb strains. J. Med. Microbiol. *5*, 243–250.

Smith, J.L., Fratamico, P.N., and Gunther, N.W. (2007). Extraintestinal pathogenic *Escherichia coli*. Foodborne Pathog. Dis. *4*, 134–163.

Sojka, W.J., Lloyd, M.K., and Sweeney, E.J. (1960). *Escherichia coli* serotypes associated with certain pig diseases. Res. Vet. Sci. *1*, 17–27.

Sonntag, A.K., Bielaszewska, M., Mellmann, A., Dierksen, N., Schierack, P., Wieler, L.H., Schmidt, M.A., and Karch, H. (2005). Shiga toxin 2e-producing *Escherichia coli* isolates from humans and pigs differ in their virulence profiles and interactions with intestinal epithelial cells. Appl. Environ. Microbiol. *71*, 8855–8863.

Staats, J.J., Chengappa, M.M., DeBey, M.C., Fickbohm, B., and Oberst, R.D. (2003). Detection of *Escherichia coli* Shiga toxin (*stx*) and enterotoxin (*estA* and *elt*) genes in fecal samples from non-diarrheic and diarrheic greyhounds. Vet. Microbiol. *94*, 303–312.

Starcic, M., Johnson, J.R., Stell, A.L., Van Der Goot, J., Hendricks, H.G., Van Vorstenbosch, C., Van Dijck, L., and Gaastra, W. (2002). Haemolytic *Escherichia coli* isolated from dogs with diarrhoea have characteristics of both uropathogenic and necrotoxigenic strains. Vet. Microbiol. *85*, 361–375.

Stordeur, P., Marlier, D., Blanco, J., Oswald, E., Biet, F., Dho-Moulin, M., and Mainil, J. (2002). Examination of *Escherichia coli* from poultry for selected adhesin genes important in disease caused by mammalian pathogenic *E. coli*. Vet. Microbiol. *84*, 231–241.

Stordeur, P., Brée, A., Mainil, J., and Moulin-Schouleur, M. (2004). Pathogenicity of *pap*-negative avian *Escherichia coli* isolated from septicaemic lesions. Microbes Infect. *6*, 637–645.

Stroemberg, N., Marklund, B.I., Lund, B., Iluer, D., Hamers, A., Gaastra, W., Karlsson, K.A., and Normak, S. (1990). Host-specificity of uropathogenic *Escherichia coli* depends on differences in binding specificity to Gal alpha 1–4 gal-containing receptors. EMBO J. *9*, 2001–2010.

Szalo, M., Goffaux, F., Pirson, V., Piérard, D., Ball, H., and Mainil, J. (2002). Presence in bovine enteropathogenic (EPEC) and enterohaemorrhagic (EHEC) *Escherichia coli* of genes coding for putative adhesins of human EHEC strains. Res. Microbiol. *153*, 653–658.

Teneberg, S., Willemsen, P.T.J., de Graaf, F.K., and Karlsson, K.A. (1990). Receptor-active glycolipids of epithelial cells of the small intestine of young and adult pigs in relation to susceptibility to infection with *Escherichia coli* K99. FEBS Lett. *263*, 10–14.

Teneberg, S., Willemsen, P.T.J., de Graaf, F.K., Stenhagen, G., Pimlott, W., Jovall, P.A., Angstrom, J., and Karlsson, K.A. (1994). Characterization of gangliosides of epithelial cells of calf small intestine, with special reference to receptor-active sequences for enteropathogenic *Escherichia coli* K99. J. Biochem. *116*, 560–574.

Timoney, J.F. (1957). Oedema disease of swine. Vet. Rec. *69*, 1160–1175.

Tivendale, K.A., Logue, C.M., Kariyawasam, S., Jordan, D., Hussein, A., Wannemuehler, Y., and Nolan, L.K. (2010). Avian pathogenic *Escherichia coli* strains are similar to neonatal meningitis *E. coli* strains and are able to cause meningitis in the rat model of human disease. Infect. Immun. *78*, 3412–3419.

Trabulsi, L.R., Keller, R., and Gomes, T.A.T. (2002). Typical and atypical enteropathogenic *Escherichia coli*. Emerg. Infect. Dis. *8*, 508–513.

Tzipori, S., Chandler, D., and Smith, M. (1983). The clinical manifestations and pathogenesis of enteritis associated with rotavirus and enterotoxigenic *Escherichia coli* infections in domestic animals. Prog. Food Nutr. Sci. *7*, 193–205.

Uber, A.P., Trabulsi, L.R., Irino, K., Beutin, L., Ghilardi, A.C., Gomes, T.A., Liberatore, A.M., de Castro, A.F., and Elias, W.P. (2006). Enteroaggregative *Escherichia coli* from humans and animals differ in major phenotypical traits and virulence genes. FEMS Microbiol. Lett. *256*, 251–257.

Valadez, A.M., Debroy, C., Dudley, E., and Cutter, C.N. (2011). Multiplex PCR detection of Shiga toxin-producing *Escherichia coli* strains belonging to serogroups O157, P103, O91, O113, O145, O111, and O26 experimentally inoculated in beef carcass swabs, beef trim, and ground beef. J. Food Prot. *74*, 228–239.

Van Bost, S., and Mainil, J. (2003). Facteurs de virulence et propriétés spécifiques des souches invasives d'*Escherichia coli*: (III) Production de toxines. Ann. Méd. Vét. *147*, 327–342.

Van Bost, S., Bâbe, M.H., Jacquemin, E., and Mainil, J. (2001). Characteristics of necrotoxigenic *Escherichia coli* isolated from septicemic and diarrheic calves between 1958 and 1970. Vet. Microbiol. *82*, 311–320.

Van Bost, S., Jacquemin, E., Oswald, E., and Mainil, J. (2003a). Multiplex PCRs for identification of necrotoxigenic *Escherichia coli* (NTEC). J. Clin. Microbiol. *41*, 4480–4482.

Van Bost, S., Roels, S., Oswald, E., and Mainil, J. (2003b). Putative role of the CNF2 and CDT-III toxins in experimental infections with necrotoxigenic *Escherichia coli* type 2 (NTEC2) strains in calves. Microbes Infect. *5*, 1189–1193.

Vandemaele, F.J., Mugasa, J.P., Vandekerchove, D., and Goddeeris, B.M. (2003). Predominance of the *pap*GII allele with high sequence homology to that of human isolates among avian pathogenic *Escherichia coli*. Vet. Microbiol. *97*, 245–257.

Veilleux, S., and Dubreuil, J.D. (2006). Presence of *Escherichia coli* carying the EAST1 toxin gene in farm animals. Vet. Res. *37*, 3–13.

Waddell, T.E., Lingwood, C.A., and Gyles, C.L. (1996). Interaction of Verotoxin 2e with pig intestine. Infect. Immun. *64*, 1714–1719.

Wales, A.D., Woodward, M.J., and Pearson, G.R. (2005). Attaching-effacing bacteria in animals. J. Comp. Pathol. *132*, 1–26.

Wang, J.Y., Wang, S.S., and Yin, P.Z. (2006). Haemolytic–uraemic syndrome caused by a non-O157:H7 *Escherichia coli* strain in experimentally inoculated dogs. J. Med. Microbiol. *55*, 23–29.

Wang, S., Niu, C., Shi, Z., Xia, Y., Yaqoob, M., Dai, J., and Lu, C. (2011a). Effects of *ibeA* deletion on virulence and biofilm formation of avian pathogenic *Escherichia coli*. Infect. Immun. *79*, 279–287.

Wang, S., Xia, Y., Dai, J., Shi, Z., Kou, Y., Li, H., Bao, Y., and Lu, C. (2011b). Novel roles for autotransporter adhesin AatA of avian pathogenic *Escherichia coli*: colonization during infection and cell aggregation. FEMS. Immun. Med. Microbiol. *63*, 328–338.

Warren, J.W. (1996). Clinical presentation and epidemiology of urinary tract infections. In Urinary Tract Infections: Molecular Pathogenesis and Clinical Management, Mobley, H.L.T., and Warren, J.W., eds. (ASM Press, Washington, DC), pp. 3–28.

Weiner, M., and Osek, J. (2004). Development of a multiplex PCR (m-PCR) test for rapid identification of genes encoding heat-labile (LTI) and heat-stable (STI and STII) toxins of enterotoxigenic *Escherichia coli* (ETEC) with internal control of amplification. Pol. J. Microbiol. *53*, 7–10.

Weinstein, D.L., Jackson, M.P., Samuel, J.E., Holmes, R.K., and O'Brien, A.D. (1988). Cloning and sequencing of a Shiga-like toxin type II variant from an *Escherichia coli* strain responsible for edema disease of swine. J. Bacteriol. *170*, 4223–4230.

Weintraub, A. (2007). Enteroaggregative *Escherichia coli*: epidemiology, virulence and detection. J. Med. Microbiol. *56*, 4–8.

White, L.J., Schukken, Y.H., Dogan, B., Green, L., Döpfer, D., Chappell, M.J., and Medley, G.F. (2010). Modelling the dynamics of intramammary *E. coli* infections in dairy cows: understanding mechanisms that distinguish transient from persistent infections. Vet. Res. *41*, 13.

W.H.O. (World Health Organization) Scientific Working Group (1980). *Escherichia coli* diarrhoea. Bull. W.H.O. *58*, 23–36.

Wieler, L.H., Bauerfeind, R., and Baljer, G. (1992). Characterization of Shiga-like toxin producing *Escherichia coli* (SLTEC) isolated from calves with and without diarrhoea. Zentralbl. Bakteriol. *276*, 243–253.

Wieler, L.H., Busse, B., Steinrück, H., Beutin, L., Weber, A., Karch, H., and Baljer, G. (2000). Enterohemorrhagic *Escherichia coli* (EHEC) strains of serogroup O118 display three distinctive clonal groups of EHEC pathogens. J. Clin. Microbiol. *38*, 2162–2169.

Wieler, L.H., Sobjinski, G., Schlapp, T., Failing, K., Weiss, R., Menge, C., and Baljer, G. (2007). Longitudinal prevalence study of diarrhoeagenic *Escherichia coli* in dairy calves. Berl. Munch. Tierarztl. Wochenschr. *120*, 296–306.

Wiles, T.J., Kulesus, R.R., and Mulvey, M.A. (2008). Origins and virulence mechanisms of uropathogenic *Escherichia coli*. Exp. Molec. Pathol. *85*, 11–19.

Willford, J., Mills, K., and Goodridge, L.D. (2009). Evaluation of three commercially available enzyme-linked immunosorbent assay kits for detection of Shiga toxin. J. Food Prot. *72*, 741–747.

Wray, C., and Woodward, M.J. (1997). *Escherichia coli* infections in farm animals. In *Escherichia coli* – Mechanisms of Virulence, Sussman, M., ed. (University Press, Cambridge, UK), pp. 49–84.

Yamamoto, S., Terai, A., Yuri, K., Kurazono, H., Takeda, Y., and Yoshida, O. (1995). Detection of urovirulence factors in *Escherichia coli* by multiplex polymerase chain reaction. FEMS Immunol. Med. Microbiol. *12*, 85–90.

Yuri, K., Nakata, K., Katae, H., Tsukamoto, T., and Hasegawa, A. (1998). Serotypes and virulence factors of *Escherichia coli* strains isolated from dogs and cats. J. Vet. Med. Sci. *61*, 37–40.

Zhang, W., Qi, W., Albert, T.J., Motiwala, A.S., Alland, D., Hyytia-Trees, E.K., Ribot, E.M., Fields, P.I., Whittam, T.S., and Swaminathan, B. (2006). Probing genomic diversity and evolution of *Escherichia coli* O157 by single nucleotide polymorphisms. Genome Res. *16*, 757–767.

Genomic Plasticity and the Emergence of New Pathogenic *Escherichia coli*

3

Shana R. Leopold, Ulrich Dobrindt, Helge Karch and Alexander Mellmann

Abstract

Changes at the genomic level of *Escherichia coli* occur both on a small scale (e.g. point mutations, insertions, deletions) as well as through lateral gene transfer and homologous recombination of larger fragments. The ability of the *E. coli* chromosome to incorporate new sequence from various sources, including mobile elements, plasmids, and other exogenous DNA, gives rise to diversity of virulence and adaptive gene content throughout the species. This plasticity of the genome enables *E. coli* to rapidly adapt, exploiting new niches and forming new pathotypes. Intra-host genome alterations have also been documented, providing models of real-time pathogen evolution. We are now learning more about this genomic flexibility as whole genome sequencing becomes more readily available.

Introduction

Several clones of emerging intestinal and extraintestinal pathogenic *Escherichia coli* have been identified over the last decades. Examples range from extraintestinal pathogenic *E. coli* (ExPEC) clone ST131 *bla*CTX M-15 (Johnson *et al.*, 2012) to intestinal enterohaemorrhagic *E. coli* (EHEC) O157:H7 (Riley *et al.*, 1983), sorbitol-fermenting (SF) EHEC O157:H⁻ (Karch *et al.*, 1993), and the newly emerged Shiga toxin 2-producing *E. coli* (STEC) clones O26:H11 (Bielaszewska *et al.*, 2013) and O104:H4 (Bielaszewska *et al.*, 2011a). Though many pathogenic *E. coli* clones have been identified only recently, phylogenetic studies show that several of these clades have been in existence for thousands of years or more (Leopold *et al.*, 2009, 2011). It is not yet fully understood why many of these pathotypes are only now being recognized, though there are many possible theories.

Modern molecular methods, such as microarray technology, RT-PCR and whole genome sequencing, have facilitated and improved detection of changes at both the genomic and individual gene expression pattern level. The feasibility of sequencing whole genomes of *E. coli* has led to a rapid increase in our knowledge and understanding of this species. In fact, several hundred *E. coli* genomes have been sequenced and deposited into GenBank to date, with countless more genome projects already under way (Hu *et al.*, 2011; Lagesen *et al.*, 2010; NCBI, 2013; Parkhill and Wren, 2011). Differences between non-pathogenic and pathogenic *E. coli*, and between different *E. coli* pathotypes in particular, are becoming clearer. This vast amount of sequence data has also been instrumental in formulating precise evolutionary models (Arnold and Jackson, 2011; Jenke *et al.*, 2012; Leopold *et al.*, 2011).

The plasticity of the genome and its resulting variability has likely made a significant contribution to the successful emergence of new pathogenic *E. coli*. Adaptation of pathogenic *E. coli* to evade host immune defences is enabled by mechanisms for rapid genome variation and diversification (Jenke *et al.*, 2012). Microbial adaptation can occur through both changes in the genomic structure and through gene expression modulation. Changes at the nucleotide level such as point mutations, insertions, deletions, rearrangements and gene duplications can confer critical changes. A single point mutation in important genes such as DNA gyrase can result in an amino acid change that confers antibiotic resistance (Turner *et al.*, 2006). Duplication of virulence-encoding genes, such as *stx*$_2$, may play an important role in genome evolution as one copy operates as a backup version, while the other is open to potentially beneficial mutations (Bielaszewska *et al.*, 2006; Yang *et al.*, 2003). Gene duplication may also aid in immune evasion or adaptation to new microenvironments by producing antigenically different proteins (Bielaszewska *et al.*, 2006). Deletion of certain genes may pave the way to enhanced virulence as well by effectively removing genes that have a negative effect on processes key to virulence (Middendorf *et al.*, 2004). An example of this is the lack of lysine decarboxylase synthesis in EHEC O111. The *cad* locus has been shown to negatively mediate *E. coli* O157:H7 and O111 adherence (Torres and Kaper, 2003; Torres *et al.*, 2005). The removal of this function has been termed a pathoadaptive mutation (Sokurenko *et al.*, 1999) because it enhances virulence. Similar pathoadaptive mutations have also been shown in other *E. coli*

pathotypes. For example, in adherent-invasive *E. coli* (AIEC) that are abnormally predominant on Crohn's disease ileal mucosa, point mutations in FimH lead to a significantly higher ability of AIEC to adhere to intestinal epithelial cells thereby increasing the ability to induce inflammation (Dreux *et al.*, 2013).

Horizontal gene transfer also enables considerable changes, with potential alterations ranging from the addition of a beneficial virulence gene to modification, gain or loss of entire genomic regions (Fig. 3.1). These rapid changes are collectively referred to as microevolution, and are manifested within a few generations in the environmental, animal or human reservoirs (Ahmed *et al.*, 2008; Brzuszkiewicz *et al.*, 2009).

Genome structure of extraintestinal and intestinal pathogenic *E. coli*

The virulence and phenotypic variability of *E. coli* is connected to its chromosomal content, which may range in size from approximately 4700 to 6500 Kbp (Hazen *et al.*, 2012). The genome is a mosaic-like structure composed of the core genomic

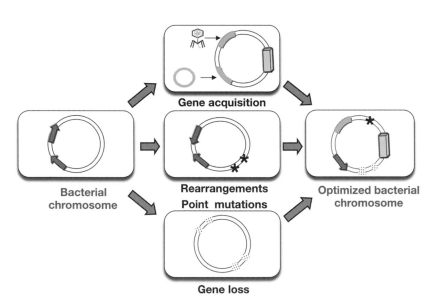

Gene acquisition

Bacterial chromosome

Rearrangements Point mutations

Optimized bacterial chromosome

Gene loss

Figure 3.1 Mechanisms contributing to bacterial genome plasticity and genome optimization. The acquisition of mobile and accessory genetic elements such as bacteriophages, plasmids and genomic islands contributes to the evolution of pathogenic *E. coli* from commensal variants. DNA rearrangements, deletions and point mutations may lead to a reduced expression of virulence genes, inactivation or loss. A colour version of this figure is available in the plate section at the back of the book.

backbone shared by all *E. coli* (housekeeping genes required for basic essential cellular processes), and the flexible genomic content (accessory and virulence properties) that varies between different pathotypes, and even between closely related strains. The accessory genes allow specific strains and pathotypes to adapt to special environmental conditions (e.g. virulence-associated factors, antibiotic resistance). The chromosome is organized so that functionally related genes are often clustered together. A relatively homogeneous guanine and cytosine (GC) content with a signature species-specific codon usage is perceptible in the chromosomal backbone. The core genome is often interrupted by the presence of DNA stretches with a different GC content and codon usage, indicative of acquisition of genes from outside (Chaudhuri and Henderson, 2012; Ochman and Jones, 2000; Touchon *et al.*, 2009; Toussaint and Chandler, 2012).

The flexible genomic gene pool consists of mobile elements such as transposons, integrons, plasmids, bacteriophages, insertion elements and genomic or pathogenicity islands (GEIs, PAIs). Virulence-encoding genes, particularly those that encode for toxins, adhesins, and other pathogenicity traits, are encoded on these accessory genetic elements (Ahmed *et al.*, 2008; Gal-Mor and Finlay, 2006; Rasko *et al.*, 2008). Varying amounts of strain- or pathotype-specific genetic information (up to 30% of the complete genome content) account for this *wide* range (Brzuszkiewicz *et al.*, 2009).

Mobile elements

Mobile and accessory genetic elements cause variable, dynamic changes of the genomic structures of *E. coli* (Brussow *et al.*, 2004; Frost *et al.*, 2005; Köhler and Dobrindt, 2011; Siefert, 2009). The mosaic nature of the chromosome is due to an on-going process of repeated rearrangements, insertions and deletions as the accessory elements are in constant flux.

Insertion elements and transposons

Insertion elements and transposons are stretches of DNA that act as jumping genes; they frequently change location in the chromosome and are also capable of undergoing horizontal transfer. Transposable elements, which often carry antibiotic resistance genes, fit into three categories: (i) insertion sequences which have small pieces of DNA with short terminal inverted repeat sequences that usually only encode information needed for their mobility, (ii) composite transposons that are flanked on both sides by insertion sequences, and (iii) non-composite transposons which contain no flanking insertion sequences (Blot, 1994; Siguier *et al.*, 2006). Integrons are other mobile genetic elements that sometimes carry determinants for antibiotic resistance and other properties. They can acquire open reading frames (ORF) from exogenous gene cassettes and confer expression. Integrons are well known because of their important role in the spread of antibiotic-resistance determinants (Mazel, 2006), but our understanding of their importance in bacterial genome evolution has broadened with the discovery of larger integron structures, termed superintegrons. These contain assemblies of genes flanked by the integron integrase targeted recombination sequence (*attC*). Superintegrons can harbour up to eight resistance cassettes (Cambray *et al.*, 2010; Hall, 2012; Naas *et al.*, 2001; Partridge *et al.*, 2009).

Plasmids and bacteriophages

Transferable genetic elements such as bacteriophages and plasmids laterally transport genetic information from one organism to another, thereby playing important roles in bacterial evolution. Bacteriophages, as viruses, can genetically modify their prokaryotic hosts by entering genomes. They might carry genes that encode new phenotypes, or modify existing traits. Additionally, the process of integration may interrupt existing chromosomal genes.

Plasmids are circular, self-replicating DNA molecules. They are autonomous molecules and exist in cells as extrachromosomal genomes. More than one can exist in a single bacterium at any time, and some also have the ability to integrate into the bacterial chromosome. Plasmids not only can exchange genetic material with their host's chromosome, they can also be exchanged laterally between *E. coli* of the same, and different, pathotypes. Additionally, some plasmid genes have been shown to up-regulate virulence genes in the chromosome. Large plasmids encoding virulence factors, such as fimbriae, serine proteases, enterotoxins or haemolysins are found in many pathogenic *E. coli* (Brunder and Karch, 2000; Brunder *et al.*, 1999, 2001; Fratamico *et al.*, 2011). Insertion sequences in these plasmids provide starting points for homologous recombination plasmid-to-plasmid or plasmid-to-chromosome. We believe that individual or entire groups of genes were integrated into existing plasmids by such transposition events. A subsequent partial deletion of insertion sequences causes immobility and thus stabilization of the virulence genes (Brzuszkiewicz *et al.*, 2009).

Homologous recombination

Homologous recombination of horizontally acquired DNA segments is a major factor in the diversity observed throughout the *E. coli* genome. Closely related *E. coli* strains can be differentiated by their horizontally acquired sequences in genomic islands, suggesting that many of these DNA segments are unstable due to successive transfers. Additionally, strains of the same patho- or serotype may have PAIs that share certain virulence traits, but with different chromosomal organization (Chaudhuri and Henderson, 2012; Leimbach *et al.*, 2013; Schmidt and Hensel, 2004). Multi-copy accessory genes in a single genome may facilitate homologous recombination within and between bacteriophages and pathogenicity islands. Individual DNA regions can be exchanged between mobile genetic elements and the chromosome with the capacity for repeated integration and excision from the bacterial chromosome. Evidence of this is several identical or closely related virulence genes found on the chromosome and/or on mobile DNA elements (Fig. 3.2) (Schubert *et al.*, 2009; Touchon *et al.*, 2009). Colicin plasmids are one example of mobile elements that have considerable sequence similarity to PAIs in *E. coli* (Starcic Erjavec *et al.*, 2003). This diversity may confer a specific advantage to particular strains, or may open the possibility of a new niche, host, or reservoir, potentially leading to the emergence of new pathogenic *E. coli* (Cascales *et al.*, 2007).

Genome alterations during human infections – a model of real-time pathogen evolution

During infection of hosts, pathogens are subjected to a variety of strong selective pressure. Genetic changes of intestinal pathogens might influence clinical outcome and have an impact on diagnosis and epidemiology. EHEC are a good example to demonstrate this.

The first suggestion of chromosomal plasticity in EHEC was suggested in the early 1990s with our demonstration of loss of Stx genes on subculture of non-O157:H7 EHEC (Karch *et al.*, 1992). This concept was resurrected when we determined several years later that not all EHEC strains from patients with bloody diarrhoea and even haemolytic–uraemic syndrome (HUS) contain Stx genes on early (i.e. soon after isolation) analysis (Bielaszewska *et al.*, 2008). Using PCR for both *stx* and *eae* to screen primary stool cultures followed by colony blot hybridization with the same PCR probes to isolate the respective strains from PCR-positive stools, we found that an appreciable subset of patients with bloody diarrhoea excrete *E. coli* that belong to EHEC serotypes (O26:H11/H⁻, O103:H2/H⁻, O145:H28/H⁻, and O157:H7/H⁻) and possess *eae*, yet the colonies on the primary agar plates (i.e. the first outgrowth following faeces emission during infection) do not contain Stx genes (Bielaszewska *et al.*, 2007a,b, 2008; Friedrich *et al.*, 2007). We observed that at the time of microbiological diagnosis a subset of patients with bloody diarrhoea or HUS no longer excrete the causative EHEC that harbour *stx*, but instead they excrete Stx-negative derivatives of EHEC that lost Stx via excision of Stx -encoding bacteriophages (Friedrich *et al.*, 2007).

Loss of a bacteriophage, even in the course of a human infection, creates a rapid change in the genomic architecture and phenotype (Mellmann *et al.*, 2009). The absence of the *stx* phage may not only increase the adaptation ability of the pathogens outside the host, but may also enable them to avoid lysis by induced Stx-encoding prophages in the event of a stress stimuli in the human gastrointestinal tract such as hydrogen peroxide or bile (Shaikh and Tarr, 2003). Excision of the Stx-phage from the EHEC chromosome

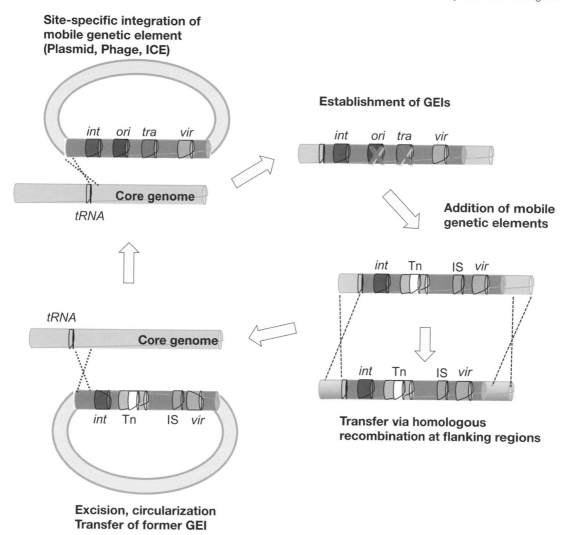

Site-specific integration of mobile genetic element (Plasmid, Phage, ICE)

int   ori   tra   vir

Core genome

tRNA

Establishment of GEIs

int   ori   tra   vir

Addition of mobile genetic elements

int   Tn   IS   vir

tRNA

Core genome

int   Tn   IS   vir

Transfer via homologous recombination at flanking regions

int   Tn   IS   vir

Excision, circularization Transfer of former GEI

Figure 3.2 Contribution of mobile genetic elements to the evolution of new pathogenic *E. coli* variants. Acquisition of mobile genetic elements (e.g. plasmids, bacteriophages, genomic islands, integrative and conjugative elements, integrons) can result in the chromosomal integration by site-specific recombination. Genomic islands are frequently inserted at the 5′-end of tRNA loci. Genome plasticity due to further gene acquisition, gene loss, or DNA rearrangements results in modification and further evolution of the genomic island. Upon conjugation and homologous recombination between flanking regions, large chromosomal regions including genomic islands can be transferred to suitable recipients. Alternatively, islands can be excised by site-specific recombination and the circularized islands can then be transferred to another recipient. A colour version of this figure is available in the plate section at the back of the book.

often reconstitutes the intact phage integration site. As a result, such strains can potentially be re-infected with Stx-phages (Mellmann *et al.*, 2008b). For example, Stx-negative EHEC serotypes O26:H11/H⁻ and SF O157:H⁻ were transduced *in vitro* by *stx₂*-phages originating from the parental EHEC strains (Bielaszewska *et al.*, 2007b). The transduced *stx₂*-phage occupied

their phage integration sites and produced active Stx in the lysogens (Bielaszewska *et al.*, 2007b; Mellmann *et al.*, 2008b). These findings demonstrate that Stx-positive and Stx-negative *E. coli* can convert their toxigenic status in both directions, yielding different pathotypes [atypical enteropathogenic *E. coli* (EPEC) and STEC], and that Stx-encoding bacteriophages are the major elements involved in this bidirectional process).

Spontaneous loss of large internal regions of genomic islands OI 43 and/or OI 48 in O157:H7 was also recently demonstrated, *in vitro* and *in vivo*, by homologous recombination between IS elements resulting in the removal of *ter* and *iha* genes in parallel with a change in colony morphology from large to small colony size (Bielaszewska *et al.*, 2011b). The intensive study of human enteric factors that induce or modulate pathogenic *E. coli* chromosome instability could open new vistas into host–microbial interactions (Mellmann *et al.*, 2009).

Genomic virulence content variability and formation of pathotypes

According to Wirth *et al.* (2006), each pathotype is composed of clones that evolved at different times, from more than one ancestral group. For example, EHEC 1 and EHEC 2 evolved from EPEC serogroups O55 and O26, respectively, and are phylogenetically unrelated. The independent import of virulence factors through horizontal gene transfer is likely responsible for this parallel evolution. EIEC clones have also emerged at different points from entirely different phylogroups of the *E. coli* species (B1 and D).

Varying presence of virulence factors in single pathogroups

Several potential adhesins have been detected in HUS-associated *E. coli* (HUSEC), a subgroup of EHEC (see www.ehec.org), but whose genes are differentially distributed. Whereas type-1 fimbriae are found in many HUSEC, Sfp fimbriae are only present in HUSEC004 and HUSEC042, two members of the HUSEC reference collection (Mellmann *et al.*, 2008a). EHEC and EPEC, both types of intestinal pathogenic *E. coli*, often possess type III secretion systems such as the intimin-encoding *eae* gene that is responsible for the pedestal formation leading to adhesion and translocation of several effector proteins and toxin. The intimin adhesin is comparatively widespread throughout the HUSEC collection, though notably absent from the 2011 O104:H4 European outbreak strain, the O104:H4 HUSEC041

reference strain isolated in 2001, and from EHEC O111:H10 isolated from HUS patients (Bielaszewska *et al.*, 2011a; Zhang *et al.*, 2007).

Such variation in the presence of potential virulence genes is seen not only with adhesins, but also with other virulence determinants. For example, the genes for subtilase cytotoxin, SubAB, are only present in two (HUSEC028, HUSEC039) of the 42 HUSEC reference strains (Karch *et al.*, 2013). Classical EPEC strains possess bundle-forming pili, absent in atypical EPEC, which allow for strong adhesion and colonization of the small bowel. EHEC, in addition, harbours phage-encoded Shiga toxins. Enterotoxigenic *E. coli* (ETEC) also have great variation, characteristically producing heat-stable and heat-labile enterotoxins as well as plasmid-encoded fimbriae, CFAI and CFAII (colonization factor antigens) in some ETEC but not all (Kaper *et al.*, 2004). Whereas enteroinvasive *E. coli* (EIEC) phenotypically resemble *Shigella* spp. and are characterized by the large pInv virulence plasmid of *Shigella*, enteroaggregative *E. coli* (EAEC) are phenotypically characterized by their characteristic aggregative adherence phenotype mediated by aggregative adherence fimbriae (AAF), which are encoded together with other virulence-associated traits on a large virulence plasmid (Chaudhuri *et al.*, 2010; Kaper *et al.*, 2004).

Evolution of EHEC – emergence of a pathotype

E. coli O157:H7 was first identified as a pathogen in 1982 and has subsequently been identified in countless local outbreaks, as well as many larger epidemics, with increasing frequency in the past three decades. Though this clade appears to be newly emerging, backbone SNP studies have shown that *E. coli* O157:H7 has likely been in existence for over 2,500 years (Leopold *et al.*, 2009). The most recently emerged cluster has been dated to be approximately 400–600 years old (Leopold *et al.*, 2009).

Although EHEC O157:H7 and SF O157:H⁻ (EHEC 1) share many virulence traits with other STEC pathogens, such as EHEC 2 serotypes O111 and O26, they are distantly related to these other groups (Reid *et al.*, 2000). Independent acquisitions of a set of virulence determinants were the first major steps in the evolution of *E. coli* O157 from a less virulent pathogen (Reid *et al.*, 2000). Many of these virulence factors were gained via mobile elements: the LEE pathogenicity island that mediates attachment to the intestinal wall (also found in EPEC), EHEC-haemolysin found on a large plasmid (Schmidt *et al.*, 1995, 1996), and Stx-encoding bacteriophages lysogenized in the chromosome (O'Brien *et al.*, 1984).

The EHEC 1 clade provides an interesting model of how horizontal gene transfer can create new pathotypes. The EHEC 1 subgroups form a model of stepwise evolution of two highly virulent pathogens (O157:H7 and SF O157:H⁻) from a less virulent ancestor, EPEC-like O55:H7 (Feng *et al.*, 1998; Jenke *et al.*, 2012) (Fig. 3.3). Two early steps in this process were the acquisition of a stx_2-bacteriophage and a homologous recombination event at the *rfb-gnd* locus, changing the *rfb* cluster encoded O-lipopolysaccharide side chain antigen (O-antigen) from type O55 to O157 (Nelson and Selander, 1994; Wang *et al.*, 2002). A hypothetical SF, Stx-carrying O157:H7 was a likely common ancestor of O157:H7 and SF O157:H⁻ (Jenke *et al.*, 2012). Investigation of the core backbone ORFs of several members of this clade suggest that the O-antigen defining recombination

was the only lasting core genome homologous recombination event in this relatively young clade (Leopold *et al.*, 2009).

The SF O157:H⁻ subgroup is known to cause bloody diarrhoea and HUS, but has to date been mainly isolated in Europe where it has caused large outbreaks (Alpers *et al.*, 2009; Ammon *et al.*, 1999; Gunzer *et al.*, 1992; Karch *et al.*, 2005). There is no known non-human reservoir; indeed this pathogen may rely on human-to-human spread (Karch *et al.*, 2005). Interestingly, there is extremely limited diversity in this group, even though an analysis of the backbone SNPs suggests that this subgroup diverged from the O157:H7 lineage approximately 7000 years ago (Leopold *et al.*, 2009).

Globally distributed O157:H7 causes no disease in its bovine reservoir, but like SF O157:H⁻ it is a highly virulent pathogen in humans. This subgroup lost the ability to ferment sorbitol early on, has mutations in the fimA promoter (Li *et al.*, 1997) and uidA (Monday *et al.*, 2001), and carries the tellurite resistance and adherence-conferring island (Friedrich *et al.*, 2005). While several clusters of *E. coli* O157:H7 have been described, only three are known to cause disease in humans. These three consecutively emerged

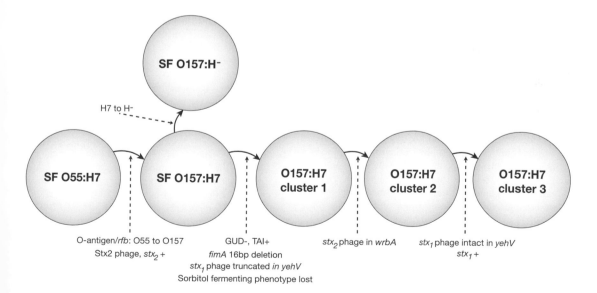

Figure 3.3 Stepwise evolution of *E. coli* O157. Each sphere represents a subgroup or cluster of EHEC 1. Important evolutionary events from published studies are represented by dotted arrows. The solid arrows represent the linear evolution of the clade. Distances are not drawn to scale. (Adapted from Leopold *et al.* 2009, Figure 1.)

clusters are characterized by the acquisition and loss of Stx bacteriophages: cluster 1 harbours a truncated stx_1 bacteriophage in *yehV*, cluster 2 organisms have the additional acquisition of a stx_2 bacteriophage in the *wrbA* gene, and in cluster 3 a full stx_1-encoding bacteriophage resides in the *yehV* gene (Shaikh *et al.*, 2007). Very few cluster 2 strains have been isolated to date, while clusters 1 and 3 account for the majority of human *E. coli* O157:H7 infections (Besser *et al.*, 2007).

Acknowledgements

AM was supported by the German Research Foundation (grant no. ME 3205/2-1), by the EU Network ERA-NET PathoGenoMics II (grant no. 0315443), by the Federal Ministry of Education and Research FBI-Zoo Network (grant no. 01KI1012B) and by the Medical Faculty of the University of Münster (grant no. BD9817044). UD was supported by the German Research Foundation (DO 789/4-1). These studies were carried out within the European Virtual Institute for Functional Genomics of Bacterial Pathogens (CEE LSHB-CT-2005-512061), and the ERA-NET PathoGenoMics I consortium 'Deciphering the intersection of commensal and extraintestinal pathogenic *E. coli*' (Federal Ministry of Education and Research (BMBF) grant no. 0313937A).

References

Ahmed, N., Dobrindt, U., Hacker, J., and Hasnain, S.E. (2008). Genomic fluidity and pathogenic bacteria: applications in diagnostics, epidemiology and intervention. Nat. Rev. Microbiol. *6*, 387–394.

Alpers, K., Werber, D., Frank, C., Koch, J., Friedrich, A.W., Karch, H., An der Heiden, M., Prager, R., Fruth, A., Bielaszewska, M., *et al.* (2009). Sorbitol-fermenting enterohaemorrhagic *Escherichia coli* O157:H⁻ causes another outbreak of haemolytic uraemic syndrome in children. Epidemiol. Infect. *137*, 389–395.

Ammon, A., Petersen, L.R., and Karch, H. (1999). A large outbreak of hemolytic uremic syndrome caused by an unusual sorbitol-fermenting strain of *Escherichia coli* O157:H⁻. J. Infect. Dis. *179*, 1274–1277.

Arnold, D.L., and Jackson, R.W. (2011). Bacterial genomes: evolution of pathogenicity. Curr. Opin. Plant Biol. *14*, 385–391.

Besser, T.E., Shaikh, N., Holt, N.J., Tarr, P.I., Konkel, M.E., Malik-Kale, P., Walsh, C.W., Whittam, T.S., and Bono, J.L. (2007). Greater diversity of Shiga toxin-encoding bacteriophage insertion sites among *Escherichia coli* O157:H7 isolates from cattle than in those from humans. Appl. Environ. Microbiol. *73*, 671–679.

Bielaszewska, M., Prager, R., Zhang, W., Friedrich, A.W., Mellmann, A., Tschäpe, H., and Karch, H. (2006). Chromosomal dynamism in progeny of outbreak-related sorbitol-fermenting enterohemorrhagic *Escherichia coli* O157:NM. Appl. Environ. Microbiol. *72*, 1900–1909.

Bielaszewska, M., Kock, R., Friedrich, A.W., von Eiff, C., Zimmerhackl, L.B., Karch, H., and Mellmann, A. (2007a). Shiga toxin-mediated hemolytic uremic syndrome: time to change the diagnostic paradigm? PLoS One *2*, e1024.

Bielaszewska, M., Prager, R., Köck, R., Mellmann, A., Zhang, W., Tschäpe, H., Tarr, P.I., and Karch, H. (2007b). Shiga toxin gene loss and transfer in vitro and in vivo during enterohemorrhagic *Escherichia coli* O26 infection in humans. Appl. Environ. Microbiol. *73*, 3144–3150.

Bielaszewska, M., Middendorf, B., Kock, R., Friedrich, A.W., Fruth, A., Karch, H., Schmidt, M.A., and Mellmann, A. (2008). Shiga toxin-negative attaching and effacing *Escherichia coli*: distinct clinical associations with bacterial phylogeny and virulence traits and inferred in-host pathogen evolution. Clin. Infect. Dis. *47*, 208–217.

Bielaszewska, M., Mellmann, A., Zhang, W., Köck, R., Fruth, A., Bauwens, A., Peters, G., and Karch, H. (2011a). Characterisation of the *Escherichia coli* strain associated with an outbreak of haemolytic uraemic syndrome in Germany, 2011: a microbiological study. Lancet Infect. Dis. *11*, 671–676.

Bielaszewska, M., Middendorf, B., Tarr, P.I., Zhang, W., Prager, R., Aldick, T., Dobrindt, U., Karch, H., and Mellmann, A. (2011b). Chromosomal instability in enterohaemorrhagic *Escherichia coli* O157:H7: impact on adherence, tellurite resistance and colony phenotype. Mol. Microbiol. *79*, 1024–1044.

Bielaszewska, M., Mellmann, A., Bletz, S., Zhang, W., Kock, R., Kossow, A., Prager, R., Fruth, A., Orth-Holler, D., Marejkova, M., *et al.* (2013). Enterohemorrhagic *Escherichia coli* O26:H11/H⁻: A new virulent clone emerges in Europe. Clin. Infect. Dis.

Blot, M. (1994). Transposable elements and adaptation of host bacteria. Genetica *93*, 5–12.

Brunder, W., and Karch, H. (2000). Genome plasticity in *Enterobacteriaceae*. Int. J. Med. Microbiol. *290*, 153–165.

Brunder, W., Schmidt, H., Frosch, M., and Karch, H. (1999). The large plasmids of Shiga toxin-producing *Escherichia coli* (STEC) are highly variable genetic elements. Microbiology *145(Pt 5)*, 1005–1014.

Brunder, W., Khan, A.S., Hacker, J., and Karch, H. (2001). Novel type of fimbriae encoded by the large plasmid of sorbitol-fermenting enterohemorrhagic *Escherichia coli* O157:H⁻. Infect. Immun. *69*, 4447–4457.

Brussow, H., Canchaya, C., and Hardt, W.D. (2004). Phages and the evolution of bacterial pathogens: from genomic rearrangements to lysogenic conversion. Microbiol. Mol. Biol. Rev. *68*, 560–602, table of contents.

Brzuszkiewicz, E., Gottschalk, G., Ron, E., Hacker, J., and Dobrindt, U. (2009). Adaptation of pathogenic

E. coli to various niches: genome flexibility is the key. Genome Dyn. *6*, 110–125.

Cambray, G., Guerout, A.M., and Mazel, D. (2010). Integrons. Annu. Rev. Genet. *44*, 141–166.

Cascales, E., Buchanan, S.K., Duche, D., Kleanthous, C., Lloubes, R., Postle, K., Riley, M., Slatin, S., and Cavard, D. (2007). Colicin biology. Microbiol. Mol. Biol. Rev. *71*, 158–229.

Chaudhuri, R.R., and Henderson, I.R. (2012). The evolution of the *Escherichia coli* phylogeny. Infect. Genet. Evol. *12*, 214–226.

Chaudhuri, R.R., Sebaihia, M., Hobman, J.L., Webber, M.A., Leyton, D.L., Goldberg, M.D., Cunningham, A.F., Scott-Tucker, A., Ferguson, P.R., Thomas, C.M., *et al.* (2010). Complete genome sequence and comparative metabolic profiling of the prototypical enteroaggregative *Escherichia coli* strain 042. PLoS One *5*, e8801.

Dreux, N., Denizot, J., Martinez-Medina, M., Mellmann, A., Billig, M., Kisiela, D., Chattopadhyay, S., Sokurenko, E., Neut, C., Gower-Rousseau, C., *et al.* (2013). Point mutations in FimH adhesin of Crohn's disease-associated adherent-invasive *Escherichia coli* enhance intestinal inflammatory response. PLoS Pathog. *9*, e1003141.

Feng, P., Lampel, K.A., Karch, H., and Whittam, T.S. (1998). Genotypic and phenotypic changes in the emergence of *Escherichia coli* O157:H7. J. Infect. Dis. *177*, 1750–1753.

Fratamico, P.M., Yan, X., Caprioli, A., Esposito, G., Needleman, D.S., Pepe, T., Tozzoli, R., Cortesi, M.L., and Morabito, S. (2011). The complete DNA sequence and analysis of the virulence plasmid and of five additional plasmids carried by Shiga toxin-producing *Escherichia coli* O26:H11 strain H30. Int. J. Med. Microbiol. *301*, 192–203.

Friedrich, A.W., Köck, R., Bielaszewska, M., Zhang, W., Karch, H., and Mathys, W. (2005). Distribution of the urease gene cluster among and urease activities of enterohemorrhagic *Escherichia coli* O157 isolates from humans. J. Clin. Microbiol. *43*, 546–550.

Friedrich, A.W., Zhang, W., Bielaszewska, M., Mellmann, A., Köck, R., Fruth, A., Tschäpe, H., and Karch, H. (2007). Prevalence, virulence profiles, and clinical significance of Shiga toxin-negative variants of enterohemorrhagic *Escherichia coli* O157 infection in humans. Clin. Infect. Dis. *45*, 39–45.

Frost, L.S., Leplae, R., Summers, A.O., and Toussaint, A. (2005). Mobile genetic elements: the agents of open source evolution. Nat. Rev. Microbiol. *3*, 722–732.

Gal-Mor, O., and Finlay, B.B. (2006). Pathogenicity islands: a molecular toolbox for bacterial virulence. Cell. Microbiol. *8*, 1707–1719.

Gunzer, F., Bohm, H., Russmann, H., Bitzan, M., Aleksic, S., and Karch, H. (1992). Molecular detection of sorbitol-fermenting *Escherichia coli* O157 in patients with hemolytic-uremic syndrome. J. Clin. Microbiol. *30*, 1807–1810.

Hall, R.M. (2012). Integrons and gene cassettes: hotspots of diversity in bacterial genomes. Ann. N.Y. Acad. Sci. *1267*, 71–78.

Hazen, T.H., Sahl, J.W., Redman, J.C., Morris, C.R., Daugherty, S.C., Chibucos, M.C., Sengamalay, N.A., Fraser-Liggett, C.M., Steinsland, H., Whittam, T.S., *et al.* (2012). Draft genome sequences of the diarrhoeagenic *Escherichia coli* collection. J. Bacteriol. *194*, 3026–3027.

Hu, B., Xie, G., Lo, C.C., Starkenburg, S.R., and Chain, P.S. (2011). Pathogen comparative genomics in the next-generation sequencing era: genome alignments, pangenomics and metagenomics. Brief. Funct. Genomics *10*, 322–333.

Jenke, C., Leopold, S.R., Weniger, T., Rothganger, J., Harmsen, D., Karch, H., and Mellmann, A. (2012). Identification of intermediate in evolutionary model of enterohemorrhagic *Escherichia coli* O157. Emerg. Infect. Dis. *18*, 582–588.

Johnson, J.R., Urban, C., Weissman, S.J., Jorgensen, J.H., Lewis, J.S., 2nd, Hansen, G., Edelstein, P.H., Robicsek, A., Cleary, T., Adachi, J., *et al.* (2012). Molecular epidemiological analysis of *Escherichia coli* sequence type ST131 (O25:H4) and *bla*CTX-M-15 among extended-spectrum-beta-lactamase-producing *E. coli* from the United States, 2000 to 2009. Antimicrob. Agents Chemother. *56*, 2364–2370.

Kaper, J.B., Nataro, J.P., and Mobley, H.L. (2004). Pathogenic *Escherichia coli*. Nat. Rev. Microbiol. *2*, 123–140.

Karch, H., Meyer, T., Russmann, H., and Heesemann, J. (1992). Frequent loss of Shiga-like toxin genes in clinical isolates of *Escherichia coli* upon subcultivation. Infect. Immun. *60*, 3464–3467.

Karch, H., Böhm, H., Schmidt, H., Gunzer, F., Aleksic, S., and Heesemann, J. (1993). Clonal structure and pathogenicity of Shiga-like toxin-producing, sorbitol-fermenting *Escherichia coli* O157:H⁻. J. Clin. Microbiol. *31*, 1200–1205.

Karch, H., Tarr, P.I., and Bielaszewska, M. (2005). Enterohaemorrhagic *Escherichia coli* in human medicine. Int. J. Med. Microbiol. *295*, 405–418.

Karch, H., Müthing, J., Dobrindt, U., and Mellmann, A. (2013). Evolution and infection biology of hemolytic-uremic syndrome (HUS) associated *E. coli* (HUSEC). Bundesgesundheitsblatt Gesundheitsforschung Gesundheitsschutz *56*, 8–14.

Köhler, C.D., and Dobrindt, U. (2011). What defines extraintestinal pathogenic *Escherichia coli*? Int. J. Med. Microbiol. *301*, 642–647.

Lagesen, K., Ussery, D.W., and Wassenaar, T.M. (2010). Genome update: the 1000th genome – a cautionary tale. Microbiology *156*, 603–608.

Leimbach, A., Hacker, J., and Dobrindt, U. (2013). *E. coli* as an all-rounder: the thin line between commensalism and pathogenicity. Curr. Top. Microbiol. Immunol. doi: 10.1007/1082_2012_1303.

Leopold, S.R., Magrini, V., Holt, N.J., Shaikh, N., Mardis, E.R., Cagno, J., Ogura, Y., Iguchi, A., Hayashi, T., Mellmann, A., *et al.* (2009). A precise reconstruction of the emergence and constrained radiations of *Escherichia coli* O157 portrayed by backbone concatenomic analysis. Proc. Natl. Acad. Sci. U.S.A. *106*, 8713–8718.

Leopold, S.R., Sawyer, S.A., Whittam, T.S., and Tarr, P.I. (2011). Obscured phylogeny and possible recombinational dormancy in *Escherichia coli*. BMC Evol. Biol. *11*.

Li, B., Koch, W.H., and Cebula, T.A. (1997). Detection and characterization of the *fimA* gene of *Escherichia coli* O157:H7. Mol. Cell. Probes *11*, 397–406.

Mazel, D. (2006). Integrons: agents of bacterial evolution. Nat. Rev. Microbiol. *4*, 608–620.

Mellmann, A., Bielaszewska, M., Köck, R., Friedrich, A.W., Fruth, A., Middendorf, B., Harmsen, D., Schmidt, M.A., and Karch, H. (2008a). Analysis of collection of hemolytic uremic syndrome-associated enterohemorrhagic *Escherichia coli*. Emerg. Infect. Dis. *14*, 1287–1290.

Mellmann, A., Lu, S., Karch, H., Xu, J.-g., Harmsen, D., Schmidt, M.A., and Bielaszewska, M. (2008b). Recycling of Shiga toxin 2 genes in sorbitol-fermenting enterohemorrhagic *Escherichia coli* O157:NM. Appl. Environ. Microbiol. *74*, 67–72.

Mellmann, A., Bielaszewska, M., and Karch, H. (2009). Intrahost genome alterations in enterohemorrhagic *Escherichia coli*. Gastroenterology *136*, 1925–1938.

Middendorf, B., Hochhut, B., Leipold, K., Dobrindt, U., Blum-Oehler, G., and Hacker, J. (2004). Instability of pathogenicity islands in uropathogenic *Escherichia coli* 536. J. Bacteriol. *186*, 3086–3096.

Monday, S.R., Whittam, T.S., and Feng, P.C. (2001). Genetic and evolutionary analysis of mutations in the *gusA* gene that cause the absence of beta-glucuronidase activity in *Escherichia coli* O157:H7. J. Infect. Dis. *184*, 918–921.

Naas, T., Mikami, Y., Imai, T., Poirel, L., and Nordmann, P. (2001). Characterization of In53, a class 1 plasmid- and composite transposon-located integron of *Escherichia coli* which carries an unusual array of gene cassettes. J. Bacteriol. *183*, 235–249.

NCBI (2013). NCBI Genomes. Available at: http://www.ncbi.nlm.nih.gov/genome/?term=Escherichia+coli (accessed on 21 February 2013).

Nelson, K., and Selander, R.K. (1994). Intergeneric transfer and recombination of the 6-phosphogluconate dehydrogenase gene (*gnd*) in enteric bacteria. Proc. Natl. Acad. Sci. U.S.A. *91*, 10227–10231.

O'Brien, A.D., Newland, J.W., Miller, S.F., Holmes, R.K., Smith, H.W., and Formal, S.B. (1984). Shiga-like toxin-converting phages from *Escherichia coli* strains that cause hemorrhagic colitis or infantile diarrhoea. Science *226*, 694–696.

Ochman, H., and Jones, I.B. (2000). Evolutionary dynamics of full genome content in *Escherichia coli*. EMBO J. *19*, 6637–6643.

Parkhill, J., and Wren, B.W. (2011). Bacterial epidemiology and biology–lessons from genome sequencing. Genome Biol. *12*, 230.

Partridge, S.R., Tsafnat, G., Coiera, E., and Iredell, J.R. (2009). Gene cassettes and cassette arrays in mobile resistance integrons. FEMS Microbiol. Rev. *33*, 757–784.

Rasko, D.A., Rosovitz, M.J., Myers, G.S., Mongodin, E.F., Fricke, W.F., Gajer, P., Crabtree, J., Sebaihia, M.,

Thomson, N.R., Chaudhuri, R., *et al.* (2008). The pangenome structure of *Escherichia coli*: comparative genomic analysis of *E. coli* commensal and pathogenic isolates. J. Bacteriol. *190*, 6881–6893.

Reid, S.D., Herbelin, C.J., Bumbaugh, A.C., Selander, R.K., and Whittam, T.S. (2000). Parallel evolution of virulence in pathogenic *Escherichia coli*. Nature *406*, 64–67.

Riley, L.W., Remis, R.S., Helgerson, S.D., McGee, H.B., Wells, J.G., Davis, B.R., Hebert, R.J., Olcott, E.S., Johnson, L.M., Hargrett, N.T., *et al.* (1983). Hemorrhagic colitis associated with a rare *Escherichia coli* serotype. N. Engl. J. Med. *308*, 681–685.

Schmidt, H., and Hensel, M. (2004). Pathogenicity islands in bacterial pathogenesis. Clin. Microbiol. Rev. *17*, 14–56.

Schmidt, H., Beutin, L., and Karch, H. (1995). Molecular analysis of the plasmid-encoded hemolysin of *Escherichia coli* O157:H7 strain EDL 933. Infect. Immun. *63*, 1055–1061.

Schmidt, H., Kernbach, C., and Karch, H. (1996). Analysis of the EHEC *hly* operon and its location in the physical map of the large plasmid of enterohaemorrhagic *Escherichia coli* O157:H7. Microbiology *142 (Pt 4)*, 907–914.

Schubert, S., Darlu, P., Clermont, O., Wieser, A., Magistro, G., Hoffmann, C., Weinert, K., Tenaillon, O., Matic, I., and Denamur, E. (2009). Role of intraspecies recombination in the spread of pathogenicity islands within the *Escherichia coli* species. PLoS Pathog. *5*, e1000257.

Shaikh, N., and Tarr, P.I. (2003). *Escherichia coli* O157:H7 Shiga toxin-encoding bacteriophages: integrations, excisions, truncations, and evolutionary implications. J. Bacteriol. *185*, 3596–3605.

Shaikh, N., Holt, N.J., Johnson, J.R., and Tarr, P.I. (2007). Fim operon variation in the emergence of enterohemorrhagic *Escherichia coli*: an evolutionary and functional analysis. FEMS Microbiol. Lett. *273*, 58–63.

Siefert, J.L. (2009). Defining the mobilome. Methods Mol. Biol. *532*, 13–27.

Siguier, P., Filee, J., and Chandler, M. (2006). Insertion sequences in prokaryotic genomes. Curr. Opin. Microbiol. *9*, 526–531.

Sokurenko, E.V., Hasty, D.L., and Dykhuizen, D.E. (1999). Pathoadaptive mutations: gene loss and variation in bacterial pathogens. Trends Microbiol. *7*, 191–195.

Starcic Erjavec, M., Gaastra, W., van Putten, J., and Zgur-Bertok, D. (2003). Identification of the origin of replications and partial characterization of plasmid pRK100. Plasmid *50*, 102–112.

Torres, A.G., and Kaper, J.B. (2003). Multiple elements controlling adherence of enterohemorrhagic *Escherichia coli* O157:H7 to HeLa cells. Infect. Immun. *71*, 4985–4995.

Torres, A.G., Vazquez-Juarez, R.C., Tutt, C.B., and Garcia-Gallegos, J.G. (2005). Pathoadaptive mutation that mediates adherence of Shiga toxin-producing *Escherichia coli* O111. Infect. Immun. *73*, 4766–4776.

Touchon, M., Hoede, C., Tenaillon, O., Barbe, V., Baeriswyl, S., Bidet, P., Bingen, E., Bonacorsi, S., Bouchier, C., Bouvet, O., *et al.* (2009). Organised genome dynamics in the *Escherichia coli* species results in highly diverse adaptive paths. PLoS Genetics *5*, e1000344.

Toussaint, A., and Chandler, M. (2012). Prokaryote genome fluidity: toward a system approach of the mobilome. Methods Mol. Biol. *804*, 57–80.

Turner, A.K., Nair, S., and Wain, J. (2006). The acquisition of full fluoroquinolone resistance in *Salmonella* Typhi by accumulation of point mutations in the topoisomerase targets. J. Antimicrob. Chemother. *58*, 733–740.

Wang, L., Huskic, S., Cisterne, A., Rothemund, D., and Reeves, P.R. (2002). The O-antigen gene cluster of *Escherichia coli* O55:H7 and identification of a new UDP-GlcNAc C4 epimerase gene. J. Bacteriol. *184*, 2620–2625.

Wirth, T., Falush, D., Lan, R., Colles, F., Mensa, P., Wieler, L.H., Karch, H., Reeves, P.R., Maiden, M.C.J., Ochman, H., *et al.* (2006). Sex and virulence in *Escherichia coli*: an evolutionary perspective. Mol. Microbiol. *60*, 1136–1151.

Yang, J., Lusk, R., and Li, W.H. (2003). Organismal complexity, protein complexity, and gene duplicability. Proc. Natl. Acad. Sci. U.S.A. *100*, 15661–15665.

Zhang, W., Mellmann, A., Sonntag, A.-K., Wieler, L., Bielaszewska, M., Tschäpe, H., Karch, H., and Friedrich, A.W. (2007). Structural and functional differences between disease-associated genes of enterohaemorrhagic *Escherichia coli* O111. Int. J. Med. Microbiol. *297*, 17–26.

Shiga Toxin-encoding Phages: Multifunctional Gene Ferries

4

Maite Muniesa and Herbert Schmidt

Abstract

Shiga toxin encoding-bacteriophages (Stx-phages) constitute a heterogeneous group of temperate lambdoid phages that harbour Shiga toxin (*stx*) genes. The *stx* genes are located in the late region of the prophage genome and expression of Stx is under phage control. In their lysogenic state, Stx-phages are incorporated in the bacterial genome at different chromosomal insertion sites. The presence of more than one Stx-phage in the same bacterial genome has implications in the expression of Stx and in the virulence of the respective host strain. Induction of Stx-phages is stimulated by activation of the lytic cycle, either by inducing agents or spontaneously, and this triggers increased production of Stx. Free Stx-phages are found in many environments, therefore being considered as vehicles that mobilize *stx* by infection and lysogenization of bacteria *in vivo* and *in vitro*. This may lead to the emergence of new bacterial pathogens. Stx-phages can acquire and mobilize foreign genes, including virulence genes present in the chromosome *of Escherichia coli* strains. The role of Stx-phages in horizontal gene transfer, and their impact on bacterial virulence, has been intensively studied, but their biological impact is not yet completely understood.

General overview on Stx-phages

Viruses that infect bacteria (bacteriophages or phages) were described independently in 1915 by Twort and in 1917 by d'Herelle (Duckworth, 1976). Until 1951, they were considered mainly as agents of bacterial lysis (Miller, 2004). Freeman (1951) was the first to demonstrate the role of phages in modification of bacterial virulence. He found that a non-toxigenic strain of *Corynebacterium diphtheriae* infected with a bacteriophage resulted in a virulent strain that produced diphtheria toxin (Freeman, 1951; Tinsley *et al.*, 2006).

The growing impact and improvement of molecular techniques, especially DNA-sequencing techniques, brought to light that many toxins and other virulence factors of Gram-positive and Gram-negative bacteria are encoded in the genome of cryptic or intact phages (Casjens, 2003). The role of phages in dynamics and evolution of bacterial chromosomes has been especially highlighted for emerging bacterial pathogens, such as Shiga toxin-producing *Escherichia coli* (STEC), where acquisition of Shiga toxin-encoding phages (Stx-phages) played a decisive role in the evolution and pathogenicity of the strains (Ogura *et al.*, 2007; Zhou *et al.*, 2010; Kyle *et al.*, 2012).

History and general description of Stx-phages

Shiga toxin was first described in 1898 by Professor Kiyoshi Shiga, from a patient with diarrhoea infected with a microbe termed *Bacillus dysenteriae* (Trofa *et al.*, 1999). The causative agent is now known as *Shigella dysenteriae*. A similar toxin (verocytotoxin, VT) was reported by Konowalchuk *et al.* (1977). This toxin, which had an irreversible damaging effect on Vero cells (Scotland *et al.*, 1985), shared the structure and biological activity to Shiga toxin (Stx) produced

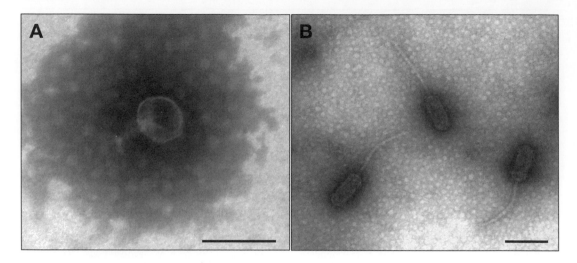

Figure 4.1 Electron micrographs of Stx-phages of *Podoviridae* morphological type corresponding to the prototype Stx2-phage 933W (A) and long-headed *Siphoviridae* corresponding to the prototype Stx1-phage, H-19 (B). Bar 100 nm.

by *S. dysenteriae* type 1. Therefore, toxins of this group were designated as Verotoxins (VT) or Shiga-like toxins (SLT). Smith and Lingwood (1971) provided evidence that the toxigenicity of a clinical diarrhoeagenic *E. coli* strain can be efficiently transmitted to a non-pathogenic strain of *E. coli*, by simply co-culturing of both strains. These findings suggested for the first time that the genes encoding Stx in STEC are located on mobile genetic elements. The genes encoding Stx toxins were then reported to be carried by temperate bacteriophages (Stx-phages) (O'Brien *et al.*, 1984; Newland *et al.*, 1985; Huang *et al.*, 1987) (Fig. 4.1). Each Stx-phage carries one *stx* operon consisting of the two subunit genes, *stxA* and *stxB*, which encode the complete AB$_5$ holotoxin. The Stx-phages encode either Shiga toxin 1 (Stx1) also known as VT1, or Shiga toxin 2 (Stx2) also known as VT2. Besides the two main variants, several *stx* variants (stx_1 (OX3), stx_{2c}, stx_{2d}, stx_{2e}, stx_{2g},) have been lately described in the genome of inducible bacteriophages in *E. coli* (Muniesa *et al.*, 2000b; Koch *et al.*, 2001; Teel *et al.*, 2002; Strauch *et al.*, 2004, 2008; García-Aljaro *et al.*, 2006, Ogura *et al.*, 2007) and *Shigella sonnei* (Strauch *et al.*, 2001). In addition, studies on flanking sequences of several *stx* variants indicated that the *stx* genes of all strains investigated are continuous with phage sequences (Unkmeir and Schmidt, 2001), although in some

cases phages could have lost their ability to be induced. To the author's knowledge up to now, no report on Shiga toxins encoded outside of phage genomes has been published. The number of Stx variants is readily growing and efforts have been done to harmonize the different toxin designations (Scheutz *et al.*, 2012)

The genetic structure of Stx-phages corresponds mainly to that of phage lambda (λ) (Schmidt, 2001; Plunkett *et al.*, 1999, O'Brien *et al.*, 1989). Within the Stx-phage genome, *stxA* and *stxB* are located directly downstream of the antiterminator sites (Neely and Friedman, 1998; Datz *et al.*, 1996). As both, *stxA* and *stxB* are located within the prophage or cryptic prophage late genes, they are transcribed when the lytic cycle is induced. The position of *stx* genes seems to be conserved among Stx-phages and they are located in the late gene region downstream of the late promoters and upstream of the lysis cassette (Neely and Friedman, 1998; Plunkett *et al.*, 1999; Unkmeir and Schmidt, 2001, Herold *et al.*, 2004; Tyler *et al.*, 2004) (Fig. 4.2). The Q protein acts at *p*R' to turn on *stx* expression, and because production of the Q protein ultimately requires transcription from the *p*R' promoter, expression of Stx should therefore require prophage induction (Fig. 4.2). Several authors have shown that derivatives of Stx2 encoding phage 933W, which

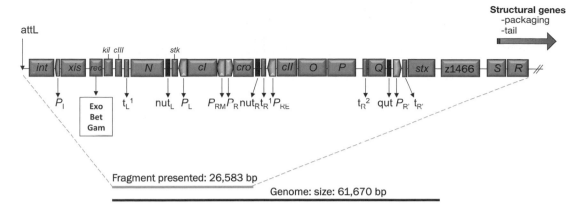

Figure 4.2 Genetic structure of the Stx-phage 933W showing the regulatory region and other relevant genes including *stx*. Not drawn to scale. The scheme was constructed using data from Tyler *et al.*, 2004. A colour version of this figure is available in the plate section at the back of the book.

have mutations in the gene *cl*, responsible of lytic cycle repression, fail to express Stx, and that agents which induce prophages can lead to an increase in Stx production (Neely and Friedman, 1998; Tyler *et al.*, 2004, Wagner *et al.*, 2001b). The operator regions in the pL promoter of Stx-phages could be different from those described for phage λ. In the 933W phage this promoter has only two operator sites while phage λ has three, although this difference has little effect on CI or Cro (activator of the lytic cycle) functions at those sites (Koudelka *et al.*, 2004).

Due to this genetic structure, which is shared by all Stx2-phages studied so far, and which is coincident with the genetic organization of phage λ, they have been assigned to the group of lambdoid phages. However, other aspects of Stx-phages biology show marked differences to phage λ, and this should be considered to avoid confusions. Among others, the morphology of phage λ is that of *Siphoviridae* with an isometric head and a long non-contractile tail, while many Stx-phages present a different morphology.

Diversity of Stx-phages

Stx-phages are highly mobile genetic elements, and they are involved in the emergence of Stx-producing bacteria, in the pathogenic profile of their bacterial hosts, in the development of genome plasticity and in the survival and dissemination of the *stx* genes in the environment. These phages

represent a heterogeneous group since basically any phage that harbours a *stx* operon is referred to as Stx-phage. Diversity of Stx-phages is characterized by the following features:

1 *Genome composition*. Phage genes or gene clusters show that Stx-phages consist of mosaic genes (Johansen *et al.*, 2001) and mosaic-like modules (Recktenwald and Schmidt, 2002), which were derived from a frequent exchange of genes with other lambdoid phages, as it will be discussed in the following sections. However, they share a general lambdoid genome of organization, as well as the location of *stx* genes in the late region.

2 *Genome size*. It varies between 42.6 and 63.4 kb (Recktenwald and Schmidt, 2002, Herold *et al.*, 2005; Plunkett *et al.*, 1999; Yokoyama *et al.*, 2000; Sato *et al.*, 2003) (Table 4.1).

3 *Morphology*. Stx-phages display different morphologies and the phages described so far belong to the families *Siphoviridae*, with isometric or long heads (Muniesa and Jofre, 2000, 2004a,b; Strauch *et al.*, 2004), *Myoviridae*, with contractile tails (Yan *et al.*, 2011), and *Podoviridae*, with icosahedral heads and short tails (Muniesa and Jofre, 2000, 2004a; Plunkett *et al.*, 1999; Yan *et al.*, 2011, for a review on morphology of phages see Ackermann, 2001.

Table 4.1 Examples of complete genome sequences of Stx1- and Stx2-phages

Stx toxin variant	Phage	GenBank accession number	Length (bp)	Reference
stx_1	Stx_1-phage	NC_004913	59,866	Sato *et al.* (2003b)
stx_1+(*nleA*)	BP-4795	AJ556162.1	57,930	Creuzburg *et al.* (2005b)
stx_1	YYZ-2008	NC_011356.1	54,896	Zhang *et al.* (2010)
stx_2	933W	AF125520.1	61,670	Plunkett *et al.* (1999)
stx_2	VT2-Sakai	NC_000902.1	60,942	Miyamoto *et al.* (1999)
stx_2	Stx_2-phage I	NC_003525.1	61,765	Sato *et al.* (2003a)
stx_2	Stx_2-phage II	NC_004914.2	62,706	Sato *et al.* (2003b)
stx_2	phage 86	NC_008464.1	60,238	Yamamoto (2009)*
stx_2	Min27	NC_010237.1	63,395	Su *et al.* (2010)
stx_2	Φ24B	HM208303.1	57,677	Smith *et al.* (2012)
stx_2	TL-2011c	JQ011318.1	60,523	L'Abée-Lund *et al.* (2012)
stx_{2c}	2851	FM180578.1	57,248	Strauch *et al.* (2008)
stx_{2c}	1717	FJ188381	62,147	Zhang *et al.* (2008)*
stx_{2e}	P27	NC_003356.1	42,575	Recktenwald and Schmidt (2002)

*Direct submission, unpublished.

4   *Bacterial host.* Stx-phages recognize different host cell receptors, such as YaeT, a highly conserved essential gene product in *Enterobacteriaceae*, or receptors that have already been described for other phages, such as LomB and FadL (Smith *et al.*, 2007; Watarai *et al.*, 1998).

5   *Superinfection immunity.* Various Stx-phages have been described that are able to circumvent superinfection immunity at high frequency. Some Stx-phages can integrate at least three times into a single host, with increasing efficiency during each successive infection. These multiple infections affect Stx production, Stx-phage release and it also might increase the opportunity for intracellular recombination. (Serra-Morreno *et al.*, 2008; Fogg *et al.*, 2011).

6   *Integration sites.* Several integration sites in the bacterial chromosome have been described for Stx-phages: *wrbA*, *yehV*, *sbcB*, *yecE*, and Z2577, the *torS-torT* intergenic region, *argW*, *ssrA*, and *prfC* (Shaik and Tarr, 2003; Besser *et al.*, 2006, Ogura *et al.*, 2007, Creuzburg *et al.*, 2005a). The use of these integration sites depends on the host strain and on the availability or presence of the preferred site.

Depending on the host strain, phages preferentially use one insertion site. For example, Stx2 phages 933W and VT-2 Sakai integrate into the *wrbA* gene, while Stx1-phages VT-1 Sakai and CP-933V integrate into the *yehV* gene (Shaik and Tarr, 2003; Serra-Moreno *et al.*, 2007; Ogura *et al.*, 2009).

7   *Host range.* Stx2-phages are mostly present in *E. coli* and *Shigella*; however, they are remarkably different with regard to the host strain used for both lysogenic and lytic infection. The host range depends on both, the phage and the host, regardless of the strain that they have been initially isolated from (Gamage *et al.*, 2003).

8   Stx-phages might *additionally carry genes acquired from other phages* and such genes (*stk, yihS, lom, bor*) might be associated with virulence, as in the case of Stx2 phage 933W (Plunkett *et al.*, 1999).

9   *Stx variants.* Different Stx variants are encoded on phage genomes. This aspect has an important effect on the infection caused by STEC since not all toxin variants have been implicated in serious human disease.

10   *Levels of Stx production.* Stx-phages can influence the level of toxin expression by the host

strains and the level of phage release and thus may be relevant to STEC pathogenesis (Wagner *et al.*, 1999; Muniesa *et al.*, 2003).

11 *Influence on the pathogenicity of the host strains.* In addition to the influence of Stx-phages in Stx production, some Stx-phages seem to be involved in up- or down-regulation of other pathogenic factors in the *E. coli* chromosome (Xu *et al.*, 2012).

Genome structure of Stx-phages

All Stx-phages known today are temperate double-stranded DNA phages. The genome sizes differ enormously and can range from ca. 42 kb up to more than 63 kb for functional phages (Recktenwald and Schmidt, 2002, Smith, 2012, Yokoyama *et al.*, 2000; Sato *et al.*, 2003; Laing *et al.*, 2012) (examples are presented in Table 4.1). Their DNA structure and gene composition is lambdoid, meaning that the sequence and regulation of genes follow largely the scheme used by phage λ and its relatives (Campbell and Botstein, 1983). In this context, lambdoid phages can be defined as temperate phages that have the choice between lysogenization and lysis. Moreover, all lambdoid phages use specific chromosomal sites where they integrate their DNA (Campbell and Botstein, 1983), and transcriptional control is achieved by transcript antitermination and lastly, the early lytic functions are transcribed to divergent transcribed operons that were negatively regulated during the lysogenic cycle by a single repressor. Between these main principles numerous variations have been observed (Campbell and Botstein, 1983).

In the early years of phage research, the modular structure of lambdoid phages was already known and a modular theory of phage evolution was proposed (Sußkind and Botstein, 1978). According to this theory, modules of different phages were exchanged between each other by recombination at specialized linker sequences. Later, phages were considered as a collection of functional modules that evolved independently in host genomes and were acquired by the phage (Campbell and Botstein, 1983). However, analysis of genome data indicated that non-homologous recombination also takes place at random points across the genome and not at specific linker sequences. In general, points of recombination are suggested outside of genes or groups of genes; however, recombination has also been observed in coding regions (for a review see Brüssow *et al.*, 2004). Therefore, a huge number of different Stx-phages can be expected, and several publications underline this assumption (Ogura *et al.*, 2007; Laing *et al.*, 2012; Unkmeir and Schmidt, 2000).

The availability of Stx-phage genome sequences facilitates the analysis of genetic relationships and recombination between phages. In one study, 51 phage genomes were aligned and analysed with SplitsTree, and a number of other phylogenetic analysis programs (Laing *et al.*, 2012). SplitsTree is able to depict phylogenetic networks and is often used in evolutionary studies (Huson and Bryant, 2005). The network-like structure obtained by SplitsTree analysis clearly demonstrated numerous recombination events in the evolution of Stx-phages. Interestingly, the Stx-phages of the German O104:H4 outbreak strains formed a phylogenetic cluster and were most closely related to a particular Stx-phage of an *E. coli* O111 strain. Furthermore, Stx-phage genomes were analysed with the software Mauve (Darling *et al.*, 2010). With Mauve, multiple genome alignments can be constructed and large-scale evolutionary events such as rearrangements and inversions can be visualized. This analysis showed a strong conservation in the organization of the *stx* region and further late phase region but the ends of the phage sequences were highly divergent (Laing *et al.*, 2012).

Phages can increase their genome size by acquiring foreign genes. Such genes are typically preceded by a transcriptional promoter sequence and/or a terminator structure, being necessary to be transcribed independently, even from a repressed prophage. Such genes have been designated morons, meaning 'more DNA than there is without the element' (Hendrix, *et al.*, 2000; Brüssow *et al.*, 2004). Moron genes have usually another nucleotide composition than the other phage genes and numerous morons have been identified in phage genomes, being located mainly in the late region. Stx-phages contain also a number of morons. One example is the type III effector gene *nleA*$_{4795}$ of the Stx1-converting phage phi4795. Transcription analysis of this gene showed that its regulation is independent from phage induction (Schwidder *et al.*, 2011).

A number of Stx-phages contain a gene encoding a serine threonine (tyrosine) kinase (*stk*). The expression of a eukaryotic-like kinase is thought to protect lysogenic bacteria from infection with another phage or even to act as a virulence factor in infection (Tyler and Friedman *et al.*, 2004; Friedman *et al.*, 2011; Robertson, 2011). The *stk* gene is mainly transcribed in the prophage state and not during the virulent cycle undermining this suggestion. The phage kinase maintains a phosphorylase activity and it has been shown that it protects a large part of the bacterial population from infection with phage HK97, and to some extent with phage HK022. After infection with one of these phages, the kinase activity increases (Tyler and Friedman, 2004; Friedman *et al.*, 2011). It has been speculated that Stk may serve as a toxin that conserves or protects a fraction of the bacterial population against severe stress (Robertson, 2011).

Homologues of the λ genes *bor* and *lom* have been identified in a number of Stx-phages (Barondess and Beckwith, 1990; 1995). These genes have been originally described as accessory sequences, which are expressed during the lysogenic state of phage λ (Barondess and Beckwith, 1990). The Lom protein has been detected in the outer membrane of the bacterial host. Lom is homologous to several virulence proteins such as Ail, PagC and RcK, which have been involved in the invasion mechanisms of *Yersinia enterocolitica*, and serum resistance of *Salmonella enterica* Serovar Typhimurium. The *bor* gene product increased the survival in animal serum. It has been suggested that these genes may confer a role in bacterial survival in the animal host (Barondess and Beckwith, 1990). The λ gene *bor* is closely related to *iss* located on the conjugative plasmid ColV, I-K94. The overall identity of both genes is 81%. It has been proposed that *iss* represents a fragment of former λ phage. It has been shown that the *iss* locus causes a 20-fold increase in survival of *E. coli* laboratory strains in animal serum *in vitro*. The *iss* locus is also a characteristic trait of avian pathogenic *E. coli* (Pfaff-McDonough *et al.*, 2000).

A large number of type III effector genes that are not located in the locus of enterocyte effacement (LEE) of enteropathogenic and enterohaemorrhagic *E. coli*, such as the cycle-inhibiting factor Cif (Marchès *et al.*, 2003; Loukiadis *et al.*, 2008), the *nle* variants or a homologue of the *S. enterica* serovar Typhimurium *avrA*, *and* the *ospB* effector protein gene (Buchrieser *et al.*, 2000) are located on prophages or prophage-like elements (Tobe *et al.*, 2006; Ogura *et al.*, 2007, Creuzburg *et al.*, 2011a,b). They were positioned in the late region of Stx-phages close to the right attachment site (Creuzburg *et al.*, 2011a).

Therefore, the effector cassette region seems to be rich in recombination events, and probably type III effector genes of members of the genera *Salmonella* and *Shigella* have been integrated into this region. The frequent occurrence of transposases underlines the impact of horizontal gene exchange for these processes. Up to now it is not clear if there is a common principle behind this accumulation of effector genes in phage genomes, or if it is at random. Interestingly, the orientation of the additional type III effector genes is not unique. Some of these genes are orientated in the identical direction to be co-transcribed with the phage genes, while other are orientated in the opposite orientation (Creuzburg *et al.*, 2011a).

The number of non-LEE encoded effector proteins identified in the genomes of pathogenic *E. coli* is readily growing and most of them seem to be encoded on prophages or prophage like-elements. (Tobe *et al.*, 2006; Ogura *et al.*, 2007). The DNA methyltransferase gene *dam* has been detected in Stx-phages (Plunkett *et al.*, 1999). Analysis of the *dam* function in BP 933W has shown that it is important for the maintenance of lysogeny (Murphy *et al.*, 2008).

Structure of the *stx* region

The structure of the *stx* region is worth mentioning. The location of *stx* between the antiterminator Q and the lysis gene S is a common principle in all Stx-phages known do date. In phage λ, this region does not contain any further genes, meaning that Q and S are contiguous (Schmidt, 2001). In Stx-phages a number of open reading frames with unknown functions are present in this region. This could be shown in earlier studies (Plunkett *et al.*, 1999; Unkmeir and Schmidt, 2000; Makino *et al.*, 1999). Besides *stx* genes, DNA sequences coding for rare tRNAs were mainly found in stx_2-carrying prophages and also a large open reading frame, the gene z1466, is present in most of the Stx2-phages. It could be shown that, upon induction with mitomycin C, the genes of this region are co-regulated with stx_2. The homology of z1466 to *nanS* and *yihs* of *E. coli* let us suggest that it could have a function in phage or host biology.

Stx-phage induction and its role in Stx production

The Stx-phages described so far share common features with phage λ and the *stx* operon does not seem to influence the general regulation of the phage genome. Located in the late gene region downstream of the late promoters and upstream of the lysis cassette, *stx* genes are strongly expressed when the lytic phage cycle is activated (Neely and Friedman, 1998, Herold *et al.*, 2004). Stx-phages regulate the transcription of Stx through the activity of phage gene promoters that are repressed during lysogeny. Moreover, Stx production increases also as a consequence of the increase in the number of *stx* gene copies produced after phage replication. Finally, the phage lysis gene products besides setting a limit on the duration of Stx production also provide a mechanism of toxin release from the cell in the process of phage-mediated bacterial lysis (Neely and Friedman, 1998).

A drastic increase in Stx- and phage production of two to three orders of magnitude within 2 to 4 h time lapse has been observed as a result of the induction of bacterial SOS response (Kimmitt *et al.*, 2000; Köhler *et al.*, 2000). The fact that the expression of *stx* genes is linked to Stx-phage induction is important because DNA-damaging agents, that induce an SOS response in bacteria, enhance Stx production and phage release (Mühldorfer *et al.*, 1996; Kimmitt *et al.*, 2000). Therefore, the conditions leading to phage induction will also increase the amount of Stx produced and phage lysis of the bacteria will release the toxin outside of the cell. This is an important aspect of STEC pathogenesis, as the Stx2 produced in the gut during the infectious process

probably influences the development of haemorrhagic colitis and haemolytic–uraemic syndrome (HUS) (Neely and Friedman, 1998; Wagner and Waldor, 2002).

The conditions that cause DNA damage or inhibition of DNA replication via the SOS response causes also prophage induction, and this allows the prophage to efficiently switch from lysogeny to induction and lytic growth when the survival of the host cell is in question (Tinsley *et al.*, 2006). Upon encountering DNA damage or inhibition of DNA synthesis, the cell initiates its SOS response. This response is a comprehensive defence mechanism in bacteria that promotes the cell survival by stimulating DNA repair via recombination, excision repair, and mutagenesis and by inhibition of cell growth. In coordination of SOS response, RecA plays a fundamental role (Little, 1996).

RecA is a key enzyme in homologous recombination. Following the phage λ model, and upon activation of SOS response, RecA catalyses self-cleavage of phage repressor CI. If the CI sufficiently decreases it will initiate the Cro expression and phage lytic growth ensues (Court *et al.*, 2007). This will lead to phage genome replication, in Stx-phages to Stx production, and phage-mediated host cell lysis (Wagner and Waldor, 2002; Court *et al.*, 2007).

Inducing agents and conditions that activate the lytic cycle of Stx-phages could affect the outcome and severity of a STEC infection. In addition, this could lead to an increase in dissemination of *stx* genes by increasing the number of free Stx2-phages that in favourable conditions might infect new susceptible hosts, and, thus, convert them into Stx-producers. An increase in Stx production and/or Stx-phage release has been observed due to certain antibiotics, host eukaryotic factors or physical factors among others:

- *Different classes of antibiotics.* Certain antibiotics used in human therapy or which had been used in animal husbandry as a growth promoters, (although these practices are banned in the EU since 2006) (ETAG, 2006), are able to stimulate the production of Stx and Stx-phages. Quinolones, which are inhibitors of bacterial DNA gyrase, accelerate Stx production in

EHEC O157 (Kimmitt *et al.*, 2000, Kohler *et al.*, 2000, Bielaszewska *et al.*, 2012). Mitomycin C, which damages DNA by cross-linking complementary strands, increases Stx production and Stx2-phage release (Mühldorfer *et al.*, 1996). Mitomycin C is commonly used in laboratory practice as an agent for Stx2-phage induction (Muniesa *et al.*, 2004a,b). Other antibiotics, such as trimethoprim, furazolidone and ciprofloxacin, are potent SOS inducers and they are also reported to induce *stx* gene transcription (Kimmitt *et al.*, 2000). Following the EHEC O104:H4 outbreak in Germany, the role of therapeutic antibiotics on Stx2-expression was investigated and it has been shown that the antibiotics meropenem, azithromycin, rifaximin, and tigecycline did not affect Stx2-production (Bielaszewska *et al.*, 2012).

- *Host eukaryotic factors.* The hormones noradrenaline and adrenaline (Pacheco and Sperandio, 2009), as well as contact with H_2O_2 or neutrophils (Wagner *et al.*, 2001a) have been implicated in increased Stx production in EHEC.

- *UV light, low dose of irradiation or high hydrostatic pressure* can potentially activate the SOS response. These treatments have also been subject of several promising studies against EHEC in foods. Even though, these treatments can reduce or inactivate bacteria (Erickson and Doyle, 2007), they also can trigger Stx-phage lytic cycle (Yamamoto *et al.*, 2003). The new emerging food technologies might potentially lead to a deposition of free Stx-phages in food (Aertsen *et al.*, 2005; Rode *et al.*, 2011).

- *Inducing factors other than antibiotics.* DNase, colicins and certain quorum sensing signals may enhance Stx2-synthesis (Toshima *et al.*, 2007; Pacheco and Sperandio, 2009). On the other hand, some bacterial metabolite products can inhibit the Stx production (de Sablet *et al.*, 2008).

A number of probiotic strains can inhibit Stx2 production *in vitro* (Carey *et al.*, 2008). In addition, probiotic bifidobacteria inhibited mitomycin C-induced Stx production in a mouse model of EHEC infection. The decrease in pH and increases in acetate concentration were proposed to mediate the repressive effect of probiotic strains in Stx2 synthesis (Asahara *et al.*, 2004).

Quorum sensing is a bacterial cell-to-cell communication system operating through the action of signal molecules when the density of the cell population reaches a threshold. Some signal molecules have been described to increase Stx production due to activation of the SOS response (Pacheco and Sperandio, 2009). In case of N-acyl-homoserine lactones (AHLs), an effect on Stx-phage induction could not be observed despite the molecule can trigger phage production in *E. coli* λ lysogens. Interestingly, the prophage induction was unaffected by RecA, suggesting that this mechanisms does not always involve SOS response (Ghosh *et al.*, 2009). This has led to a new paradigm of RecA-independent phage induction. Quorum sensing could also be responsible of the differential activation of Stx-phage lytic cycle within a bacterial population, although there is not clear experimental evidence.

There are reports about Stx2-phage induction independently of RecA, observed when inducing Stx-phages from a *recA*-negative strain lysogenic for a Stx-phage (Muniesa *et al.*, 2004; Imamovic and Muniesa, 2012). Similarly, this fact has been described for phage λ. Under certain conditions, specific *in vitro* cleavage of highly purified λ repressor protein can take place in the absence of RecA protein. The autodigestion reaction cleaves the same alanine–glycine bond, as it happens in the RecA–dependent cleavage reaction. The repressor substrate could be broken down spontaneously in the presence of EDTA, and EDTA has been described to induce Stx-phages in a RecA negative background (Imamovic and Muniesa, 2012). The cleavage was additionally stimulated by the presence of Ca^{2+}, Co^{2+}, or Mg^{2+} and alkaline pH (Little, 1984).

Stx1, but not Stx2, production is also affected by iron (Calderwood *et al.*, 1987; Sung *et al.*, 1990). Stx1-phages contain a binding site for Fur protein which complexes iron. In the presence of iron, Fur binds to DNA and blocks transcription of stx_1. Therefore, Stx1- production could be regulated in response to host-derived signals, such as low iron concentration. Nevertheless, for Stx1 release out of the cell, the activation of phage lysis machinery seems to be required (Wagner *et al.*,

2002). In contrast, Zn inhibits Stx expression at both the protein and the RNA level (Crane *et al.*, 2011). Zinc's inhibitory effects on Stx have been observed in strains expressing Stx1 or Stx2 only, as well as in strains expressing both the toxins types. Zinc inhibits both the basal as well as the antibiotic-induced Stx expression. However, the mechanism of Zn influence on Stx-phages has not been clearly defined.

Expression of Stx mainly depends on the inducibility of the phage carrying the *stx* genes. If the phage is inducible, an increase in the Stx production will be observed independently of the Stx variant (de Sablet *et al.*, 2008), although sometimes spontaneous induction of Stx-phages does not allow comparison of the induction in the presence or absence of an inducing agent. In some other occasions, a point mutation can produce defective toxin, that could not be properly monitored by immunoassays or cell toxicity assays, suggesting that no toxin is produced after phage induction, and leading to contradictory results.

Some reports indicate that efficiency of formation of progeny phages is inhibited by citrate or amino acid starvation (Nejman-Faleńczyk *et al.*, 2012). This effect seems not to be connected to the early stages of phage induction, which is not down-regulated by the presence of citrate. The treatment of Stx-phages with citrate causes an increase of Stx expression, but it will lead to a general reduction of the number of functional phage particles, which are responsible for the spreading of the *stx* genes.

Stx transduction and the emergence of new pathogens

A major driving force in the emergence and evolution of pathogenic bacteria is horizontal gene transfer and acquisition of virulence factors. Several mobile genetic elements, including insertion sequences, plasmids and pathogenicity islands have been implicated in the horizontal transfer of virulence genes, however, the role that bacteriophages play in this process has often been underestimated (Casjens *et al.*, 2000). As a growing number of intact prophages and remnant prophages has been discovered and mapped onto bacterial genomes, it becomes clear that bacteriophages constitute one of the most significant entities involved in bacterial evolution, with up to 20% of some bacterial genomes being composed of bacteriophage genes (Casjens *et al.*, 2000).

Stx-phages are highly mobile genetic elements, which are able to transmit the *stx* genes to other members of the family *Enterobacteriaceae*. Lysogeny with toxin-encoding phage therefore has important implications in the evolution of new Stx-producing strains. Several studies *in vitro* confirmed the ability of these phages to transfer toxin genes to *E. coli* or *Shigella* spp. (Muniesa *et al.*, 2004a; James *et al.*, 2001; Schmidt *et al.*, 1999). Furthermore, intestinal dissemination of these phages and *stx* transduction has been demonstrated (Acheson *et al.*, 1998; Tóth *et al.*, 2003; Cornick *et al.*, 2006; Sekse *et al.*, 2008b). Different studies support *stx* transduction that has been demonstrated in food and water matrices (Imamovic *et al.*, 2009), under dairy processes conditions (Picozzi *et al.*, 2012) and within biofilms (Solheim *et al.*, 2013). Furthermore, several external conditions are found to enhance *stx* transduction, such as a combination of UV and high temperatures (Yue *et al.*, 2012).

New pathogenic serotypes, outbreaks in Norway and Germany

An unfortunate example of the emergence of a new virulent strain can be found in the outbreak that occurred in Germany in May 2011 caused by *E. coli* serotype O104:H4 (Buchholz *et al.*, 2011) and which also affected France (King *et al.*, 2011). The most unusual fact concerning the O104:H4 strain is a genetic background that classifies it as an enteroaggregative *E. coli* strain (EAEC) and, unlike the rest of EAEC, its ability to produce Stx2 (Rohde *et al.*, 2011, Bielaszewska *et al.*, 2011). It seems feasible that incorporation of *stx* has been the differential fact for the generation of this new strain. Combination of Stx production and an optimal adherence caused by the aggregative characteristics of the strain, encoded in a plasmid similar to pAA plasmid found in typical EAEC, has led to its exceptional virulence and the high incidence of HUS reported in this outbreak

(Rasko *et al.*, 2011). This strain is a good example of the clear mosaic of *E. coli* genomes, composed of different mobile genetic elements (phages, plasmids, pathogenicity islands, etc.) that converge to generate more virulent clones (Bielaszewska *et al.*, 2011, Mellmann *et al.*, 2011, Karch *et al.*, 2012).

This is not the only example, an *E. coli* strain of serotype O103:H25 caused an outbreak in Norway in 2006 (Schimmer *et al.*, 2008), with a high incidence of HUS. In the two *stx*-positive O103:H25 strains isolated from two patients in this outbreak, Stx2-phages were well conserved (Sekse *et al.*, 2008a). Other outbreak-associated isolates, including all food isolates, were *stx*-negative, and carried a different phage replacing the Stx2 phage. This phage was of similar size to the Stx2-phage, but had a distinctive early phage region and no *stx* gene (L'Abée-Lund *et al.*, 2012).

It is remarkable that the nucleotide identity between the Stx2-phages from the Norwegian and German outbreak strains was about 90% (L'Abée-Lund *et al.*, 2012), and this suggests a common origin for both phages. The identity includes a 1 bp silent nucleotide mutation in the *stx2A* gene (Bielaszewska *et al.*, 2011), which is rare in other *stx2* genes. In contrast, there are remarkable differences between the Stx-phages in *E. coli* O104:H4 strains from different outbreaks in Germany, Norway and Georgia (Beutin *et al.*, 2012).

The impact of multiple Stx-phages in the same genome

The lytic and lysogenic cycles of stx_2-phages are controlled by the analogous regulator genes *cI* and *cro*, (Fig. 4.2) similar to the regulation in phage λ. The levels of expression of *cI* and *cro* depend on the cellular environment. As an example, cell stress, DNA damage or certain antibiotics present in the extracellular environment would promote the lytic pathway, while high cAMP concentration or low proteases concentration in the cytoplasm would lead to the lysogenic pathway (Herskowitz and Hagen, 1980).

CI and Cro are DNA-binding proteins that share the same DNA target site. They are both able to bind to the operator sites in pR and pL promoters. If CI binds to these sites, *cro* expression will be inhibited, while *cI* expression will be activated. Therefore, CI is known as the lytic repressor and Cro as the anti-repressor. Early *cro* and *cI* expression levels depend on other factors; for instance, in the absence of CI its expression will be determined by CII.

CII protein is easily removed from the cytoplasm by the action of cellular proteases, such as HflA in *E. coli*. CII will only be stable if CIII is bound to it, protecting CII from the proteolytic effect. Hence, depending on CII and CIII, CI and Cro levels will vary, favouring lysogeny or lysis. CII also promotes proper expression of *int*. *int* is only expressed by its proper promoter (pI) under CII stimulation, avoiding the synthesis of altered *int* messenger RNA. Phage integrase recognize homologous sequences in both bacterial chromosome and phage genome (*attB* and *attP*, respectively) to allow phage DNA integration (Herskowitz and Hagen, 1980).

It is not known why the phage integrases have evolved to select specific insertion loci for Stx-prophage integration or if integration in these loci confers biological advantage for the phage and/ or the bacteria. Presumably insertion sites were selected since they are conserved and this could guarantee phage integration. But it could also be speculated that this is not the main reason. For example, insertion in a gene that protects against oxidative stress, as *wrbA* or *Z2577*, could reduce the oxidative stress response, and some oxygen derivatives that induce oxidative stress responses, have been shown to have an effect on Stx-phage induction (Wagner *et al.*, 2001b). Truncation of these loci would therefore increase the presence of the Stx-phage in its prophage state by reducing the induction of lytic cycle. Nevertheless, this has never been confirmed experimentally.

In lysogens harbouring an Stx-phage, the CI repressor prevents the expression of phage genes (Dodd *et al.*, 2005). Repression is not complete and there is always a small fraction of lysogens that are induced. Any loss of prophage repression that, in the absence of a known inducing agent, leads to expression of a phage encoded product, even if it does not result in phage production or host lethality, is known as *spontaneous phage induction* (Livny and Freidman, 2004), although the mechanisms that promote spontaneous induction are poorly understood. One of the differences of Stx-phages with phage λ is that it has been demonstrated that most Stx2 prophages induce spontaneously more readily than lambdoid prophages that do not carry *stx* genes (Livny and Friedman, 2004; Shimizu *et al.*, 2009). When observing laboratory lysogens, Stx-phages could be induced at higher rates in an *E. coli* C600 than in *recA*-negative strains, such as *E. coli* DH5α or a *recA*-mutant (Imamovic and Muniesa, 2012). The number of spontaneously induced phages decreases when the *recA* mutation is complemented, suggesting a role of SOS response on spontaneous induction.

Theoretically, the presence of one Stx-phage should confer immunity protection similar to the situation in phage λ (Herskowitz and Hagen, 1980). However, immunity profiles of different Stx-phages were not in concordance with the model established for phage λ, and there are several descriptions of double lysogeny by the same Stx-phage, so Stx-phages are able of superinfecting bacterial cells containing the same or very similar prophages (Allison *et al.*, 2003; Serra-Moreno *et al.*, 2007, 2008). Significantly, the rate at which these multiple infections occur is extremely high, increasing in frequency with each successive superinfection (Fogg *et al.*, 2011, Serra-Moreno *et al.*, 2008). These results are in concordance with the large proportion of the STEC strains isolated from the environment that carry more than one Stx-prophage (Bielaszewska *et al.*, 2006; Johansen *et al.*, 2001; Teel *et al.*, 2002; Unkmeier and Schmidt, 2000). This is remarkable, since *attP* and *attB* share a common sequence necessary for recognition, so for a given phage, its corresponding *attB* must be present in each new insertion locus (Schmidt, 2001). In principle, the specificity of the integrases would not allow integration of a second phage if its integration locus is already occupied. For Stx-phages this is not necessarily true, since it has been shown that phage integrases can use secondary integration sites for Stx-phages if the primary site is truncated or occupied (Serra-Moreno *et al.*, 2007). Similarly, multiple integration sites have been described for other Stx-phages (Fogg *et al.*, 2007). Although laboratory experiments were performed under antibiotic selection, the clear generation of double lysogens of the same phage, and the amount of wild-type strains harbouring more than one Stx-phage indicates a favourable selection for the insertion of more than one phage without a plausible biological meaning. One can speculate that the generation of a double lysogeny could be a strategy for maintenance of the prophages as such within the cells. It could be observed that in some cases there is a reduction on phage induction when the same Stx-phages are present twice in the same bacterial genome. This fact would be beneficial for the bacteria, because they keep two copies of the virulence gene *stx*, while the possibilities of activation of the lytic cycle and as a consequence the chances to be killed by lysis are reduced. But also there are advantages for the phages, since they can remain stable in a prophage state that could be suggested as their 'preferred' state.

The presence of two Stx-phages in the same bacterial genome has direct implications for the pathogenicity of the host strains. An increased virulence could be inferred from the presence of multiple copies of the *stx* gene in a single cell with the consequent higher expression of Stx as reported by some authors (Hayashi *et al.*, 2001; Wagner *et al.*, 1999). However, it has been proposed that the presence of two Stx-phages could decrease induction of the phages and hence decreasing virulence potential (Bielaszewska *et al.*, 2006; Cornick *et al.*, 2002; Muniesa *et al.*, 2003; Serra-Moreno *et al.*, 2008). The observed reduction of Stx-phages production and the lower levels of Stx expression could be due to diverse possibilities. The first explanation proposed is that CI repressors operating in *trans* could cause the establishment of lysogeny of the two phages in a single bacterial genome. This would regulate and maintain the lysogenic state of both phages, thus inhibiting the induction and reducing Stx expression (Serra-Moreno *et al.*, 2008). A second possibility could be that in the presence of two phages, the gene expression levels of phage late genes increase, as demonstrated by qPCR, but the resources within an infected cell are not sufficient and are simply overwhelmed upon prophage induction, causing reduction on Stx production in some double lysogens (Fogg *et al.*, 2012).

Stx2-phages in the environment, difficulties to evaluate the presence of lytic Stx2-phages

While the lysogeny with Stx-phages has important implications for the evolution of new pathogenic strains, Stx-phages induction contribute to the release of the free phage particles in the environment. The environmental *stx* gene pool is an additional risk factor for the emergence of new STEC types.

Environmental viruses are undoubtedly the largest reservoir of genetic diversity on the planet (Suttle, 2007). They structure microbial communities, cause the lysis of a large proportion of the ocean's biomass on a daily basis, transfer genetic material among host organisms, and shunt nutrients between particulate and dissolved phases of the biosphere. Gene transfer occurs in the natural microbial community, however the scale of the process, the benefit to the host and virus, and the implications for the evolution of the organism involved are poorly understood (Rohwer *et al.*, 2009). Viruses not only have the ability to move genes between hosts and themselves, but they are also capable of moving genetic material between ecosystems.

Free Stx-phages can be found in rivers and sewage systems, and it can be assumed that they originate from contamination with faecal matter containing both free Stx-phages and STEC strains which may later release Stx-phages (Muniesa and Jofre, 1998, 2000; Muniesa *et al.*, 2004b; Tanji *et al.*, 2003; Dumke *et al.*, 2006; Imamovic *et al.*, 2010a). These phages persist longer than the Shiga toxin-producing bacteria in the environment when the treatments commonly used for the elimination of microorganisms are used. Chlorination and pasteurization decreased bacterial densities $4 \log_{10}$ units, in contrast to a $1 \log_{10}$ unit decrease in Stx-phage densities (Muniesa *et al.*, 1999). In contrast, Stx-phages have been shown to be inactivated during composting process at temperatures exceeding 60°C (Johannessen *et al.*, 2005). This was expected since these experiments evaluated Stx-phages persistence after 40 days of treatment (Johannessen *et al.*, 2005).

The survival of bacteriophages and their hosts in river water and sewage has been studied through *in situ* survival experiments (Muniesa *et al.*, 1999). Bacteriophages that infect *E. coli* O157:H7 and carry the stx_2 gene survived equally to somatic coliphages and all other bacteriophages which infect *E. coli* O157:H7, and more successfully than naturally occurring *E. coli*. Hence it would appear that naturally occurring bacteriophages, which infect *E. coli* O157:H7 and carry the stx_2 gene, persist in the aquatic environment more successfully than their bacterial host. The levels of bacteriophages and bacteria were also enumerated over a 2-week period, with bacterial numbers decreasing by a total of 3 logs, and bacteriophages by only 1–$2 \log_{10}$ (Muniesa *et al.*, 1999).

The concept that Stx-phages are stable in the environment for a long period of time and moreover, that they are able to survive processes that eradicate their bacterial hosts can allow them to access new environments and new hosts. In the case that the new host possesses the traits to colonize the mammalian gut, a new pathogen could be created. Therefore, it has been suggested that Stx-phages played a crucial role in the step-wise emergence of EHEC O157 from its EPEC O55 ancestor (Feng *et al.*, 1998; Zhang *et al.*, 2007; Kyle *et al.*, 2012). The environmental persistence of phages, due to their structural characteristics, is greater than the persistence shown by free DNA (either linear fragments or plasmids) (Zhu, 2006).

Studies on the occurrence of Stx-phages and their susceptible hosts can help to evaluate the potential risk for transduction. However, the environmental pool of Stx-phages might be underestimated due to the fact that the classical technique for their enumeration relies on the detection of the host cell lysis *in vitro* (plaque assay). Direct visualization of plaques is complicated since Stx-phages produce poorly visible ones (Muniesa *et al.*, 2004a). For a more accurate evaluation, different approaches have been assayed, among others plaque blot hybridization with a *stx* specific probe (Muniesa *et al.*, 2000, 2004a) or the use of alternative plaquing methods with addition of antibiotics (Santos *et al.*, 2009).

Besides the difficulty of observation of plaques produced by Stx-phages, they can also conduct lysogenic conversion of the host strain (Schmidt *et al.*, 1999; James *et al.*, 2001), instead of causing its lysis. It should be also considered that the host strain can be resistant to the phage infection or that uses strategies to avoid infection, e.g. abortive infection (Barksdale and Arden, 1974; Fineran *et al.*, 2009). Therefore, there is a need for the development of culture-independent phage enumeration techniques.

For this purpose, genetic methodologies based on conventional PCR and more recently on quantitative real-time PCR (qPCR) have been developed (Imamovic *et al.*, 2010b; Rooks *et al.*, 2010). The culture-independent methods have the advantage to overpass the limitations for Stx-phage enumeration, but it should be considered that the number of particles calculated by means of a qPCR does not give a good indication of the infectivity of the phages, which is required for subsequent bacterial conversion. A qPCR will detect non-infectious or defective phage particles in the same way that the infectious ones. Comparison on the enumeration of Stx-phages by qPCR showed that values detected are close to the calculations of phage particles performed by transmission electron microscopy (Imamovic *et al.*, 2010b). The comparison of the number of particles with the number of infectious phages detected by plaque blot hybridization onto suitable bacterial host strains showed, as expected, lower values of infectious particles, with variable correlations among cultivable and non-cultivable techniques that depend on the Stx-phage and on the host strain assayed (Imamovic *et al.*, 2010b).

The concept that phage-encoded virulence genes may be stable in the environment for long periods of time is further supported through the examination of other virulence genes, such as cholera toxin (*ctx*) Staphylococcus enterotoxin A (*sea*) and diphtheria toxin (*dtx*) recovered from different environments, including water, soil and sediments. The data show that phage-encoded exotoxin genes are relatively common in the environment with one of each of the four phage-encoded virulence genes found in 10% of the samples screened, each in association with bacteriophage from both aquatic and terrestrial environments where occurrence of the associated bacterial disease itself was rare (Casas *et al.*, 2006). The ability of the genes encoding exotoxins to survive processes that eradicate their host cell, may

impact upon their distribution, allowing them to access new environments and novel hosts. If the host cell possesses the necessary traits to colonize the mammalian gut, a new pathogen may ensue. For example, Stx-phage transduction has been proposed between STEC strains and *E. coli* O26. Acquisition of stx_2-encoding bacteriophages could confer *E. coli* O26 with greater pathogenic potential for humans, and possession of stx_2 may alter the bovine immune response and lead to longer shedding and to larger quantities of the shed bacterium, resulting in greater exposure to humans (Chase-Topping *et al.*, 2012). Moreover, infection of a commensal gut species could result in *in vivo* transduction and the transient production of toxin in the gut (Acheson *et al.*, 1998; Tóth *et al.*, 2003; Cornick *et al.*, 2006; Sekse *et al.*, 2008b).

Biological impact of Stx-phages

One of the most critical questions concerning the evolutionary implication of phage biology is: What is the advantage for *E. coli* to harbour an Stx-phage?

Answering to this question could indeed be of help in explaining the immense spread of Stx-phages in the entire *E. coli* population. Although the biological meaning of the symbiotic–parasitic relationship between Stx-phage and STEC is not clear, it is clear that bacteria benefits from the incorporation of a Stx-phage. However, the costs for the bacteria should not be neglected. By carrying an Stx-phage, bacteria also carries the potential to be eliminated if the lytic cycle of the phage is activated, mostly due to environmental factors. But the presence of more than one phage in a bacterial genome is a common fact that gives hints that somehow there is some benefit for the host. A non-inducible defective phage would be more desirable for the host, but then there would be no spread of Stx-phages and therefore potential conversion of new hosts to STEC. It seems feasible to imagine that bacterial populations have found an equilibrium among the increased virulence due to Stx-phage lysogeny and the potential death caused by phage lysis, in order to keep the advantages provided by both, without being likely to eliminate the whole bacterial population. During the process of infection, this equilibrium could be reached, although it should imply a sort of 'altruistic suicide', in which part of the population would remain stable and do not activate phage lytic cycle, but producing low amounts of toxin, while other part of the population increases its virulence by activating the phage lytic cycle, and in the meanwhile producing drastic high amounts of Stx. This biological agreement would allow an increase of virulence, spread of Stx-phages and prevalence of the population.

In accordance with this hypothesis, it has been shown that lysogeny with an Stx2-phage represses type III secretion in EHEC (Xu *et al.*, 2012), proposing a model in which the Stx phage regulates bacterial colonization and persistence making it dependent on the bacteriophage, with a subpopulation of bacteria that can then induce the type III secretion system when the appropriate niche signal is detected.

The symmetrical question to be discussed is what would be the benefit for the bacteriophages to carry additional genes. Interesting answers were given in the early times of phage research. Assuming that *stx* genes have been incorporated in a non-Stx-phage during a relatively recent evolutionary event, probably by specialized transduction, why this fact has been positively selected? Following the idea of Darwinian biology, the acquisition and maintenance of foreign genes on a phage genome makes only sense when the lysogenic host confers advantage in a given ecological niche. In a general point of view, all genes that are supporting host's growth, should also ensure phage propagation and therefore benefit the phage. One could assume that phages may pay a rent to the bacterial host to get not excised and deleted from the population. Of course, the benefit of carrying and expressing *stx* is yet not clear.

A possible explanation comes from experiments with parasites. It could be suspected that human infection with STEC is coincidental and that Stx and other virulence factors have evolved for other reasons. One possibility could be that these factors facilitate STEC survival in ruminants. It has been shown that some virulence factors, including *stx*, promote STEC

colonization of the cattle. Another scenario could be protection against bactivorous protozoa such as *Tetrahymena*. Ciliate protozoa are organisms commonly found in the rumen of ruminants (Veira, 1986). Since ciliates are able to eat bacteria, predation by bactivorous protozoa may shape bacterial communities, Stx production could be a defence mechanism against protozoan predation and there are a number of studies underlining this hypothesis (Steinberg and Levin, 2007; Lainhart *et al.*, 2009).

Conclusions

Bacteriophages are probably the most abundant organisms on earth with an estimated amount of 10^{31} (Hendrix *et al.*, 2000). The high diversity of Stx-phages and the widely distributed occurrence in vertebrates as well as in different environments suggests that the Stx-phage is a very successful entity in terms of biological evolution. Since humans are one of the few species, which get sick upon infection with Stx-producing bacteria while the most of other species do not, the question to be answered is whether human infection is accidental. Although highly mobile and heterogeneous, all known Stx-phages characterized from 1982 up to now have a common feature, the *stx* region, having a similar structure and that could be considered as the 'heart' of the phage. This observation suggests that the release of toxin upon phage induction may be of importance for either the phage or the host strain biology. Stx-phage harbouring *E. coli* populations show different range of lysogeny, but to the authors' knowledge, Stx-phages are never completely virulent. Lysogeny, or pseudolysogeny, is in fact necessary to avoid complete elimination of the bacterial population (Barksdale and Arden, 1974).

Conclusively, further study of Stx-phages, their impact to the bacterial host physiology and population biology, and therefore for pathogenicity in humans is important to combat the EHEC-threat that has distributed worldwide. New tools, such as next generation sequencing, labelling and cell culture techniques, and *in vivo* imaging techniques may be of great help in clarifying the role of Stx-phages in bacterial biology.

References

Acheson, D.W., Reidl, J., Zhang, X., Keusch, G.T., Mekalanos, J.J., and Waldor, M.K. (1998). *In vivo* transduction with Shiga toxin 1-encoding phage. Infect. Immun. *66*, 4496–4498.

Ackermann, H.W. (2001). Frequency of morphological phage descriptions in the year 2000. Brief review. Arch. Virol. *146*, 843–857.

Aertsen, A., Faster, D., and Michiels, C.W. (2005). Induction of Shiga toxin-converting prophage in *Escherichia coli* by high hydrostatic pressure. Appl. Environ. Microbiol. *71*, 1155–1162.

Allison, H.E., Sergeant, M.J., James, C.E., Saunders, J.R., Smith, D.L., Sharp, R.J., Marks, T.S., and McCarthy, A.J. (2003). Immunity profiles of wild-type and recombinant Shiga-like toxin-encoding bacteriophages and characterization of novel double lysogens. Infect. Immun. *71*, 3409–3418.

Asahara, T., Shimizu, K., Nomoto, K., Hamabata, T., Ozawa, A., and Takeda, Y. (2004). Probiotic bifidobacteria protect mice from lethal infection with Shiga toxin-producing *Escherichia coli* O157:H7. Infect. Immun. *72*, 2240–2247.

Barksdale, L., and Arden, S.B. (1974). Persisting bacteriophages infections, lysogeny adn phage conversions. Annu. Rev. Microbiol. *28*, 265–300.

Barondess, J.J., and Beckwith, J. (1990). A bacterial virulence determinant encoded by lysogenic coliphage lambda. Nature *346*, 871–874.

Barondess, J.J., and Beckwith, J. (1995). *bor* gene of phage lambda, involved in serum resistance, encodes a widely conserved outer membrane lipoprotein. J. Bacteriol. *177*, 1247–1253.

Besser, T.E., Shaikh, N., Holt, N.J., Konkel, M.E., Malik-Kale, P., Walsh, C.W., Whittman, T., and Bono, J.L. (2006). Greater diversity of Shiga toxin-encoding bacteriophage insertion sites among *Escherichia coli* O157:H7 isolates from cattle than from humans. Appl. Environ. Microbiol. *73*, 671–679.

Beutin, L., Kaulfuss, S., Herold, S., Oswald, E., and Schmidt, H. (2005). Genetic analysis of enteropathogenic and enterohemorrhagic *Escherichia coli* serogroup O103 strains by molecular typing of virulence and housekeeping genes and pulsed-field gel electrophoresis. J. Clin. Microbiol. *43*, 1552–1563.

Beutin, L., Hammerl, J.A., Strauch, E., Reetz, J., Dieckmann, R., Kelner-Burgos, Y., Martin, A., Miko, A., Strockbine, N.A., Lindstedt, B.A., *et al.* (2012). Spread of a distinct Stx2-encoding phage prototype among *E. coli* O104:H4 strains from outbreaks in Germany, Norway and Georgia. J. Virol. *86*, 10444–10455.

Bielaszewska, M., Prager, R., Zhang, W., Friedrich, A.W., Mellmann, A., Tschäpe, H., and Karch, H. (2006). Chromosomal dynamism in progeny of outbreak-related sorbitol-fermenting enterohemorrhagic *Escherichia coli* O157:NM. Appl. Environ. Microbiol. *72*, 1900–1999.

Bielaszewska, M., Mellmann, A., Zhang, W., Köck, R., Fruth, A., Bauwens, A., Peters, G., and Karch, H. (2011). Characterisation of the *Escherichia coli* strain associated with an outbreak of haemolytic uraemic syndrome in Germany, 2011: a microbiological study. Lancet Infect. Dis. *11*, 671–676.

Bielaszewska, M., Idelevich, E.A., Zhang, W., Bauwens, A., Schaumburg, F., Mellmann, A., Peters, G., and Karch, H. (2012). Effects of antibiotics on Shiga toxin 2 production and bacteriophage induction by epidemic *Escherichia coli* O104:H4 strain. Antimicrob. Agents Chemother. *56*, 3277–3282.

Brüssow, H., Canchaya, C., and Hardt, W.D. (2004). Phages and the evolution of bacterial pathogens: from genomic rearrangements to lysogenic conversion. Microbiol. Mol. Biol. Rev. *68*, 560–602.

Buchholz, U., Bernard, H., Werber, D., Böhmer, M.M., Remschmidt, C., Wilking, H., Deleré, Y., an der Heiden, M., Adlhoch, C., Dreesman, J., *et al.* (2011). German outbreak of *Escherichia coli* O104:H4 associated with sprouts. N. Engl. J. Med. *365*, 1763–1770.

Buchrieser, C., Glaser, P., Rusniok, C., Nedjari, H., D'Hauteville, H., Kunst, F., Sansonetti, P., and Parsot, C. (2000). The virulence plasmid pWR100 and the repertoire of proteins secreted by the type III secretion apparatus of *Shigella flexneri*. Mol. Microbiol. *38*, 760–771.

Calderwood, S.B., and Mekalanos, J.J. (1987). Iron regulation of Shiga-like toxin expression in *Escherichia coli* is mediated by the fur locus. J. Bacteriol. *169*, 4759–4764.

Campbell, A., and Botstein, D. (1983). Evolution of the lambdoid phages. In Lambda 11, Hendrix, R.W., Roberts, J.W., Stahl, F.W., and Weisberg, R.A., eds. (Cold Spring Harbor Laboratory, NY, USA), pp. 365–380.

Carey, C.M., Kostrzynska, M., Ojha, S., and Thompson, S. (2008). The effect of probiotics and organic acids on Shiga-toxin 2 gene expression in enterohemorrhagic *Escherichia coli* O157:H7. J. Microbiol. Methods. *73*, 125–132.

Casas, V., Miyake, J., Balsley, H., Roark, J., Telles, S., Leeds, S., Zurita, I., Breitbart, M., Bartlett, D., Azam, F., and Rohwer, F. (2006). Widespread occurrence of phage-encoded exotoxin genes in terrestrial and aquatic environments in Southern California. FEMS Microbiol. Lett. *261*, 141–149.

Casjens, S. (2003). Prophages and bacterial genomics: what have we learned so far? Mol. Microbiol. *49*, 277–300.

Chase-Topping, M.E., Rosser, T., Allison, L.J., Courcier, E., Evans, J., McKendrick, I.J., Pearce, M.C., Handel, I., Caprioli, A., Karch, H., *et al.* (2012). Pathogenic potential to humans of bovine *Escherichia coli* O26, Scotland. Emerg. Infect. Dis. *18*, 439–448.

Cornick, N.A., Jelacic, S., Ciol, M.A., and Tarr, P.I. (2002). *Escherichia coli* O157:H7 infections: discordance between filterable fecal Shiga toxin and disease outcome. J. Infect. Dis. *186*, 57–63.

Cornick, N.A., Helgerson, A.F., Mai, V., Ritchie, J.M., and Acheson, D.W. (2006). *In vivo* transduction of an Stx-encoding phage in ruminants. Appl. Environ. Microbiol. *72*, 5086–5088.

Court, D.L., Oppenheim, A.B., and Adhya, S.L. (2007). A new look at bacteriophage lambda genetic networks. J. Bacteriol. *189*, 298–304.

Crane, J.K., Byrd, I.W., and Boedeker, E.C. (2011). Virulence inhibition by zinc in shiga-toxigenic *Escherichia coli*. Infect. Immun. *79*, 1696–1705.

Creuzburg, K., Köhler, B., Hempel, H., Schreier, P., Jacobs, E., and Schmidt, H. (2005a) Genetic structure and chromosomal integration site of the cryptic prophage CP-1639 encoding Shiga toxin 1. Microbiology. *151*, 941–950.

Creuzburg, K., Recktenwald, J., Kuhle, V., Herold, S., Hensel, M., and Schmidt, H. (2005b). The Shiga toxin 1-converting bacteriophage BP-4795 encodes an NleA-like type III effector protein. J. Bacteriol. *187*, 8494–8498.

Creuzburg, K., Heeren, S., Lis, C.M., Kranz, M., Hensel, M., and Schmidt, H. (2011a) Genetic background and mobility of variants of the gene *nleA* in attaching and effacing *Escherichia coli*. Appl. Environ. Microbiol. *77*, 8705–8713.

Creuzburg, K., Middendorf, B., Mellmann, A., Martaler, T., Holz, C., Fruth, A., Karch, H., and Schmidt, H. (2011b) Evolutionary analysis and distribution of type III effector genes in pathogenic *Escherichia coli* from human, animal and food sources. Environ. Microbiol. *13*, 439–452.

Darling, A.E., Mau, B., and Perna, N.T. (2010). Progressive Mauve: Multiple Genome Alignment with Gene Gain, Loss, and Rearrangement. PLoS One *5*, e11147.

Datz, M., Janetzki-Mittmann, C., Franke, S., Gunzer, F., Schmidt, H., and Karch, H. (1996). Analysis of the enterohemorrhagic *Escherichia coli* O157 DNA region containing lambdoid phage gene p and Shiga-like toxin structural genes. Appl. Environ. Microbiol. *62*, 791–797.

Dodd, I.B., Shearwin, K.E., and Egan, J.B. (2005). Revisited gene regulation in bacteriophage lambda. Curr. Opin. Genet. Dev. *15*, 145–152.

Duckworth, D.H. (1976). Who discovered bacteriophage?' Bacteriol. Rev. *40*, 793–802.

Dumke, R., Schröter-Bobsin, U., Jacobs, E., and Röske, I. (2006). Detection of phages carrying the Shiga toxin 1 and 2 genes in waste water and river water samples. Lett. Appl. Microbiol. *42*, 48–53.

Erickson, M.C., and Doyle, M.P. (2007). Food as a vehicle for transmission of Shiga toxin-producing *Escherichia coli*. J. Food Prot. *70*, 2426–2449.

European Technology Assessment Group ETAG. (2006). Antibiotic resistance. European Parliament. Policy Department. Economic and Scientific Policy. IP/A/STOA/ST/(2006)–4. Brussels, Belgium.

Feng, P., Lampel, K.A., Karch, H., and Whittam, T.S. (1998). Genotypic and phenotypic changes in the emergence of *Escherichia coli* O157:H7. J. Infect. Dis. *177*, 1750–1753.

Fineran, P.C., Blower, T.R., Foulds, I.J., Humphreys, D.P., Lilley, K.S., and Salmond, G.P. 2009. The phage abortive infection system, ToxIN, functions as a protein-RNA toxin–antitoxin pair. Proc. Natl. Acad. Sci. U.S.A. *106*, 894–899.

Fogg, P.C., Gossage, S.M., Smith, D.L., Saunders, J.R., McCarthy, A.J., and Allison, H.E. (2007). Identification of multiple integration sites for Stx-phage Phi24B in the *Escherichia coli* genome, description of a novel integrase and evidence for a functional anti-repressor. Microbiology. *153*, 4098–4110.

Fogg, P.C., Rigden, D.J., Saunders, J.R., McCarthy, A.J., and Allison, H.E. (2011). Characterization of the relationship between integrase, excisionase and antirepressor activities associated with a superinfecting Shiga toxin encoding bacteriophage. Nucleic Acids Res. *39*, 2116–2129.

Fogg, P.C., Saunders, J.R., McCarthy, A.J., and Allison, H.E. (2012). Cumulative effect of prophage burden on Shiga toxin production in *Escherichia coli*. Microbiology. *158*, 488–497.

Freeman, V.J. (1951). Studies on the virulence of bacteriophage-infected strains of *Corynebacterium diphtheriae*. J. Bacteriol. *61*, 675–688.

Friedman, D.I., Mozola, C.C., Beeri, K., Ko, C.C., and Reynolds, J.L. (2011). Activation of a prophage-encoded tyrosine kinase by a heterologous infecting phage results in a self-inflicted abortive infection. Mol. Microbiol. *82*, 567–577.

Gamage, S.D., Strasser, J.E., Chalk, C.L., and Weiss, A.A. (2003). Non-pathogenic *Escherichia coli* can contribute to the production of Shiga toxin. Infect. Immun. *71*, 3107–3115.

García-Aljaro, C., Muniesa, M., Jofre, J., and Blanch, A.R. (2006). Newly identified bacteriophages carrying the stx_{2g} Shiga toxin gene isolated from *Escherichia coli* strains in polluted waters. FEMS Microbiol. Lett. *258*, 127–135.

Ghosh, D., Roy, K., Williamson, K.E., Srinivasiah, S., Wommack, K.E., and Radosevich, M. (2009). Acyl-homoserine lactones can induce virus production in lysogenic bacteria: an alternative paradigm for prophage induction. Appl. Environ. Microbiol. *75*, 7142–7152.

Hardt, W.D., and Galán, J.E. (1997). A secreted Salmonella protein with homology to an avirulence determinant of plant pathogenic bacteria. Proc. Natl. Acad. Sci. U.S.A. *94*, 9887–9892.

Hayashi, T., Makino, K., Ohnishi, M., Kurokawa, K., Ishii, K., Yokoyama, K., Han, C.G., Ohtsubo, E., Nakayama, K., Murata, T., et al. (2001). Complete genome sequence of enterohemorrhagic *Escherichia coli* O157:H7 and genomic comparison with a laboratory strain K-12. DNA Res. *8*, 11–22.

Hendrix, R.W. (2005). Bacteriophage HK97: assembly of the capsid and evolutionary connections. Adv. Virus Res. *64*, 1–14.

Hendrix, R.W., Lawrence, J.G., Hatfull, G.F., and Casjens, S. (2000). The origins and ongoing evolution of viruses. Trends Microbiol. *8*, 504–508.

Herold, S., Karch, H., and Schmidt, H. (2004). Shiga toxin-encoding bacteriophages: genomes in motion. Int. J. Med. Microbiol. *294*, 115–121.

Herold, S., Siebert, J., Huber, A., and Schmidt, H., (2005). Global expression of prophage genes in *Escherichia coli* O157:H7 strain EDL933 in response to norfloxacin. Antimicrob. Agents Chemother. *49*, 931–944.

Herskowitz, I., and Hagen, D. (1980). The lysis-lysogeny decision of phage lambda: explicit programminng and responsiveness. Ann. Rev. Genet. *14*, 399–445.

Huang, A., Friesen, J., and Brunton, J.L. (1987). Characterization of a bacteriophage that carries the genes for production of Shiga-like toxin 1 in *Escherichia coli*. J. Bacteriol. *169*, 4308–4312.

Huson, D., H., and Bryant, D. (2005). Application of phylogenetic networks in evolutionary studies. Mol. Biol. Evol. *23*, 254–267.

Imamovic, L., and Muniesa, M. (2012). Characterizing RecA-independent induction of Shiga toxin2-encoding phages by EDTA treatment. PLoS One 7, e32393.

Imamovic, L., Jofre, J., Schmidt, H., Serra-Moreno, R., and Muniesa, M. (2009). Phage-mediated Shiga toxin 2 gene transfer in food and water. Appl. Environ. Microbiol. *75*, 1764–1768.

Imamovic, L., Ballesté, E., Jofre, J., and Muniesa, M. (2010a). Quantification of Shiga toxin-converting bacteriophages in wastewater and in fecal samples by real-time quantitative PCR. Appl. Environ. Microbiol. *76*, 5693–5701.

Imamovic, L., Serra-Moreno, R., Jofre, J., and Muniesa, M. (2010b). Quantification of Shiga toxin 2-encoding bacteriophages, by real-time PCR and correlation with phage infectivity. J. Appl. Microbiol. *108*, 1105–1114.

James, C.E., Stanley, K.N., Allison, H.E., Flint, H.J., Stewart, C.S., Sharp, R.J., Saunders, J.R., and McCarthy A.J. (2001). Lytic and lysogenic infection of diverse *Escherichia coli* and *Shigella* strains with a verocytotoxigenic bacteriophage. Appl. Environ. Microbiol. *67*, 4335–4337.

Johannessen, G.S., James, C.E., Allison, H.E., Smith, D.L., Saunders, J.R., and McCarthy, A.J. (2005). Survival of a Shiga toxin-encoding bacteriophage in a compost model. FEMS Microbiol. Lett. *245*, 369–375.

Johansen, B.K., Wasteson, Y., Granum, P.E., and Brynestad, S. (2001). Mosaic structure of Shiga-toxin-2-encoding phages isolated from *Escherichia coli* O157:H7 indicates frequent gene exchange between lambdoid phage genomes. Microbiol. *147*, 1929–1936.

Karch, H., Denamur, E., Dobrindt, U., Finlay, B.B., Hengge, R., Johannes, L., Ron, E.Z., Tønjum, T., Sansonetti, P.J., and Vicente, M. (2012). The enemy within us: lessons from the 2011 European *Escherichia coli* O104:H4 outbreak. EMBO Mol. Med. *4*, 841–848.

Kimmitt, P.T., Harwood, C.R., and Barer, M.R. (2000). Toxin gene expression by Shiga toxin-producing *Escherichia coli*: the role of antibiotics and the bacterial SOS response. Emerg. Infect. Dis. *6*, 458–465.

King, L.A., Nogareda, F., Weill, F.X., Mariani-Kurkdjian, P., Loukiadis, E., Gault, G., Jourdan-Dasilva, N., Bingen, E., Macé, M., Thevenot, D., *et al.* (2012). Outbreak of Shiga toxin-producing *Escherichia coli* O104:H4 associated with organic fenugreek sprouts, France, June 2011. Clin. Infect. Dis. *54*, 1588–1594.

Koch, C., Hertwig, S., Lurz, R., Appel, B., and Beutin, L. (2001). Isolation of a lysogenic bacteriophage carrying the stx(1(OX3)) gene, which is closely associated with Shiga toxin-producing *Escherichia coli* strains from sheep and humans. J. Clin. Microbiol. *39*, 3992–3998.

Köhler, B., Karch, H., and Schmidt, H. (2000). Antibacterials that are used as growth promoters in animal husbandry can affect the release of Shiga-toxin-2-converting bacteriophages and Shiga toxin 2 from *Escherichia coli* strains. Microbiology. *146*, 1085–1090.

Konowalchuk, J., Speirs, J.I., and Stavric, S. (1977). Vero response to a cytotoxin of *Escherichia coli*. Infect. Immun. *18*, 775–779.

Koudelka, A.P., Hufnagel, L.A., and Koudelka, G.B. (2004). Purification and characterization of the repressor of the Shiga toxin encoding bacteriophage 933W: DNA binding, gene regulation and autocleavage. J. Bacteriol. *186*, 7659–7669.

Kyle, J.L., Cummings, C.A., Parker, C.T., Quiñones, B., Vatta, P., Newton, E., Huynh, S., Swimley, M., Degoricija, L., Barker, M., *et al.* (2012). *Escherichia coli* serotype O55:H7 diversity supports parallel acquisition of bacteriophage at Shiga toxin phage insertion sites during evolution of the O157:H7 lineage. J. Bacteriol. *194*, 1885–1896.

L'Abée-Lund, T.M., Jørgensen, H.J., O'Sullivan, K., Bohlin, J., Ligård, G., Granum, P.E., and Lindbäck, T. (2012). The highly virulent 2006 Norwegian EHEC O103:H25 outbreak strain is related to the 2011 German O104:H4 outbreak strain. PLoS One 7, e31413.

Laing, C.R., Zhang, Y., Gilmour, M.W., Allen, V., Johnson, R., Thomas, J.E., and Gannon, V.P. (2012). A comparison of Shiga-toxin 2 bacteriophage from classical enterohemorrhagic *Escherichia coli* serotypes and the German *E. coli* O104:H4 outbreak strain. PLoS One 7, e37362.

Lainhart, W., Stolfa, G., and Koudelka, G.B. (2009). Shiga toxin as a bacterial defense against a eukaryotic predator, *Tetrahymena thermophila*. J. Bacteriol. *191*, 5116–5122.

Little, J.W. (1984). Autodigestion of lexA and phage λ repressors. Proc. Natl. Acad. Sci. U.S.A. *81*, 1375–1379.

Livny, J., and Friedman, D.I. (2004). Characterizing spontaneous induction of Stx-encoding phages using a selectable reporter system. Mol. Microbiol. *51*, 1691–1704.

Loukiadis, E., Nobe, R., Herold, S., Tramuta, C., Ogura, Y., Ooka, T., Morabito, S., Kérourédan, M., Brugère, H., Schmidt, H., Hayashi, T., and Oswald, E. (2008). Distribution, functional expression, and genetic organization of Cif, a phage-encoded type III-secreted effector from enteropathogenic and enterohaemorrhagic *Escherichia coli*. J. Bacteriol. *190*, 275–285.

Makino, K., Yokoyama, K., Kubota, Y., Yutsudo, C.H., Kimura, S., Kurokawa, K., Ishii, K., Hattori, M., Tatsuno, I., Abe, H., *et al.* (1999). Complete nucleotide sequence of the prophage VT2-Sakai carrying the verotoxin 2 genes of the enterohaemorrhagic *Escherichia coli* O157:H7 derived from the Sakai outbreak. Genes Genet. Syst. *74*, 227–239.

Marchès, O., Ledger, T.N., Boury, M., Ohara, M., Tu, X., Goffaux, F., Mainil, J., Rosenshine, I., Sugai, M., De Rycke, J., and Oswald, E. (2003): Enteropathogenic and enterohaemorrhagic *Escherichia coli* deliver a novel effector called Cif, which blocks cell cycle G2/M transition. Mol. Microbiol. *50*, 1553–1567.

Mellmann, A., Harmsen, D., Cummings, C.A., Zentz, E.B., Leopold, S.R., Rico, A., Prior, K., Szczepanowski, R., Ji, Y., Zhang, W., *et al.* (2011). Prospective genomic characterization of the German enterohemorrhagic *Escherichia coli* O104:H4 outbreak by rapid next generation sequencing technology. PLoS One *6*, e22751.

Miller, R.V. (2004). Bacteriophage-mediated transduction: an engine for change and evolution. In Microbial evolution, gene establishment survival and exchange, Miller, R.V., and Day, M.J., eds. (ASM Press, Washington, DC), pp. 144–156.

Miyamoto, H., Nakai, W., Yajima, N., Fujibayashi, A., Higuchi, T., Sato, K., and Matsushiro, A. (1999). Sequence analysis of Stx2-converting phage VT2-Sa shows a great divergence in early regulation and replication regions. DNA Res. *6*, 235–240.

Muhldorfer, I., Hacker, J., Keusch, G.T., Acheson, D.W.K., Tschaepe, H., Kane, A.V., Ritter, A., Olschläger, T., and Donohue-Rolfe, A. (1996). Regulation of the Shiga-like toxin II operon in *Escherichia coli*. Infect. Immun. *64*, 495–502.

Muniesa, M., and Jofre, J. (1998). Abundance in sewage of bacterioophages that infect *E. coli* O157:H7 and that carry the Shiga toxin 2 gene. Appl. Environ. Microbiol. *64*, 2443–2448.

Muniesa, M., and Jofre, J. (2000). Occurrence of phages infecting *Escherichia coli* O157:H7 carrying the Stx$_2$ gene in sewage from different countries. FEMS Microbiol. Lett. *183*, 197–200.

Muniesa, M., Lucena, F., and Jofre, J. (1999). Comparative survival of free shiga toxin 2-encoding phages and *Escherichia coli* strains outside the gut. Appl. Environ. Microbiol. *65*, 5615–5618.

Muniesa, M., Recktenwald, J., Bielaszewska, M., Karch, H., and Schmidt, H. (2000). Characterization of a shiga toxin 2e-converting bacteriophage from an *Escherichia coli* strain of human origin. Infect. Immun. *68*, 4850–4855.

Muniesa, M., de Simón, M., Prats, G., Ferrer, D., Pañella, H., and Jofre, J. (2003). Shiga toxin 2-converting bacteriophages associated with clonal turnover in *Escherichia coli* O157:H7 strains of human origin isolated from a single outbreak. Infect. Immun. *71*, 4554–4562.

Muniesa, M., Blanco, J.E., de Simón, M., Serra-Moreno, R., Blanch, A.R., and Jofre, J. (2004a). Diversity of *stx$_2$* converting bacteriophages induced from Shiga-toxin-producing *Escherichia coli* strains isolated from cattle. Microbiology. *150*, 2959–2971.

Muniesa, M., Serra-Moreno, R., and Jofre, J. (2004b). Free Shiga toxin bacteriophages isolated from sewage showed diversity although the *stx* genes appeared conserved. Environ. Microbiol. *6*, 716–725.

Murphy, K.C., Ritchie, J.M., Waldor, M.K., Løbner-Olesen, A., and Marinus, M.G. (2008). Dam methyltransferase is required for stable lysogeny of the Shiga toxin (Stx2)-encoding bacteriophage 933W of enterohemorrhagic *Escherichia coli* O157:H7. J. Bacteriol. *190*, 438–441.

Neely, M.N., and Friedman, D.I. (1998). Functional and genetic analysis of regulatory regions of coliphage H-19B: location of shiga-like toxin and lysis genes suggest a role for phage functions in toxin release. Mol. Microbiol. *28*, 1255–1267.

Nejman-Faleńczyk, B., Golec, P., Maciąg, M., Wegrzyn, A., and Węgrzyn, G. (2012). Inhibition of development of Shiga toxin-converting bacteriophages by either treatment with citrate or amino acid starvation. Foodborne Pathog. Dis. *9*, 13–19.

Newland, J.W., Strockbine, N.A., Miller, S.F., O'Brien, A.D., and Holmes, R.K. (1985). Cloning of Shiga-like toxin structural genes from a toxin converting phage of *Escherichia coli*. Science *230*, 179–181.

O'Brien, A.D., Newland, J.W., Miller, S.F., Holmes, R.K., Smith, H.W., and Formal, S.B. (1984). Shiga-like toxin-converting phages from *Escherichia coli* strains that cause hemorrhagic colitis or infantile diarrhoea. Science *226*, 694–696.

O'Brien, A.D., Marques, L.R., Kerry, C.F., Newland, J.W., and Holmes, R.K. (1989). Shiga-like toxin converting phage of enterohemorrhagic *Escherichia coli* strain 933. Microb. Pathog. *6*, 381–390.

Ogura, Y., Ooka, T., Asadulghani, M., Terajima, J., Nougayrède, J.P., Kurokawa, K., Tashiro, K., Tobe, T., Nakayama, K., Kuhara, S., *et al.* (2007). Extensive genomic diversity and selective conservation of virulence-determinants in enterohemorrhagic *Escherichia coli* strains of O157 and non-O157 serotypes. Genome Biol. *8*, R138.

Ogura, Y., Ooka, T., Iguchi, A., Toh, H., Asadulghani, M., Oshima, K., Kodama, T., Abe, H., Nakayama, K., Kurokawa, K., *et al.* (2009). Comparative genomics reveal the mechanism of the parallel evolution of O157 and non-O157 enterohemorrhagic *Escherichia coli*. Proc. Natl. Acad. Sci. U.S.A. *106*, 17939–17944.

Pacheco, A.R., and Sperandio, V. (2009). Inter-kingdom signaling: chemical language between bacteria and host. Curr. Opin. Microbiol. *12*, 192–198.

Pfaff-McDonough, S.J., Horne, S.M., Giddings, C.W., Ebert, J.O., Doetkott, C., Smith, M.H., and Nolan, L.K. (2000). Complement resistance-related traits among *Escherichia coli* isolates from apparently healthy birds and birds with colibacillosis. Avian Dis. *44*, 23–33.

Picozzi, C., Volponi, G., Vigentini, I., Grassi, S., and Foschino, R. (2012). Assessment of transduction of *Escherichia coli* Stx2-encoding phage in dairy process conditions. Int. J. Food Microbiol. *153*, 388–394.

Plunkett, G., Rose, J.D., Durfee, T.J., and Blattner, F.R. (1999). Sequence of Shiga toxin 2 phage 933W from *Escherichia coli* O157:H7: Shiga toxin as a phage late-gene product. J. Bacteriol. *181*, 1767–1778.

Rasko, D.A., Webster, D.R., Sahl, J.W., Bashir, A., Boisen, N., Scheutz, F., Paxinos, E.E., Sebra, R., Chin, C.S., Iliopoulos, D., *et al.* (2011). Origins of the *E. coli* strain causing an outbreak of hemolytic-uremic syndrome in Germany. N. Engl. J. Med. *365*, 709–717.

Recktenwald, J., and Schmidt, H. (2002). The nucleotide sequence of Shiga toxin (Stx) 2e-encoding phage phiP27 is not related to other Stx phage genomes, but the modular genetic structure is conserved. Infect. Immun. *70*, 1896–1908.

Robertson, E.S. (2011). Survival of the fittest: a role for phage-encoded eukaryotic-like kinases. Mol. Microbiol. *82*, 539–541.

Rode, T.M., Axelsson, L., Granum, P.E., Heir, E., Holck, A., and L'Abée-Lund, T.M. (2011). High stability of Stx2 phage in food and under food-processing conditions. Appl. Environ. Microbiol. *77*, 5336–5341.

Rohde, H., Qin, J., Cui, Y., Li, D., Loman, N.J., Hentschke, M., Chen, W., Pu, F., Peng, Y., Li, J., *et al.* (2011). *E. coli* O104:H4 genome analysis crowd-sourcing consortium. Open-source genomic analysis of Shiga toxin-producing *E. coli* O104:H4. N. Engl. J. Med. *365*, 718–724.

Rohwer, F., Prangishvili, D., and Lindell, D. (2009). Roles of viruses in the environment. Environ. Microbiol. *11*, 2771–2774.

Rooks, D.J., Yan, Y., McDonald, J.E., Woodward, M.J., McCarthy, A.J., and Allison, H.E. (2010). Development and validation of a qPCR-based method for quantifying Shiga toxin-encoding and other lambdoid bacteriophages. Environ. Microbiol. *12*, 1194–1204.

de Sablet, T., Bertin, Y., Vareille, M., Girardeau, J.P., Garrivier, A., Gobert, A.P., and Martin, C. (2008). Differential expression of stx_2 variants in Shiga toxin-producing *Escherichia coli* belonging to seropathotypes A and C. Microbiology. *154*, 176–186.

Samba-Louaka, A., Nougayrède, J.P., Watrin, C., Oswald, E., and Taieb, F. (2009). The enteropathogenic *Escherichia coli* effector Cif induces delayed apoptosis in epithelial cells. Infect. Immun. *77*, 5471–5477.

Santos, S.B., Carvalho, C.M., Sillankorva, S., Nicolau, A., Ferreira, E.C., and Azeredo, J. (2009). The use of antibiotics to improve phage detection and enumeration by the double-layer agar technique. BMC Microbiol. *9*, 148–157.

Sato, T., Shimizu, T., Watarai, M., Kobayashi, M., Kano, S., Hamabata, T., Takeda, Y., and Yamasaki, S. (2003a). Distinctiveness of the genomic sequence of Shiga toxin 2-converting phage isolated from *Escherichia coli* O157:H7 Okayama strain as compared to other Shiga toxin 2-converting phages. Gene *309*, 35–48.

Sato, T., Shimizu, T., Watarai, M., Kobayashi, M., Kano, S., Hamabata, T., Takeda, Y., and Yamasaki, S. (2003b). Genome analysis of a novel Shiga toxin 1 (Stx1)-converting phage which is closely related to Stx2-converting phages but not to other Stx1-converting phages. J. Bacteriol. *185*, 3966–3971.

Scheutz, F., Teel, L.D., Beutin, L., Piérard, D., Buvens, G., Karch, H., Mellmann, A., Caprioli, A., Tozzoli, R., Morabito, S., *et al.* (2012). Multicenter evaluation of a sequence-based protocol for subtyping shiga toxins and standardizing stx nomenclature. J. Clin. Microbiol. *50*, 2951–2963.

Schimmer, B., Nygard, K., Eriksen, H.M., Lassen, J., Lindstedt, B.A., Brandal, L.T., Kapperud, G., and Aavitsland, P. (2008). Outbreak of haemolytic uraemic syndrome in Norway caused by stx2-positive *Escherichia coli* O103:H25 traced to cured mutton sausages. BMC Infect. Dis. *8*, 41.

Schmidt H. (2001). Shiga toxin converting bacteriophages. Res. Microbiol. *152*, 687–695.

Schmidt, H., Bielaszewska, M., and Karch, H. (1999). Transduction of enteric *Escherichia coli* isolates with a derivative of Shiga-toxin 2-encoding bacteriophage 3538 isolated from *Escherichia coli* O157:H7. Appl. Environ. Microbiol. *65*, 3855–3861.

Schwidder, M., Hensel, M., and Schmidt, H. (2011). Regulation of nleA in Shiga toxin-producing *Escherichia coli* O84:H4 strain 4795/97. J. Bacteriol. *193*, 832–841.

Scotland, S.M., Smith, H.R., and Rowe, B. (1985). Two distinct toxins active on Vero cells from *Escherichia coli* O157. Lancet *2*, 885–886.

Sekse, C., Muniesa, M., and Wasteson, Y. (2008a). Conserved Stx2 phages from *Escherichia coli* O103:H25 isolated from patients suffering from hemolytic uremic syndrome. Foodborne Pathog. Dis. *5*, 801–810.

Sekse, C., Solheim, H., Urdahl, A.M., and Wasteson, Y. (2008b). Is lack of susceptible recipients in the intestinal environment the limiting factor for transduction of Shiga toxin-encoding phages? J. Appl. Microbiol. *105*, 1114–1120.

Serra-Moreno, R., Jofre, J., and Muniesa, M. (2007). Insertion site occupancy by stx_2-bacteriophages depends on the locus availability of the host strain chromosome. J. Bacteriol. *189*, 6645–6654.

Serra-Moreno, R., Jofre, J., and Muniesa, M. (2008). The CI repressors of Shiga toxin-converting prophages are involved in coinfection of *Escherichia coli* strains, which causes a down-regulation in the production of Shiga toxin 2. J. Bacteriol. *190*, 4722–4735.

Shaikh, N., and Tarr, P.I. (2003). *Escherichia coli* O157:H7 Shiga toxin-encoding bacteriophages: integrations, excisions, truncations, and evolutionary implications. J. Bacteriol. *185*, 3596–3605.

Shimizu, T., Ohta, Y., and Noda, M. (2009). Shiga toxin 2 is specifically released from bacterial cells by two different mechanisms. Infect. Immun. *77*, 2813–2823.

Smith, D.L., James, C.E., Sergeant, M.J., Yaxian, Y., Saunders, J.R., McCarthy, A.J., and Allison, H.E. (2007). Short-tailed Stx phages exploit the conserved YaeT protein to disseminate Shiga toxin genes among enterobacteria. J. Bacteriol. *189*, 7223–7233.

Smith, D.L., Rooks, D.J., Fogg, P.C., Darby, A.C., Thomson, N.R., McCarthy, A.J., and Allison, H.E. (2012). Comparative genomics of Shiga toxin encoding bacteriophages. BMC Genomics *13*, 311.

Smith, H.W., and Linggood, M.A. (1971). The transmissible nature of enterotoxin production in a human enteropathogenic strain of *Escherichia coli*. J. Med. Microbiol. *4*, 301–305.

Solheim, H.T., Sekse, C., Urdahl, A.M., Wasteson, Y., and Nesse LL. 2013. Biofilm as an environment for dissemination of *stx* genes by transduction. Appl Environ Microbiol. *79*, 896–900.

Steinberg, K.M., and Levin, B.R. (2007). Grazing protozoa and the evolution of the *Escherichia coli* O157:H7 Shiga toxin-encoding prophage. Proc. Biol. Sci. *274*, 1921–1929.

Strauch, E., Lurzand, R., and Beutin, L. (2001). Characterization of a Shiga Toxin-encoding temperate bacteriophage of *Shigella sonnei*. Infect. Immun. *69*, 7588–7595.

Strauch, E., Schaudinn, C., and Beutin, L. (2004). First-time isolation and characterization of a bacteriophage encoding the Shiga toxin 2c variant, which is globally spread in strains of Escherichia coli O157. Infect. Immun. 72, 7030–7039.

Strauch, E., Hammerl, J.A., Konietzny, A., Schneiker-Bekel, S., Arnold, W., Goesmann, A., Pühler, A., and Beutin, L. (2008). Bacteriophage 2851 is a prototype phage for dissemination of the Shiga toxin variant gene 2c in Escherichia coli O157:H7. Infect. Immun. 76, 5466–5477.

Su, L.K., Lu, C.P., Wang, Y., Cao, D.M., Sun, J.H., and Yan, Y.X. (2010). Lysogenic infection of a Shiga toxin 2-converting bacteriophage changes host gene expression, enhances host acid resistance and motility. Mol. Biol. 44, 54–66.

Sung, L.M., Jackson, M.P., O'Brien, A.D., and Holmes, R.K. (1990). Transcription of the Shiga-like toxin type II and Shiga-like toxin type II variant operons of Escherichia coli. J. Bacteriol. 172, 6386–6395.

Susskind, M.M., and Botstein, D. (1978). Molecular genetics of bacteriophage P22. Microbiol. Rev. 42, 385–413.

Suttle, C.A. (2007). Marine viruses: major players in the global ecosystem. Nat. Rev. Microbiol. 5, 801–812.

Tanji, Y., Mizoguchi, K., Yoichi, M., Morita, M., Kijima, N., Kator, H., and Unno, H. (2003). Seasonal change and fate of coliphages infected to Escherichia coli O157:H7 in a wastewater treatment plant. Water Res. 37, 1136–1142.

Teel, L.D., Melton-Celsa, A.R., Schmitt, C.K., and O'Brien, A.D. (2002). One of two copies of the gene for the activatable shiga toxin type 2d in Escherichia coli O91:H21 strain B2F1 is associated with an inducible bacteriophage. Infect. Immun. 70, 4282–4291.

Tinsley, C.R., Bille, E., and Nassif, X. (2006). Bacteriophages and pathogenicity: more than just providing a toxin? Microbes Infect. 8, 1365–1371.

Tobe, T., Beatson, S.A., Taniguchi, H., Abe, H., Bailey, C.M., Fivian, A., Younis, R., Matthews, S., Marches, O., Frankel, G., et al. (2006). An extensive repertoire of type III secretion effectors in Escherichia coli O157 and the role of lambdoid phages in their dissemination. Proc. Natl. Acad. Sci. U.S.A. 103, 14941–14946.

Toshima, H., Yoshimura, A., Arikawa, K., Hidaka, A., Ogasawara, J., Hase, A., Masaki, H., and Nishikawa, Y. (2007). Enhancement of Shiga toxin production in enterohemorrhagic Escherichia coli serotype O157:H7 by DNase colicins. Appl. Environ. Microbiol. 73, 7582–7588.

Tóth, I., Schmidt, H., Dow, M., Malik, A., Oswald, E., and Nagy, B. (2003). Transduction of porcine enteropathogenic Escherichia coli with a derivative of a shiga toxin 2-encoding bacteriophage in a porcine ligated ileal loop system. Appl. Environ. Microbiol. 69, 7242–7247.

Trofa, A.F., Ueno-Olsen, H., Oiwa, R., and Yoshikawa, M. (1999). Dr. Kiyoshi Shiga: discoverer of the dysentery bacillus. Clin. Infect. Dis. 29, 1303–1306.

Tyler, J.S., and Friedman, D.I. (2004). Characterization of a eukaryotic-like tyrosine protein kinase expressed by the Shiga toxin-encoding bacteriophage 933W. J. Bacteriol. 186, 3472–3479.

Tyler, J.S., Mills, M.J., and Friedman, D.I. (2004). The operator and early promoter region of the Shiga toxin type 2-ecoding bacteriophage 933W and control of toxin expression. J. Bacteriol. 186, 7670–7679.

Unkmeir, A., and Schmidt, H. (2000). Structural analysis of phage-borne stx genes and their flanking sequences in shiga toxin-producing Escherichia coli and Shigella dysenteriae type 1 strains. Infect. Immun. 68, 4856–4864.

Veira, D.M. (1986). The role of ciliate protozoa in nutrition of the ruminant. J. Anim. Sci. 63, 1547–1560.

Wagner, P.L., and Waldor, M.K. (2002). Bacteriophage control of bacterial virulence. Infect. Immun. 70, 3985–3993.

Wagner, P.L., Acheson, D.W.K., and Waldor, M.K. (1999). Isogenic lysogens of diverse Shiga toxin 2-encoding bacteriophages produce markedly different amounts of Shiga toxin. Infect. Immun. 67, 6710–6714.

Wagner, P.L., Acheson, D.W.K., and Waldor, M.K. (2001a). Human neutrophils and their products induce Shiga toxin production by enterohemorrhagic Escherichia coli. Infect. Immun. 69, 1934–1937.

Wagner, P.L., Neely, M.N., Zhang, X., Acheson, D.W., Waldor, M.K., and Friedman, D.I. (2001b) Role for a phage promoter in Shiga toxin 2 expression from a pathogenic Escherichia coli strain. J. Bacteriol. 183, 2081–2085.

Watarai, M., Sato, T., Kobayashi, M., Shimizu, T., Yamasaki, S., Tobe, T., Sasakawa, C., and Takeda, Y. (1998). Identification and characterization of a newly isolated Shiga toxin 2-converting phage from shiga toxin-producing Escherichia coli. Infect. Immun. 66, 4100–4107.

Xu, X., McAteer, S.P., Tree, J.J., Shaw, D.J., Wolfson, E.B., Beatson, S.A., Roe, A.J., Allison, L.J., Chase-Topping, M.E., Mahajan, A., et al. (2012). Lysogeny with Shiga Toxin 2-encoding bacteriophages represses Type III secretion in enterohemorrhagic Escherichia coli. PLoS Pathog. 8, e1002672.

Yamamoto, T., Kojio, S., Taneike, I., Nakagawa, S., Iwakura, N., and Wakisaka-Saito, N. (2003). ^{60}Co irradiation of Shiga toxin (Stx)-producing Escherichia coli induces Stx phage. FEMS Microbiol. Lett. 222, 115–121.

Yan, Y., Shi, Y., Cao, D., Meng, X., Xia, L., and Sun, J. (2011). Prevalence of Stx phages in environments of a pig farm and lysogenic infection of the field Escherichia coli O157 isolates with a recombinant converting phage. Curr. Microbiol. 62, 458–464.

Yokoyama, K., Makina, K., Kubota, Y., Watanabe, M., Kimura, S., Yutsudo, C.H., Kurokawa, K., Ishii, K., Hattori, M., Tatsuno, I., et al. (2000). Complete nucleotide sequence of the prophage VT1-Sakai carrying the Shiga toxin 1 genes of enterohemorragic Escherichia coli O157:H7 strain derived from the Sakai outbreak. Gene 258, 163–177.

Yue, W.F., Du, M., and Zhu, M.J. (2012). High temperature in combination with UV irradiation enhances horizontal transfer of stx2 gene from E. coli O157:H7 to non-pathogenic E. coli. PLoS One 7, e31308.

Zhang, Y., Laing, C., Steele, M., Ziebell, K., Johnson, R., Benson, A.K., Taboada, E., and Gannon, V.P. (2007). Genome evolution in major *Escherichia coli* O157:H7 lineages. BMC Genomics 8, 121.

Zhou, Z., Li, X., Liu, B., Beutin, L., Xu, J., Ren, Y., Feng, L., Lan, R., Reeves, P.R., and Wang, L. (2010). Derivation of *Escherichia coli* O157:H7 from its O55:H7 precursor. PLoS One 5, e8700.

Zhu, B. (2006). Degradation of plasmid and plant DNA in water microcosms monitored by natural transformation and real-time polymerase chain reaction PCR. Water Res. *40*, 3231–3238.

Shiga Toxins

Samuel Juillot and Winfried Römer

5

Abstract

Shiga toxins are virulence factors produced by *Shigella dysenteriae* and certain *Escherichia coli* strains called STEC. The toxins affect target cells already by binding to the plasma membrane through induction of signalling cascades that mainly lead to apoptosis. Furthermore, they inhibit protein biosynthesis by inactivating the 60S subunit of ribosomes. In order to reach their cytoplasmic target, Shiga toxins are endocytozed and transported through the retrograde trafficking pathway towards the endoplasmic reticulum (ER), from where the catalytically active subunit is retro-translocated to the cytosol with the help of ER chaperons and translocon components. Even though Shiga toxins are still a threat to human health, the receptor-binding subunit of Shiga toxins represents a powerful tool to study the mechanisms of intracellular transport and may be exploited as biomedical tool in immunotherapy and tumour imaging.

Discovery of Shiga toxins

Shiga toxins, also well-known as verotoxins or Shiga-like toxins belong to a group of structurally and functionally related exotoxins comprising Shiga toxin from *Shigella dysenteriae* serotype 1 and Shiga toxins that are produced from entero-haemorrhagic strains of *Escherichia coli* (EHEC). The use of diverse terms to describe very similar toxins has historical reasons. In 1897, the Japanese microbiologist Kiyoshi Shiga characterized the dysentery bacillus – *Shigella dysenteriae* (Shiga, 1898; Trofa *et al.*, 1999). In 1903, the presence of a toxic agent (Shiga toxin) in autolysates from cultured *S. dysenteriae* was demonstrated. It was first classified as 'neurotoxin' because parenteral inoculation in rabbits resulted in limb paralysis, cerebral and spinal cord haemorrhages followed by death (Conradi, 1903; Flexner and Sweet, 1906). In 1972, Keusch *et al.* (1972) reported the first partial purification of this '*Shigella* neurotoxin' and demonstrated that the cytotoxic and enterotoxic effects were due to a single toxin – Shiga toxin (Keusch and Jacewicz, 1975). In two different studies using volunteers and a monkey model, it was suggested that *S. dysenteriae* produce the toxin to influence the severity of dysentery (Levine *et al.*, 1973; Fontaine *et al.*, 1988). In 1977, Konowalchuk and colleagues discovered a group of *E. coli* isolates associated with infant diarrhoea that produced a toxin, different from the *E. coli* heat-labile enterotoxin, which was able to kill Vero (African green monkey kidney) cells in culture (Konowalchuk *et al.*, 1977). The toxin was termed 'Vero cell toxin', and the bacteria were termed verotoxin-producing *E. coli* (VTEC). A few years later, Shiga toxin was purified to homogeneity (Olsnes and Eiklid, 1980; O'Brien *et al.*, 1980) and reported as cytotoxic for tissue culture (HeLa cells), enterotoxic in rabbit ileal loops and lethal when injected into rabbits or mice (O'Brien *et al.*, 1980). These biological activities can be neutralized by a specific antiserum derived by immunization with toxoid. Using neutralizing antibodies, O´Brien and colleagues demonstrated in the early 1980s that some isolated *E. coli* strains produced a toxin very similar to Shiga toxin and named it Shiga-like toxin. The bacteria were termed Shiga-like toxin-producing *E. coli* (STEC) (O'Brien *et al.*, 1982, 1984) and were later associated with haemolytic uraemic syndrome (HUS) (Karmali *et al.*, 1983, 1985). Moreover, as neutralizing antibodies against purified Shiga-like toxin were also efficient on VTEC (O'Brien *et al.*, 1983), researchers realized that they were probably studying the same family of highly related Shiga toxins (Stxs).

Indeed, Shiga toxin from *S. dysenteriae* is nearly identical to Shiga-like toxin 1 (Stx1) produced by *E. coli*, differing only by a single amino acid in the catalytic A-subunit of the toxin (Strockbine *et al.*, 1988). STEC strains can produce two types of Shiga-like toxins – Stx1 (with three subtypes Stx1a, Stx1c and Stx1d) and Stx2 (with seven subtypes, Stx2a to Stx2g). Although Stx1 and Stx2 share a common receptor and exert the same intracellular mechanism of action, Stx2 cannot be neutralized by antibodies against either Stx or Stx1 (Strockbine *et al.*, 1986; Calderwood *et al.*, 1987), most probably since they are only 56% identical at the amino acid sequence level (Jackson *et al.*, 1987). Some Stx-producing bacteria express only one type, while others express a combination of variants of one or both types (Karch *et al.*, 2005). However, severe disease was epidemiologically linked to the presence of Stx2 (Boerlin *et al.*, 1999).

Genetic and environmental regulation of Shiga toxin production

Shiga toxins are encoded by genes that localize on the genomes of functional or defective lambdoid bacteriophages (O'Brien *et al.*, 1984). *S. dysenteriae* originally carried a Stx-encoding lambdoid prophage, which became defective due to loss of bacteriophage sequences after IS element insertions and rearrangements (McDonough and Butterton, 1999; Unkmeir and Schmidt, 2000). Hence, the *stx* genes in *S. dysenteriae* are considered to be chromosomally encoded and not transmissible (O'Brien *et al.*, 1992), whereas genes encoding the production of Stx produced by *E. coli* are located on toxin-converted lambdoid phages (O'Brien *et al.*, 1984; Newland *et al.*, 1985). Two different phages encode genes for the two immunologically distinct toxin sub-types, Stx1 and Stx2 (O'Brien *et al.*, 1984; Strockbine

et al., 1986), except for Stx2e (associated with the piglet oedema disease) that is usually chromosomally encoded (most likely an integrated phage) (O'Brien and Kaper, 1998).

These Stx-phages, which are defined by the presence of the Shiga toxin operon, are highly mobile genetic elements. Bacterial host cells can carry several Stx-phages, allowing them to produce several Shiga toxin variants. The presence of several Stx-phages and cryptic prophages within a single cell increases the likelihood for recombination between phage sequences, with the potential for creating novel phages and for genome diversification (Johansen *et al.*, 2001; Herold *et al.*, 2004). Even though Stx-phages are genetically diverse, the *stx* genes are located in the late gene region downstream of the late promoters and upstream of the lysis cassette and are highly expressed when the lytic cascade of the phage is activated (Wagner *et al.*, 2002; Herold *et al.*, 2004; Tyler *et al.*, 2004). Phages regulate Stx production through the amplification of gene copy number, activity of phage gene promoters, and through toxin release (Neely and Friedman, 1998; Tyler *et al.*, 2004). After expression, folding and assembly of toxin subunits, Stx is located in the periplasm of Gram-negative bacteria (Donohue-Rolfe and Keusch, 1983). Since the release of the toxin seems to coincide with phage-induced bacterial lysis (Neely and Friedman, 1998), there is no absolute necessity for a specific secretion system. However, there is evidence for at least one secretion system capable of releasing Stx2 from viable *E. coli* cells (Shimizu *et al.*, 2009). Another possibility might be the destruction of bacterial cells by the human immune system that leads to release of Stx in the absence of phage-mediated lysis (Wagner *et al.*, 2002).

The production of Stx from *S. dysenteriae* and Stx1 from *E. coli* are temperature-regulated, being maximal at 37°C, and are repressed in conditions with a high iron concentration (Dubos and Geiger, 1946; Weinstein *et al.*, 1988). Indeed, Stx and Stx1 possess a binding site for the Fur protein on the operator region upstream of the gene for the A-subunit. The Fur protein forms a complex with iron, binds to DNA and thus blocks transcription (Calderwood and Mekalanos, 1987; Svinarich and Pachaudhuri, 1992). These two environmental conditions suggest that the production of Stx can be regulated in response to host-derived signals such as body temperature and a low iron environment, which would be expected in the distal small intestine and colon. In contrast, Stx2 does not appear to be regulated by environmental factors (Mühldorfer *et al.*, 1996).

Shiga toxin structure

Shiga and Shiga-like toxins belong to the family of AB_5 toxins, which e.g. also comprises the cholera toxin and the heat-labile toxins from enterotoxigenic *E. coli* (Stein *et al.*, 1992; Fraser *et al.*, 1994). Shiga toxins are composed of an enzymatically active monomeric A-subunit (StxA) with a molecular mass of 32.2 kDa and a non-toxic

B-subunit (StxB) that binds specifically to cell-surface receptors (O'Brien *et al.*, 1992; Stein *et al.*, 1992) (Fig. 5.1a). The B-subunit consists of five identical monomers (7.7 kDa) that form a dough-nut-shaped pentameric structure with a central pore into which the C-terminus of StxA inserts (Stein *et al.*, 1992; Fraser *et al.*, 1994; Richardson *et al.*, 1997) (Fig. 5.1a and b).

As initially described for heat-labile entero-toxins from *E. coli*, StxA and StxB fragments are expressed with N-terminal signal peptides targeting the protein for transport to the bacterial periplasm, where the signal peptides are removed and the fragments assemble non-covalently into the mature holotoxin (Hirst *et al.*, 1984). By using recombinant toxin subunits, Austin *et al.* found that intact toxin A- and B-subunits have the ability

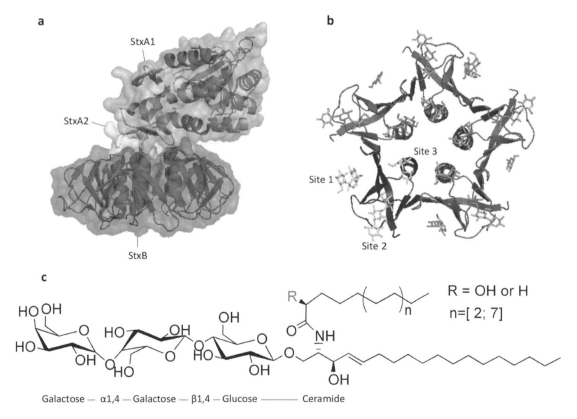

Figure 5.1 The structure of Shiga toxin and its cellular receptor globotriaosylceramide (Gb3). (a) A ribbon diagram of Shiga toxin, consisting of one A-subunit (StxA), which can be cleaved into the fragments A1 (orange) and A2 (yellow), and the homopentameric B-subunit (StxB) that consists of five B-fragments (red). (b) A ribbon diagram of the StxB subunit from the membrane-oriented surface, highlighting one B-fragment (blue) with three Gb3-binding sites. Gb3 is shown as a stick representation. Note the central pore that is lined by α–helices. (c) Chemical structure of the glycosphingolipid Gb3 composed of Galα1–4Galβ1–4GlcCer. A colour version of this figure is available in the plate section at the back of the book.

to form the holotoxin *in vitro* (Austin *et al.*, 1994). However, the A1-fragment is unable to combine with the B-subunit to form the holotoxin, suggesting that the A2-fragment is essential for toxin assembly. Indeed, site-directed mutagenesis revealed that nine amino acids of the A2-fragment, which form an α-helix, penetrate the pore of the StxB pentamer (Haddad and Jackson, 1993).

In *S. dysenteriae*, the structural genes encoding the A- and B-fragments of Shiga toxin are present on a single transcriptional unit (Strockbine *et al.*, 1988). The biosynthetic mechanism of holotoxin assembly remains unclear as the fragments are transcribed at a 1:1 ratio (Hale, 1991). However, Habib and Jackson identified and characterized a second B-subunit gene promoter in the *stx* operon, which is less efficient, but not repressed by iron. These findings suggest that independent transcription of the *stxB* gene may regulate the overproduction of the B-fragments and may contribute to the 1:5 stoichiometry of the holotoxin (Habib and Jackson, 1992). Moreover, the transcript of B-fragments was found to bind more avidly to ribosomes than that of A-fragments, resulting in an increased translation of B-fragments (Habib and Jackson, 1993).

The first X-ray crystallographic studies of Shiga holotoxin (Fraser *et al.*, 1994) and the B-subunit of Stx1 (Stein *et al.*, 1992) have been conducted in the absence of receptor molecules. Interestingly, in the absence of StxA, StxB alone still adopts a pentameric structure that is functionally equivalent to the holotoxin in receptor binding (Donohue-Rolfe *et al.*, 1989). In 1998, the crystal structure of the Stx1 B-subunit was determined in complex with a soluble trisaccharide analogue of its receptor Gb3, demonstrating three potential carbohydrate-binding sites per B-fragment (Ling *et al.*, 1998). Each B-fragment owns two carbohydrate binding sites and a third one is formed in the cleft between the β-sheets of two adjacent B-fragments (Fig. 5.1b) leading to a total number of 15 potential receptor-binding sites per pentameric B-subunit. All binding sites face the same direction thereby defining the membrane interaction surface. Mutational analysis indicated that two of the binding sites (site 1 and site 2) provide high-affinity receptor binding and are most relevant for cell cytotoxicity, whereas the third

(site 3) recognizes Gb3 at lower affinity (Bast *et al.*, 1999; Soltyk *et al.*, 2002). Optimal binding involves all three binding sites per monomer (Soltyk *et al.*, 2002), supported by a recent model suggesting that high-density receptor clustering induced by Shiga toxin leads to the formation of tubular membrane invaginations on cells and synthetic membrane systems (Römer *et al.*, 2007; Windschiegl *et al.*, 2009; Johannes and Römer, 2010; Safouane *et al.*, 2010).

During the intracellular transport of Shiga toxins (most probably at the stage of early endosomes), StxA is proteolytically cleaved by the membrane-associated endoprotease furin into the enzymatically active A1-fragment (~ 27.5 kDa) and the much smaller A2-fragment (~ 4.5 kDa) (Garred *et al.*, 1995). This cleavage is important to increase the enzymatic activity of Stx (Reisbig *et al.*, 1981). However, after furin-induced cleavage, the A1-fragment still remains bound to the A2-fragment via a disulfide bond between Cys242 in StxA1 and Cys261 in StxA2. Furin recognizes a specific sequence motif in the protein loop formed by the disulfide bond and cleaves the A-fragment at site Arg251-Met252 in Shiga toxin and Shiga-like toxin 1, and between Arg250 and Ala251 in Shiga-like toxin 2 (Garred *et al.*, 1995; Fagerquist and Sultan, 2010). Finally, the A1-fragment is released upon exposure to the reducing conditions in the lumen of the endoplasmic reticulum (ER) (Olsnes *et al.*, 1981; O'Brien and Holmes, 1987; Garred *et al.*, 1995, 1997; Lea *et al.*, 1999; Kurmanova *et al.*, 2007) and translocated into the cytosol where it exerts its cytotoxic activity. Even cells lacking furin can cleave and activate Shiga toxins, also less efficiently and at a later stage of the transport (Garred *et al.*, 1997).

Shiga toxin receptor(s)

The receptor for Shiga toxin in human cells is the neutral glycosphingolipid globotriaosylceramide (Gb3), also known as CD77 or the P^k blood group antigen, which is present in the extracellular leaflet of target cell plasma membranes (Jacewicz *et al.*, 1986; Lingwood *et al.*, 1987). Gb3 is regarded as the only functional receptor for Shiga toxins in most mammals (Okuda *et al.*, 2006). The interaction of Shiga toxin with Gb3 is complex and still far from being completely understood. Gb3 consists of a ceramide moiety (which is composed of a relatively invariable mono-unsaturated sphingosine chain of 18 carbon atoms and a fatty acid chain with variations in length and saturation) and a carbohydrate moiety (Gb3 trisaccharide) to which Shiga toxin is specifically binding (Fig. 5.1c). Despite the low affinity of Gb3 molecules for individual Shiga toxin binding sites (with binding constants in the order of $10^3/M$), Gb3 binding to multiple binding sites of the pentameric B-subunit remarkably increases the binding affinity of the toxin for cells (with binding constants in the range of $10^9/M$) (Fuchs et al., 1986; St. Hilaire et al., 1994). The binding of Shiga toxin is not only influenced by the number of receptors, but also by the availability and bio-distribution of Gb3 isoforms (varying in fatty acid content, hydroxylation, chain length and degree of unsaturation) (Pellizzari et al., 1992; Boyd et al., 1994; Kiarash et al., 1994; Arab and Lingwood, 1996; Nakajima et al., 2001; Binnington et al., 2002; Mahfoud et al., 2009) and the local membrane environment of the receptor. Increased cholesterol levels inhibited Stx1 binding, but not of Stx2, while increased sphingomyelin levels decreased binding of Stx2, but not the binding of Stx1 (Tam et al., 2008). Moreover, the Gb3 fatty acid content may play a key role in the intracellular trafficking and sorting of the toxin to the cytosol (Kiarash et al., 1994; Sandvig et al., 1996). Please refer to the following reviews for a more detailed description of these topics (Lingwood et al., 2010a,b).

Fuchs et al. were the first to provide unequivocal evidence that Stx bound to a receptor molecule on mammalian cells (Fuchs et al., 1986). Treatment of HeLa cells with tunicamycin, a compound that inhibits N-linked glycosylation, abrogates the cytotoxic effect of Stx, indicating that the toxin bound to a carbohydrate-containing receptor on these cells. Several groups subsequently identified Gb3 as receptor using different approaches, e.g. using radiolabelled toxin on thin-layer chromatograms containing separated glycolipids and binding of toxin to glycolipid-coated micro-titre plates (Jacewicz et al., 1986; Lindberg et al., 1987; Waddell et al., 1990). Contrary to the other Stx variants that all bind to Gb3, Stx2e uses the glycosphingolipid globotetraosylceramide (Gb4),

which contains an additional terminal $\beta[1-3]$-linked N-acetylgalactosamine residue, as cellular receptor in pigs (DeGrandis et al., 1989). Expression of Gb3 and exposure at the plasma membrane of target cells is essential for Stx invasiveness since inhibition of Stx-binding by Gb3 analogues or monoclonal antibodies towards Gb3 (Jacewicz et al., 1986; Lingwood et al., 1987), digestion of Gb3 with α-galactosidase (Lingwood et al., 1987; Mobassaleh et al., 1989) and inhibition of the biosynthetic pathways, which also impair Gb3 expression (Jacewicz et al., 1994), result in the loss of cytotoxicity for cultured epithelial cells. In addition, the incorporation of exogenous Gb3 into the plasma membrane of toxin-resistant cells, such as Daudi Burkitt lymphoma cells, using liposomes containing Gb3, results in a substantial increase of sensitivity to the toxin (Jacewicz et al., 1986, 1994; Waddell et al., 1990).

The expression of Gb3 is also developmentally regulated. For instance, suckling rabbit ileum is resistant to the effect of Shiga toxin, but after weaning, Gb3 receptors are expressed in villus epithelial cells, which parallels the increase in susceptibility to the enterotoxic and cytotoxic effects of Shiga toxin (Mobassaleh et al., 1989). Similarly, the expression of Gb3 can be induced in Caco-2 cells by incubating the cells with butyrate, which stimulates the biosynthetic pathways responsible for the production of Gb3 and confers toxin sensitivity on previously resistant cells (Jacewicz et al., 1995). Although Gb3 expression is a prerequisite for Shiga toxin sensitivity, it is in itself not sufficient as toxicity generally requires efficient retrograde transport of the toxin (Sandvig et al., 1992, 1994; Jacewicz et al., 1994).

Toxin internalization and retrograde transport

Already the binding of Shiga toxin to the plasma membrane activates various kinases, e.g. the serine/threonine kinase PKCδ (Torgersen et al., 2007), the mitogen-activated protein kinase (MAPK) p38α (Wälchli et al., 2008) and the tyrosine kinases Syk (Lauvrak et al., 2006), Lyn (Mori et al., 2000) and Yes (Katagiri et al., 1999). Shiga toxin-induced signalling seems to be crucial for its cellular uptake and endosome-to-Golgi transport. The question how Shiga toxin that binds to the carbohydrate moiety of the glycosphingolipid Gb3 in the extracellular leaflet of the plasma membrane, communicates with cytosolic signalling molecules still remains challenging and largely unresolved. One might speculate that Gb3 interacts with transmembrane-spanning proteins that in turn might affect the membrane dynamics of Gb3. Recently, it has been shown that Toll-like receptor 4 (TLR4) is involved in the binding of Shiga toxin in several cell lines (Torgersen et al., 2011) supporting the above-mentioned hypothesis. Moreover, actin, CD44, cytokeratin, ezrin, focal adhesion kinases, paxilin, α- and γ-tubulins, and vimentin get redistributed upon binding of Shiga toxin to the cell membrane leading to cytoskeleton remodelling (Takenouchi et al., 2004).

After binding to the cell surface, Shiga toxin is very efficiently internalized into target cells by different endocytic mechanisms (Sandvig et al., 2010) (Fig. 5.2, left), dependent and independent of coat proteins.

Shiga toxin was the first lipid-binding ligand that was described to use clathrin-coated pits for its endocytosis (Sandvig et al., 1989). When interfering with proteins involved in clathrin-mediated endocytosis, e.g. clathrin heavy chain, dynamin, epsin and Eps15 (Nichols et al., 2001; Lauvrak et al., 2004; Saint-Pol et al., 2004), Shiga toxin internalization is only partially blocked suggesting that other endocytic uptake mechanisms distinct to clathrin-mediated endocytosis exist. Since down-regulation of one pathway might be compensated by others, one has to be extremely careful when concluding about the proportions of different uptake mechanisms in unperturbed cells.

It has been mentioned above (see 'Shiga toxin structure') that Shiga toxin has the ability to induce negative membrane curvature through the clustering of Gb3 receptor molecules, leading to the formation of tubular plasma membrane invaginations as initial step of its entry (Fig. 5.2, left). This new mechanism was demonstrated either by using energy-depleted cells or cells in which dynamin or actin function was inhibited by several methods (Römer et al., 2007). The reconstitution of Shiga toxin-induced lipid clustering on supported lipid bilayers (Windschiegl et al., 2009) and the

formation of tubular membrane invaginations on cytosol-free giant unilamellar liposomes (Römer et al., 2007) is a fruitful synthetic approach to elucidate the first step of the cellular uptake of Shiga toxin that seems to occur without the help of the cytosolic machinery reviewed in (Eierhoff et al., 2012). Shiga toxin-induced membrane invaginations are then processed by the cellular machinery involving cholesterol-dependent actin polymerization that leads to membrane reorganization and scission of Stx-positive transport vesicles driven by membrane boundary forces, in concomitance or not with dynamin (Römer et al., 2010).

Shiga toxin is known to use the retrograde transport pathway from the early endosomes to the Golgi and the endoplasmic reticulum (ER) (Sandvig et al., 1989, 1991, 1992; Donta et al., 1995). Once internalized into cells, Shiga toxin localizes to early and recycling endosomes (Sandvig et al., 1989), then bypasses the late endocytic pathway to be directed towards the trans-Golgi network (TGN) (Mallard et al., 1998), and finally to the ER. Interestingly, StxB alone is enough to traffic to the ER (Sandvig et al., 1994; Johannes and Goud, 1998). The retrograde trafficking route that links the early endocytic pathway to the TGN is also used by several cellular proteins with diverse functions ranging from cell signalling and glucose transport to morphogen trafficking and tissue remodelling (Johannes and Popoff, 2008). But how does Shiga toxin escape from the early endocytic pathway? Does a signal sequence exist for the retrograde pathway? For instance, Pseudomonas exotoxin A (Jackson et al., 1999) contains a sequence in its carboxy-terminus end, termed KDEL (-Lys-Asp-Glu-Leu-) sequence, which is important for retrograde trafficking (Sandvig and van Deurs, 1996). StxB does not contain the KDEL sequence, and probably as a consequence of this, it might depend on cellular proteins like clathrin (Lauvrak et al., 2004; Saint-Pol et al., 2004) and the retromer complex (Bujny et al., 2007; Popoff et al., 2007; Utskarpen et al., 2007) that are of crucial importance for the retrograde transport of the toxin. Although the exact mechanism of Shiga toxin-containing retrograde carrier formation on the early endosome is not yet understood, clathrin and retromer probably function in a sequential manner (Popoff et al., 2007):

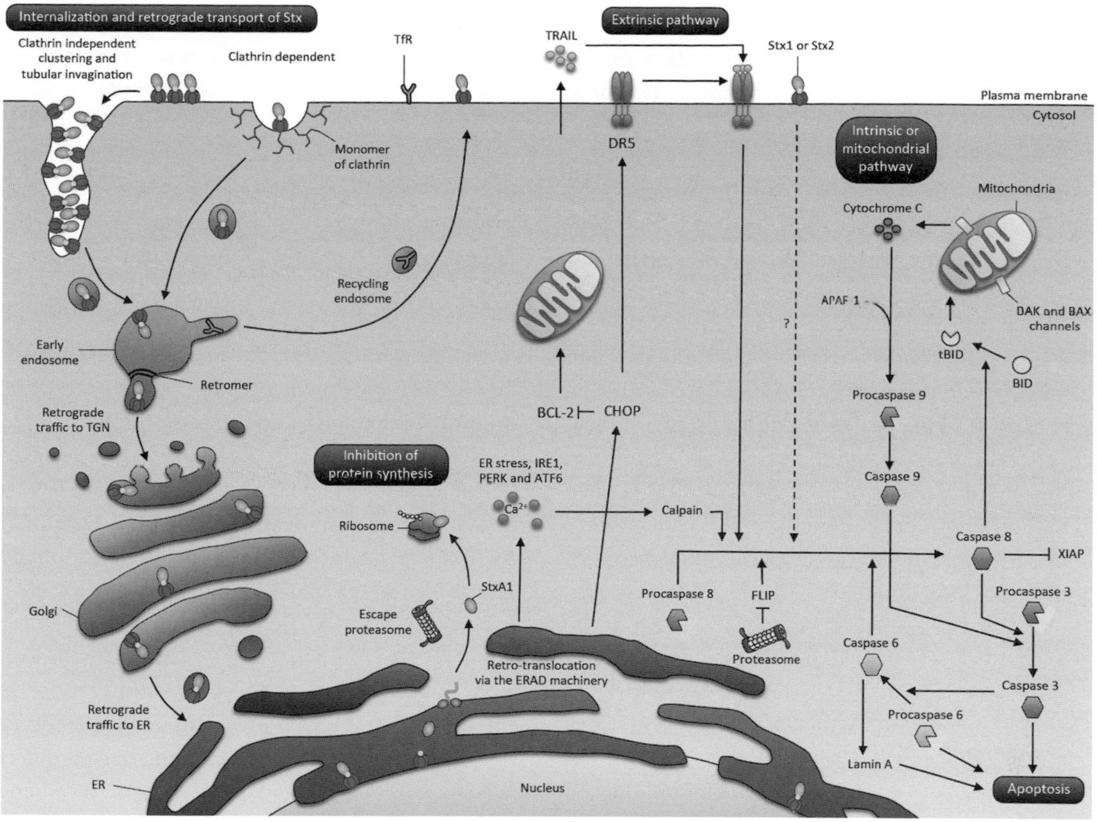

Figure 5.2 (Left part) Intracellular trafficking of Shiga toxins. After binding to its receptor, Stx induces clathrin-dependent and –independent endocytosis. The toxin preferentially localizes to retrograde tubules in early endosomes that are formed in a clathrin-dependent manner and then processed by scission in a retromer-dependent manner. Contrary to the transferrin receptor (TfR) that is recycled back to the plasma membrane, Stx bypasses the late endocytic pathway and is transferred to the *trans*–Golgi network (TGN) and then to the endoplasmic reticulum (ER). There, the A1-fragment (StxA1) utilizes the ER-associated degradation (ERAD) machinery to facilitate its retro-translocation into the host cell cytosol where it inhibits the protein synthesis. (Right part) Apoptosis pathways induced by Shiga toxins. This sketch sums up findings from different cell types, where the apoptosis initiator caspase 8 is activated, resulting in the activation of the executioner caspase 3. In most cases, the enzymatic activity of Shiga holotoxins is required for the induction of apoptosis. APAF-1, apoptotic protease-activating factor 1; ATF6, cyclic AMP-dependent transcription factor 6; BAK, BCL-2-homologous agonist/killer; BCL-2, B cell lymphoma 2; BID, BH3-interacting domain; CHOP, C/EBP-homologous protein; DR5, death receptor 5; ER, endoplasmic reticulum; FLIP, FLICE-like inhibitory protein; Gb3, globotriaosylceramide; PERK, Protein kinase-like ER kinase; tBID, truncated BID; TRAIL, TNF-related apoptosis-inducing ligand; XIAP, X-linked inhibitor of apoptosis protein. A colour version of this figure is available in the plate section at the back of the book.

Clathrin that is recruited by a host of proteins, including the phosphatidylinositol-4-phosphate-binding protein Epsin-related protein (EpsinR), for which a role in retrograde transport of Shiga toxin has been shown (Saint-Pol. *et al.*, 2004), is required for the generation of initial membrane curvature on the early endosome. The retromer complex is composed of a membrane curvature-inducing subcomplex (consisting of SNX1, SNX2, SNX5 and SNX6) and a cargo-recognition subcomplex consisting of the three vacuolar protein sorting-associated proteins, VPS26, VPS29 and VPS35 (reviewed in Bonifacinoa and Hurley, 2008). The membrane domain containing Shiga toxin would interact with clathrin adaptors and the cargo-recognition subcomplex of retromer to localize to the sites of transport intermediate formation. Retromer may also have a role during the scission of nascent retrograde tubules into retrograde transport intermediates. This model is discussed in detail in (Johannes and Popoff, 2008).

The transport of Shiga toxin from the endosomes to the *trans*-Golgi network/Golgi apparatus is also dependent on several Rab proteins, which are proteins that belong to the Ras superfamily of monomeric G proteins, e.g. Rab11a (Wilcke *et al.*, 2000), Rab6 and Rab6' (Mallard *et al.*, 2002; Del Nery *et al.*, 2006) and Rab33b (Starr *et al.*, 2010), and various Rab GTPase-activating proteins (GAPs) including TBC1D10A-C and EVI5 have been identified as specific regulators of Shiga toxin transport (Fuchs *et al.*, 2007). Two SNARE complexes (Soluble NSF Attachment Protein) that are involved in the final stages of vesicle docking and fusion have been discussed to be involved in the transport of Shiga toxin to the *trans*-Golgi network. One consists of VAMP3/VAMP4 v-SNARE that is interacting with syntaxin16/syntaxin6/Vti1a t-SNARE (Mallard *et al.*, 2002), the other one is the complex of the v-SNARE GS15 and the t-SNARE syntaxin 5/Ykt6/GS28 (Tai *et al.*, 2004).

Recently, it has been reported that the cycling Golgi protein GPP130 interacts with StxB, which is sorted to lysosomes when interfering with the function of this protein, e.g. by treating cells with small amounts of manganese (Mukhopadhyay *et al.*, 2010; Mukhopadhyay and Linstedt, 2012).

Although the precise mechanisms that underlie the transport of Shiga toxin between the TGN and the ER still remain largely unexplored, Shiga toxin seems to use a cytosolic coat protein complex I (COPI)-independent pathway, which is specifically regulated by Rab6a' and dependent on the small GTPase Cdc42, N-WASP, the ERGIC-localized protein Yip1A, microtubules, actin and the motor protein myosin II (Girod *et al.*, 1999; White *et al.*, 1999; Luna *et al.*, 2002; Duran *et al.*, 2003; Del Nery *et al.*, 2006; Johannes and Popoff, 2008; Kano *et al.*, 2009; McKenzie *et al.*, 2009).

Retro-translocation

Retrograde trafficking is a crucial step for intoxication with Shiga toxin, since some cell lines resistant to the toxin were able to bind and internalize Stx, but no transport to the Golgi was observed (Sandvig *et al.*, 1992, 1996; Sandvig and van Deurs, 1996; Falguières *et al.*, 2001). The essential question comes back again: How to enter the retrograde trafficking pathway? Experimental evidence suggests that the composition of the fatty acid chain of the Gb3 receptor molecules, rather than the toxin itself, might define such a trafficking signal (Sandvig *et al.*, 1994). Indeed, after addition of butyrate to A431 cells, which largely increases the expression of Gb3 receptors and the export to the cell surface, intoxication of previously resistant cells was observed after incubation with Shiga holotoxin (Sandvig *et al.*, 1992). Whereas the carbohydrate moiety was similar to other resistant cell lines, the fatty acid chain length was increased (Arab and Lingwood, 1998). Sandvig *et al.* (1996) confirmed this mechanism showing that changing Gb3 fatty acid composition or chain length induces major alterations in intracellular trafficking of the toxin. Further studies based on cholesterol extraction showed nicely the importance of lipid rafts for the retrograde transport of Shiga toxin (Falguières *et al.*, 2001).

Upon its arrival in the ER, the A1-fragment of the toxin is liberated by reduction of the disulfide bond in the ER lumen and subsequently released to the cytosol. Shiga toxin is known to inhibit protein synthesis through its highly specific RNA N-glycosidase activity, which cleaves an adenine base of 28S ribosomal RNA (rRNA) of eukaryotic ribosomes (Endo *et al.*, 1988). In order to gain access to its target in the cytosol, the enzymatic A1-fragment of the toxin has to translocate across the ER membrane. The translocation process, which relies on the host cell machinery, is now starting to be understood. StxA has been co-immunoprecipitated with the ER chaperons HEDJ (also known as ERdj3), BiP and the 94 kDa glucose-regulated protein (GRP94) known to be involved in protein retro-translocation from the ER to the cytosol (Yu and Haslam, 2005). Moreover, HEDJ-associated Stx was found to interact with the integral membrane Sec61 translocon core unit (Yu and Haslam, 2005), which is part of the ER-quality control of misfolded proteins. It has been proposed that a hydrophobic C-terminal domain of the A1-fragment is recognized as a misfolded peptide domain by the ER export mechanism (LaPointe *et al.*, 2005). These findings suggest that Stx exploits the ER-associated protein degradation (ERAD) pathway by interacting with HEDJ and other ER-localized chaperones and translocon components for its release from the ER lumen to the cytosol (Meunier *et al.*, 2002; LaPointe *et al.*, 2005; Yu and Haslam, 2005; Falguières and Johannes, 2006). The observation that siRNA-mediated knockdown of Sec61B protects against Shiga intoxication (Moreau *et al.*, 2011) strongly supports the crucial role of Sec61 for the retro-translocation of Shiga toxin. After its translocation, the A1-fragment of Shiga toxin escapes the proteasome that degrades misfolded proteins, probably due to the lack of lysine residues, which may minimize ubiquitination, together with its ability to refold rapidly (Hazes and Read, 1997). However, a fraction of the translocated A1-fragment seems to be degraded since inhibition of the proteasome increases the amount of cytosolic A1-fragment and thereby enhances cytotoxicity (Wesche *et al.*, 1999; Deeks *et al.*, 2002).

Shiga toxin-induced apoptosis

Besides the capacity of Shiga toxins to inhibit protein synthesis by catalytic inactivation of eukaryotic ribosomes (see section below about 'Inhibition of protein synthesis'), Shiga toxins are also known to trigger apoptosis (Tesh, 2010). Apoptosis is the 'programmed cell death' characterized by many morphological modifications such as cytoplasmic vacuolation, blebbing, chromatin condensation and DNA fragmentation. These complex mechanisms are triggered by activation of a series of aspartyl-specific cysteine proteases called caspases. Shiga toxin-induced apoptosis can be induced through different mechanisms in different cell types (Fig. 5.2, right) (Fujii *et al.*, 2003, 2008; Lee *et al.*, 2005, 2008). In many cell types, the enzymatic activity of Stx, which results in the inhibition of protein synthesis, induces an apoptotic signalling mechanism, called the ER stress response that is triggered by the accumulation of unfolded or misfolded proteins in the ER. ER stress leads to Ca^{2+} release from ER stores and

the activation of the Ca^{2+}-dependent cysteine protease calpain, which contributes to the cleavage of caspase 8 (Lee *et al.*, 2008). Furthermore, Stx incubation increases the activation of the ER membrane-associated stress sensors inositol-requiring ER to nucleus signal kinase 1 (IRE1), protein kinase-like ER kinase (PERK) and cyclic AMP-regulator dependent transcription factor 6 (ATF6) and also the expression of transcriptional regulator C/EBP-homologous protein (CHOP) and death receptor 5 (DR5) at mRNA and protein levels (Fig. 5.2). Besides, the level of the survival factor B cell lymphoma 2 (BCL-2) decreases whereas the secretion of TNF-related apoptosis-inducing ligand (TRAIL) increases. All these factors lead to the induction of apoptosis, which can be categorized into two major pathways: the extrinsic or death receptor pathway and the intrinsic or mitochondrial pathway. The extrinsic pathway is induced when death-inducing ligands, such as TRAIL, interact with their receptors, here DR5. Following the activation of DR5, caspase 8 is cleaved, which either may directly trigger the activation of caspase 3 leading to apoptosis, or initiate apoptosis through the mitochondrial pathway by cleavage of the inactive form of BH3-interacting domain (BID) to the active form, called truncated BID (tBID). tBid translocates to mitochondria, facilitates Bax/Bak-induced pore formation and increases mitochondrial outer membrane permeability. Pore formation results in the release of cytochrome c which, with procaspase 9, dATP and apoptosis protease activating factor (APAF)-1, triggers the formation of a macromolecular complex called the apoptosome. Apoptosome formation leads to caspase 9 activation and cleavage of procaspase 3 into caspase 3, which in turn cleaves many downstream substrates that contribute to apoptosis. Depending on the cell line and the Stx variant, one of these apoptosis pathways could occur. For instance in HeLa cells, Stx1 induces apoptosis through the activation of caspase 8, caspase 6 and caspase 3 but not caspase 9 (Fujii *et al.*, 2003). Indeed, the toxin increases the expression of the caspase 9 inhibitor, called X-linked inhibitor of apoptosis protein (XIAP) in these cells. Moreover, apoptosis induced by Stx1 can be associated with enhanced expression of the pro-apoptotic protein BAX (Physiol et al., 2000) and with the inhibition of the expression of the anti-apoptotic BCL-2 family member MCL1 (Erwert et al., 2003). Overexpression of BCL-2 has been shown to protect cells against Stx1-induced cell death (Suzuki et al., 2000). In human brain microvascular endothelial cells (HBMEC), Stx2 induces DNA fragmentation, and activation of caspase 3, caspase 6, caspase 8 and caspase 9 is mediated by CHOP up-regulation and the complete degradation of the anti-apoptotic protein FLICE-like inhibitory protein (FLIP), which in turn enhances the activation of caspase 8 (Fujii et al., 2008).

Many pathogens have evolved mechanisms to manipulate apoptosis signalling in order to favour pathogen survival or dissemination. In the case of Shiga toxin, many evidences suggest that apoptosis is crucial for the development of vascular lesions and tissue damage followed by translocation of the toxins into the bloodstream.

Inhibition of protein synthesis and intoxication

Early reports indicated that Stx inhibits protein synthesis (reviewed in O'Brien and Holmes, 1987). Subsequent experimentation revealed a highly specific mechanism of action. Using a cell-free system, Reisberg et al. (1981) demonstrated that Stx inactivates the large 60S ribosomal subunit that is ubiquitous in eukaryotic cells. These large ribosomal subunits comprise three fragments of 5S, 5.8S and 28S ribosomal RNA (rRNA) subunits. Trypsin-activated Stx that was incubated with ^{14}C-phenylalanine in a reticulocyte lysate inhibited the synthesis of polyphenylalanine chains. This inhibition has been identified to act on the 60S ribosomal subunits and to be independent of peptidyl-transferase activity, which is required to elongate the peptide chains of newly synthesized polypeptides. Instead, Stx inhibits peptide chain elongation by blocking amino-acyl tRNA binding to the acceptor site on rRNA (Fig. 5.3) (Brown et al., 1986; Obrig et al., 1987). In vitro treatment of rat ribosomes with Stx1 or 2 caused the release of an adenine base from the 28S rRNA at position 4324 by cleaving a specific N-glycosidic bond (Endo et al., 1988). Thus,

the precise target of StxA1 is the docking site of amino-acyl tRNA on the ribosome. When the protein synthesis is disrupted, JUN N-terminal kinase (Jnk) proteins and the mitogen-activated protein kinase (MAPK) p38 are activated and extracellular signal-regulated kinase 1 (ERK1)- and ERK2-signalling is altered (Ikeda et al., 2000; Foster and Tesh, 2002; Smith et al., 2003; Jandhyala et al., 2008). This mechanism is called ribotoxic stress.

In addition to its effect on eukaryotic cells, Stx has a similar effect on bacterial ribosomes, resulting in decreased proliferation of susceptible bacteria such as E. coli (Suh et al., 1998). This raises the possibility that Stx may facilitate bacterial survival by inhibiting the growth of potential competitors in the lumen of the gastrointestinal tract. The N-glycosidase activity of Stx is very similar to that of the plant toxin ricin, implying some evolutionary conservation within these toxins.

Shiga toxin-induced cytokine synthesis

During STEC infection, changes in cytokine levels can be ascribed to the cytotoxic effect of Shiga toxin on different cell types (Hughes et al., 1998b, 2000, 2001; Paton and Paton, 1998). While many epithelial and endothelial cells in the kidney and central nervous system are highly sensitive to Shiga toxin-induced cytotoxicity, others like peripheral blood monocytes have been reported to be resistant to intoxication despite Gb3 expression, Shiga toxin binding and internalization (Ramegowda and Tesh, 1996; Harrison et al., 2004). These toxin-resistant cells respond by synthesizing and releasing pro-inflammatory cytokines that in turn stimulate Gb3 synthesis on many endothelial cells and sensitize them to cytotoxic action of Shiga toxins in a process that often involves signalling through the p38 pathway (Stricklett et al., 2005; Stone et al., 2008). For instance, Shiga toxins can induce the release of IL-1, IL-6 and tumour necrosis factor (TNF) from macrophages, human glomerular epithelial cells and human proximal tubule cells (Hughes et al., 1998b, 2001; Paton and Paton, 1998). Besides, Stx1 combined with lipopolysaccharide (LPS) stimulated production of all three cytokines to a greater extent than the

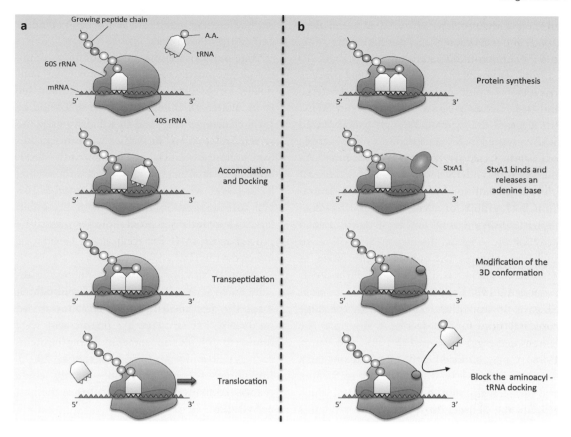

Figure 5.3 Shiga toxin-induced inhibition of protein synthesis. (a) A simplified view on protein synthesis. The 60S- and the 40S-ribosomal subunits assemble around the mRNA and recognize an amino-acyl tRNA specific of the codon. After the accommodation and the docking, the amino-acyl moiety on the tRNA will bind to the growing peptide chain triggered by GTP and elongation factors. Finally, the ribosome moves forward on the mRNA from 5′ to 3′, releasing a free tRNA. (b) StxA1 inhibits the elongation during protein synthesis by binding to the 28S rRNA on the large 60S-subunit of the ribosome and cleaving an adenine base at the position 4324. Thus, the folding of the ribosome changes, modifying the docking site of the amino-acyl tRNA. As consequence, no amino-acyl tRNA could be added to the growing peptide chain resulting in the inhibition of protein synthesis. A colour version of this figure is available in the plate section at the back of the book.

toxin alone (Hughes *et al.*, 2001). Accordingly, IL-1 stimulates the expression of Gb3 on the surface of human proximal tubular (Hughes *et al.*, 1998a), umbilical endothelial, and saphenous vein endothelial cells (van de Kar *et al.*, 1992; Kaye *et al.*, 1993) and up-regulates cell sensitivity to Stx cytotoxicity. Similarly, TNF increases Gb3 expression and Stx1 cytotoxicity in glomerular, umbilical, and saphenous vein endothelial cells (van de Kar *et al.*, 1992; Kaye *et al.*, 1993; van Setten *et al.*, 1997; Warnier *et al.*, 2006). One may speculate that inflammatory cytokines enhance renal injury in post-diarrhoeal haemolytic–uraemic syndrome (HUS).

Pathophysiology of organ damage

Production of Shiga toxins is essential for many of the pathological as well as life-threatening features of STEC infection. Normally, infections with Shiga toxin-producing bacteria start with diarrhoea and sometimes progress into dysentery and haemorrhagic colitis. Primarily, young children and elderly people are affected (Lingwood, 1996). In some cases, the disease can further develop into haemolytic uraemic syndrome (HUS), which is characterized by acute renal failure, thrombocytopenia and microangiopathic haemolytic anaemia (Melton-Celsa *et al.*, 2012), and neurological complications and death (Proulx *et al.*, 2001; Nathanson *et al.*, 2010).

The pathogenesis is a multistep process, involving complex interactions between bacterial and host factors. *Shigella dysenteriae* and STEC are ingested orally, often through contaminated food or water in low initial doses. The bacteria must initially survive the harsh environment of the stomach and then compete with other gut microorganisms to establish intestinal colonization. Both groups of microbial organisms cause severe intestinal damage leading to diarrhoea and dehydration. *Shigella dysenteriae* first colonizes and then invades the epithelium, resulting in necrosis of the mucosa (reviewed in Hale, 1991; Sansonetti, 1992). In contrast, STEC are non-invasive. Instead, they associate with the large intestinal mucosa and secrete toxins into the lumen of the gut (reviewed in Paton and Paton, 1998). Once produced in the lumen, toxins must be first absorbed by the intestinal epithelium and then translocated to the bloodstream, which permits the delivery to specific toxin receptors on target cell surfaces inducing both local and systemic effects. Conceivably, the production of Shiga toxins, which induce ischaemic necrosis of the gut, may facilitate the bacterial survival by causing bleeding into the bowel lumen. Some strains of STEC also carry a plasmid-encoded enterohaemolysin, which has been found to lyse sheep erythrocytes. It has been suggested that haemoglobin released by the action of STEC may provide a source of iron, thereby stimulating the growth of the bacteria in the gut (Nataro and Kaper, 1998; Paton and Paton, 1998). These pathogenic factors may provide essential nutrients and other growth promoters, which allow for the growth of bacteria in a rather hostile environment.

antibodies, StxA-inhibitors, synthetic Stx binders and transport inhibitors.

The most promising treatment is currently the development of human monoclonal antibodies for passive immunotherapy (Mukherjee *et al.*, 2002a,b). Indeed, when administered to toxin-challenged mice and to a piglet model, the selected monoclonal antibodies for both Stx1 and Stx2 prolonged survival and prevented the development of fatal neurological signs and cerebral lesions, even when antibodies were given 6–12 h after toxin application. Their Stx-neutralizing ability could potentially be used clinically to passively protect against HUS development in humans infected with STEC.

Another strategy focuses on the enzymatically active StxA-subunit. Using high-throughput screening, few small molecule inhibitors based on guanine-like structure like pteroic acid, were found to inhibit ribosome-inhibiting proteins such as ricin and Shiga toxin (Miller *et al.*, 2002). Inhibition of toxin enzymatic activity could reduce the ribotoxic stress and enhance the host cell viability.

Moreover, several promising therapeutic approaches are based on inhibiting the binding of StxB to the plasma membrane of target cells before its cellular uptake into host cells. Multivalent high avidity inhibitors against StxB have been developed, which markedly inhibit cytotoxic activities of Shiga toxins (Table 5.1).

Some small molecule inhibitors that are mostly used to elucidate the retrograde trafficking pathway have shown the potential to block the cytotoxic activity of Shiga toxins (Table 5.2). However, most of these compounds show dramatic effects on the integrity of the Golgi apparatus, which questions their further development as therapeutic agents.

Intervention strategies

Since conventional antimicrobial therapies are limited and may even be counterproductive, as killing of bacteria by antibiotics may lead to the release of higher toxin quantities (Proulx *et al.*, 1992; Walterspiel *et al.*, 1992), several therapeutic approaches have been developed to counteract the threat displayed by Shiga toxin including monoclonal

Biomedical applications of Shiga toxins

Shiga toxin and its derivatives are promising, biomedical tools for targeted therapy and imaging of tumours and tumour vasculature since their receptor, the glycosphingolipid Gb3, shows a relatively restricted expression in human tissues, and is highly enriched in certain types of

Table 5.1 Synthetic Shiga toxin binders

Name	Structure	Action	Impact on Stx activity	Reference(s)
P^k ligands	P^k water-soluble trisaccharides	Bind to one StxB binding site	Millimolar inhibitory activity; weak affinity to StxB	Armstrong et al. (1995), Kitov et al. (2000), Paton et al. (2000), Nishikawa et al. (2002), Watanabe et al. (2004), Isobe et al. (2007), Neri et al. (2007)
STARFISH	Oligovalent and water-soluble carbohydrate ligand	Formation of face-to-face complexes between two copies of Stx	Binding affinity increases 1,000,000-fold; subnanomolar inhibitory activity in vitro, but not optimal in vivo.	Kitov et al. (2000), Mulvey et al. (2003)
DAISY	Optimized version of STARFISH	Formation of face–to–face complexes between two copies of Stx	Protects mice against Stx; does not interfere with the immune system to produce Stx-specific antibodies	Mulvey et al. (2003)
Synsorb-P^k	Trisaccharides coupled to an inert silicon-based matrix (Chromosorb-P)	Bind to many Stx	Protection of cells during in vitro treatment, but the agent failed in clinical trials	Takeda et al. (1999), Trachtman et al. (2003)
Supramolecular templating	Multivalent carbohydrate ligands like STARFISH, coupled with a complementary templating receptor	Supramolecular scaffolding between inhibitors and toxins	Enhancement of the inhibitory activity of STARFISH-type P^k-dendrimers	Kitov et al. (2008)
Probiotic bacteria	Non-pathogenic strains of E. coli designed to display a mimic of the Shiga toxin receptor on its surface	High affinity binding to Stx, which is adsorbed and neutralized	Efficient protection of mice from lethal toxin doses	Paton et al. (2000), Pinyon et al. (2004)

tumours (reviewed in Johannes and Römer, 2010). Carcinogenesis is characterized by an aberrant glycosylation pattern on the cell membrane, which may alter cell signalling, growth, adherence and motility (Hakomori, 1989, 1996; Hakomori and Zhang, 1997). In 1984, Gb3 was first identified as a fibrosarcoma-associated antigen in rats (Ito et al., 1984) as well as in Burkitt's lymphoma cell lines (Nudelman et al., 1983) and in four out of eight primary Burkitt-like B cell lymphomas (Wiels et al., 1981). Later, Gb3 was found to be expressed in a large proportion of several other types of B-cell lymphoma (Murray et al., 1985; LaCasse et al., 1996, 1999). However, colorectal carcinoma is the cancer type for which the highest number of patients having high level of Gb3 expression has been reported (Kovbasnjuk et al., 2005; Falguières et al., 2008; Distler et al., 2009). As mentioned above (see 'Shiga toxin-induced

cytokine synthesis'), several cytokines have been shown to induce Gb3 expression in endothelial cells, suggesting that overexpression of Gb3 might be due to the host response towards a tumour (van de Kar et al., 1992; van Setten et al., 1997; Lingwood, 1999).

As StxB itself is sufficient for the binding to Gb3 receptors, targeted imaging of cancers has been performed by injection of labelled StxB with several types of imaging agents (Janssen et al., 2006; Viel et al., 2008). Indeed, Janssen et al. have shown that it is possible to detect tumours in the digestive tract of mice using confocal laser endoscopy after force-feeding of the animals with a single dose of 300–500 mg fluorescence-labelled StxB, or after retro-orbital injection of ^{18}F-labelled StxB followed by PET (positron emission tomography) imaging at 1–2 h after injection (Janssen et al., 2006). Later, the same group reported on

Table 5.2 Small molecule inhibitors affecting the retrograde transport of Stx

Name	Protein affected	Action	Impact on Stx trafficking	Reference(s)
Mn²⁺ (manganese)	GPP130	Blocks endosome-to-Golgi trafficking of Stx	Stx is transported to lysosomes	Mukhopadhyay and Linstedt (2012)
Dynasore	Dynamin	Blocks scission of clathrin coated pits during endocytosis; May block endosome-to-Golgi trafficking	Blocks Stx at multiple levels	Lauvrak et al. (2004), Macia et al. (2006)
Pitstop	Clathrin	Blocks clathrin mediated-endocytosis; May block endosome-to-Golgi trafficking	Blocks Stx at multiple levels	Lauvrak et al. (2004)
Retro-1/Retro-2	Syntaxin 5	Block endosome-to-Golgi trafficking	Block Stx in early endosomes	Stechmann et al. (2010)
AIF4 (aluminium fluoride)	n/k	Disrupts the Golgi apparatus	Blocks Stx in recycling endosomes	McKenzie et al. (2009)
BFA (brefeldin A)	GTP exchange factor that activates Arf1p	Disrupts the Golgi apparatus	Blocks Stx in tubular parts of early endosomes	Donta et al. (1995)
GCA (golgicide A)	Arf-GEF, GBF1	Disrupts the Golgi apparatus	NS	Saenz et al. (2009)
Exo2/LG186	Arf-GEF GBF1	Disrupt the Golgi apparatus	Inhibit the delivery of Stx to the ER	Spooner et al. (2008), Boal et al. (2010)
75	NK	May inhibit transport at an early stage of endocytosis	Blocks Stx in early endosomes	Saenz et al. (2007)
134	NK	May inhibit transport at a post-recycling endosome stage	Blocks Stx in recycling endosomes	Saenz et al. (2007)

NK, not known; NS, not specified.

the imaging of tumours that were subcutaneously implanted in mice (specimen from human colon cancer) from 5 h to 9 days after oral, intraperitoneal or intravenous administration of 50 mg of fluorescence-labelled StxB (Viel et al., 2008).

With its ability to specifically target and bind cancer cells, Shiga toxin could also be used in targeted drug delivery. There are two different options, using either Shiga holotoxin, or the B-subunit coupled to therapeutic drugs, each one having its therapeutic effects as well as different side effects. As Stx can kill cells very efficiently, the use of the holotoxin at low doses with one or few rounds of treatment could be advantageous compared to conventional chemotherapeutic drugs. In different studies, it was found that intratumoural or intraperitoneal injection of Stx1 inhibited tumour growth in a murine metastatic fibrosarcoma model (Farkas-Himsley et al., 1995), as well

as in mouse xenograft models of human malignant meningiomas (Salhia et al., 2002), atypical human bladder carcinoma with endothelial characteristics (Heath-Engel and Lingwood, 2003), human renal carcinoma (Ishitoya et al., 2004) and human astrocytoma (Arab et al., 1999). Complete regression of the tumours was observed in the two latter cases, at 7–10 days after a single intratumoural injection of Stx1 while no side effects were reported (Arab et al., 1999; Ishitoya et al., 2004). Although it might also be interesting to use the holotoxin in cancer therapy, further studies are needed to determine if this treatment is tolerated by humans and if the tumour-unspecific A-subunit is neutralized and degraded – if not, it might induce other unexpected toxic effects on the organism.

Instead of the holotoxin, the non-toxic B-subunit of Shiga toxin could also be used as drug-delivery tool. Indeed, El Alaoui and colleagues coupled StxB to the active part of the topoisomerase I inhibitor camptothecin 11, a cytotoxic compound, and tested its effects on colorectal cancer cells (El Alaoui et al., 2007). The drug was conjugated to StxB in form of a prodrug through a disulfide linkage. Retrograde transport and reducing conditions in the Golgi apparatus and the ER were essential for the cleavage and the release of the active compound that was highly cytotoxic to HT29 colorectal carcinoma cells (El Alaoui et al., 2007). More studies are required to assess the short- and long-term effects of Stx, StxB and the various Stx derivatives in animals and humans.

Conclusions and future directions

Although Shiga toxins are still considered as a threat to human health, these toxins have great potential to develop into valuable tools to fight disease. They may be exploited as vectors to bring proteins and peptides into cells, with the possibility of inducing immune responses to virus or cancer cells. Moreover, they may be used as constituents of immunotoxins and other targeted chimeric molecules. Clearly, further studies are necessary to acquire increasing and more profound knowledge about Shiga toxins and their interactions with target cells in order to clarify the potential of these agents in medicine. Already nowadays, Shiga toxins are highly appreciated as tools in cell biology to elucidate intracellular transport pathways.

Acknowledgements

This work was supported in part by the Excellence Initiative of the German Research Foundation (EXC 294 and GSC-4) and by a starting grant of the European Research Council to W.R. (Programme 'Ideas' – call identifier: ERC-2011-StG 282105-lec&lip2invade).

References

Arab, S., and Lingwood, C.A. (1996). Influence of phospholipid chain length on verotoxin/globotriaosyl ceramide binding in model membranes: comparison of a supported bilayer film and liposomes. Glycoconj. J. *13*, 159–166.

Arab, S., and Lingwood, C. (1998). Intracellular targeting of the endoplasmic reticulum/nuclear envelope by retrograde transport may determine cell hypersensitivity to verotoxin via globotriaosyl ceramide fatty acid isoform traffic. J. Cell Physiol. *177*, 646–660.

Arab, S., Rutka, J., and Lingwood, C.A. (1999). Verotoxin induces apoptosis and the complete, rapid, long-term elimi- nation of human astrocytoma xenografts in nude mice. Onc. Res. *11*, 33–39.

Armstrong, G., Rowe, P., Goodyer, P., Orrbine, E., Klassen, T., Wells, G., MacKenzie, A., Lior, H., Blanchard, C., and Auclair, F. (1995). A phase-I study of chemically synthesized verotoxin (Shiga-like toxin) Pk-trisaccharide receptors attached to chromosorb for preventing hemolytic-uremic syndrome. J. Infect. Dis. *171*, 1042–1045.

Austin, P.R., Jablonski, P.E., Bohach, G.A., Dunker, A.K., and Hovde, C.J. (1994). Evidence that the A2 fragment of Shiga-like toxin type I is required for holotoxin integrity. Infect. Immun. *62*, 1768–1775.

Bast, D.J., Banerjee, L., Clark, C., Read, R.J., and Brunton, J.L. (1999). The identification of three biologically relevant globotriaosyl ceramide receptor binding sites on the Verotoxin 1 B subunit. Mol. Microbiol. *32*, 953–960.

Binnington, B., Lingwood, D., Nutikka, A., and Lingwood, C.A. (2002). Effect of globotriaosyl ceramide fatty acid α-hydroxylation on the binding by verotoxin 1 and verotoxin 2. Neurochem. Res. *27*, 807–813.

Boal, F., Guetzoyan, L., Sessions, R.B., Zeghouf, M., Spooner, R.A., Lord, J.M., Cherfils, J., Clarkson, G.J., Roberts, L.M., and Stephens, D.J. (2010). LG186: An inhibitor of GBF1 function that causes Golgi disassembly in human and canine cells. Traffic *11*, 1537–1551.

Boerlin, P., McEwen, S.A., Boerlin-Petzold, F., Wilson, J.B., Johnson, R.P., and Gyles, C.L. (1999). Associations between virulence factors of Shiga toxin-producing *Escherichia coli* and disease in humans. J. Clin. Microbiol. *37*, 497–503.

Bonifacinoa, J.S., and Hurley, J.H. (2008). Retromer. Curr. Opin. Cell Biol. *20*, 427–436.

Boyd, B., Magnusson, G., Zhiuyan, Z., and Lingwood, C.A. (1994). Lipid modulation of glycolipid receptor function: Availability of Gal(a1–4)Gal disaccharide for verotoxin binding in natural and synthetic glycolipids. Eur. J. Biochem *223*, 873–878.

Brown, J.E., Obrig, T.G., Ussery, M. a, and Moran, T.P. (1986). Shiga toxin from *Shigella dysenteriae* 1 inhibits protein synthesis in reticulocyte lysates by inactivation of aminoacyl-tRNA binding. Microb. Pathog. *1*, 325–334.

Bujny, M.V., Popoff, V., Johannes, L., and Cullen, P.J. (2007). The retromer component sorting nexin-1 is required for efficient retrograde transport of Shiga toxin from early endosome to the trans Golgi network. J. Cell Sci. *120*, 2010–2021.

Calderwood, S.B., and Mekalanos, J.J. (1987). Iron regulation of Shiga-like toxin expression in *Escherichia coli* is mediated by the fur locus. J. Bacteriol. *169*, 4759–4764.

Calderwood, S.B., Auclairt, F., Donohue-rolfet, A., and Keuscht, G.T. (1987). Nucleotide sequence of the Shiga-like toxin genes. Proc. Natl. Acad. Sci. U.S.A. *84*, 4364–4368.

Conradi, H. (1903). Ueber lösliche, durch aseptische Autolyse erhaltene Giftstoffe von Ruhr- und Typhus-Bazillen. Dtsch. Med. Wochenschr *29*, 26–28.

DeGrandis, S., Law, H., Brunton, J., Gyles, C., and Lingwood, C. a (1989). Globotetraosylceramide is recognized by the pig edema disease toxin. J. Biolog. Chem. *264*, 12520–12525.

Deeks, E.D., Cook, J.P., Day, P.J., Smith, D.C., Roberts, L.M., and Lord, J.M. (2002). The low lysine content of ricin A chain reduces the risk of proteolytic degradation after translocation from the endoplasmic reticulum to the cytosol. Biochemistry *41*, 3405–3413.

Distler, U., Souady, J., Hulsewig, M., Drmic-Hofman, I., Haier, J., and Friedrich, A.W. (2009). Shiga toxin receptor Gb3Cer/CD77: tumor- association and promising thera- peutic target in pancreas and colon cancer. PLoS ONE *4*, e6813.

Donohue-Rolfe, A., and Keusch, G.T. (1983). *Shigella dysenteriae* 1 cytotoxin: periplasmic protein releasable by polymyxin B and osmotic shock. Infect. Immun. *39*, 270–274.

Donohue-Rolfe, A., Jacewicz, M., and Keusch, G.T. (1989). Isolation and characterization of functional Shiga toxin subunits and renatured holotoxin. Mol. Microbiol. *3*, 1231–1236.

Donta, S.T., Tomicic, T.K., and Donohue-Rolfe, A. (1995). Inhibition of Shiga-like toxins by brefeldin A. J. Infect. Dis. *171*, 721–724.

Dubos, R.J., and Geiger, J.W. (1946). Preparation and properties of Shiga toxin and toxoid. J. Exp. Med. *84*, 143–156.

Duran, J.M., Valderrama, F., Castel, S., Magdalena, J., Hosoya, H., Renau-piqueras, J., Malhotra, V., and Egea, G. (2003). Myosin motors and not actin comets are mediators of the actin-based Golgi-to-endoplasmic reticulum protein transport. Mol. Biol. Cell *14*, 445–459.

Eierhoff, T., Stechmann, B., and Römer, W. (2012). Pathogen and toxin entry – how pathogens and toxins induce and harness endocytotic mechanisms. In Molecular Regulation of Endocytosis, Ceresa, B., ed. (In Tech, Rijeka), pp. 249–276.

El Alaoui, A., Schmidt, F., Amessou, M., Sarr, M., Decaudin, D., Florent, J.-C., and Johannes, L. (2007). Shiga toxin-mediated retrograde delivery of a topoisomerase I inhibitor prodrug. Angewandte Chemie *46*, 6469–6472.

Endo, Y., Tsurugi, K., Yutsudo, T., Takeda, Y., Ogasawara, T., and Igarashi, K. (1988). Site of action of a Vero toxin (VT2) from *Escherichia coli* 0157 : H7 and of Shiga toxin on eukaryotic ribosomes. RNA N-glycosidase activity of the toxins. Eur. J. Biochem *171*, 45–50.

Erwert, R.D., Eiting, K.T., Tupper, J.C., Winn, R.K., Harlan, J.M., and Bannerman, D.D. (2003). Shiga toxin induces decreased expression of the anti-apoptotic protein Mcl-1 concomitant with the onset of endothelial apoptosis. Microb. Pathog. *35*, 87–93.

Fagerquist, C.K., and Sultan, O. (2010). Top-down proteomic identification of furin-cleaved α-subunit of Shiga toxin 2 from *Escherichia coli* O157:H7 using MALDI-TOF-TOF-MS/MS. J. Biomed. Biotechnol. *123460*.

Falguières, T., and Johannes, L. (2006). Shiga toxin B-subunit binds to the chaperone BiP and the nucleolar protein B23. Biol. Cell *98*, 125–134.

Falguières, T., Mallard, F., Baron, C., Hanau, D., Lingwood, C., Goud, B., Salamero, J., and Johannes, L. (2001). Targeting of Shiga toxin B-subunit to retrograde transport route in association with detergent-resistant membranes. Mol. Biol. Cell *12*, 2453–2468.

Falguières, T., Maak, M., von Weyhern, C., Sarr, M., Sastre, X., and Poupon, M.F. (2008). Human colorectal tumors and metastases express Gb3 and can be targeted by an intestinal pathogen-based delivery tool. Mol. Cancer Ther. *7*, 2498–2508.

Farkas-Himsley, H., Hill, R., Rosen, B., Arab, S., and Lingwood, C.A. (1995). The bacterial colicin active against tumor cells *in vitro* and *in vivo* is verotoxin 1. Proc. Natl. Acad. Sci. U.S.A. *92*, 6996–7000.

Flexner, S., and Sweet, J.E. (1906). The pathogenesis of experimental colitis and the relation of colitis in animal and man. J. Exp. Med. *8*, 514–535.

Fontaine, A., Arondel, J., and Sansonetti, P.J. (1988). Role of Shiga toxin in the pathogenesis of bacillary dysentery, studied by using a Tox(−) mutant of *Shigella dysenteriae* 1. Infect. Immun. *56*, 3099–3109.

Foster, G.H., and Tesh, V.L. (2002). Shiga toxin 1-induced activation of c-Jun NH 2 -terminal kinase and p38 in the human monocytic cell line THP-1 : possible involvement in the production of TNF- alpha. J. Leukocyte Biol. *71*.

Fraser, M.E., Chernaia, M.M., Kozlov, Y.V., and James, M.N. (1994). Crystal structure of the holotoxin from *Shigella dysenteriae* at 2.5 Å resolution. Nature Struct. Biol. *1*, 59–64.

Fuchs, G., Mobassaleh, M., Donohue-Rolfe, a, Montgomery, R.K., Grand, R.J., and Keusch, G.T. (1986). Pathogenesis of *Shigella* diarrhoea: rabbit intestinal cell microvillus membrane binding site for *Shigella* toxin. Infect. Immun. *53*, 372–377.

Fuchs, E., Haas, A.K., Spooner, R. a, Yoshimura, S., Lord, J.M., and Barr, F.A. (2007). Specific Rab GTPase-activating proteins define the Shiga toxin and epidermal growth factor uptake pathways. J. Cell Biol. *177*, 1133–1143.

Fujii, J., Matsui, T., Heatherly, D.P., Schlegel, K.H., Lobo, P.I., Yutsudo, T., Ciraolo, G.M., Morris, R.E., and Obrig, T. (2003). Rapid Apoptosis Induced by Shiga Toxin in HeLa Cells. Infect. Immun. *71*, 2724–2735.

Fujii, J., Wood, K., Matsuda, F., Carneiro-Filho, B. a, Schlegel, K.H., Yutsudo, T., Binnington-Boyd, B., Lingwood, C. a, Obata, F., Kim, K.S., *et al.* (2008). Shiga toxin 2 causes apoptosis in human brain microvascular endothelial cells via C/EBP homologous protein. Infect. Immun. *76*, 3679–3689.

Garred, O., van Deurs, B., and Sandvig, K. (1995). Furin induced cleavage and activation of shiga toxin. J. Biol. Chem. *270*, 10817–10821.

Garred, O., Dubinina, E., Polesskaya, a, Olsnes, S., Kozlov, J., and Sandvig, K. (1997). Role of the disulfide bond in Shiga toxin A-chain for toxin entry into cells. J. Biol. Chem. 272, 11414–11419.

Girod, A., Storrie, B., Simpson, J., Johannes, L., Goud, B., Roberts, L., Lord, J., Nilsson, T., and Pepperkok, R. (1999). Evidence for a COP-I-independent transport route from the Golgi complex to the endoplasmic reticulum. Nature Cell Biol. 1, 423–430.

Habib, N.F., and Jackson, M.P. (1992). Identification of a B subunit gene promoter in the Shiga toxin operon of Shigella dysenteriae 1. J. Bacteriol. 174, 6498–6507.

Habib, N.F., and Jackson, M.P. (1993). Roles of a ribosome-binding site and mRNA secondary structure in differential expression of Shiga toxin genes. J. Bacteriol. 175, 597–603.

Haddad, J.E., and Jackson, M.P. (1993). Identification of the Shiga toxin A-subunit residues required for holotoxin assembly. J. Bacteriol. 175, 7652–7657.

Hakomori, S. (1989). Aberrant glycosylation in tumors and tumor-associated carbohydrate antigens. Adv. Cancer Res. 52, 257–331.

Hakomori, S. (1996). Tumor malignancy defined by aberrant glycosylation and sphingo (glyco) lipid metabolism tumor Malignancy defined by aberrant glycosylation and sphingo (glyco) lipid metabolism. Cancer Res. 5309–5318.

Hakomori, S., and Zhang, Y. (1997). Glycosphingolipid antigens and cancer therapy. Chem. Biol. 4, 97–104.

Hale, T.L. (1991). Genetic basis of virulence in Shigella species. Microbiol. Rev. 55, 206–224.

Harrison, L.M., van Haaften, W.C.E., and Tesh, V.L. (2004). Regulation of Proinflammatory Cytokine Expression by Shiga Toxin 1 and/or Lipopolysaccharides in the Human Monocytic Cell Line THP-1. Infect. Immun. 72, 2618–2627.

Hazes, B., and Read, R.J. (1997). Accumulating evidence suggests that several AB-toxins subvert the endoplasmic reticulum-associated protein degradation pathway to enter target cells. Biochemistry 36, 4–7.

Heath-Engel, H.M., and Lingwood, C.A. (2003). Verotoxin sensitivity of ECV304 cells in vitro and in vivo in a xenograft tumour model: VT1 as a tumour neovascular marker. Angiogenesis 6, 129–141.

Herold, S., Karch, H., and Schmidt, H. (2004). Shiga toxin- encoding bacteriophages – genomes in motion. Int. J. Med. Microbiol. 294, 115–121.

Hirst, T.R., Sanchez, J., Kaper, J.B., Hardy, S.J., and Holmgren, J. (1984). Mechanism of toxin secretion by Vibrio cholerae investigated in strains harboring plasmids that encode heat-labile enterotoxins of Escherichia coli. Proc. Natl. Acad. Sci. U.S.A. 81, 7752–7756.

Hughes, A.K., Stricklett, P.K., and Kohan, D.E. (1998a). Cytotoxic effect of Shiga toxin-1 on human proximal tubule cells. Kidney Int. 54, 426–437.

Hughes, A.K., Stricklett, P.K., and Kohan, D.E. (1998b). Shiga toxin-1 regulation of cytokine production by human proximal tubule cells. Kidney Int. 54, 1093–1106.

Hughes, A.K., Stricklett, P.K., Schmid, D., and Kohan, D.E. (2000). Cytotoxic effect of Shiga toxin-1 on human glomerular epithelial cells. Kidney Int. 57, 2350–2359.

Hughes, A.K., Stricklett, P.K., and Kohan, D.E. (2001). Shiga toxin-1 regulation of cytokine production by human glomerular epithelial cells. Nephron 88, 14–23.

Ikeda, M., Gunji, Y., Yamasaki, S., and Takeda, Y. (2000). Shiga toxin activates p38 MAP kinase through cellular Ca(2+) increase in Vero cells. FEBS Letters 485, 94–98.

Ishitoya, S., Kurazono, H., Nishiyama, H., Nakamura, E., Kamoto, T., Habuchi, T., Terai, A., Ogawa, O., and Yamamoto, S. (2004). Verotoxin induces rapid elimination of human renal tumor xenografts in SCID mice. J. Urol. 171, 1309–1313.

Isobe, H., Cho, K., Solin, N., Werz, D., Seeberger, P., and Nakamura, E. (2007). Synthesis of fullerene glycoconjugates via a copper-catalyzed Huisgen cycloaddition reaction. Org. Lett. 9, 4611–4614.

Ito, M., Suzuki, E., Naiki, M., Sendo, F., and Arai, S. (1984). Carbohydrates as antigenic determinants of tumor-associated antigens recognized by monoclonal anti-tumor antibodies produced in a syngeneic system. Int. J. Cancer 34, 689–697.

Jacewicz, M., Clausen, H., Nudelman, E., Donohue-rolfe, A., and Keusch, G.T. (1986). Pathogenesis of Shigella Diarrhea XI. Isolation of a Shigella toxin-binding glycolipid from rabbit jejunum and HeLa cells and its identification as globotriaosylceramide. J. Exp. Med. 163, 1391–1404.

Jacewicz, M.S., Mobassaleh, M., Gross, S.K., Balasubramanian, K.A., Daniel, P.F., Raghavan, S., McCluer, R.H., and Keusch, G.T. (1994). Pathogenesis of Shigella diarrhoea: XVII. A mammalian cell membrane glycolipid, Gb3 is required but not sufficient to confer sensitivity to shiga toxin. Rev. Infect. Dis. 169, 538–546.

Jacewicz, M.S., Acheson, D.W.K., Mobassaleh, M., Donohue-rolfe, A., Balasubramanian, K.A., and Keusch, G.T. (1995). Maturational regulation of globotriaosylceramide, the Shiga-like toxin i receptor, in cultured human gut epithelial Cells. J. Clin. Invest. 96, 1328–1335.

Jackson, M.E., Simpson, J.C., Girod, A., Pepperkok, R., Roberts, L.M., and Lord, J.M. (1999). The KDEL retrieval system is exploited by Pseudomonas exotoxin A, but not by Shiga-like toxin-1, during retrograde transport from the Golgi complex to the endoplasmic reticulum. J. Cell Sci. 112, 467–475.

Jackson, M.P., Newland, J.W., Holmes, R.K., and O'Brien, a D. (1987). Nucleotide sequence analysis of the structural genes for Shiga-like toxin I encoded by bacteriophage 933J from Escherichia coli. Microb. Pathog. 2, 147–153.

Jandhyala, D.M., Ahluwalia, A., Obrig, T., and Thorpe, C.M. (2008). ZAK: a MAP3Kinase that transduces Shiga toxin- and ricin-induced proinflammatory cytokine expression. Cell. Microbiol. 10, 1468–1477.

Janssen, K.-P., Vignjevic, D., Boisgard, R., Falguières, T., Bousquet, G., Decaudin, D., Dollé, F., Louvard, D., Tavitian, B., Robine, S., et al. (2006). In vivo tumor targeting using a novel intestinal pathogen-based delivery approach. Cancer Res. *66*, 7230–7236.

Johannes, L., and Goud, B. (1998). Surfing on a retrograde wave: how does Shiga toxin reach the endoplasmic reticulum? Trends Cell Biol. *8*, 158–162.

Johannes, L., and Popoff, V. (2008). Tracing the retrograde route in protein trafficking. Cell *135*, 1175–1187.

Johannes, L., and Römer, W. (2010). Shiga toxins--from cell biology to biomedical applications. Nature Rev. Microbiol. *8*, 105–116.

Johansen, B.K., Wasteson, Y., Granum, P.E., and Brynestad, S. (2001). Mosaic structure of Shiga-toxin-2-encoding phages isolated from *Escherichia coli* O157:H7 indicates frequent gene exchange between lambdoid phage genomes. Microbiology *147*, 1929–1936.

Kano, F., Yamauchi, S., Yoshida, Y., Watanabe-Takahashi, M., Nishikawa, K., Nakamura, N., and Murata, M. (2009). Yip1A regulates the COPI-independent retrograde transport from the Golgi complex to the ER. J. Cell Sci. *122*, 2218–2227.

van de Kar, N.C., Monnens, L.A., Karmali, M.A., and van Hinsbergh, V.W. (1992). Tumor necrosis factor and interleukin-1 induce expression of the verocytotoxin receptor globotriaosylceramide on human endothelial cells: implications for the pathogenesis of the hemolytic uremic syndrome. Blood *80*, 2755–2764.

Karch, H., Tarr, P.I., and Bielaszewska, M. (2005). Enterohaemorrhagic *Escherichia coli* in human medicine. Int. J. Med. Microbiol. *295*, 405–418.

Karmali, M.A., Steele, B.T., Petric, M., and Lim, C. (1983). Sporadic cases of haemolytic-uraemic syndrome associated with faecal cytotoxin and cytotoxin-producing *Escherichia coli* in stools. Lancet *1*, 619–620.

Karmali, M.A., Petric, M., Lim, C., Fleming, P.C., Arbus, G.S., and Lior, H. (1985). The association between idiopathic hemolytic uremic syndrome and infection with verotoxin producing *Escherichia coli*. J. Infect. Dis. *151*, 775–782.

Katagiri, Y.U., Mori, T., Nakajima, H., Katagiri, C., Taguchi, T., Takeda, T., Kiyokawa, N., and Fujimoto, J. (1999). Activation of Src family kinase yes induced by Shiga toxin binding to globotriaosyl ceramide (Gb3/CD77) in low density, detergent-insoluble microdomains. J. Biol. Chem. *274*, 35278–35282.

Kaye, S. a, Louise, C.B., Boyd, B., Lingwood, C. a, and Obrig, T.G. (1993). Shiga toxin-associated hemolytic uremic syndrome: interleukin-1 beta enhancement of Shiga toxin cytotoxicity toward human vascular endothelial cells in vitro. Infect. Immun. *61*, 3886–3891.

Keusch, G.T., and Jacewicz, M. (1975). Relationship of Shiga enterotoxin, neurotoxin and cytotoxin. J. Infect. Dis. *131*, S33–S39.

Keusch, G.T., Grady, G.F., Mata, I.J., and McIver, J. (1972). The pathogenesis of *Shigella* diarrhea. J. Clin. Invest. *51*, 1212–1218.

Kiarash, A., Boyd, B., and Lingwood, C.A. (1994). Glycosphingolipid receptor function is modified by fatty acid content. Verotoxin 1 and verotoxin 2c preferentially recognize different globotriaosyl ceramide fatty acid homologues. J. Biol. Chem. *269*, 11138–11146.

Kitov, P.I., Sadowska, J.M., Mulvey, G., Armstrong, G.D., Ling, H., Pannu, N.S., Read, R.J., and Bundle, D.R. (2000). Shiga-like toxins are neutralized by tailored multivalent carbohydrate ligands. Nature *403*, 669–672.

Kitov, P.I., Mulvey, G.L., Griener, T.P., Lipinski, T., Solomon, D., Paszkiewicz, E., Jacobson, J.M., Sadowska, J.M., Suzuki, M., Yamamura, K.-I., et al. (2008). In vivo supramolecular templating enhances the activity of multivalent ligands: a potential therapeutic against the *Escherichia coli* O157 AB5 toxins. Proc. Natl. Acad. Sci. U.S.A. *105*, 16837–16842.

Konowalchuk, J., Speirs, J.I., and Starvric, S. (1977). Vero Response to a Cytotoxm of *Escherichia coli*. Infect. Immun. *18*, 775–779.

Kovbasnjuk, O., Mourtazina, R., Baibakov, B., Wang, T., Elowsky, C., and Choti, M.A. (2005). The glycosphingolipid globotriaosylceramide in the metastatic transformation of colon cancer. Proc. Natl. Acad. Sci. U.S.A. *102*, 19087–19092.

Kurmanova, A., Llorente, A., Polesskaya, A., Garred, O., Olsnes, S., Kozlov, J., and Sandvig, K. (2007). Structural requirements for furin-induced cleavage and activation of Shiga toxin. Biochem. Biophys. Res. Comm. *357*, 144–149.

LaCasse, E.C., Saleh, M.T., Patterson, B., Minden, M.D., and Gariepy, J. (1996). Shiga-like toxin purges human lymphoma from bone marrow of severe combined immunodeficient mice. Blood *88*, 1561–1567.

LaCasse, E.C., Bray, M.R., Patterson, B., Lim, W.M., Perampalam, S., and Radvanyi, L.G. (1999). Shiga-like toxin-1 receptor on human breast cancer, lymphoma, and myeloma and absence from CD34(+) hematopoietic stem cells: implications for *ex vivo* tumor purging and autologous stem cell transplantation. Blood *94*, 2901–2910.

LaPointe, P., Wei, X., and Gariépy, J. (2005). A role for the protease-sensitive loop region of Shiga-like toxin 1 in the retrotranslocation of its A1 domain from the endoplasmic reticulum lumen. J. Biol. Chem. *280*, 23310–23318.

Lauvrak, S.U., Torgersen, M.L., and Sandvig, K. (2004). Efficient endosome-to-Golgi transport of Shiga toxin is dependent on dynamin and clathrin. J. Cell Sci. *117*, 2321–2331.

Lauvrak, S.U., Wälchli, S., Iversen, T., Slagsvold, H.H., Torgersen, M.L., Spilsberg, B., and Sandvig, K. (2006). Shiga toxin regulates its entry in a Syk-dependent Manner. Mol. Biol. Cell *17*, 1096–1109.

Lea, N., Lord, J.M., and Roberts, L.M. (1999). Proteolytic cleavage of the A subunit is essential for maximal cytotoxicity of *Escherichia coli* O157:H7 Shiga-like toxin-1. Microbiology *145*, 999–1004.

Lee, S., Cherla, R.P., Caliskan, I., Tesh, V.L., Al, L.E.E.E.T., and Mmun, I.N.I. (2005). Shiga toxin 1 induces apoptosis in the human myelogenous leukemia cell line THP-1 by a caspase-8-dependent, tumor necrosis factor receptor-independent mechanism. Infect. Immun. *73*, 5115–5126.

Lee, S.-Y., Lee, M.-S., Cherla, R.P., and Tesh, V.L. (2008). Shiga toxin 1 induces apoptosis through the endoplasmic reticulum stress response in human monocytic cells. Cell. Microbiol. *10*, 770–780.

Levine, M.M., DuPont, H.L., Formal, S.B., Hornick, R.B., Takeuchi, A., Gangarosa, E.J., Snyder, J.D., and J.P., L. (1973). Pathogenesis of *Shigella dysenteriae* 1 (Shiga) dysentery. J. Infect. Dis. *127*, 261–270.

Lindberg, a a, Brown, J.E., Strömberg, N., Westling-Ryd, M., Schultz, J.E., and Karlsson, K. a (1987). Identification of the carbohydrate receptor for Shiga toxin produced by *Shigella dysenteriae* type 1. J. Biol. Chem. *262*, 1779–1785.

Ling, H., Boodhoo, a, Hazes, B., Cummings, M.D., Armstrong, G.D., Brunton, J.L., and Read, R.J. (1998). Structure of the shiga-like toxin I B–pentamer complexed with an analogue of its receptor Gb3. Biochemistry *37*, 1777–1788.

Lingwood, C.A. (1996). Role of verotoxin receptors in pathogenesis. Trends Microbiol. *4*, 147–153.

Lingwood, C.A. (1999). Verotoxin/globotriaosyl ceramide recognition: angiopathy, angiogenesis and antineoplasia. Biosci. Rep. *19*, 345–354.

Lingwood, C.A., Law, H., Richardson, S., Bruntonqglili, J.L., Grandis, S.D., and Karmali, M. (1987). Glycolipid Binding of Purified and Recombinant *Escherichia coli* Produced Verotoxin *in vitro*. J. Biol. Chem. *262*, 8834–8839.

Lingwood, C.A., Binnington, B., Manis, a, and Branch, D.R. (2010a). Globotriaosyl ceramide receptor function – where membrane structure and pathology intersect. FEBS Lett. *584*, 1879–1886.

Lingwood, C.A., Manis, A., Mahfoud, R., Khan, F., Binnington, B., and Mylvaganam, M. (2010b). New aspects of the regulation of glycosphingolipid receptor function. Chem. Phys. Lipids *163*, 27–35.

Luna, A., Matas, O.B., Martı, A., Dura, J.M., Way, M., and Egea, G. (2002). Regulation of protein transport from the Golgi complex to the endoplasmic reticulum by CDC42 and. Mol. Biol. Cell *13*, 866–879.

McDonough, M.A., and Butterton, J.R. (1999). Spontaneous tandem amplification and deletion of the shiga toxin operon in *Shigella* dysenteriae 1. Mol. Microbiol. *34*, 1058–1069.

Macia, E., Ehrlich, M., Massol, R., Boucrot, E., Brunner, C., and Kirchhausen, T. (2006). Dynasore, a cell-permeable inhibitor of dynamin. Dev. Cell *10*, 839– 850.

McKenzie, J., Johannes, L., Taguchi, T., and Sheff, D. (2009). Passage through the Golgi is necessary for Shiga toxin B subunit to reach the endoplasmic reticulum. FEBS J. *276*, 1581–1595.

Mahfoud, R., Manis, A., and Lingwood, C. a (2009). Fatty acid-dependent globotriaosyl ceramide receptor function in detergent resistant model membranes. J. Lipid Res. *50*, 1744–1755.

Mallard, F., Antony, C., Tenza, D., Salamero, J., Goud, B., and Johannes, L. (1998). Direct pathway from early/recycling endosomes to the Golgi apparatus revealed through the study of shiga toxin B-fragment transport. J. Cell Biol. *143*, 973–990.

Mallard, F., Tang, B.L., Galli, T., Tenza, D., Saint-Pol, A., Yue, X., Antony, C., Hong, W., Goud, B., and Johannes, L. (2002). Early/recycling endosomes-to-TGN transport involves two SNARE complexes and a Rab6 isoform. J. Cell Biol. *156*, 653–664.

Melton-Celsa, A., Mohawk, K., Teel, L., and O'Brien, A. (2012). Pathogenesis of Shiga-toxin producing *Escherichia coli*. Curr. Top. Microbiol. Immunol. *357*, 67–103.

Meunier, L., Usherwood, Y., Chung, K.T., and Hendershot, L.M. (2002). A Subset of Chaperones and Folding Enzymes Form Multiprotein Complexes in Endoplasmic Reticulum to Bind Nascent Proteins. Mol. Biol. Cell *13*, 4456–4469.

Miller, D., Ravikumar, K., Shen, H., Suh, J., Kerwin, S., and Robertus, J. (2002). Structure-based design and characterization of novel platforms for ricin and shiga toxin inhibition. J. Med. Chem. *45*, 90–98.

Mobassaleh, M., Gross, S.K., McCluer, R.H., Donohue-Rolfe, A., and Keusch, G.T. (1989). Quantitation of the rabbit intestinal glycolipid receptor for shiga toxin. Further evidence for the developmental regulation of globotriaosylceramide in microvillus membranes. Gastroenterology *97*, 384–391.

Moreau, D., Kumar, P., Wang, S.C., Chaumet, A., Chew, S.Y., Chevalley, H., and Bard, F. (2011). Genome-wide RNAi screens identify genes required for Ricin and PE intoxications. Develop. Cell *21*, 231–244.

Mori, T., Kiyokawa, N., Katagiri, Y.U., Taguchi, T., Suzuki, T., Sekino, T., Sato, N., Ohmi, K., Nakajima, H., Takeda, T., *et al.* (2000). Globotriaosyl ceramide (CD77/Gb3) in the glycolipid-enriched membrane domain participates in B-cell receptor-mediated apoptosis by regulating lyn kinase activity in human B cells. Exper. Hematol. *28*, 1260–1268.

Mühldorfer, I., Hacker, J., Keusch, G.T., Acheson, D.W., Tschäpe, H., Kane, a V., Ritter, a, Olschläger, T., and Donohue-Rolfe, a (1996). Regulation of the Shiga-like toxin II operon in *Escherichia coli*. Infect. Immun. *64*, 495–502.

Mukherjee, J., Chios, K., Fishwild, D., Hudson, D., Donnell, S.O., Rich, S.M., Donohue-rolfe, A., and Tzipori, S. (2002a). Human Stx2-Specific Monoclonal Antibodies Prevent Systemic Complications of *Escherichia coli* O157 : H7 Infection. Infect. Immun. *70*, 612–619.

Mukherjee, J., Chios, K., Fishwild, D., Hudson, D., Donnell, S.O., Rich, S.M., Donohue-rolfe, A., and Tzipori, S. (2002b). Production and characterization of protective human antibodies against Shiga toxin 1. Infect. Immun. *70*, 5896–5899.

Mukhopadhyay, S., and Linstedt, A.D. (2012). Manganese blocks intracellular trafficking of shiga toxin and protects against shiga toxicosis. Science *335*, 332–335.

Mukhopadhyay, S., Bachert, C., Smith, D.R., and Linstedt, A.D. (2010). Manganese-induced trafficking and turnover of the cis-Golgi glycoprotein GPP130. Mol. Biol. Cell *21*, 1282–1292.

Mulvey, G.L., Marcato, P., Kitov, P.I., Sadowska, J., Bundle, D.R., and Armstrong, G.D. (2003). Assessment in mice of the therapeutic potential of tailored, multivalent Shiga toxin carbohydrate ligands. J. Infect. Dis. *187*, 640–649.

Murray, L.J., Habeshaw, J.A., Wiels, J., and Greaves, M.F. (1985). Expression of Burkitt lymphoma-associated antigen (defined by the monoclonal antibody 38.13) on both normal and malignant germinal-centre B cells. Int. J. Cancer 36, 561–565.

Nakajima, H., Kiyokawa, N., Katagiri, Y.U., Taguchi, T., Suzuki, T., Sekino, T., Mimori, K., Ebata, T., Saito, M., Nakao, H., et al. (2001). Kinetic analysis of binding between Shiga toxin and receptor glycolipid Gb3Cer by surface plasmon resonance. J. Biol. Chem. 276, 42915–42922.

Nataro, J.P., and Kaper, J.B. (1998). Diarrheagenic Escherichia coli. Clin. Microbiol. Rev. 11, 142–201.

Nathanson, S., Kwon, T., Elmaleh, M., Charbit, M., Launay, E.A., Harambat, J., Brun, M., Ranchin, B., Bandin, F., Cloarec, S., et al. (2010). Acute neurological involvement in diarrhoea-associated hemolytic uremic syndrome. CJASN 5, 1218–1228.

Neely, M.N., and Friedman, D.I. (1998). Functional and genetic analysis of regulatory regions of coliphage H-19B: location of shiga-like toxin and lysis genes suggest a role for phage functions in toxin release. Mol. Microbiol. 28, 1255–1267.

Neri, P., Tokoro, S., Yokoyama, S.-I., Miura, T., Murata, T., Nishida, Y., Kajimoto, T., Tsujino, S., Inazu, T., Usui, T., et al. (2007). Monovalent Gb3-/Gb2-derivatives conjugated with a phosphatidyl residue: a novel class of Shiga toxin-neutralizing agent. Biol. Pharm. Bull. 30, 1697–1701.

Del Nery, E., Miserey-Lenkei, S., Falguières, T., Nizak, C., Johannes, L., Perez, F., and Goud, B. (2006). Rab6A and Rab6A' GTPases play non-overlapping roles in membrane trafficking. Traffic 7, 394–407.

Newland, J.W., Strockbine, N. a, Miller, S.F., O'Brien, a D., and Holmes, R.K. (1985). Cloning of Shiga-like toxin structural genes from a toxin converting phage of Escherichia coli. Science 230, 179–181.

Nichols, B.J., Kenworthy, a K., Polishchuk, R.S., Lodge, R., Roberts, T.H., Hirschberg, K., Phair, R.D., and Lippincott-Schwartz, J. (2001). Rapid cycling of lipid raft markers between the cell surface and Golgi complex. J. Cell Biol. 153, 529–541.

Nishikawa, K., Matsuoka, K., Kita, E., Okabe, N., Mizuguchi, M., Hino, K., Miyazawa, S., Yamasaki, C., Aoki, J., Takashima, S., et al. (2002). A therapeutic agent with oriented carbohydrates for treatment of infections by Shiga toxin-producing Escherichia coli O157:H7. Proc. Natl. Acad. Sci. U.S.A. 99, 7669–7674.

Nudelman, E., Kannagi, R., Hakomori, S., Parsons, M., Lip- inski, M., and Wiels, J. (1983). A glycolipid antigen associated with Burkitt lymphoma defined by a monoclonal antibody. Science 220, 509–511.

Obrig, T.G., Moran, T.P., and Brown, J.E. (1987). The mode of action of Shiga toxin on peptide elongation of eukaryotic protein synthesis. Biochem. J. 244, 287–294.

Okuda, T., Tokuda, N., Numata, S., Ito, M., Ohta, M., Kawamura, K., Wiels, J., Urano, T., Tajima, O., Furukawa, K., et al. (2006). Targeted disruption of Gb3/CD77 synthase gene resulted in the complete deletion of globo-series glycosphingolipids and loss of sensitivity to verotoxins. J. Biol. Chem. 281, 10230–10235.

Olsnes, S., and Eiklid, K. (1980). Isolation and Characterization of Shigella shigue Cytotoxin. J. Biol. Chem. 255, 284–289.

Olsnes, S., Reisbig, R., and Eiklid, K. (1981). Subunit structure of Shigella cytotoxin. J. Biol. Chem. 256, 8732–8738.

O'Brien, A.D., and Holmes, R.K. (1987). Shiga and Shiga-Like Toxins. Microbiol. Rev. 51, 206–220.

O'Brien, A.D., and Kaper, J.B. (1998). Escherichia coli and other Shiga Toxin-Producing E. coli Strains (American Society for Microbiology, Washington, DC).

O'Brien, A.D., LaVeck, G.D., Griffin, D.E., and Thompson, M.R. (1980). Characterization of Shigella dysenteriae 1 (Shiga) toxin purified by anti-Shiga toxin affinity chromatography. Infect. Immun. 30, 170–179.

O'Brien, A.D., LaVeck, G.D., Thompson, M.R., and Formal, S.B. (1982). Production of Shigella dysenteriae 1 like cytotoxin by Escherichia coli. J. Infect. Dis. 146, 763–769.

O'Brien, A.D., Lively, T.A., Chen, M.E., Rothman, S.W., and Formal, S.B. (1983). Escherichia coli O157:H7 strains associated with haemorrhagic colitis in the United States produce a Shigella dysenteriae 1 (SHIGA) like cytotoxin. Lancet 1, 702.

O'Brien, A.D., Newland, J.W., Miller, S.F., Holmes, R.K., Smith, H.W., and Formal, S.B. (1984). Shiga-Like Toxin-Converting Phages from Eschenrchia coli Strains That Cause Hemorrhagic Colitis or Infantile Diarrhea. Science 226, 694–696.

O'Brien, A.D., Tesh, V.L., Donohue-Rolfe, A., Jackson, M.P., Olsnes, S., Sandvig, K., Lindberg, A.A., and Keusch, G.T. (1992). Shiga toxin: biochemistry, genetics, mode of action and role in pathogenesis. Curr. Top. Microbiol. Immunol. 180, 65–94.

Paton, J.C., and Paton, A.W. (1998). Pathogenesis and diagnosis of Shiga toxin-producing Escherichia coli infections. Clin. Microbiol. Rev. 11, 450–479.

Paton, A., Morona, R., and Paton, J.C. (2000). A new biological agent for treatment of Shiga toxigenic Escherichia coli infections and dysentery in humans. Nature Med. 6, 265–270.

Pellizzari, A., Pang, H., and Lingwood, C.A. (1992). Binding of verocytotoxin 1 to its receptor is influenced by differences in receptor fatty acid content. Biochemistry 31, 1363–1370.

Physiol, A.J., Liver, G., Jones, N.L., Islur, A., Haq, R., Mascarenhas, M., Karmali, M.A., Perdue, M.H., Zanke, B.W., and Sherman, P.M. (2000). Escherichia coli Shiga toxins induce apoptosis in epithelial cells that is regulated by the Bcl-2 family Escherichia coli Shiga toxins induce apoptosis in epithelial cells that is regulated by the Bcl-2 family. Am. J. Physiol. Gastrointest. Liver Physiol. 278, 811–819.

Pinyon, R. a, Paton, J.C., Paton, A.W., Botten, J. a, and Morona, R. (2004). Refinement of a therapeutic Shiga toxin-binding probiotic for human trials. J. Infect. Dis. 189, 1547–1555.

Popoff, V., Mardones, G. a, Tenza, D., Rojas, R., Lamaze, C., Bonifacino, J.S., Raposo, G., and Johannes, L. (2007). The retromer complex and clathrin define an early endosomal retrograde exit site. J. Cell Sci. 120, 2022–2031.

Proulx, F., Turgeon, J., Delage, G., Lafleur, L., and Chicoine, L. (1992). Randomized, controlled trial of antibiotic therapy for *Escherichia coli* O157:H7 enteritis. J. Pediatr. *121*, 299–303.

Proulx, F., Seidman, E.G., and Karpman, D. (2001). Pathogenesis of Shiga toxin-associated hemolytic uremic syndrome. Pediatr. Res. *50*, 163–171.

Ramegowda, B., and Tesh, V.L. (1996). Differentiation-associated toxin receptor modulation, cytokine production, and sensitivity to Shiga-like toxins in human monocytes and monocytic cell lines. Infect. Immun. *64*, 1173–1180.

Reisbig, R., Olsnes, S., and Eiklid, K. (1981). The cytotoxic activity of *Shigella* toxin. Evidence for catalytic inactivation of the 60S ribosomal subunit. J. Biol. Chem. *256*, 8739–8744.

Richardson, J.M., Evans, P.D., Homans, S.W., and Donohue-Rolfe, A. (1997). Solution structure of the carbohydrate-binding B subunit homopentamer of verotoxin VT-1 from *E. coli*. Nat. Genet. *4*, 190–193.

Römer, W., Berland, L., Chambon, V., Gaus, K., Windschiegl, B., Tenza, D., Aly, M.R.E., Fraisier, V., Florent, J.-C., Perrais, D., *et al*. (2007). Shiga toxin induces tubular membrane invaginations for its uptake into cells. Nature *450*, 670–675.

Römer, W., Pontani, L.-L., Sorre, B., Rentero, C., Berland, L., Chambon, V., Lamaze, C., Bassereau, P., Sykes, C., Gaus, K., *et al*. (2010). Actin dynamics drive membrane reorganization and scission in clathrin-independent endocytosis. Cell *140*, 540–553.

Saenz, J.B., Doggett, T.A., and Haslam, D.B. (2007). Identification and characterization of small molecules that inhibit intracellular toxin transport. Infect. Immun. *75*, 4552–4561.

Saenz, J.B., Sun, W.J., Chang, J.W., Li, J., Bursulaya, B., Gray, N.S., and Haslam, D.B. (2009). Golgicide A reveals essential roles for GBF1 in Golgi assembly and function. Nat. Chem. Biol. *5*, 157–165.

Safouane, M., Berland, L., Callan-Jones, A., Sorre, B., Römer, W., Johannes, L., Toombes, G.E.S., and Bassereau, P. (2010). Lipid cosorting mediated by shiga toxin induced tubulation. Traffic *11*, 1519–1529.

Saint-Pol, A., Yélamos, B., Amessou, M., Mills, I., Dugast, M., Tenza, D., Schu, P., Antony, C., McMahon, H., Lamaze, C., *et al*. (2004). Clathrin adaptor epsinR is required for retrograde sorting on early endosomal membranes. Dev. Cell *6*, 525–538.

Salhia, B., Rutka, J.T., Lingwood, C.A., Nutikka, A., and Van Furth, W.R. (2002). The treatment of malignant meningioma with verotoxin. Neoplasia *4*, 304–311.

Sandvig, K., and van Deurs, B. (1996). Endocytosis, intracellular transport, and cytotoxic action of Shiga toxin and ricin. Physiol. Rev. *76*, 949–966.

Sandvig, K., Olsnes, S., Brown, J.E., Petersen, O.W., and van Deurs, B. (1989). Endocytosis from coated pits of Shiga toxin: a glycolipid-binding protein from *Shigella dysenteriae* 1. J. Cell Biol. *108*, 1331–1343.

Sandvig, K., Prydz, K., Ryd, M., and van Deurs, B. (1991). Endocytosis and intracellular transport of the glycolipid-binding ligand Shiga toxin in polarized MDCK cells. J. Cell Biol. *113*, 553–562.

Sandvig, K., Garred, O., Prydz, K., Kozlov, J., Hansen, S., and van Deurs, B. (1992). Retrograde transport of endocytosed Shiga toxin to the endoplasmic reticulum. Nature *358*, 510–512.

Sandvig, K., Ryd, M., Garred, O., Schweda, E., Holm, P.K., and van Deurs, B. (1994). Retrograde transport from the Golgi complex to the ER of both Shiga toxin and the nontoxic Shiga B-fragment is regulated by butyric acid and cAMP. J. Cell Biol. *126*, 53–64.

Sandvig, K., Garred, O., van Helvoort, a, van Meer, G., and van Deurs, B. (1996). Importance of glycolipid synthesis for butyric acid-induced sensitization to shiga toxin and intracellular sorting of toxin in A431 cells. Mol. Biol. Cell *7*, 1391–1404.

Sandvig, K., Bergan, J., Dyve, A.-B., Skotland, T., and Torgersen, M.L. (2010). Endocytosis and retrograde transport of Shiga toxin. Toxicon *56*, 1181–1185.

Sansonetti, P.J. (1992). Molecular and cellular biology of *Shigella flexneri* invasiveness: from cell assay systems to Shigellosis. Curr. Top. Microbiol. Immunol. *180*, 1–19.

van Setten, P.A., van Hinsbergh, V.W., van der Velden, T.J., van de Kar, N.C., Vermeer, M., and Mahan, J.D. (1997). Effects of TNF alpha on verocytotoxin cytotoxicity in purified human glomerular microvascular endothelial cells. Kidney Int. *51*, 1245–1256.

Shiga, K. (1898). Ueber den Dysenterie-bacillus (*Bacillus dysenteriae*). Centralblatt Für Bakteriologie, Parasitenkunde Und Infektionskrankheiten, Erste Abteilung *24*, 913–918.

Shimizu, T., Ohta, Y., and Noda, M. (2009). Shiga toxin 2 is specifically released from bacterial cells by two different mechanisms. Infect. Immun. *77*, 2813–2823.

Smith, W.E., Kane, A.V., Campbell, S.T., Acheson, D.W.K., Cochran, B.H., and Thorpe, C.M. (2003). Shiga toxin 1 triggers a ribotoxic stress response leading to p38 and JNK activation and induction of apoptosis in intestinal epithelial cells. Infect. Immun. *71*, 1497–1504.

Soltyk, A.M., MacKenzie, C.R., Wolski, V.M., Hirama, T., Kitov, P.I., Bundle, D.R., and Brunton, J.L. (2002). A mutational analysis of the globotriaosylceramide-binding sites of verotoxin VT1. J. Biol. Chem. *277*, 5351–5359.

Spooner, R.A., Watson, P., Smith, D.C., Boal, F., Amessou, M., Johannes, L., Clarkson, G.J., Lord, J.M., Stephens, D.J., and Roberts, L.M. (2008). The secretion inhibitor Exo2 perturbs trafficking of Shiga toxin between endosomes and the trans-Golgi network. Biochem. J. *414*, 471–484.

St. Hilaire, P.M., Boyd, M.K., and Toone, E.J. (1994). Interaction of the Shiga-like toxin type 1 B-subunit with its carbohydrate receptor. Biochemistry *33*, 14452–14463.

Starr, T., Sun, Y., Wilkins, N., and Storrie, B. (2010). Rab33b and Rab6 are functionally overlapping regulators of Golgi homeostasis and trafficking. Traffic *11*, 626–636.

Stechmann, B., Bai, S., Gobbo, E., Lopez, R., Merer, G., Pinchard, S., Panigai, L., Beaumelle, B., Sauvaire, D., Gillet, D., *et al*. (2010). Inhibition of retrograde transport protects mice from lethal ricin challenge. Cell *141*, 231–242.

Stein, P.E., Boodhoo, A., Tyrrell, G.J., Brunton, J.L., and Read, R.J. (1992). Crystal structure of the cell-binding B oligomer of verotoxin-1 from *E. Coli*. Nature *335*, 748–750.

Stone, M.K., Kolling, G.L., Lindner, M.H., and Obrig, T.G. (2008). P38 mitogen-activated protein kinase mediates lipopolysaccharide and tumor necrosis factor alpha induction of Shiga toxin 2 sensitivity in human umbilical vein endothelial cells. Infect. Immun. *76*, 1115–1121.

Stricklett, P.K., Hughes, A.K., and Kohan, D.E. (2005). Inhibition of p38 mitogen-activated protein kinase ameliorates cytokine up-regulated shigatoxin-1 toxicity in human brain microvascular endothelial cells. J. Infect. Dis. *191*, 461–471.

Strockbine, N.A., Marques, L.R.M., Newland, J.W., Smith, H.W., Holmes, R.K., and Brienl, A.D.O. (1986). Two Toxin-Converting Phages from *Escherichia coli* O157 : H7 Strain 933 Encode Antigenically Distinct Toxins with Similar Biologic Activities. Infect. Immun. *53*, 135–140.

Strockbine, N. a, Jackson, M.P., Sung, L.M., Holmes, R.K., and O'Brien, a D. (1988). Cloning and sequencing of the genes for Shiga toxin from *Shigella dysenteriae* type 1. J. Bacteriol. *170*, 1116–1122.

Suh, J.K., Hovde, C.J., and Robertus, J.D. (1998). Shiga toxin attacks bacterial ribosomes as effectively as eucaryotic ribosomes. Biochemistry *37*, 9394–9398.

Suzuki, A., Doi, H., Matsuzawa, F., Aikawa, S., Takiguchi, K., Kawano, H., Hayashida, M., and Ohno, S. (2000). Bcl-2 antiapoptotic protein mediates verotoxin II – induced cell death : possible association between Bcl-2 and tissue failure by *E. coli* O157 : H7. Genes Develop. *14*, 1734–1740.

Svinarich, B.M., and Pachaudhuri, S. (1992). Regulation of the SLT-1A toxin operon by ferric uptake regulatory protein in toxinogenic strains of *Shigella dysenteriae* type 1. J. Diarrhoeal. Dis. Res. *10*, 139–145.

Tai, G., Lu, L., Wang, T.L., Tang, B.L., Goud, B., Johannes, L., and Hong, W. (2004). Participation of the syntaxin 5/Ykt6/GS28/GS15 SNARE complex in transport from the early/recycling endosome to the trans-Golgi network. Mol. Biol. Cell *15*, 4011–4022.

Takeda, T., Yoshino, K., Adachi, E., Sato, Y., and Yamagata, K. (1999). In vitro assessment of a chemically synthesized Shiga toxin receptor analogue attached to chromosorb P (Synsorb Pk) as a specific absorbing agent of Shiga toxin 1 and 2. Microbiol. Immunol. *43*, 331–337.

Takenouchi, H., Kiyokawa, N., Taguchi, T., Matsui, J., Katagiri, Y.U., Okita, H., Okuda, K., and Fujimoto, J. (2004). Shiga toxin binding to globotriaosyl ceramide induces intracellular signals that mediate cytoskeleton remodeling in human renal carcinoma-derived cells. J. Cell Sci. *117*, 3911–3922.

Tam, P., Mahfoud, R., Nutikka, A., Khine, A.A., Binnington, B., Paroutis, P., and Lingwood, C. (2008). Differential intracellular transport and binding of verotoxin 1 and verotoxin 2 to globotriaosylceramide-containing lipid assemblies. J. Cell. Physiol. *216*, 750–763.

Tesh, V.L. (2010). Induction of apoptosis by Shiga toxins. Future Microbiol. *5*, 431–453.

Torgersen, M.L., Wälchli, S., Grimmer, S., Skånland, S.S., and Sandvig, K. (2007). Protein kinase Cdelta is activated by Shiga toxin and regulates its transport. J. Biol. Chem. *282*, 16317–16328.

Torgersen, M.L., Engedal, N., Pedersen, A.-M.G., Husebye, H., Espevik, T., and Sandvig, K. (2011). Toll-like receptor 4 facilitates binding of Shiga toxin to colon carcinoma and primary umbilical vein endothelial cells. FEMS Immunol. Med. Microbiol. *61*, 63–75.

Trachtman, H., Cnaan, A., Christen, E., Gibbs, K., Zhao, S., Acheson, D., Weiss, R., Kaskel, F., Spitzer, A., and Hirschman, G. (2003). Effect of an oral shiga toxin-binding agent on diarrhea-associated hemolytic uremic syndrome in children. JAMA *290*, 1337–1344.

Trofa, a F., Ueno-Olsen, H., Oiwa, R., and Yoshikawa, M. (1999). Dr. Kiyoshi Shiga: discoverer of the dysentery bacillus. Clin. Infect. Dis. *29*, 1303–1306.

Tyler, J.S., Mills, M.J., and Friedman, D.I. (2004). The operator and early promoter region of the Shiga toxin type 2-encoding bacteriophage 933w and control of toxin expression. J. Bacteriol. *186*, 7670–7679.

Unkmeir, a, and Schmidt, H. (2000). Structural analysis of phage-borne stx genes and their flanking sequences in shiga toxin-producing *Escherichia coli* and *Shigella dysenteriae* type 1 strains. Infect. Immun. *68*, 4856–4864.

Utskarpen, A., Slagsvold, H., Dyve, A., Skånland, S., and K., S. (2007). SNX1 and SNX2 mediate retrograde transport of Shiga toxin. Biochem. Biophys. Res. Commun. *358*, 566–570.

Viel, T., Dransart, E., Nemati, F., Henry, E., Thézé, B., Decaudin, D., Boisgard, R., Johannes, L., and Tavitian, B. (2008). *In vivo* tumor targeting by the B-subunit of Shiga toxin. Mol. Imag. *7*, 23–31.

Waddell, T., Cohen, A., and Lingwood, C.A. (1990). Induction of verotoxin sensitivity in receptor-deficient cell lines using the receptor glycolipid globotriosylceramide. Proc. Natl. Acad. Sci. U.S.A. *87*, 7898–7901.

Wagner, P.L., Livny, J., Neely, M.N., Acheson, D.W.K., Friedman, D.I., and Waldor, M.K. (2002). Bacteriophage control of Shiga toxin 1 production and release by *Escherichia coli*. Mol. Microbiol. *44*, 957–970.

Walterspiel, J., Ashkenazi, S., Morrow, A., and Cleary, T. (1992). Effect of subinhibitory concentrations of antibiotics on extracellular Shiga-like toxin I. Infection *20*, 25–29.

Warnier, M., Römer, W., Geelen, J., Lesieur, J., Amessou, M., van den Heuvel, L., Monnens, L., and Johannes, L. (2006). Trafficking of Shiga toxin/Shiga-like toxin-1 in human glomerular microvascular endothelial cells and human mesangial cells. Kidn. Int. *70*, 2085–2091.

Watanabe, M., Matsuoka, K., Kita, E., Igai, K., Higashi, N., Miyagawa, A., Watanabe, T., Yanoshita, R., Samejima, Y., Terunuma, D., *et al.* (2004). Oral therapeutic agents with highly clustered globotriose for treatment of Shiga toxigenic *Escherichia coli* infections. J. Infect. Dis. *189*, 360–368.

Weinstein, D.L., Holmes, R.K., and O'Brien, A.D. (1988). Effects of iron and temperature on Shiga-like toxin I production by *Escherichia coli*. Infect. Immun. *56*, 106–111.

Wesche, J., Rapak, a, and Olsnes, S. (1999). Dependence of ricin toxicity on translocation of the toxin A-chain from the endoplasmic reticulum to the cytosol. J. Biol. Chem. *274*, 34443–34449.

White, J., Johannes, L., Mallard, F., Girod, a, Grill, S., Reinsch, S., Keller, P., Tzschaschel, B., Echard, a, Goud, B., *et al.* (1999). Rab6 coordinates a novel Golgi to ER retrograde transport pathway in live cells. J. Cell Biol. *147*, 743–760.

Wiels, J., Fellous, M., and Tursz, T. (1981). Monoclonal antibody against a Burkitt lymphoma-associated antigen. Proc. Natl. Acad. Sci. U.S.A. *78*, 6485–6488.

Wilcke, M., Johannes, L., Galli, T., Mayau, V., Goud, B., and Salamero, J. (2000). Rab11 regulates the compartmentalization of early endosomes required for efficient transport from early endosomes to the trans-Golgi network. J. Cell Biol. *151*, 1207–1220.

Windschiegl, B., Orth, A., Römer, W., Berland, L., Stechmann, B., Bassereau, P., Johannes, L., and Steinem, C. (2009). Lipid reorganization induced by Shiga toxin clustering on planar membranes. PloS One *4*, e6238.

Wälchli, S., Skånland, S.S., Gregers, T.F., Lauvrak, S.U., Torgersen, M.L., Ying, M., Kuroda, S., Maturana, A., and Sandvig, K. (2008). The Mitogen-activated Protein Kinase p38 Links Shiga Toxin-dependent Signaling and Trafficking. Mol. Biol. Cell *19*, 95–104.

Yu, M., and Haslam, D.B. (2005). Shiga Toxin Is Transported from the Endoplasmic Reticulum following Interaction with the Luminal Chaperone HEDJ/ERdj3. Infect. Immun. *73*, 2524–2532.

Escherichia coli Subtilase Cytotoxin: Structure, Function and Role in Disease

6

Adrienne W. Paton, Hui Wang, Valeria Michelacci, Stefano Morabito and James C. Paton

Abstract

Subtilase cytotoxin (SubAB) is a recently discovered AB5 toxin family produced by certain strains of pathogenic *Escherichia coli*, particularly Shiga toxigenic *E. coli* (STEC) strains that lack the locus of enterocyte effacement (LEE). Its A subunit is a serine protease belonging to the Peptidase_S8 (subtilase) family, while the pentameric B subunit binds to cell surface receptor glycans terminating in the sialic acid *N*-glycolylneuraminic acid. Receptor binding triggers internalization of the holotoxin and retrograde trafficking to the endoplasmic reticulum (ER), where the A subunit cleaves its only known substrate, the essential Hsp70 family chaperone BiP (GRP78). BiP is a highly conserved master regulator of ER function, which is essential for survival of eukaryotes from simple yeasts to higher organisms such as mammals. Thus, SubAB-mediated BiP cleavage has devastating consequences for the cell, triggering a severe and unresolved ER stress response, ultimately leading to apoptosis. Interestingly, intraperitoneal injection of SubAB is lethal for mice and induces pathological features overlapping those seen in the haemolytic uraemic syndrome, a life-threatening complication of Shiga toxigenic *E. coli* infection in humans. However, an unequivocal role for SubAB in human disease pathogenesis is yet to be established and carefully designed molecular epidemiological investigations are required to resolve this issue.

Introduction

AB5 toxins are key virulence factors for several major bacterial pathogens, including Shiga toxigenic and enterotoxigenic *Escherichia coli* (STEC and ETEC, respectively). AB5 toxins are characterized by enzymatic A subunits capable of disrupting critical host cell functions, non-covalently linked to pentameric B subunits that bind to specific glycan receptors on target eukaryotic cells triggering toxin uptake (Fan *et al.*, 2000; Beddoe *et al.*, 2010). There are three well-characterized AB5 toxin families, namely the Shiga toxins (Stx) (also called Verocytotoxins [VT]) produced by STEC and *Shigella dysenteriae*, cholera toxin (Ctx) and the closely related labile enterotoxins (LT) produced by *Vibrio cholerae* and ETEC, respectively, and pertussis toxin (Ptx) produced by *Bordetella pertussis*. The A subunits of Stx toxins are RNA-*N*-glycosidases which cleave a specific adenine base in 28S rRNA, thereby inhibiting eukaryotic protein synthesis. The A subunits of Ctx/LT and Ptx are ADP-ribosyltransferases which modify distinct host G proteins, resulting in alteration of intracellular cAMP levels and dysregulation of ion transport mechanisms (Beddoe *et al.*, 2010). Subtilase cytotoxin (SubAB) is the prototype of a fourth AB5 toxin family, with distinct A subunit enzymatic activity (Paton *et al.*, 2004) and in this chapter we summarize the molecular and cellular biology and role in disease pathogenesis of this recently discovered toxin.

Biological characterization

SubAB was discovered in a serotype O113:H21 STEC strain (designated 98NK2) that caused an outbreak of haemolytic uraemic syndrome (HUS) in South Australia (Paton *et al.*, 1999). HUS is a life-threatening systemic complication of STEC infection, characterized by a triad of microangiopathic haemolytic anaemia, thrombocytopenia and renal failure. These features are a direct manifestation of endothelial damage believed to be caused by Stx after absorption into the circulation from the gut lumen (Paton and Paton, 1998). O113:H21 is a prominent STEC serotype frequently associated with serious human disease, and was among the first STEC serotypes to be causally associated with HUS (Karmali *et al.*, 1985). However, unlike other prominent HUS-associated STEC serotypes such as the infamous O157:H7, O113:H21 strains lack the locus of enterocyte effacement (LEE), which encodes important accessory virulence traits promoting intestinal colonization and gastrointestinal pathology. The clinical presentation was also unusual in this particular outbreak, as the affected patients exhibited more marked neurological involvement than in previous HUS cases seen at the same hospital. This led to the hypothesis that O113:H21 and perhaps some other virulent STEC strains might produce an additional cytotoxin capable of either augmenting the effects of Stx or causing pathology in its own right.

A novel AB5 toxin operon was subsequently isolated from a cosmid gene bank of 98NK2 genomic DNA. The encoded toxin had a distinct cytopathic effect (CPE) on Vero cells to that of Stx; it was maximal after 3–4 days incubation and featured rounding of cells, detachment from the substratum, and loss of viability (Paton *et al.*, 2004). The operon consists of two closely linked genes, designated *subA* and *subB*, located on the O113:H21 megaplasmid pO113. The *subA* gene encodes a 347 amino acid protein with similarity to members of the Peptidase_S8 (subtilase) family of serine proteases (pfam00082.8). Its closest bacterial relative is the BA_2875 gene product of *Bacillus anthracis*. SubA contains a 'catalytic triad' comprising conserved Asp, His and Ser domains characteristic of members of the subtilase family (Siezen and Leunissen, 1997). It matches consensus sequences for these domains at 11/12, 10/11 and 10/11 positions, respectively, including the known critical active site residues Asp_{52}, His_{89} and Ser_{272} (Paton *et al.*, 2004). The *subB* gene is 16 nucleotides downstream of *subA* and encodes a 141 amino acid protein with significant

similarities to putative exported proteins from *Yersinia pestis* (YPO0337; 56% identity, 79% similarity over 136 amino acids) and *Salmonella Typhi* (STY1891; 50% identity, 68% similarity over 117 amino acids). STY1891 has significant similarity (30% identity over 101 amino acids) to the S2 subunit of Ptx, but there is only 18% identity between SubB and the latter (Paton *et al.*, 2004).

Both *subA* and *subB* genes are required for expression of cytotoxicity in *E. coli*, and this can also be abolished by mutagenesis of any one of the three critical A subunit active site residues. Thus, serine protease activity is fundamental to the cytotoxic mechanism. The holotoxin is highly toxic for a range of cell types suggesting widespread expression of its receptor. Concentrations as low as 1 pg/ml are sufficient to cause a detectable CPE in Vero cells after 3 days (Paton *et al.*, 2004). However, an additional vacuolating activity of SubAB, which was due to SubB alone has also been described (Morinaga *et al.*, 2007). This phenotype was dependent on vacuolar ATPase activity and occurred at early time points in Vero cells, before the SubA-dependent, protease-mediated cytotoxicity became apparent. Moreover, vacuolation required very high toxin doses (> 1 µg/ml). A similar phenotype in SubAB-treated HeLa cells, with formation of numerous vacuoles derived from elements of the ER, Golgi and probably also the mitochondria, coupled with appearance of lipid droplets in the cytoplasm has also been described (Lass *et al.*, 2008). Notably, however, these changes were dependent on the proteolytic activity of SubA, as they were not seen in cells treated with a SubAB derivative with an active site Ser_{272}-Ala mutation ($SubA_{A272}B$).

Cytotoxic mechanism

Proteomic comparison of Vero cells treated with SubAB or $SubA_{A272}B$ showed that the only cellular substrate of SubA was BiP (GRP78), a Hsp70 family chaperone located principally in the endoplasmic reticulum (ER). The toxin cleaves a di-leucine motif (Leu_{416}-Leu_{417}) in the hinge region connecting the N-terminal ATPase and C-terminal protein-binding domains of BiP, releasing 44 kDa and 28 kDa fragments. The protease is extremely specific, as high doses of purified

SubA or SubAB were incapable of cleaving even the most closely related Hsp70 family chaperones *in vitro* (Paton *et al.*, 2006). This is consistent with the crystal structure of SubA, which reveals an unusually deep active site cleft, relative to other subtilase family proteases (Paton *et al.*, 2006).

BiP has a variety of critical functions in the ER compartment. One of these is to mediate correct folding of nascent secretory proteins in the ER lumen by binding to exposed hydrophobic regions via its C-terminus. Subsequent release is coupled with ATP hydrolysis and a series of binding and release events folds the proteins into their correct conformation. BiP also maintains the permeability barrier of the ER membrane, and targets terminally mis-folded proteins via the Sec61 apparatus for degradation by the proteasome. One of its most critical functions, however, is as a master regulator of ER stress signalling responses, and it plays a crucial role in triggering the unfolded protein response (UPR). It also exhibits anti-apoptotic properties through interference with caspase activation (Gething, 1999; Hendershot, 2004). Thus, disablement of BiP by SubA-mediated proteolysis would be expected to have serious consequences for cell survival. Significantly, transfected Vero cells co-expressing a SubA protease-resistant BiP derivative ($Leu_{416}Asp$) were resistant to SubAB-mediated cytotoxicity, directly confirming the central role of BiP cleavage in the lethal mechanism (Paton *et al.*, 2006). This mechanism of action is unique amongst bacterial toxins.

Although SubAB-mediated cell death may take up to 3 days, early events are critical. Washing cell monolayers within a few minutes of initial exposure to the toxin will not rescue the cells from inevitable death (Paton *et al.*, 2004). However, much remains to be learnt regarding the intervening molecular events. One early consequence of SubAB intoxication is induction of a rapid and severe ER stress response (Wolfson *et al.*, 2008). This involves a series of changes in cellular activity designed to restore ER homeostasis so that the cells can recover (Boyce and Yuan, 2006). One of the major arms of the ER stress response is the UPR, which involves transcriptional up-regulation of ER chaperones (including BiP) to boost the folding capacity of the ER, activation

of proteasome-dependent ER-associated degradation (ERAD) to remove unfolded proteins from the ER lumen, and transient inhibition of translation to slow down the traffic of nascent proteins into the ER compartment that require folding (Boyce and Yuan, 2006). In mammals, ER stress responses are triggered by activation of three sentinel proteins [PKR-like ER kinase (PERK), inositol-requiring enzyme 1 (IRE1) and activating transcription factor 6 (ATF6)] that span the ER membrane and interact with BiP via their luminal domains. However, accumulation of unfolded proteins in the ER lumen competitively displaces BiP, thereby activating the signalling pathways. Activated PERK phosphorylates $eIF2\alpha$, inhibiting general protein synthesis, yet still permitting translation of some mRNAs such as ATF4, which activates genes that ultimately assist in re-establishing ER homeostasis (Szegezdi *et al.*, 2006). When BiP releases ATF6, it traffics to the Golgi and is cleaved, releasing a 50 kDa activated form. This translocates to the nucleus, where it binds to the ER stress response element, inducing genes encoding ER chaperones such as BiP, GRP94 and protein disulfide isomerase, as well as the transcription factors C/EBP homologous protein (CHOP) and X-box-binding protein 1 (XBP1). Activation of IRE1 splices XBP1 mRNA such that it is translated into XBP1 protein, which then up-regulates genes encoding ER chaperones and the HSP40 family member $P58^{IPK}$. $P58^{IPK}$ provides a feedback loop by binding and inhibiting PERK, thereby relieving the $eIF2\alpha$-mediated translational block (Szegezdi *et al.*, 2006).

Treatment of Vero cells with SubAB activated all three ER stress signalling pathways (Wolfson *et al.*, 2008). Active PERK-dependent phosphorylation of $eIF2\alpha$ occurred within 30 min of toxin treatment, and correlated with inhibition of translation. Activation of IRE1 was demonstrated by splicing of XBP1 mRNA, while ATF6 activation was demonstrated by depletion of the 90 kDa un-cleaved form, and appearance of the 50 kDa cleaved form. SubAB treatment activated the PERK and IRE1 pathways quite rapidly, before all the cellular pool of BiP had been depleted, and presumably before significant levels of unfolded proteins would have accumulated in the ER lumen. This suggests that cleavage by the toxin causes BiP to immediately dissociate from the signalling molecules. However, ATF6 activation in response to SubAB treatment appeared to be markedly slower. Interestingly, BiP has been reported to form a stable interaction with ATF6, with dissociation requiring direct triggering mediated by an ER stress-responsive sequence in the ATF6 luminal domain (Shen *et al.*, 2005). Thus, accumulation of unfolded proteins in SubAB-treated cells may be required before the ATF6 pathway is activated. During the following 24 h period, further downstream consequences of BiP cleavage were detected, including up-regulation of GRP94, ATF4, EDEM, CHOP, and GADD34. BiP itself was also up-regulated at the mRNA level, but at the protein level, it continued to be degraded by SubAB in the ER lumen, presumably preventing restoration of ER homeostasis (Wolfson *et al.*, 2008). Thus, SubAB treatment induced a severe and sustained ER stress response in Vero cells, and at 30 h, there was evidence of apoptosis, as judged by DNA fragmentation. These *in vitro* findings are consistent with our observation of CHOP induction in the liver, as well as evidence of apoptosis in the kidneys, spleen and liver of SubAB-treated mice (Paton *et al.*, 2006; Wang *et al.*, 2007).

There are well-established links between UPR signalling and cell death pathways via Bcl-2 family proteins (Li *et al.*, 2006). Activated IRE1 directly facilitates Bax/Bak oligomerization at the ER membrane, and SubAB-induced apoptosis in mouse embryo fibroblasts and HeLa cells has been shown to be Bax/Bak-dependent (May *et al.*, 2010; Yahiro *et al.*, 2010). Conversely, ATF6- and PERK-mediated up-regulation of CHOP did not appear to be essential, at least in HeLa cells (Yahiro *et al.*, 2010). Nevertheless, the PERK pathway appears to play an important role in triggering apoptosis through phospho-eIF2α-mediated translational blockade of anti-apoptotic proteins (Yahiro *et al.*, 2012). Interestingly, SubAB-mediated activation of the PERK and IRE1 pathways in renal epithelial cells has been shown to result in phosphorylation of JNK, ERK, and p38 MAP kinases, leading to activation of AP-1 and induction of AP-1-dependent transcription (Zhao *et al.*, 2011). Such MAP kinase activation has previously been associated with Stx-induced HUS in

humans. SubAB has also been reported to induce cell cycle arrest in G1 phase, possibly through down-regulation of cyclin D1 due to a combination of translational inhibition and proteasomal degradation (Morinaga *et al.*, 2008).

Receptor binding

Of course, before SubAB can cleave BiP and trigger the above responses, it must bind to cognate glycan receptors on target cells, then be internalized and transported to the ER compartment. Receptor specificity is critical for the pathogenic process, as it determines host susceptibility, tissue tropism, and the nature and spectrum of the resultant pathology (Beddoe *et al.*, 2010). The B subunits of both Stx and Ctx bind to host cell glycolipids (Gb_3 and GM1, respectively), whereas the S2 subunit of Ptx binds to sialated glycoproteins (Beddoe *et al.*, 2010). SubB shares about 18% amino acid identity with Ptx S2 and binds to N-linked glycans displayed on several glycoproteins on the surface of Vero and HeLa cells, including α2β1integrin, which is known to be heavily sialated. Moreover, RNAi knock-down of β1integrin in Vero cells abrogated the vacuolating activity of the toxin, although it did not block SubA-dependent cytotoxicity (Yahiro *et al.*, 2006). This is probably because RNAi knockdown of integrin expression was less than 100% efficient and binding to other surface glycoproteins may also have contributed to toxin uptake. Lack of involvement of glycolipid receptors is also supported by studies in knock out mice with defects in biosynthesis of a range of glycosphingolipids and gangliosides, none of which were protected from SubAB (Kondo *et al.*, 2009).

Glycan array analysis has shown that SubB has a high specificity for glycans terminating with α2–3-linked N-glycolylneuraminic acid (Neu5Gc). Unlike the situation for CtxB and StxB, the nature of the sub-terminal sugars had little impact on binding affinity. However, roughly 20-fold weaker binding was seen with otherwise identical glycans that terminated in α2–3-linked N-acetylneuraminic acid (Neu5Ac), which differs by one hydroxyl group from Neu5Gc. Binding was reduced over 30-fold if the Neu5Gc linkage was changed from α2–3 to α2–6, and 100-fold if the terminal sialic acid was removed. This high specificity for Neu5Gc-terminating glycans is, to the best of our knowledge, unique amongst bacterial toxins (Byres *et al.*, 2008).

Structural analysis of SubB confirmed that it forms a homopentameric ring, like CtxB and StxB. There was also a high degree of structural similarity with the last 100 amino acids of Ptx S2, with the two structures almost superimposing. This is in spite of relatively modest (18%) amino acid sequence identity. Neu5Gc bound to a shallow pocket halfway down the sides of the SubB pentamer, whereas identical experiments using Neu5Ac failed to show any binding (Byres *et al.*, 2008). The Ptx S2 sialic acid binding site is also shallow and in the same location (Stein *et al.*, 1994). In contrast, the B subunits of Stx, Ctx and LT, whose receptors are glycolipids rather than glycoproteins, have deep receptor binding pockets located on the base of the pentamer, juxtaposed to the membrane (Merritt *et al.*, 1994; Merritt *et al.*, 1997; Ling *et al.*, 1998). In the SubB–Neu5Gc complex, Neu5Gc makes key interactions with the side chains of Asp_8, Ser_{12}, Glu_{36} and Tyr_{78}. Neu5Gc differs from Neu5Ac by the addition of a hydroxyl on the methyl group of the *N*-Acetyl moiety, which makes additional crucial interactions with SubB; namely the extra hydroxyl points towards and interacts with Tyr_{78}^{OH} and also hydrogen bonds with the main chain of Met_{10}. These key interactions could not occur with Neu5Ac, thus explaining the marked preference for Neu5Gc (Byres *et al.*, 2008). The biological relevance of the structural analysis was confirmed by mutagenesis of key glycan-interacting residues in SubB. Mutagenesis of either Ser_{12}, Glu_{36} or Tyr_{78} significantly reduced cell binding and specific cytotoxicity of the respective holotoxin. Of these, the Ser_{12} mutation had the greatest impact, reducing cytotoxicity by 99.98%. Importantly, mutagenesis of Tyr_{78}, which interacts only with the OH group unique to Neu5Gc, reduced cytotoxicity by 96.9% (Byres *et al.*, 2008).

Intracellular trafficking

Receptor binding by AB5 toxins triggers internalization and intracellular trafficking, such that the catalytic A subunit has access to its substrate.

Like Stx and Ctx, SubAB is trafficked from the cell surface via the Golgi to the ER via a retrograde pathway. However, SubAB internalization and trafficking is exclusively clathrin-dependent, whereas Stx or Ctx can also engage the lipid raft transport pathway (Chong *et al.*, 2008). The route through the Golgi is also distinct, with SubAB exploiting a novel p115/golgin-84-independent, COG/Rab6/COPI-dependent mechanism, and unlike Stx, retrograde transport is not dependent on the endosomal sorting nexins SNX1 and SNX2 (Smith *et al.*, 2009). Trafficking of the other AB_5 toxins also differs from SubAB because their substrates are located in the cytoplasm, while that for SubAB is confined to the ER lumen. Thus, the catalytic subunits of the other toxins must also be retro-translocated across the ER membrane, by subversion of the Sec61 translocon (Lencer and Tsai, 2003; Yu and Haslam, 2005). Interestingly, at least for StxA, retro-translocation is believed to occur following interaction with BiP and another chaperone HEDJ/ERdj3 (Yu and Haslam, 2005). SubAB is also known to inhibit ERAD, presumably through reduced Sec61-mediated trafficking of substrates (Lass *et al.*, 2008). Thus, it is possible that SubAB-mediated BiP cleavage might interfere with entry of StxA into the cytosol, and modulate the *in vivo* consequences of Stx intoxication in patients infected with a bacterial strain producing both toxins.

Strain distribution of SubAB and identification of allelic variants

PCR screening of strain collections has shown that the *subAB* operon is widely distributed and is present in STEC isolates belonging to over 30 O-serogroups emanating from Australia, Japan, Europe, North America and South America. So far, *subAB* has been detected almost exclusively in LEE-negative STEC, and there appears to be an association between presence of *subAB* and STEC carrying stx_2, or $stx_1 + stx_2$, rather than stx_1 alone (Paton *et al.*, 2004; Paton and Paton, 2005; Izumiya *et al.*, 2006; Khaitan *et al.*, 2007; Cergole-Novella *et al.*, 2008; Karama *et al.*, 2008; Wolfson *et al.*, 2009; Gerhardt *et al.*, 2009; Slanec *et al.*, 2009; Bugarel *et al.*, 2010; Irino *et al.*, 2010;

Tozzoli *et al.*, 2010; Orden *et al.*, 2011). However, given that at least in O113:H21, the megaplasmid that carries the *subAB* operon is capable of conjugative transmission (Srimanote *et al.*, 2002), there is potential for wider dissemination amongst other *E. coli* pathotypes and possibly other Enterobacteriaceae.

Interestingly, SubAB production has now been demonstrated in two *E. coli* strains that do not produce Stx. These strains were isolated from unrelated cases of mild diarrhoea in Italy, and were detected as a consequence of the routine use of Vero cell cytotoxicity assays for diagnosis of STEC infection. Both strains (designated ED 32 and ED 591) induced a CPE resembling that produced by Stx, but showing a different time course. The CPE was not neutralized by antisera raised against the major Stx types, Stx1 and Stx2 and furthermore, PCR analyses showed that the strains were negative for all the known *stx*-related genes. The genetic determinants responsible for the CPE were identified by random transposon mutagenesis. One such mutant had lost cytotoxic activity and sequence analysis subsequently demonstrated that it had a transposon insertion in the *subA* gene (Tozzoli *et al.*, 2010). The nucleotide sequences of the *subA* and *subB* genes of the two *E. coli* strains were identical to each other, but were only 90% identical to the prototype genes encoded on the pO113 plasmid of STEC strain 98NK2, suggesting the existence of an allelic variant of the *subAB* operon (Tozzoli *et al.*, 2010). The presence of *subAB* sequence variants with about 90% identity to the prototype had also been reported amongst STEC isolates from Japan (Izumiya *et al.*, 2006). However, it is unclear whether these are identical to those from Italy, as the sequences themselves are not publicly available.

Given the lack of *stx* genes and the variant *subAB* allele in the Italian strains, ED 32 and ED 591 were tested for the presence of accessory virulence genes, such as *saa*, which is considered a hallmark for the presence of the pO113 plasmid, as well as for genetic markers associated with other *E. coli* pathogroups. However, this yielded negative results for all the genes tested, suggesting that the two *E. coli* strains belonged to a new pathogroup. Moreover, even though both the strains carried a large plasmid, the variant

subAB-operon was located on the chromosome (Tozzoli *et al.*, 2010). Further analysis revealed that it was located next to a gene, termed *tia*, encoding an invasion factor previously described in some ETEC strains (Fleckenstein *et al.*, 1996). The physical association of two chromosomally located virulence-related genes suggested the existence of a pathogenicity island, a mobile genetic element often implicated in the transmission of virulence genes. Thus, the *subAB-tia* locus may represent an alternative genetic vehicle to pO113 for the dissemination of the SubAB-encoding genes. An identical chromosomal region carrying *subAB* and *tia* has also been identified in other *subAB*-positive STEC strains (Tozzoli *et al.*, 2010), and the gene content of the putative PAI, designated Subtilase-Encoding PAI (SE-PAI) has recently been determined (Michelacci *et al.*, 2013). Apart from the *subAB* and *tia* genes, it contains the genetic determinant for an integrase, possibly involved in the PAI mobilization machinery, an unnamed gene encoding a sulfatase and the *shiA* gene, described in strains of *Shigella flexneri* (Ingersoll and Zychlinsky, 2006). The absence of virulence genes associated with other *E. coli* pathotypes and simultaneous presence of three putative virulence determinants in SE-PAI (*subAB*, *shiA* and *tia*) supports the hypothesis that this element contributes to disease pathogenesis in its own right in the non-STEC *E. coli* strains in which it was first described. However, like the prototype *subAB* allele, the SE-PAI has been identified in STEC strains lacking the LEE locus. This finding may indicate that a synergistic action with Stx is required for these strains to cause significant disease and that ED 32 and ED 591 may have lost the Stx-encoding genes at some stage during the infection. Indeed, whereas both *subAB* alleles seem to be well represented in LEE-negative STEC strains causing disease in humans (Michelacci *et al.*, 2013), ED 32 and ED 591 remain the only SE-PAI-containing non-STEC strains reported to date from human cases of diarrhoea.

The finding that the *subAB*-operon seems to be restricted to STEC lacking the LEE locus implies that these strains deploy colonization mechanisms other than the attaching and effacing lesion (see Chapter 10). Interestingly, both the allelic variants of the *subAB* operon are transmitted by MGEs together with determinants encoding known or putative colonization factors. The prototype *subAB*-operon in STEC O113:H21 strain 98NK2 is located on a self-transmissible megaplasmid pO113 together with genes encoding Saa, an auto-agglutinating adhesin belonging to the trimeric autotransporter (TA) family that has been shown to play a role in adherence to HEp-2 cell monolayers (Paton *et al.*, 2001), as well as Sab, another TA family protein required for biofilm formation (Herold *et al.*, 2009). This megaplasmid also carries genes encoding additional putative virulence determinants including the enterohaemolysin EhxA, two distinct extracellular serine proteases EspP and EpeA, and the adherence-promoting ferric siderophore receptor Iha (Newton *et al.*, 2009). On the other hand, the new allelic variant of the *subAB*-operon co-localizes in the SE-PAI with at least two genes whose products may act as colonization factors. The product of the *tia* gene mediates invasion of intestinal epithelial cell lines by ETEC (Fleckenstein *et al.*, 1996) and could act in synergism with the product of the *shiA* gene, apparently involved in the down-regulation of the inflammatory response following infection with strains of *Shigella flexneri*, another pathogen that uses cellular invasion to colonize the gastrointestinal tract of the host.

Most studies investigating the distribution of the *subAB* operon among *E. coli* strains of human and animal origin have used PCR assays that do not distinguish between the two allelic variants, thus hindering the assessment of their relative distribution. However, a recent study found a distinct distribution of the two variants in different animal species (Orden *et al.*, 2011). The variants were distinguished on the basis of the efficiency of the PCR reaction and direct sequencing of the amplicons, with the prototype operon from pO113 and the variant from SE-PAI designated *subAB1* and *subAB2*, respectively. This study showed that the *subAB* genes were present in about 17% and > 90% of LEE-negative STEC isolated from cattle and small ruminants (sheep and goats), respectively. Interestingly, *subAB1* was the principal variant present in bovine strains, while *subAB2* was the most prevalent variant in the strains from small ruminants (Orden *et al.*, 2011). Another recent study used an allele-specific PCR strategy to test a

large panel of LEE-negative STEC strains isolated from human cases of diarrhoea in Denmark (162 strains) and healthy sheep in Spain (108 strains) (Michelacci *et al.*, 2012). The authors found that 72% and 86% of the human and ovine strains, respectively, harboured the *subAB* operon. Interestingly, almost all of them (98%) possessed the *subAB2* allelic variant regardless the source of isolation. Additionally, the *subAB2* genes were always found in association with the *tia* gene, suggesting the presence of the SE-PAI in these strains (Michelacci *et al.*, 2012). These findings strongly support a role for SubAB in the pathogenesis of disease in humans. They also point to a possible zoonotic origin for human infections with small ruminants as the likely animal reservoir of *subAB2*-positive, LEE-negative STEC. Furthermore, the finding that almost all the *subAB* operons identified belonged to the *subAB2* variant indicates that, at least in parts of Europe, the SE-PAI seems to be the prevalent mobile genetic element for transmission of the genes encoding this cytotoxin.

Pathological features and immune modulation

Studies on the *in vivo* effects of SubAB have so far been confined to mouse models. Infection of streptomycin-treated mice with recombinant *E. coli* carrying the *subAB* operon on a low copy-number plasmid did not cause obvious diarrhoea, but the mice lost about 15% of body weight over the following six days. In contrast, mice colonized with a clone expressing the non-toxic mutant $subA_{A272}B$ operon continued to thrive. Interestingly, the mice infected with the active toxin-producing clone appeared to recover and gained weight after day 6, which correlated with sero-conversion against the toxin (Paton *et al.*, 2004). Immunization with purified $SubA_{A272}B$ also protected mice from weight loss induced by subsequent colonization with *E. coli* expressing active SubAB (Talbot *et al.*, 2005). The effects of intraperitoneal injection of purified SubAB, however, were far more dramatic with doses as low as 200 ng being invariably fatal (Paton *et al.*, 2004). Principal pathological manifestations in the toxin-treated mice included microangiopathic haemolytic anaemia, thrombocytopenia and renal impairment, a triad of features that defines Stx-induced HUS in humans (Wang *et al.*, 2007). There was extensive microvascular thrombosis and other histological damage in the brain, kidneys and liver, as well as dramatic splenic atrophy (Fig. 6.1). A more recent study also detected inflammation and severe haemorrhage in the small intestine of SubAB-injected mice (Furukawa *et al.*, 2011).

Interestingly, BiP has been reported to inhibit the activation of tissue factor (TF), the major initiator of extrinsic coagulation. Thus, we hypothesized that the apparent prothrombotic effect of SubAB *in vivo* may involve the stimulation of TF-dependent procoagulant activity (PCA). *In vitro* treatment of human macrophage (U937) cells and primary human umbilical vein endothelial cells (HUVECs) with SubAB (but not SubA$_{A272}$B) significantly increased TF-dependent PCA and induced TF mRNA expression, suggesting that the procoagulant effect of SubAB may be dependent on both the up-regulation of TF expression as well as the activation of TF via BiP cleavage (Wang *et al.*, 2010).

Following IP injection, SubAB bound to murine peritoneal leucocytes (including T and B lymphocytes, neutrophils and macrophages) and elicited profound leukocytosis, which peaked at 24 h. SubAB also increased neutrophil activation in the blood and peritoneal cavity. It also induced a marked redistribution of leucocytes between the three compartments, with increases in leucocyte subpopulations in the blood and peritoneal cavity coinciding with a significant decline in splenic cells (Wang *et al.*, 2011), consistent with the earlier observation of gross splenic atrophy (Wang *et al.*, 2007). SubAB-treatment also elicited a significant increase in the apoptosis rate of CD4+ T cells, B lymphocytes and macrophages (Wang *et al.*, 2011).

Further examples of immune modulation by SubAB include preferential inhibition of secretion of immunoglobulins by activated murine B lymphocytes, whilst leaving cytokine secretion relatively unscathed. SubAB preferentially cleaved newly synthesized BiP in these cells, and the C-terminal BiP fragment remained tightly bound to nascent immunoglobulin light chains, trapping them in the ER compartment (Hu *et al.*, 2009). SubAB also has pro-inflammatory properties and caused transient phosphorylation of Akt and activation of NF-κB in rat renal tubular epithelial cells, which was mediated via the ATF6 branch of the UPR (Yamazaki *et al.*, 2009). Activation of NF-κB is believed to play an important role in HUS and renal injury. However, at subcytotoxic concentrations, SubAB has actually been shown to inhibit LPS-mediated NF-κB activation in a murine macrophage cell line, and to protect mice from LPS-induced endotoxic lethality and experimental arthritis (Harama *et al.*, 2009). The above findings indicate that apart from direct cytotoxic effects, SubAB has more subtle effects on the host, interacting with cellular components of both the innate and adaptive arms of the immune system, with potential consequences for disease pathogenesis.

Role in disease

Although the above findings suggest that SubAB could directly contribute to pathogenesis of disease in humans, direct extrapolation from mice to humans is problematic. The SubB pentamer exhibits a high degree of specificity for receptor glycan structures terminating in α2–3-linked Neu5Gc (Byres *et al.*, 2008). However, unlike most other mammals including the great apes, humans cannot make this sugar due to a mutation in the *cmah* gene, which encodes the CMP-*N*-acetylneuraminic acid hydroxylase that converts CMP-Neu5Ac to CMP-Neu5Gc (Varki, 2001). This suggests that humans might be genetically hyposusceptible to the toxin, since toxin binding would be dependent on lower affinity interactions with Neu5Ac glycans on target cell surfaces. However, humans can assimilate dietary Neu5Gc and incorporate it into glycoconjugates expressed on epithelial and endothelial surfaces (Tangvoranuntakul *et al.*, 2003). This would confer full susceptibility to SubAB, and *in vitro* binding of SubAB to human gut epithelium and microvascular endothelium has been demonstrated (Byres *et al.*, 2008). The *cmah* mutation also means that humans do not express Neu5Gc glycans on glycoproteins present in serum or in intestinal mucus. Paradoxically, these glycoconjugates could have a protective effect in *cmah*-positive species, by competing with otherwise identical glycans displayed

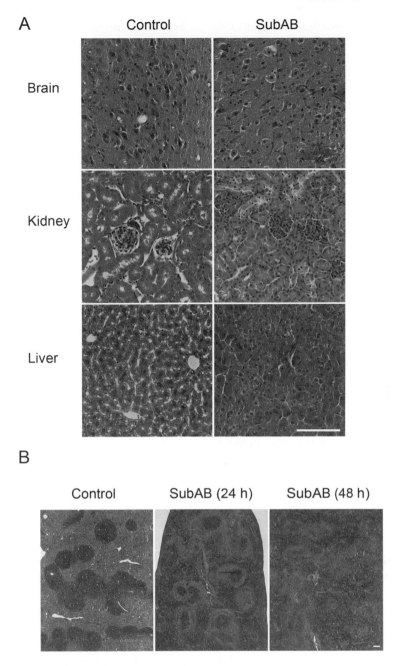

Figure 6.1 SubAB-induced pathology. (A) Haematoxylin–eosin (HE)-stained brain (medulla), kidney and liver sections showing microthrombi and haemorrhage after SubAB injection. Mice were injected intra-peritoneally with 5 μg SubAB or PBS (control) and examined 72 h after treatment (scale bar, 0.1 mm). (B) HE-stained mouse spleen sections showing leucocyte depletion from the white pulp at 24 and 48 h after SubAB injection (scale bar, 0.1 mm). A colour version of this figure is available in the plate section at the back of the book.

on target cell surfaces for binding to SubAB. Indeed, we have shown that chimpanzee serum (which contains Neu5Gc glycoproteins) inhibits binding of SubAB to human renal tissue sections *in vitro*, whereas human serum does not (Byres *et al.*, 2008). Lack of protective Neu5Gc-glyco- proteins in serum is also thought to account for the unexpected susceptibility of *cmah*-knock-out mice to injected SubAB (Byres *et al.*, 2008). Thus, if humans consume Neu5Gc-rich foods, significant levels of the sugar are incorporated into key target tissues, and in the absence of protective Neu5Gc-glycoproteins in serum or mucus, they may actually become hyper- rather than hypo-susceptible to SubAB cytotoxicity. In a further ironic twist, the richest dietary sources of Neu5Gc are red meat and dairy products, and these are also amongst the commonest sources of STEC contamination in the human food chain. Thus, we have a unique paradigm of bacterial pathogenesis, whereby humans may directly contribute to disease through dietary choices, simultaneously exposing themselves to the risk of STEC infection and sensitizing their tissues to SubAB.

Notwithstanding the devastating effect of purified SubAB in mice, the contribution of SubAB to pathogenesis of disease caused by STEC strains that produce it is very difficult to assess. STEC virulence is multifactorial, involving a diverse array of accessory virulence factors encoded on pathogenicity islands or on plasmids. This includes factors required for efficient colonization of the host intestinal tract, possibly in a species-dependent fashion. Unfortunately, although useful for comparing the potency of toxins, existing animal models are generally unsuitable for STEC colonization studies. Human volunteer studies cannot be undertaken, because of the potentially life-threatening nature of STEC disease. Thus, involvement of a given virulence factor in human disease can only be inferred by epidemiological associations, but such studies are complicated by the genetic diversity of STEC strains. To date, *subAB* has been detected almost exclusively in STEC strains that lack the LEE type III secretion system required for generation of attaching effacing lesions, as well as delivery of other potentially deleterious effector proteins into host cells. Thus, LEE-negative STEC strains have been considered to be less virulent than LEE-positive strains such as O157:H7. Nevertheless, many LEE-negative strains are capable of causing severe disease in humans, and production of SubAB may be one of the factors that compensates for the absence of LEE and increases the likelihood of life-threatening complications such as HUS (Wickham *et al.*, 2006). To address the key question, there needs to

be a comprehensive comparison of the frequency of *subAB* in LEE-negative STEC isolates from cases of uncomplicated diarrhoea, haemorrhagic colitis or HUS in humans, using large STEC collections derived from multiple geographical regions. Comparison with isolates from potential environmental sources such as livestock or meat samples may not be helpful in addressing the role of SubAB, as there will be compounding factors, such as linkage with genes required for colonization of the human gut. Analyses also need to take account of the precise sequence subtype(s) of *stx* gene(s) present in the strain, as an association between production of particular Stx2 subtypes and propensity of a given STEC strain to cause HUS has already been reported (Bielaszewská *et al.*, 2006; Friedrich *et al.*, 2002; Persson *et al.*, 2007; Scheutz *et al.*, 2012). Superimposed on this is the recent recognition of two distinct *subAB* alleles linked to MGEs carrying distinct sets of accessory virulence genes. Furthermore, although amino acid sequence differences between SubAB1 and SubAB2 are not in residues known to be required for A subunit catalytic activity or B-subunit receptor interaction (Tozzoli *et al.*, 2010), the effect of these changes on other properties such as intracellular trafficking efficiency or holotoxin stability are not yet known.

Conclusions

A substantial body of knowledge of the biological properties of SubAB and its interactions with host cells has been generated in the 8 years since its initial discovery. These studies have provided two new paradigms in bacterial pathogenesis. SubAB is the first example of a bacterial cytotoxin whose mechanism of action involves covalent modification (in this case proteolytic cleavage) of a host chaperone protein (in this case BiP). It is also the first example of a bacterial toxin whose uptake by host cells (at least in the case of humans) is dependent upon dietary assimilation of a glycan (Neu5Gc) that the host cannot synthesize. The cytotoxic potency of SubAB underscores the critical role that BiP plays in host cell survival. Moreover, the exquisite specificity of the SubA protease for BiP has provided a powerful cell biological tool for examining the role of this essential chaperone and ER stress signalling in diverse cellular processes, including ERAD (Lass *et al.*, 2008), T-cell activation and inflammatory responses of a variety of cell types (Du *et al.*, 2009; Hayakawa *et al.*, 2008, 2009; Okamura *et al.*, 2009; Takano *et al.*, 2007; Huang *et al.*, 2009), ER Ca^{2+} permeability (Schäuble *et al.*, 2012) and multiple stages of virus infection of cells (Buchkovich *et al.*, 2008, 2009; Wati *et al.*, 2009; Goodwin *et al.*, 2011). BiP is also critical for survival of cancer cells and hence appropriately targeted SubA has potential as a cancer therapeutic (Backer *et al.*, 2009; Martin *et al.*, 2011; Ray *et al.*, 2012).

Notwithstanding the above advances, our understanding of the role of SubAB in disease in both humans and animals is limited. The situation is complicated by the fact that with only two exceptions, SubAB has been detected only in strains of *E. coli* that also produce Stx, raising fascinating questions regarding the relative contributions to pathogenesis of these two highly potent AB5 cytotoxins. Extensive epidemiological investigations such as those referred to above are required to determine whether production of SubAB correlates with increased risk of severe complications of STEC disease. Secondly, it is important to determine whether Stx and SubAB act in synergism, or whether they antagonize one another. Both toxins inhibit protein synthesis; Stx acts directly by modification of eukaryotic 28S rRNA, while SubAB acts indirectly through activation of eIF2α, as described above. Conversely, SubAB-mediated BiP cleavage might interfere with retro-translocation of StxA into the cytosol, limiting access to its ribosomal substrate. Both toxins also have distinct effects on inflammatory signalling pathways, which may play a significant role in pathogenesis of STEC disease. Understanding these events may be very important for clinical management of affected patients.

Acknowledgements

Research in the authors' laboratory is supported by the National Health and Medical Research Council of Australia (Program Grant 565526 and Project Grant 1002792) and the Australian Research Council (Discovery Grants DP1095420 and DP120103178). A.W.P. is an Australian Research Council DORA Fellow; J.C.P. is a National Health and Medical Research Council Australia Fellow.

References

Backer, J.M., Krivoshein, A., Hamby, C.V., Pizzonia, J., Gilbert, K., Ray, Y.S., Brand, H., Paton, A.W., Paton, J.C., and Backer, M.V. (2009). Chaperone-targeting cytotoxin and ER stress-inducing drug synergize to kill cancer cells. Neoplasia *11*, 1165–1173.

Beddoe, T., Paton, A.W., Le Nours, J., Rossjohn, J., and Paton, J.C. (2010). Structure, biological functions and applications of the AB₅ toxins. Trends Biochem. Sci. *35*, 411–418.

Bielaszewská, M., Friedrich, A.W., Aldick, T., Schurk-Bulgrin, R., and Karch, H. (2006). Shiga toxin activatable by intestinal mucus in *Escherichia coli* isolated from humans: predictor for a severe clinical outcome. Clin. Infect. Dis. *13*, 1160–1167.

Boyce, M., and Yuan, J. (2006). Cellular response to endoplasmic reticulum stress: A matter of life or death. Cell Death Differ. *13*, 363–373.

Buchkovich, N.J., Maguire, T.G., Yu, Y., Paton, A.W., Paton, J.C., and Alwine, J.C. (2008). Human Cytomegalovirus specifically controls the levels of the endoplasmic reticulum chaperone BiP/GRP78, which is required for virion assembly. J. Virol. *82*, 31–39.

Buchkovich, N.J., Maguire, T.G., Paton, A.W., Paton, J.C., and Alwine, J.C. (2009). The endoplasmic reticulum chaperone BiP/GRP78 is important in the structure and function of the HCMV assembly compartment. J. Virol. *83*, 11421–11428.

Bugarel, M., Beutin, L., Martin, A., Gill, A., and Fach, P. (2010). Micro-array for the identification of Shiga toxin-producing *Escherichia coli* (STEC) seropathotypes associated with Hemorrhagic Colitis and Hemolytic Uremic Syndrome in humans. Int. J. Food Microbiol. *142*, 318–329.

Byres, E., Paton, A.W., Paton, J.C., Löfling, J.C., Smith D.F., Wilce, M.C.J., Talbot, U.M., Chong, D.C., Yu, H., Huang, S., *et al.* (2008). Incorporation of a non–human glycan mediates human susceptibility to a bacterial toxin. Nature *456*, 648–652.

Cergole-Novella, M.C., Nishimura, L.S., dos Santos, L.F., Irino, K., Vaz, T.M.I., Bergamini, A.M.M., and Guth, B.E.C. (2008). Distribution of virulence profiles related to new toxins and putative adhesins in Shiga toxin-producing *Escherichia coli* isolated from diverse sources in Brazil. FEMS Microbiol. Lett. *274*, 329–334.

Chong, D.C., Paton, J.C., Thorpe, C.M., and Paton, A.W. (2008). Clathrin-dependent trafficking of subtilase cytotoxin, a novel AB₅ toxin that targets the endoplasmic reticulum chaperone BiP. Cell. Microbiol. *10*, 795–806.

Du, S., Hiramatsu, N., Hayakawa, K., Kasai, A., Okamura, M., Huang, T., Yao, J., Takeda, M., Araki, I., Sawada, N., Paton, A.W., Paton, J.C., and Kitamura, M. (2009). Suppression of NF-κB by cyclosporin A and tacrolimus (FK506) via induction of the C/EBP family: Implication for unfolded protein response. J. Immunol. *182*, 7201–7211.

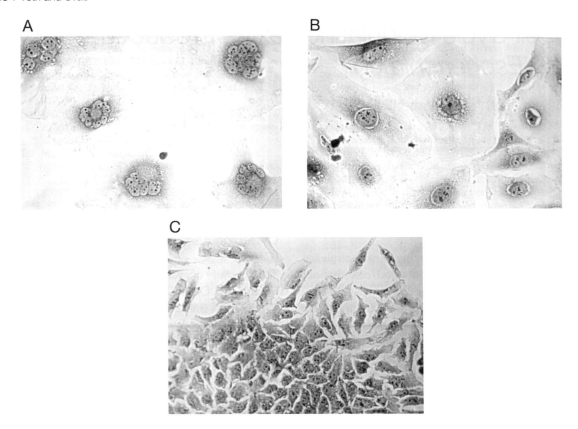

Figure 7.1 CNF- and CDT-specific cytopathic effects on HeLa cells. HeLa cell monolayers were infected with (A) CNF-1 and (B) CDT-IV-producing *E. coli* strains. (C) Untreated HeLa cells served as control. After 3 days' incubation, infecting material was removed, HeLa cells were washed and stained with Giemsa. Morphological changes were investigated by light microscope. The original magnification was 40. Photographs were taken by I. Tóth. A colour version of this figure is available in the plate section at the back of the book.

in rabbits causes skin necrosis and subcutaneous inoculation into the abdomen of guinea pigs and leads to widespread haemorrhage (Caprioli *et al.*, 1983; De Rycke *et al.*, 1999).

CNF also induces an extensive reorganization of the cytoskeleton characterized by irreversible formation of actin stress fibres and focal adhesions (Fiorentini *et al.*, 1988; Oswald *et al.*, 1994). *In vitro* experiments have shown that the effect of CNF on the cell cytoskeleton may trigger the entry of non-invasive bacteria into the epithelial cells (Falzano *et al.*, 1993).

Experimental infections also support the role of CNFs as a virulence factor in human and animal infections. Infections of neonatal calves (Smith, 1974) and neonatal pigs (Wray *et al.*, 1993) have shown that orally inoculated CNF-producing *E. coli* caused septicaemia, enteritis as well as

histological changes characteristic of toxaemic effects in the brain, heart, liver, and kidney. Similar lesions were reported after intravenous inoculation of purified CNF1 in lambs (De Rycke and Plassiart, 1990). CNF1- and CNF2-producing strains are involved in extraintestinal human infections (Caprioli *et al.*, 1987) and intestinal colibacillosis (Caprioli *et al.*, 1983; Blanco *et al.*, 1992), but their phenotypes and necrotic effects are different. CNF1 cytotoxin causes extensive multinucleation and enlargement of cells in HeLa cell culture assays, moderate necrosis in the rabbit skin test, and absence of necrosis in the mouse footpad test. CNF2 cytotoxin triggers moderate multinucleation and elongation of HeLa cells, and intense necrotic response in both the rabbit skin test and the mouse footpad test (De Rycke *et al.*, 1990).

Cytoskeleton alteration is regulated by the small Rho GTPase proteins (Oswald et al., 1994; Fiorentini et al., 1995). CNF1 is the paradigm of bacterial toxins activating host Rho GTPases. CNF catalyses the permanent activation of Rho proteins by post-translational modification of GTPases (reviewed by Lemonnier et al., 2007). The activated Rho GTPases regulate a great variety of signal transduction cascades associated with fundamental cellular processes, first of all with cell-cycle progression and actin and microtubular dynamics (reviewed by Fabbri et al., 2010).

Epidemiology of CNF-producing *E. coli*

After the initial discovery of CNF (CNF1), De Rycke et al. (1987) identified from the faeces of diarrhoeic calves another serologically related necrotoxin, CNF2. CNF2 was first called Vir-cytotoxin (Oswald et al., 1989), since the toxic effect was encoded with one of the previously described plasmids (Oswald et al., 1994), associated with the virulence of septicaemic *E. coli* strains (Smith, 1974; Smith and Huggins, 1976). *E. coli* strains producing CNF2 (NTEC-2) and containing the Vir plasmid have been involved in diseases of cattle, dairy cows, and lambs, and they are found in the normal faecal flora of calves (Orden et al., 1999; Kadhum et al., 2008). Recently, Orden et al. (2007) have identified CNF3-producing *E. coli* strains in small ruminants.

In addition, CNF1-producing *E. coli* was first reported from infant enteritis cases, and CNF1 is produced by a large proportion of *E. coli* causing extraintestinal infections in humans (Caprioli et al., 1987; Blanco et al., 1992) as well as in septicaemic and diarrhoeic pigs (Dozois et al., 1997; Blanco et al., 1988, Tóth et al., 2000).

Strains producing CNF2 have been isolated from diarrhoeal, septic and healthy calves (Oswald et al., 1991; Orden et al., 1999, 2002; Osek, 2001). On the other hand, CNF3 strains are frequently recognized from small ruminants: from a representative percentage of healthy lambs, and sporadically from healthy adult sheep and goat kids as well as from diarrhoeic calves, lamb and goat kids (Orden et al., 2007).

In NTEC strains, CNFs are closely associated with further specific virulence genes. Accordingly,

CNF1 is associated with alpha-haemolysin (Caprioli et al., 1987) and P fimbriae, which are located on a pathogenicity island (PAI), termed PAI II$_{J96}$ (Blum et al., 1995). CNF is also associated with cytolethal distending toxin (CDT) types I and IV (Tóth et al., 2003), as well as with P, S fimbriae encoding genes (Tóth et al., 2000; Salvarani et al., 2012). Accumulation of an exceptional repertoire of extraintestinal pathogenic virulence-associated genes including at least 1 toxin gene with carcinogenic potential (colibactin, cytolethal distending toxins, or cytotoxic necrotizing factor) was detected in *E. coli* strains from human prostatitis (Krieger et al., 2011). In the same way as in human isolates, a great number of virulence genes were detected in UPEC strains isolated from dogs and cats as well. The high correlation between *cnf1*, *hlyA*, *sfa*, and pyelonephritis confirms the presence of pathogenicity islands (Hacker et al., 1997) in these strains (Tramuta et al., 2011). The ExPEC isolated from prostatitis and the isolates from animal pyelonephritis uniformly belong to phylogenetic group B2 (Krieger et al., 2011; Tramuta et al., 2011). Also, a pathogenicity island II$_{J96}$-like domain (Blum et al., 1995; Bingen-Bidois et al., 2002) was identified in a life-threatening 'flesh-eating' strain of *E. coli* (Grimaldi et al., 2010), which is a typical trait of isolates from urinary tract infection and neonatal meningitis (Houdouin et al., 2002).

The association of CNF2 with F17-b fimbriae (Oswald et al., 1991) and CDT-III (Pérès et al., 1997) has been known since the discovery of CNF2, but the complete nucleotide sequence analysis of Vir plasmid from a bovine NTEC2 strain revealed the co-existence of further virulence factors (Johnson et al., 2010). The sequenced pVir plasmid (pVir68) of 138.7 kb length belongs to the RepFIB and RepFIIA replicon types, and within this plasmid an approximately 60 kb PAI was identified. The pVir68 PAI harbours multiple virulence factors within distinct genetic regions of lower G + C content bounded by inverted repeats. The pVir68 PAI encodes putative virulence factors, including F17b fimbrial genes, genes of a novel fimbrial operon, *tibAC*, haemolysins, and the *cnf2* and *cdt* toxin-encoding genes. Intensive genetic screening of 96 CNF-2-positive strains for the virulence- and replication-associated genes of the pVir plasmid revealed that the most prevalent genes among this collection included *repA* (RepFIB), *cnf2*, an *ompP* homologue, and the *tib-AC* genes encoding for aggregation and biofilm formation (Johnson *et al.*, 2010). Interestingly, pVir68 harbours only part of the *hlyCABD* operon (Emődy *et al.*, 2003); however, the analysis of haemolytic abilities of the DH10:pVir transconjugant indicated that pVir68 encodes partial haemolytic activity towards sheep red blood cells (Johnson *et al.*, 2010).

CNF3-positive *E. coli* strains uniformly have *eae* and most of them carry *ehxA* (Orden *et al.*, 2007) while none of them are positive for the above-mentioned virulence markers associated with CNF1 or CNF2. The most prevalent serotypes of *cnf3* NETEC strains are O4:H11, O153:H11 and O177:H11, which are not represented between CNF1 and CNF2 strains. Interestingly, among the serotyped *cnf3*, *eae*, *enhxA* strains typical human EPEC and EHEC serotypes of O26:NM and O111:NM are also identified. The association of *cnf3* with beta-1 type of *eae* and *ehxA* as well as their presence in typical EPEC and EHEC serotypes suggests that CNF3 NTEC strains might have zoonotic potential (Orden *et al.*, 2007).

Genetics and structure of CNF

The family of CNFs consists of structurally and functionally similar protein toxins. CNFs are synthesized as a single-chain polypeptide organized into three functional domains (Lemichez *et al.*, 1997).

CNF1 and CNF3 are encoded by chromosomal genes, CNF2 via the Vir plasmid. The ORF encoding CNF1, *cnf1*, exhibits an overall G + C content (36.5%) lower than that of the *E. coli* genome, suggesting that this determinant could have only recently come to reside in the *E. coli* chromosome. The ORF corresponds to a protein of 1,014 amino acids with a predicted molecular mass of 113.7 kDa (Falbo *et al.*, 1993). CNF2 is a plasmid-encoded 114.7 kDa protein (De Rycke *et al.*, 1990; Oswald *et al.*, 1994) and the most recent sequence analysis of the Vir plasmid revealed that *cnf2* is part of a pathogenicity island (Johnson *et al.*, 2010). Similarly, the *cnf1* gene together with the *hlyCABD* operon is also located on a PAI in the chromosome of uropathogenic *E. coli* strains

(Falbo *et al.*, 1992; Blum *et al.*, 1995). The *cnf3* gene corresponds to 1013 amino acids and the calculated CNF3 mass is 114.5 KDa. Analysis of the *cnf3* locus revealed the presence of mobile genetic elements, suggesting the mobility of *cnf3*. The low G + C content also suggests that *cnf3* was acquired by horizontal gene transfer (Orden *et al.*, 2007). CNF1 and CNF2 share 85% identical residues over the whole proteins (Falbo *et al.*, 1993; Oswald *et al.*, 1994). The predicted CNF3 is highly similar to CNF1 and CNF2 (70.1 and 69.9% identity, respectively; Orden *et al.*, 2007). The homology between CNF1, 2 and 3 proteins is evenly distributed through the entire sequence. *E. coli* CNFs share two variable regions in positions 411–427 and 565–568 amino acids, respectively. Furthermore, CNF3 shows four exclusive regions of dissimilarity, but the cysteine and histidine residues essential for the biological activity (Schmidt *et al.*, 1998) are conserved in the three CNFs (Orden *et al.*, 2007).

CNF1 is the best characterized member of the dermonecrotic toxins. It is organized into three functional domains: the N-terminal part containing the cell-binding domain, a putative central membrane-spanning region, and a C-terminal catalytic region (Lemichez *et al.*, 1997). The binding region is restricted to the first 190 amino acids, and hydrophilic amino acids 53–75 are strictly necessary for cell receptor recognition. Additionally, the first 48 codons of *cnf1* are involved in the translational regulation of CNF1 synthesis. This regulation consists of both a positive (codons 6–20) and a negative (codons 45–48) control (Fabbri *et al.*, 1999). CNF1 exhibits a translocation domain containing two predicted hydrophobic helices (H1–2) (aa 350–412) separated by a short peptidic loop (aa 373–386) with acidic residues. The catalytic domain corresponds to amino acids 720 to 1014 (Lemichez *et al.*, 1997). Functional (Schmidt *et al.*, 1998) and crystal structure (Buetow *et al.*, 2001) analyses revealed that CNF1 has a Val833-Cys866-His881 catalytic triad reminiscent of enzymes belonging to the catalytic triad superfamily.

There is no evidence for a bacterial signal peptide sequence at any of the potential start codons, which agrees with the lack of toxin secretion reported for CNF1-producing strains (Caprioli *et* al., 1987; De Rycke et al., 1989). However, it has been reported that CNF1 was released and transferred from an UPEC strain in a complex with outer membrane vesicles (OMVs). CNF1-bearing vesicles kept the biological activity and attenuated the phagocyte function of polynuclear leucocytes (Davis et al., 2005).

CNF-triggered cellular and molecular effects

CNFs bind to their cell-surface receptors, are internalized by endocytosis and reach the endosomal compartment. The acidic conditions allow the transfer of the catalytic domain of the necrotoxins into the cytosol where it deamidates a glutamine residue of small GTPases of the Rho proteins (glutamine 63 of Rho and glutamine 61 of Rac and Cdc42) into glutamic acid (Flatau et al., 1997; Schmidt et al., 1997; Lerm et al., 1999). CNF1 is the first bacterial toxin described that uses both a clathrin-independent endocytic mechanism and an acidic-dependent membrane translocation step in its delivery of the catalytic domain to the cell cytosol (Contamin et al., 2000).

These modifications of glutamine 63/61 trigger permanent activation of Rho proteins (reviewed by Nougayrède et al., 2005; Lemonnier et al., 2007). Rho GTPases are molecular switches that control a wide variety of signal transduction pathways in all eukaryotic cells including cytoskeletal dynamics, cell proliferation, shape, motility, transformation, apoptosis, and oncogenesis (Etienne-Manneville and Hall, 2002).

The Rho GTPases are positive regulators of cell cycle progression. Rho GTPases contribute to G1 progression mainly through cyclin D1 up-regulation and through down-regulation of the cycle-dependent kinase inhibitor p21[cip1] (Assoian and Schwartz, 2001). At the same time, CNF1-associated mitotic block in HeLa cells and delayed lethality in the uroepithelial cell line T24 were also reported (DeRycke et al., 1996; Falzano et al., 2006). CNF1 induced the accumulation of T24 cells in the G2/M phase by sequestering cyclin B1 in the cytoplasm and down-regulating its expression (Falzano et al., 2006).

CNF1 promotes transcription and release of tumour necrosis factor alpha, gamma interferon, interleukin-6 (IL-6), and IL-8 proinflammatory cytokines, and increases the production of reactive oxygen species (ROS) in uroepithelial T24 cells (Falzano et al., 2003).

Activation of Rho proteins is followed by their ubiquitination and proteasome-mediated degradation. Degradation of Rho proteins dampens the inflammatory responses resulting from Rho protein activation and could accentuate the invasive capacities of uropathogenic E. coli producing this toxin (Doye et al., 2002; Munro et al., 2004). At the same time, CNF1 could protect host cells from apoptosis. CNF1-induced up-regulation of signal transduction leads to an increased level of anti-apoptotic Bcl-2 proteins leading to mitochondrial homeostasis (Fiorentini et al., 1998).

CNF as a virulence factor

In addition to epidemiological data and the historic experimental infection studies mentioned above, in vivo studies conducted with isogenic pairs underline the virulence feature of CNF.

Infection of rat prostates with CNF1-positive UPEC strain caused more inflammation-mediated morphological and histological tissue damage than did infection with isogenic CNF1-negative mutants (Rippere-Lampe et al., 2001a). Similarly, CNF1 proved to be a virulence factor in a murine model as well. All the investigated wild-type strains colonized the bladder and kidney better than did the CNF1-negative mutants, and the CNF1-positive strains consistently induced deeper and more extensive inflammation than the isogenic mutants. A CNF1-positive strain showed increased resistance to neutrophils, which in turn permits these bacteria to gain access to deeper tissues and persist better in the lower urinary tract (Rippere-Lampe et al., 2001b).

Besides inflammatory responses and delayed neutrophil transmigration (Hofman et al., 1998), CNF1 also modulates the immune response. In mice co-fed toxin and the soluble protein antigen ovalbumin (OVA), CNF1 elicited both systemic and mucosal adjuvanticity anti-OVA responses. In contrast, the catalytic inactive mutant CNF1-Cys866Ser demonstrated no effects (Munro et al., 2005). Dermonecrotic toxin from Bordetella has the same adjuvant property that is also related to its catalytic domain.

It is important that CNF1 expression partially

protects the bacterium against the antimicrobial properties of polymorphonuclear leucocytes (PMNs). Accordingly, CNF1-positive UPEC down-regulates phagocytosis, limits the release of reactive oxygen species, and therefore allows the bacteria to survive the acute inflammatory response and facilitates their enhanced replication (Davis *et al.*, 2005).

Cytolethal distending toxins

Discovery and main characteristics

Cytolethal distending toxin (CDT) was the first inhibitory cyclomodulin described on the basis of a novel phenotype observed in tissue cultures infected with pathogenic *E. coli* strains isolated from children with diarrhoea (Johnson and Lior, 1988; Fig. 7.1). Since then five genetic variants have been identified in *E. coli*, and CDTs have been found in several Gram-negative pathogenic bacteria. CDT is a tripartite AB_2-type holotoxin, the active subunit protein (CdtB) causes DNA damage in the target cells, leading to a characteristic distended morphology and cell cycle arrest in the G2/M phase in a broad range of mammalian cell lines. CDT might have a role of slowing the regenerative processes, and modulating the host's immune response by inhibiting the proliferation of rapidly growing cell populations.

Nomenclature

CDT was initially described in *E. coli*, but because of the existence of numerous variants and its wide dissemination in multiple species (*Escherichia albertii; Campylobacter jejuni, C. lari, C. fetus, C. upsaliensis, C. coli; Haemophilus ducreyi, H. parasuis; Aggregatibacter actinomycetemcomitans; Helicobacter hepaticus, H. bilis, H. mastomyrinus, H. cinaedi, H. canis, H. pullorum, H. winghamensis; Shigella boydii, S. dysenteriae; Salmonella enterica* serovar Typhi, reviewed by Jinadasa *et al.*, 2011) it is necessary to use a nomenclature. In this chapter we will use the most recent one established by Jinadasa *et al.* (2011). Accordingly, the host organism is designated by the first letter of its genus name and the first three letters of its species name. Therefore, the CDT produced by *E. coli* is designated as

EcolCDT in this system, AactCDT stands for the CDT produced by *Aggregatibacter actinomycetemcomitans*, etc. As in this chapter we will mainly overview EcolCDT, in the following section the genetic variants of this toxin will be referred to simply as CDT-I to CDT-V, respectively.

Epidemiology of CDT-producing *E. coli*

Since the discovery of cytolethal-distending toxin, EcolCDTs have been detected in a high number of pathogenic *E. coli* strains, and so far five genetic variants have been identified: CDT-I (Scott and Kaper, 1994), CDT-II (Pickett *et al.*, 1994), CDT-III (Pérès *et al.*, 1997), CDT-IV (Tóth *et al.*, 2003), and CDT-V (Janka *et al.*, 2003). All five CDT variants have been reported from strains of various sero-pathotypes. However, only a few stronger associations between the different genetic variants and the sero- or pathotypes of the host strains have been established.

Five types of cytolethal distending toxin (CDT-I to CDT-V) have been identified in *E. coli* strains originating from intestinal and extraintestinal infections of humans as well as different animal species. CDT-I and CDT-II have been detected in *E. coli* strains that mainly cause human gastrointestinal infections. CDT-III and CDT-IV were identified in human- and animal-pathogenic *E. coli* strains of intestinal and extraintestinal origin, and these strains frequently produced cytotoxic necrotizing factors CNF2 and CNF1, respectively, as well as haemolysin. CDT-V was detected in EHEC *E. coli* O157 isolates and non-O157 STEC strains from human patients as well as in atypical bovine O157 strains.

CDT-producing *E. coli* strains represent various pathotypes, such as STEC/EHEC (Bielaszewska *et al.*, 2004; Orth *et al.*, 2006), EPEC (Pérès *et al.*, 1997; Tóth *et al.*, 2003), UPEC (Tóth *et al.*, 2003; Dubois *et al.*, 2010), NTEC (CNF 2 Van Bost *et al.*, 2003), NMEC (Johnson *et al.*, 2002), ExPEC (Tóth *et al.*, 2003, Bielaszewska *et al.*, 2004) and also a number of strains which do not fit any of the established pathotypes (Hinenoya *et al.*, 2009; Tóth *et al.*, 2009b). While the majority of strains investigated in these works originate from human patients, strains of veterinary significance also produce CDT, as the toxin has also been detected from septicaemic cattle (Van Bost *et al.*, 2003; Tóth *et al.*, 2003; Ghanbarpour and Oswald, 2009), pork and lamb (Tóth *et al.*, 2003).

CDT-I is associated mainly with the EPEC pathotype, particularly with the O127:H7 and O86a:H34 serotypes (Pandey *et al.*, 2003; Asakura *et al.*, 2007; Kim *et al.*, 2009), although UPEC (Tóth *et al.*, 2003; Dubois *et al.*, 2010) and STEC (Bouzari *et al.*, 2005) strains were also found to produce this variant. Similarly, CDT-II strains have been detected in *E. coli* strains that mainly cause human gastrointestinal infections (Pickett *et al.*, 1994; Hinenoya *et al.*, 2009).

CDT-III is characteristic of pathogenic strains isolated from bovine source, particularly of the O127 serogroup (Pérès *et al.*, 1997; Clark *et al.*, 2002; Tóth *et al.*, 2003; Orth *et al.*, 2006; Borriello *et al.*, 2012) and its presence is associated with the gene encoding the CNF2 (Van Bost *et al.*, 2003; Mainil *et al.*, 2003; Borriello *et al.*, 2012) being both harboured on the same plasmid (Pérès *et al.*, 1997; Johnson *et al.*, 2010). CDT-IV was identified in human- and animal-pathogenic *E. coli* strains of intestinal and extraintestinal origin, and these strains frequently produced cytotoxic necrotizing factor 1 (CNF1) as well as haemolysin (Tóth *et al.*, 2003). CDT-V has been detected in EHEC *E. coli* O157 isolates and non-O157 STEC strains from human patients as well as in atypical bovine O157 strains. CDT-V is frequently associated with sorbitol-fermenting (SF) O157:NM STEC strains (Janka *et al.*, 2003; Bielaszewska *et al.*, 2004; Osek, 2005; Kim *et al.*, 2009; Wu *et al.*, 2010), and while this variant has also been detected in various strains of other sero- and pathotypes (Tóth *et al.*, 2009b; Hinenoya *et al.*, 2009) the SF O157:NM STEC strains so far did not harbour any other variant of CDT. CDT-V in O157:H7 strains is strongly, but not exclusively associated with phage types 14 and 34 (Friedrich *et al.*, 2006). O91:H21 STEC strains harboured CDT-V exclusively and with high prevalence (70%, Bielaszewska *et al.*, 2009).

There are some cases where CDT is the only established virulence factor of the given strain, which has led to the introduction of the term 'cytolethal distending toxin producing *E. coli*',

abbreviated as CLDTEC (Okeke *et al.*, 2000) or CTEC (Hinenoya *et al.*, 2009). A common characteristic of the CTEC strains in these studies was that they were all isolated from children with diarrhoea, typically 5 years of age or younger (Okeke *et al.*, 2000; Hinenoya *et al.*, 2009).

CDT genes and proteins

CDT is a tripartite AB_2-type holotoxin, where CdtB is the active subunit and the two other subunits, CdtA and CdtC are required for the transport and binding of the active subunit to the target cell (Lara-Tejero and Galán, 2001). The subunit proteins are encoded by the *cdtABC* operon, which consists of the three adjacent, slightly overlapping genes encoding the three subunits. In the case of EcolCDT, the length of *cdtA*, *cdtB* and *cdtC* varies between 714–777, 810–822, and 546–573 bp, respectively (GenBank accession numbers: *cdt-I* U03293.1, *cdt-II* U04208.1, *cdt-III* U89305.1, *cdt-IV* AY578329.1, *cdt-V* AJ508930.1) and their encoded proteins are of 27, 29 and 20 kDa, respectively (Scott and Kaper, 1994).

Sequence analysis revealed the homology of CdtB with mammalian DNAse I (Elwell and Dreyfus, 2000), and its DNAse activity was shown by multiple studies (Lee *et al.*, 2003; Haghjoo, 2004). The resolution of the crystal structure of AactCDT has revealed that CdtB shares conserved residues of the active site of the mammalian DNAse I (Nesić *et al.*, 2004).

The sequence of CdtA contains two characteristic conserved regions. One of them is a large aromatic cluster of eight side chains, which is thought to aid the binding of the toxin to the cell surface. The other conserved region is the interaction surface with CdtC (Nesić *et al.*, 2004). Furthermore, in AactCdtA (Mayer *et al.*, 1999), a signal peptidase II recognition site was found, which is a characteristic motif for lipoproteins. This suggests that CdtA anchors to the bacterial outer membrane (Heywood *et al.*, 2005).

In the case of CDT produced by *Haemophilus ducreyi* (HducCDT), the N terminus of CdtC blocks the active site of CdtB, which is possibly a self-regulatory mechanism by which the DNAse activity of CdtB is probably restricted until it has reached its target (Nesić *et al.*, 2004).

Pathogenesis: *in vitro* and *in vivo* models

As CDT is the first described inhibitory cyclo-modulin, its mode of action at the cellular and molecular level has been extensively studied, and also *in vivo* models have been established to understand its role in pathogenesis (reviewed in Ge *et al.*, 2008).

Pandey *et al.* (2003) tested the effective titre of CDT on tissue cultures, and found higher titres in *E. coli* strains originating from patients with more severe symptoms (bloody diarrhoea), suggesting the contribution of the toxin to pathogenesis. The characteristic distended morphology and the cell cycle arrest at the G2/M phase were demonstrated on multiple human endothelial cell lines for CDT-I (Comayras *et al.*, 1997; Sert *et al.*, 1999). Furthermore, the bacteria may use the toxin to block the proliferation of the crypt cells, aiding the colonization of the intestinal epithelium by the CDT-producing *E. coli* (Comayras *et al.*, 1997). Similar effects were demonstrated on various cell lines in the case of CDT-III (Pérès *et al.*, 1997), CDT-IV (Tóth *et al.*, 2009a), and CDT-V (Bielaszewska *et al.*, 2005).

While the effects of CDT have been studied extensively on the cellular level *in vitro*, currently there are very few data regarding the potential or actual role of CDT in the pathogenesis of *E. coli* strains. In a review, Heywood *et al.* (2005) proposed that CDT-producing *E. coli* might be opportunistic pathogens, and the effect of CDT does not directly support the intestinal infection. In the case of pathogenic *E. coli*, it is possible that the effects of other virulence factors make it difficult to observe the *in vivo* effects of CDT, or its role in the pathogenic process (Oswald *et al.*, 2005). However, the fact that CDT has been detected in a wide range of pathogenic strains isolated from patients with serious symptoms, underlines the pathogenic potential of this toxin.

Currently there are few data regarding the role of CDT in the acute phase of infection, and studies so far show that the toxin does not play a significant role in these models (Lewis *et al.*, 2001; Young *et al.*, 2001; Van Bost *et al.*, 2003). There are, however, several animal models, which demonstrate that CDT has a significant role in chronic infections. CjejCDT enhances bacterial invasion in severe combined immunodeficient mice (Purdy *et al.*, 2000), in mice deficient in necrosis factor-κB complex (Fox *et al.*, 2004), as well as in mucin-deficient mice (McAuley *et al.*, 2007). In these chronic models CDT promotes pro-inflammatory response. The inflammatory response serves the host's defence against the invading bacteria but at the same time it could increase shedding and therefore promote the spreading of the pathogen (Jinadasa *et al.*, 2011). The persistence of CDT-associated chronic inflammation in this case might also be associated with potential carcinogenesis (Guerra *et al.*, 2011).

Studies have revealed that tissue cultures of haematopoietic origin are more susceptible to CDT than those of mesenchymal origin (Cortes-Bratti *et al.*, 2001; Shenker *et al.*, 2001). CDT causes apoptosis more rapidly (between 24 and 72 h) in cell lines of haematopoietic origin as opposed to over 96 h in mesenchymal cell lines (Cortes-Bratti *et al.*, 2001; Shenker *et al.*, 2001). Caspase-dependent apoptosis induced by CDT could be observed in the case of Jurkat and MOULT-4 cell lines intoxicated with AactCDT (Shenker *et al.*, 2001; Ohara *et al.*, 2004). U937 monocytic cells (Rabin *et al.*, 2009) and immature dendritic cells (Li *et al.*, 2002) are also reported to be highly susceptible to AactCDT. This higher susceptibility observed in cell lines of haematopoietic origin, more specifically in T and B lymphocytes (Gelfanova *et al.*, 1999; Svensson *et al.*, 2001), led to the suggestion that by inhibiting cell proliferation, CDT may have its most severe effect on rapidly growing cell populations (Heywood *et al.*, 2005; Guerra *et al.*, 2011).

CDT may have an immunomodulatory role as well, particularly in the case of chancroid caused by *H. ducreyi* and in periodontitis caused by *A. actinomycetemcomitans*. HducCDT, besides inducing apoptosis in B lymphocytes and dendritic cells (Gelfanova *et al.*, 1999; Cortes-Bratti *et al.*, 2001; Xu *et al.*, 2004), also inhibits proliferation and interferon-γ secretion in T-lymphocytes (Svensson *et al.*, 2001).

The inhibition of cell proliferation can be utilized differently by the diverse CDT-producing pathogenic bacteria (Heywood *et al.*, 2005; Oswald *et al.*, 2005). In the case of *Campylobacter*-related enterocolitis, CDT seems to aid the invasion of the host organism by the bacteria (Purdy *et al.*, 2000; Fox *et al.*, 2004; McAuley *et al.*, 2007), while in the case of *H. ducreyi*, HducCDT seems to have a role in aggravating chancroid, by dampening the immune response (Xu *et al.*, 2004) and slowing the regenerative processes (Ohara *et al.*, 2011). The possible immunomodulatory role of *A. actinomycetemcomitans* could have similar functions in the case of periodontitis (reviewed by Guerra *et al.*, 2011).

Mode of action

The specific receptor(s) of CDTs could not be identified to date. This can be due to the finding that several CDTs are species and type specific concerning the factors that they interact with and the cell lines which are susceptible to them (Eshraghi *et al.*, 2010). A few key molecules that can play a role in the binding of CDT to the target cell have been identified in recent works. Fucose may have a role in binding CDT-II to the target cell (McSweeney and Dreyfus, 2005), and it has also been shown that this CDT variant is secreted in outer membrane vesicles (OMVs, Berlanda Scorza *et al.*, 2008). The binding of CDT-III to Chinese hamster ovary (CHO) cells is cholesterol dependent, while fucose and glycosphingolipids also play a role in the binding of CDT-III to this cell line (Eshraghi *et al.*, 2010). The binding of AactCDT and HducCDT to the cell surface requires intact lipid rafts (Guerra *et al.*, 2005; Boesze-Battaglia *et al.*, 2006). Haploid human cell lines with sphingomyelin or transmembrane protein 181 (TMEM181) deficiency are resistant to CDT. This finding suggests that TMEM181 may have a role in the internalization of the toxin complex (Carette *et al.*, 2009).

CdtB is retrogradely transported to the nucleus through the Golgi apparatus and the endoplasmic reticulum (Cortes-Bratti *et al.*, 2000; Guerra *et al.*, 2005, 2009), aided by the nuclear localization signal sequences found at the carboxy-terminal region of CdtB-II (Nishikubo *et al.*, 2003; McSweeney and Dreyfus, 2004).

Once CdtB reaches the nucleus of the target cell, it causes double-strand breaks in the DNA, which in turn cause the intoxicated cells to activate the full pattern of DNA damage responses

(DDRs). Human epithelial and mesenchymal cells intoxicated with CDT are arrested in the G2 phases of the cell cycle, which results in 4N DNA content which has been shown by flow cytometry in several works (Pérès *et al.*, 1997; Comayras *et al.*, 1997; Cortes-Bratti *et al.*, 2001; Bielaszewska *et al.*, 2005; Tóth *et al.*, 2009a). The cell cycle arrest caused by CDT is characterized by the activation of the ataxia telangiectasia mutated (ATM) kinase (Cortes-Bratti *et al.*, 2001) together with the proto-oncogene c-Myc (Guerra *et al.*, 2010) and, in turn, the activation of tumour suppressor p53, which promotes the transcription of the cyclin-dependent kinase inhibitor p21. The activated ATM phosphorylates the checkpoint kinase 2 (Chk2), which in turn phosphorylates cdc25, making it unable to dephosphorylate the cyclin B–cyclin-dependent kinase 1 (CDK1) complex. This cascade leads to cell cycle arrest at the G2 phase (Comayras *et al.*, 1997; Sert *et al.*, 1999; Cortes-Bratti *et al.*, 2001; Guerra *et al.*, 2010).

Through the activation of ATM, CDT also causes the appearance of actin stress fibres in the surviving intoxicated cells (Cortes-Bratti *et al.*, 1999; Frisan *et al.*, 2003) due to Net1-mediated activation of RhoA, which is a key factor in the assembly of stress fibres and cell survival (Frisan *et al.*, 2003; Guerra *et al.*, 2008). The ATM kinase also phosphorylates the histone, which accumulates at the site of DNA double-strand breaks (DSB), as they have a role in repairing damaged DNA. The accumulation of H2AX was used to demonstrate CDT activity in several studies (Hassane *et al.*, 2003; Bielaszewska *et al.*, 2005; Tóth *et al.*, 2009a; Liyanage *et al.*, 2010).

The cell cycle of fibroblast cells is characteristically arrested in the G1 phase by CDT (Hassane *et al.*, 2003; Belibasakis *et al.*, 2004; Wising *et al.*, 2005). This stoppage is caused by the dephosphorylation of the cyclin E–CDK2 complex by p21 (reviewed by Ge *et al.*, 2008).

There are exceptions to the tripartite toxin structure, the most notable being StypCdtB, encoded by *S. enterica* serovar Typhi, where only a *cdtB* gene could be found (Parkhill *et al.*, 2001). In this species, the products of two other genes from the same genomic island, *pltB* (pertussis-like toxin B) and *pltA* (pertussis-like toxin A) are likely required to transport CdtB out to the extracellular medium (Spanò *et al.*, 2008). Another example for a sole *cdtB* without the genes *cdtA* and *cdtC* was found in a bacteriophage that infects *Candidatus* Hamiltonella defensa, a mutualistic endosymbiont of aphids (Degnan and Moran, 2008). The function and potential role of the toxin in this bacterium have not yet been tested. It has been reported that in the case of AactCDT, an incomplete complex lacking CdtA is also capable of causing cell-cycle arrest (Akifusa *et al.*, 2001), which was also observed in HducCDT (Deng *et al.*, 2001), while on the other hand, an AactCDT complex lacking CdtC also retained some of its toxicity (Saiki *et al.*, 2001).

Genetics and evolutionary aspects of EcolCDT

Sequence analysis of individual *cdt* genes indicates that genetic variants of EcolCDT fall in two clusters, with CDT-I and IV being in one cluster, and CDT-II, III and V forming another group (Tóth *et al.*, 2009a). The sequence of *cdtB* proved to be remarkably more conserved than the sequences of the two genes encoding the other subunits, which suggests selective pressure for the structure of the active subunit, indicating its importance as a virulence factor. So far all the *cdt* operons are either part of mobile genetic elements or associated to them.

The *cdt-I* operon in O127:H7 and O142:H6 human EPEC strains is located in the genome of a 60 kb long inducible converting lambdoid prophage. The phage also carries a truncated version of another virulence gene, the cell cycle inhibiting factor *(cif)*, in its virulence module (Asakura *et al.*, 2007).

The *cdt-III* operon is located on the large plasmid pVir68, as part of a large pathogenicity island encoding several virulence-related genes (Pérès *et al.*, 1997; Johnson *et al.*, 2010; details are given in the CNF part).

The *cdt-IV* operon is framed with lambdoid prophage genes very similar to the prophage harbouring *cdt-I*, although some sequence heterogeneity and the loss of a few ORFs directly upstream of the *cdt* operon can be observed (Tóth *et al.*, 2009a). The most complete information regarding the integration site of the CDT-IV-carrying lambdoid prophages comes from the whole sequenced genome of the avian pathogenic *E. coli* APEC O1 strain, where the *cdt-IV* operon is flanked by two prophages, tRNA genes and genes associated with pathogenicity islands. Interestingly, the structural genes of these lambdoid prophages are similar to those of *stx2* phages; however, the integration site of the *cdt* cluster is different from that of *stx2* genes (Johnson *et al.*, 2007).

The *cdt-V* operon is flanked by P2-like prophage sequences in O157:NM STEC (Janka *et al.*, 2003) and CTEC strains of various serotypes (Hinenoya *et al.*, 2009). An inducible CDT-V carrying P2-like prophage was also isolated from STEC strains of the O157:H7, O91:H21 and other serotypes found in the environment. However, the P2-like phages do not seem to be universally inducible in CDT-V-producing *E. coli* strains (Allué-Guardia *et al.*, 2011). This discrepancy is similar to the case of lambdoid prophages flanking the CDT-I gene cluster, suggesting a degree of sequence heterogeneity in the CDT-V-carrying P2-like prophages. This is also supported by the observations of Friedrich *et al.* (2006), that there is a considerable clonal diversity within CDT-V-producing O157:H7 strains. This might be the result of the different phage susceptibility of the strains, due to the presence of phage-specific receptors or genomic integration sites (Friedrich *et al.*, 2006). This notion is supported by the fact that *E. coli* of phage type 14, which is susceptible to all the test phages (Ahmed *et al.*, 1987), was among the carriers of *cdt-V* (Friedrich *et al.*, 2006).

The foreign origin of *cdt* operon in *E. coli* is further supported by the 41–44% average G + C content of the operons, which is markedly lower than the average genomic G + C content of *E. coli* (50.8%; Touchon *et al.*, 2009). The above data also suggest that based on their genetic background and similarly to their sequence–based relations, *cdt* operons could form at least two clusters, with *cdt-I* and *cdt-IV* being in one cluster, both carried by lambdoid phages. This relationship is further confirmed by the length of the individual subunit genes. The size of *cdtA*, *cdtB* and *cdtC* are 714, 822 and 573 bp in *cdt-I* and *cdt-IV*. Their respective lengths in *cdt-II*, *cdt-III* and *cdt-V* are 777, 810 and 546 bp. The sequence heterogeneity of the prophages (Asakura *et al.*, 2007) suggests that the cdt-I and -IV operons might have been acquired from a common ancestor by phage transduction and, after integrating into the bacterial host genome, the phages began to adapt to the different hosts, leading to changes in sequences and, in some cases, the apparent temperation of the phage. All these findings indicate that horizontal gene transfer played a significant role in the present distribution of CDT variants, and the genetic diversity of both the CDT operons and its flanking sequences is probably the result of adaptation to the different hosts.

Cycle inhibiting factors (Cif)

Cycle inhibiting factors (Cifs) are type III secretion system (TTSS) secreted effectors produced by EPEC, EHEC as well as by *Yersinia pseudotuberculosis*, *Photorhabdus luminescens*, *Photorhabdus asymbiotica* and *Burkholderia pseudomallei*. Cif induces a progressive cytopathic effect characterized by intensive actin rearrangement, cell and nucleus enlargement, increased DNA content, cell cycle arrest, and apoptosis. Cif proteins inhibit the eukaryotic host cell-cycle progression at G1/S and G2/M transitions by accumulation of the cyclin-dependent kinase inhibitors p21 and p27 (Taieb *et al.*, 2011).

Discovery and main characteristics of Cif

The Cif-associated irreversible cytopathic effect (CPE) was first reported by De Rycke *et al.* (1997) in HeLa cultures infected with rabbit *E. coli* O103 EPEC (REPEC) strains and also by human EPEC isolates. This CPE is characterized by progressive recruitment of focal adhesion plaques leading to the assembly of stress fibres, dramatic cell enlargement and inhibition of the cell cycle G2/M phase transition (De Rycke *et al.*, 1997; Nougayrède *et al.*, 1999, 2001). The Cif-associated effects increase with time, and cause cell death about 5 days after the interaction. Elegant experiments have revealed that the CPE is linked to the locus of enterocyte effacement (LEE) pathogenicity island (De Rycke *et al.*, 1997). Besides EPEC, EHEC also produce Cif and induce cytoskeletal alterations and cell cycle block at the G2/M phase transition due to the functional LEE type

III secretion machinery but not to intimin or Tir (Nougayrède *et al.*, 1999; Marchès *et al.*, 2003). This novel cell cycle inhibitory effector molecule has been identified and designated as 'cycle inhibitory factor' (Marches *et al.*, 2003).

Epidemiology of Cif-producing *E. coli*

The distribution and functional expression of *cif* were investigated in a collection of more than 5000 *E. coli* strains of human, animal and environmental origin. This study has revealed that 2.3% (115/5049) of *E. coli* isolates from diverse origins and geographic locations harboured the internal region of the *cif* gene (Loukiadis *et al.*, 2008). In harmony with previous reports (De Rycke *et al.*, 1997; Marchès *et al.*, 2003) all the 115 *cif*-positive isolates also carried the *eae* gene, confirming the strong association between *cif* and LEE pathogenicity island. The highly representative collection contained only 24 *eae*-positive and *cif*-negative isolates. The *cif*-positive strains were not restricted to specific *eae* gene or phylogenetic groups. The *cif*-positive strains belonged to phylogenetic groups A, B1 and B2. As many as 71% (34/48) of EPEC and EHEC strains of human and animal origin carried an internal *cif* gene, while only 38% (13/34) of them induced the Cif-related cytopathic effect. The lack of CPE in these strains was due to frameshift mutations or insertion of an insertion sequence (IS) element into the *cif* gene (Loukiadis *et al.*, 2008; Taieb *et al.*, 2011). Interestingly, half (12/24) of the EPEC but only 10% (1/10) of *cif*-positive EHEC strains carried functional *cif*. This phenomenon could be due to the low number of EHEC strains investigated or could suggest an evolutionary counter-selection of functional Cif in Shiga-toxin producing EHEC strains (Loukiadis *et al.*, 2008; Taieb *et al.*, 2011).

In another molecular epidemiological study, the dissemination of cyclomodulins was investigated in uroseptic *E. coli* strains of human origin. Among the urosepsis-associated strains (n = 146) two *cif*-harbouring strains were observed and none of the *E. coli* control strains isolated from faeces (n = 51) carried the *cif* gene. These *cif*-positive strains belonged to phylogroup B1 and harboured no other cyclomodulin-encoding gene. Further analysis confirmed that *cif*-positive strains proved to be EPEC strains by detection of the *eae* gene but not of *stx* genes. All the other cyclomodulins were observed with higher incidence (Dubois *et al.*, 2010).

A set of *E. coli* strains from calves and dogs with and without diarrhoea or presenting cystitis was investigated for the presence of cyclomodulins and for their functional expression. As many as 21.4% (80/374) of 374 *E. coli* isolates were positive for at least one cyclomodulin. One-fourth of these cyclomodulin-positive strains carried *cif* gene. All the 20 *cif*-positive isolates were associated with the *eae* gene ($P < 0.01$), 16 of which were identified as EPEC and 4 as EHEC. Fifteen *cif*-positive strains showed characteristic CPE on HeLa cells, while in the remaining strains the deletion of one adenine (in four isolates) or the addition of one adenine (in one isolate) was identified. These modifications occurred at the 3'-end of the gene that generated a stop codon and a non-functional truncated Cif protein. Isolates *cif*-positive and carrying additional intestinal virulence genes, were found in isolates causing diarrhoea and mainly belonged to the B1 phylogenetic group, while *cnf*- and *cdt*-positive *E. coli* isolates carried extraintestinal virulence factors, and belonged predominantly to phylogroups B2 and D, regardless of their origin (Salvarani *et al.*, 2012). Accordingly, Cif is an exclusive virulence factor of EPEC and EHEC causing attaching and effacing lesions.

Genetics and structure of Cif and Cif homologues

The entire *cif* gene is 846 bp in length and encodes a 282 amino acid protein with a predicted molecular mass of 32 kDa (Marchès *et al.*, 2003). The amino acid sequence comparison of Cif does not show significant matches with any available proteins, except with hypothetical proteins encoded by ORFs in four other species, the human pathogens *Yersinia pseudotuberculosis*, *Burkholderia pseudomallei* and in entomopathogens *Photorabdus luminescens* and *Photorabdus asymbiotica* (Jubelin *et al.*, 2009; Taieb *et al.*, 2011). Furthermore the alignment, as well as the crystal structures, of *E. coli* Cif and *E. coli* Cif homologues (ECCH) reveals well-conserved amino acid positions at the C terminal of the proteins (Jubelin

et al., 2009). The structure of Cifs contains functional domains: the N-terminal region, referred to as the tail, that contains the translocation and substrate binding site, and the C-terminal, which forms the head domain harbouring the enzymatic site. The first 16 amino acids (aa) of Cif are sufficient and necessary to mediate translocation into the host cells. Similarly, the first 20 aa of the effector proteins Map, EspF, and Tir, which are encoded in the same region as the TTSS, mediate secretion and translocation in a type III-dependent but chaperone-independent manner. The truncated form of Cif lacking its first 20 aa was no longer secreted and translocated, but fusion with the first 20 aa of Tir, Map, or EspF restores both secretion and translocation (Charpentier and Oswald, 2004). The Cif proteins are members of the papain subfamily of cysteine proteases, acetyl transferases, deamidases and transglutamases (Yao *et al.*, 2009).

The catalytic triad consists of Cys109-His165-Gln185 residues that are identical in cysteine proteases and acetyltransferases. Mutation of these conserved active site residues abolishes the ability of Cif to block cell-cycle progression (Hsu *et al.*, 2008; Crow *et al.*, 2009).

The integrity of the catalytic triad is also essential for the induction of cell death by apoptosis (Chavez *et al.*, 2010). Interestingly, irreversible cysteine protease inhibitors do not abolish the cytopathic effect of Cif, suggesting that another enzymatic activity may underlie the biological activity of this virulence factor (Hsu *et al.*, 2008).

Although linked to the LEE (De Rycke *et al.*, 1997), Cif is not encoded by LEE but by a lambda-like prophage integrated into the bacterial chromosome (Marchès *et al.*, 2003; Loukiadis *et al.*, 2008). The Cif-encoding prophage is present in most of the EPEC and EHEC serovars. However, the *cif* gene is absent or truncated in EHEC reference strain O157:H7 Sakai and in EPEC O127:H7 strain E2348/69. In the EPEC O103 E22 genome, the Cif-encoding prophage is integrated near the *bio* operon, which is part of the core genome present in every *E. coli* organism, and the prophage is similar to the Sp3 cryptic prophage of EHEC O157:H7 Sakai and the EDL933 prophage (Marchès *et al.*, 2003). *In vitro* experiments have revealed that the *cif* gene is located on an inducible lambdoid prophage spread between EPEC and EHEC strains (Loukiadis *et al.*, 2008). Importantly, in all bacteria harbouring Cif protein homologues the *cif* genes are located in prophages or in the neighbourhood of mobile genetic elements, which also carry type III secretion systems genes (Jubelin *et al.*, 2009). The G + C content of *cif* genes in known cif-positive bacteria is lower by an average of 10% than the average genomic G + C content of the respective species, indicating horizontal acquisition of the genes (Jubelin *et al.*, 2009). In the case of EPEC, the gene cluster encoding the type III secretion system has similar G + C content (Jubelin *et al.*, 2009), suggesting that the acquisition of these virulence determinants possibly happened at the same time (GenBank Numbers AAQ07241 [*cif* cluster] and NC_011601 [EPEC strain E2348/69]).

Cellular and molecular pathogenesis

All the above-mentioned epidemiological data and cell biological results clearly show that Cif is a virulence or fitness factor of *E. coli* harbouring functional TTSS. Upon translocation into the host cell Cif triggers a progressive cytopathic effect characterized by stress fibre and focal adhesion formation, nucleus enlargement and by cell cycle arrest at both the G2/M and G1/S transitions, depending on the stage of cells in the cell cycle during infection (Marchès *et al.*, 2003; Samba-Louaka *et al.*, 2008). Cells transformed by Cif accumulate with a DNA content of 4N and reinitiate DNA without division leading to an 8–16N DNA content (Oswald *et al.*, 2005). The cell cycle inhibition is not a consequence of cytoskeleton alteration (Nougayrède *et al.*, 2001), since the inhibition of stress fibre formation by Rho inhibitors does not prevent the G2 arrest of epithelial cells. The cell cycle inhibition correlates with the stabilization of cyclin-dependent kinase (CDK) inhibitors, the $p21^{waf1}$ and $p27^{kip1}$ proteins that regulate the host cell cycle (Samba-Louaka *et al.*, 2008). Cif-induced cell cycle arrest that, like CDT- and colibactin-associated cell cycle arrest, is associated with the inactive phosphorylated state of CDK1 (Marchès *et al.*, 2003; Taieb *et al.*, 2006). However, in contrast to genotoxic CDT and colibactin, Cif does not cut DNA, nor does

it activates the DNA damage response pathway. The Cif-induced G2 phase inhibition correlates with the accumulation of p21 regulator protein and the G1 inhibition is due to the accumulation of p21 and p27 (Samba-Louaka *et al.*, 2008). p21 and p27 inhibit CDK2-cyclin E and A complexes whose activation is required for G1/S and S phase progression (Taieb *et al.*, 2011). The accumulation of Cif-induced cycle-dependent kinase inhibitors is independent of transcriptional activation, but relies on their reduced ubiquitin-mediated proteolysis (Samba-Louka *et al.*, 2008). Ubiquitination is the main post-translational modification that regulates proteasome-mediated degradation of proteins including the key cell cycle regulators. Ubiquitination involves three successive steps mediated by the E1–E2–E3 enzyme cascade (Pickart and Eddins, 2004). E3 is the ubiquitin ligase that modulates the specific target proteins including cell cycle regulators. Cullin-RING ligases (CRLs) comprise the largest ubiquitin E3 subclass, in which a central cullin subunit links a substrate-binding adaptor with an E2-binding RING (Petroski and Deshaises, 2005). Covalent attachment of the ubiquitin-like protein NEDD8 (neural precursor cell expressed, developmentally down-regulated 8) to CRL (neddylation) stimulates ubiquitination activity and prevents binding of the inhibitor CAND1 (Duda *et al.*, 2008). Cif inhibits the E3 ligase activity of neddylated cullins by catalysing the deamidation of a specific glutamine residue (Gln40) in NEDD8 and the related protein ubiquitin. This modification prevents recycling of neddylated cullin-RING ligases leading to stabilization of related targets including cycle-dependent kinase inhibitors (Cui *et al.*, 2010; Crow *et al.*, 2012).

Since CLRs play regulatory roles in diverse cellular functions, the Cif-mediated inhibitions could explain the stabilization of Rho proteins inducing stress fibres, as well as the inhibition and suppression of inflammatory cascades, respectively. Last but not least, the Cif-mediated cell cycle inhibition could slow down the multiplication of intestinal progenitors and the delayed epithelial cell renewal could promote bacterial colonization (reviewed by Taieb *et al.*, 2011). Furthermore, the elimination of p21 and p27 does not overcome the Cif-induced cell cycle arrest, suggesting the existence of further pathway(s) controlling the cell cycle progression (Samba-Louka *et al.*, 2008).

Like the Cif-producing *E. coli*, all the known *E. coli* Cif homologues (ECCH) producing pathogenic or symbiotic bacteria induce cytopathic effects identical to those observed with Cif from pathogenic *E. coli* (Jubelin *et al.*, 2009; Chavez *et al.*, 2010).

Evolutionary aspects of Cif-harbouring bacteria

Phages from the Cif prototype EPEC O103 E22 strain can be induced and spread by transduction. Interestingly, phage E22 was inserted in the recombinant K12 strain to the same site as in the parental E22 (Loukiadis *et al.*, 2008). In all the other *cif*-positive bacteria *cif* genes are also associated with mobile genetic elements. In *Photorabdus* strains the *cif* genes are located downstream of a prophage. Interestingly, the *Photorabdus* genomes contain several copies of this prophage that encodes several other putative virulence factors including a YopT homologue TTSS effector. In *B. pseudomallei*, the *cif* gene is located between two vestigial transposase genes and the same locus also encodes one of three TTSSs carried by the strain. In *Yersinia pseudotuberculosis*, *cif* is located at a highly variable chromosomal locus close to the *yrs* region, which is a site-specific recombination target for filamentous bacteriophages. Accordingly, all these xenologues are part of flexible genetic modules suggesting the acquisition of *cif* genes by horizontal gene transfer in all these pathogenic bacteria, presumably early in their evolution (Jubelin *et al.*, 2009).

Colibactin

Discovery and main characteristics

Colibactin was discovered in *E. coli* by Nougayrède *et al.* (2006). A set of ExPEC and commensal phylogenetic group B2 strains carried a genetic island that was associated with CPE characterized by cell body and nucleus enlargement and cell cycle arrest in the G2/M phase of the eukaryotic tissue cultures. The genetic island termed *pks* encodes non-ribosomal peptide megasynthases, polyketide megasynthases and other accessory enzymes. *E. coli* possessing *pks* island are capable of inducing double-strand DNA breaks and lasting chromosome aberrations in infected cells (Nougayrède *et al.*, 2006).

Epidemiology and virulence potential of colibactin

Nougayrède *et al.* (2006) reported for the first time that colibactin production could characterize ExPEC and commensal faecal *E. coli* isolates belonging to the B2 phylogenetic group, but not the intestinal pathogenic isolates. In their original paper 53% of ExPEC and 34% of strains from healthy humans carried *pks* island genes, while such genes were carried by none of the intestinal pathogenic *E. coli* isolates. Similarly, as many as 45% of the ExPEC strains isolated from blood carried *pks* island genes (Johnson *et al.*, 2008). Colibactin was observed in 72.2% of the *E. coli* isolates from healthy young adults with acute prostatitis, and colibactin-positive isolates belonged to the B2 phylogenetic group (Krieger *et al.*, 2011). In a highly representative study, more than 1500 strains representing 24 species and 14 genera were screened for the presence of *pks* island genes. In harmony with previous studies among *E. coli* isolates, colibactin was detected in ExPEC and in faecal commensal isolates but none of the intestinal pathogenic strains carried *pks* island gene. Nearly all (101/104) of colibactin-positive isolates belonged to phylogenetic group B2, while the other three strains belonged to group B1 with extended spectrum β-lactamase positivity. Interestingly, the colibactin genes were detected not only in *E. coli* but also in *Klebsiella pneumoniae* (5/141) and *Enterobacter aerogenes* (3/11) strains and also in a *C. koseri* reference isolate (ATCC BAA-895). These colibactin island-positive strains are clinical extraintestinal pathogenic isolates (Putze *et al.*, 2009).

Recently, the relationship between carriage of the *pks* island and the capacity of *E. coli* strains to persist in the gut microbiota of infants has been investigated. The study revealed that strains harbouring the *pks* island were significantly more prevalent among long-term colonisers than among either intermediate-term or transient strains. As observed previously (Nougayrède *et al.*, 2006; Johnson *et al.*, 2008; Putze *et al.*, 2009), the *pks* island exclusively characterized the B2 strains and the carriage of the *pks* island was linked to the carriage of other pathogenicity traits, including six PAI markers and the *papC*, *fimH*, and *hlyA* genes. Therefore, the *pks* island may not in itself confer long-term colonization ability, but rather it may be one of the multiple pathogenicity traits that have accumulated in B2 strains and that collectively enhance the colonization capacities of these strains (Nowrouzian and Oswald, 2012) as it was reported for uropathogenic isolates (Dubois *et al.*, 2010).

Pathogenesis – *in vivo* and *in vitro* experiments

Colibactin-producing strains cause DNA double-strand breaks in the nucleus of the target cell, this blocks the cell cycle in the G2/M phase (Nougayrède *et al.*, 2006; Cuevas-Ramos *et al.*, 2010). DNA damage response deficient cells undergo apoptosis when exposed to colibactin-positive *E. coli*, and normal cells also die when they are infected with a high number of colibactin-positive bacteria. Cells receiving a low dose of infective colibactin-carrying bacteria tend to survive intoxication, but propagate lasting chromosome aberrations: γH2AX foci, lagging chromosomes and multipolar mitosis could be observed after 72 h. These aberrations can result in numerical instability in the chromosomes: aneuploidia and tetraploidia were also observed in the surviving cells. Through these mechanisms, the mutagenic potential of colibactin was also demonstrated (Cuevas-Ramos *et al.*, 2010).

Genetics and evolutionary aspects

In addition to colibactin, there are only two known non-ribosomal peptidase and polyketide/non-ribosomal peptide hybrids in *Enterobacteriaceae*, the iron chelators enterobactin and yersiniabactin, respectively (Crosa and Walsh, 2002; Pfeifer *et al.*, 2003). In contrast to colibactin, none of these iron chelators cause cytopathic effect on eukaryotic cells *in vitro*. Furthermore, a correlation was observed between the presence of the colibactin determinant and the high pathogenicity island (HPI) coding for the siderophore system yersiniabactin (Carniel, 2001). Accordingly, all colibactin-positive strains were also yersiniabactin

positive, but the presence of the HPI was not always associated with colibactin genes in *E. coli*, *K. pneumoniae* and *E. aerogenes*. The *pks* island sequences are highly conserved (99% identity), suggesting their recent integration into the different strains and species. The fact that the G + C content of the *pks* island and the average genomic G + C content of the colibactin-producing *C. koseri* is the same (53%), advocates that this species could be the source of the *pks* island (Putze *et al.*, 2009).

Conclusion

Since the discovery of the first cell cycle manipulating toxin (CNF) thirty years ago, research generated significant insight into the mechanisms of cyclomodulins produced by pathogenic *E. coli* originating from human and animal infections as well as from healthy individuals. Fundamental molecular and cellular observations have revealed that the members of this emerging functional toxin family have their own ways for manipulating the eukaryotic host cell function presumably for their own needs. Epidemiological data, experimental infections and cell biological studies indicate that cyclomodulins are emerging virulence and fitness factors that are disseminated among several *E. coli* clones and a wide range of bacterial species. In view of the fact that all the so far known cyclomodulins of *E. coli* are encoded by mobile genetic elements, the appearance of cyclomodulin toxin producing strains representing a novel geno- and/ or pathotype is only a question of time.

Acknowledgement

This work was supported by the Hungarian Research Fund OTKA (K81252).

References

Ahmed, R., Bopp, C., Borczyk, A., and Kasatiya, S. (1987). Phage-typing scheme for *Escherichia coli* O157:H7. J. Infect. Dis. *155*, 806–809.

Akifusa, S., Poole, S., Lewthwaite, J., Henderson, B., and Nair, S.P. (2001). Recombinant *Actinobacillus actinomycetemcomitans* cytolethal distending toxin proteins are required to interact to inhibit human cell cycle progression and to stimulate human leukocyte cytokine synthesis. Infect. Immun. *69*, 5925–5930.

Allué-Guardia, A., García-Aljaro, C., and Muniesa, M. (2011). Bacteriophage-encoding cytolethal distending toxin type V gene induced from nonclinical *Escherichia coli* isolates. Infect. Immun. 79, 3262–3272.

Asakura, M., Hinenoya, A., Alam, M.S., Shima, K., Zahid, S.H., Shi, L., Sugimoto, N., Ghosh, A.N., Ramamurthy, T., Faruque, S.M., *et al.* (2007). An inducible lambdoid prophage encoding cytolethal distending toxin (Cdt-I) and a type III effector protein in enteropathogenic *Escherichia coli*. Proc. Natl. Acad. Sci. U.S.A. 104, 14483–14488.

Assoian, R.K., and Schwartz, M.A. (2001). Coordinate signaling by integrins and receptor tyrosine kinases in the regulation of G1 phase cell-cycle progression. Curr. Opin. Genet. Dev. 11, 48–53.

Belibasakis, G.N., Mattsson, A., Wang, Y., Chen, C., and Johansson, A. (2004). Cell cycle arrest of human gingival fibroblasts and periodontal ligament cells by *Actinobacillus actinomycetemcomitans:* involvement of the cytolethal distending toxin. APMIS 112, 674–685.

Berlanda Scorza, F., Doro, F., Rodríguez-Ortega, M.J., Stella, M., Liberatori, S., Taddei, A.R., Serino, L., Gomes Moriel, D., Nesta, B., Fontana, M.R., *et al.* (2008). Proteomics characterization of outer membrane vesicles from the extraintestinal pathogenic *Escherichia coli* DeltatoIR IHE3034 mutant. Mol. Cell Proteomics 7, 473–485.

Bielaszewska, M., Fell, M., Greune, L., Prager, R., Fruth, A., Tschäpe, H., Schmidt, M.A., and Karch, H. (2004). Characterization of cytolethal distending toxin genes and expression in shiga toxin-producing *Escherichia coli* strains of non-O157 serogroups. Infect. Immun. 72, 1812–1816.

Bielaszewska, M., Sinha, B., Kuczius, T., and Karch, H. (2005). Cytolethal distending toxin from Shiga toxin-producing *Escherichia coli* O157 causes irreversible G2/M arrest, inhibition of proliferation, and death of human endothelial cells. Infect. Immun. 73, 552–562.

Bielaszewska, M., Stoewe, F., Fruth, A., Zhang, W., Prager, R., Brockmeyer, J., Mellmann, A., Karch, H., and Friedrich, A.W. (2009). Shiga toxin, cytolethal distending toxin, and hemolysin repertoires in clinical *Escherichia coli* O91 isolates. J. Clin. Microbiol. 47, 2061–2066.

Bingen-Bidois, M., Clermont, O., Bonacorsi, S., Terki, M., Brahimi, N., Loukil, C., Barraud, D., and Bingen, E. (2002). Phylogenetic analysis and prevalence of urosepsis strains of *Escherichia coli* bearing pathogenicity island-like domains. Infect. Immun. 70, 3216–3226.

Blanco, J., González, E.A., García, S., Blanco, M., Regueiro, B., and Bernárdez, I. (1988). Production of toxins by *Escherichia coli* strains isolated from calves with diarrhoea in Galicia (north-western Spain). Vet. Microbiol. 18, 297–311.

Blanco, J., Blanco, M., Alonso, M.P., Blanco, J.E., González, E.A., and Garabal, J.I. (1992). Characteristics of haemolytic *Escherichia coli* with particular reference to production of cytotoxic necrotizing factor type 1 (CNF1). Res. Microbiol. 143, 869–978.

Blum, G., Falbo, V., Caprioli, A., and Hacker, J. (1995). Gene clusters encoding the cytotoxic necrotizing factor type 1, Prs-fimbriae and alpha-hemolysin form the pathogenicity island II of the uropathogenic *Escherichia coli* strain J96. FEMS Microbiol. Lett. 126, 189–195.

Boesze-Battaglia, K., Besack, D., McKay, T., Zekavat, A., Otis, L., Jordan-Sciutto, K., and Shenker, B.J. (2006). Cholesterol-rich membrane microdomains mediate cell cycle arrest induced by *Actinobacillus actinomycetemcomitans* cytolethal-distending toxin. Cell. Microbiol. 8, 823–836.

Borriello, G., Lucibelli, M.G., De Carlo, E., Auriemma, C., Cozza, D., Ascione, G., Scognamiglio, F., Iovane, G., and Galiero, G. (2012). Characterization of enterotoxigenic *E. coli* (ETEC), Shiga-toxin producing *E. coli* (STEC) and necrotoxigenic *E. coli* (NTEC) isolated from diarrhoeic Mediterranean water buffalo calves (*Bubalus bubalis*). Res. Vet. Sci. 93, 18–22.

Bouzari, S., Oloomi, M., and Oswald, E. (2005). Detection of the cytolethal distending toxin locus cdtB among diarrhoeagenic *Escherichia coli* isolates from humans in Iran. Res. Microbiol. 156, 137–144.

Buetow, L., Flatau, G., Chiu, K., Boquet, P., and Ghosh, P. (2001). Structure of the Rho-activating domain of *Escherichia coli* cytotoxic necrotizing factor 1. Nat. Struct. Biol. 8, 584–588.

Caprioli, A., Falbo, V., Roda, L.G., Ruggeri, F.M., and Zona, C. (1983). Partial purification and characterization of an *Escherichia coli* toxic factor that induces morphological cell alterations. Infect. Immun. 39, 1300–1306.

Caprioli, A., Donelli, G., Falbo, V., Possenti, R., Roda, L.G., Roscetti, G., and Ruggeri, F.M. (1984). A cell division-active protein from *E. coli*. Biochem. Biophys. Res. Commun. 118, 587–593.

Caprioli, A., Falbo, V., Ruggeri, F.M., Baldassarri, L., Bisicchia, R., Ippolito, G., Romoli, E., and Donelli, G. (1987). Cytotoxic necrotizing factor production by hemolytic strains of *Escherichia coli* causing extraintestinal infections. J. Clin. Microbiol. 25, 146–149.

Carette, J.E., Guimaraes, C.P., Varadarajan, M., Park, A.S., Wuethrich, I., Godarova, A., Kotecki, M., Cochran, B.H., Spooner, E., Ploegh, H.L., and Brummelkamp, T.R. (2009). Haploid genetic screens in human cells identify host factors used by pathogens. Science 326, 1231–1235.

Carniel, E. (2001). The *Yersinia* high-pathogenicity island: an iron-uptake island. Microbes Infect. 3, 561–569.

Charpentier, X., and Oswald, E. (2004). Identification of the secretion and translocation domain of the enteropathogenic and enterohemorrhagic *Escherichia coli* effector Cif, using TEM-1 beta-lactamase as a new fluorescence-based reporter. J. Bacteriol. 186, 5486–5495.

Chavez, C.V., Jubelin, G., Courties, G., Gomard, A., Ginibre, N., Pages, S., Taieb, F., Girard, P.A., Oswald, E., Givaudan, A., Zumbihl, R., and Escoubas, J.M. (2010). The cyclomodulin Cif of *Photorhabdus luminescens* inhibits insect cell proliferation and triggers host cell death by apoptosis. Microbes Infect. 14–15, 1208–1218.

Clark, C.G., Johnson, S.T., Easy, R.H., Campbell, J.L., and Rodgers, F.G. (2002). PCR for detection of cdt-III and the relative frequencies of cytolethal distending toxin variant-producing *Escherichia coli* isolates from humans and cattle. J. Clin. Microbiol. *40*, 2671–2674.

Comayras, C., Tasca, C., Pérès, S.Y., Ducommun, B., Oswald, E., and De Rycke, J. (1997). *Escherichia coli* cytolethal distending toxin blocks the HeLa cell cycle at the G2/M transition by preventing cdc2 protein kinase dephosphorylation and activation. Infect. Immun. *65*, 5088–5095.

Contamin, S., Galmiche, A., Doye, A., Flatau, G., Benmerah, A., and Boquet, P. (2000). The p21 Rho-activating toxin cytotoxic necrotizing factor 1 is endocytosed by a clathrin-independent mechanism and enters the cytosol by an acidic-dependent membrane translocation step. Mol. Biol. Cell *11*, 1775–1787.

Cortes-Bratti, X., Chaves-Olarte, E., Lagergård, T., and Thelestam, M. (1999). The cytolethal distending toxin from the chancroid bacterium *Haemophilus ducreyi* induces cell-cycle arrest in the G2 phase. J. Clin. Invest. *103*, 107–115.

Cortes-Bratti, X., Chaves-Olarte, E., Lagergård, T., and Thelestam, M. (2000). Cellular internalization of cytolethal distending toxin from *Haemophilus ducreyi*. Infect. Immun. *68*, 6903–6911.

Cortes-Bratti, X., Karlsson, C., Lagergård, T., Thelestam, M., and Frisan, T. (2001). The *Haemophilus ducreyi* cytolethal distending toxin induces cell cycle arrest and apoptosis via the DNA damage checkpoint pathways. J. Biol. Chem. *276*, 5296–5302.

Crosa, J.H., and Walsh, C.T. (2002). Genetics and assembly line enzymology of siderophore synthesis in bacteria. Microbiol. Mol. Biol. Rev. *66*, 223–249.

Crow, A., Race, P.R., Jubelin, G., Varela Chavez, C., Escoubas, J.M., Oswald, E., and Banfield, M.J. (2009). Crystal structures of Cif from bacterial pathogens *Photorhabdus luminescens* and *Burkholderia pseudomallei*. PLoS One *4*, e5582.

Crow, A., Hughes, R.K., Taieb, F., Oswald, E., and Banfield, M.J. (2012). The molecular basis of ubiquitin-like protein NEDD8 deamidation by the bacterial effector protein Cif. Proc. Natl. Acad. Sci. U.S.A. *109*, E1830–1838.

Cuevas-Ramos, G., Petit, C.R., Marcq, I., Boury, M., Oswald, E., and Nougayrède, J. (2010). *Escherichia coli* induces DNA damage *in vivo* and triggers genomic instability in mammalian cells. Proc. Natl. Acad. Sci. U.S.A. *107*, 11537–11542.

Cui, J., Yao, Q., Li, S., Ding, X., Lu, Q., Mao, H., Liu, L., Zheng, N., Chen, S., and Shao, F. (2010). Glutamine deamidation and dysfunction of ubiquitin/NEDD8 induced by a bacterial effector family. Science *329*, 1215–1218.

Davis, J.M., Rasmussen, S.B., and O'Brien, A.D. (2005). Cytotoxic necrotizing factor type 1 production by uropathogenic *Escherichia coli* modulates polymorphonuclear leukocyte function. Infect. Immun. *73*, 5301–5310.

Degnan, P.H., and Moran, N.A. (2008). Diverse phage-encoded toxins in a protective insect endosymbiont. Appl. Environ. Microbiol. *74*, 6782–6791.

Deng, K., Latimer, J.L., Lewis, D.A., and Hansen, E.J. (2001). Investigation of the interaction among the components of the cytolethal distending toxin of *Haemophilus ducreyi*. Biochem. Biophys. Res. Commun. *285*, 609–615.

De Rycke, J., and Plassiart, G. (1990). Toxic effects for lambs of cytotoxic necrotising factor from *Escherichia coli*. Res. Vet. Sci. *49*, 349–354.

De Rycke, J., Guillot, J.F., and Boivin, R. (1987). Cytotoxins in non-enterotoxigenic strains of *Escherichia coli* isolated from feces of diarrheic calves. Vet. Microbiol. *15*, 137–150.

De Rycke, J., Phan-Thanh, L., and Bernard, S. (1989). Immunochemical identification and biological characterization of cytotoxic necrotizing factor from *Escherichia coli*. J. Clin. Microbiol. *27*, 983–988.

De Rycke, J., González, E.A., Blanco, J., Oswald, E., Blanco, M., and Boivin, R. (1990). Evidence for two types of cytotoxic necrotizing factor in human and animal clinical isolates of *Escherichia coli*. J. Clin. Microbiol. *28*, 694–699.

De Rycke, J., Mazars, P., Nougayrède, J.P., Tasca, C., Boury, M., Herault, F., Valette, A., and Oswald, E. (1996). Mitotic block and delayed lethality in HeLa epithelial cells exposed to *Escherichia coli* BM2–1 producing cytotoxic necrotizing factor type 1. Infect. Immun. *64*, 1694–1705.

De Rycke, J., Comtet, E., Chalareng, C., Boury, M., Tasca, C., and Milon, A. (1997). Enteropathogenic *Escherichia coli* O103 from rabbit elicits actin stress fibers and focal adhesions in HeLa epithelial cells, cytopathic effects that are linked to an analog of the locus of enterocyte effacement. Infect. Immun. *65*, 2555–2563.

De Rycke, J., Milon, A., and Oswald, E. (1999). Necrotoxic *Escherichia coli* (NTEC): two emerging categories of human and animal pathogens. Vet. Res. *30*, 221–233.

Dozois, C.M., Clément, S., Desautels, C., Oswald, E., and Fairbrother, J.M. (1997). Expression of P, S, and F1C adhesins by cytotoxic necrotizing factor 1-producing *Escherichia coli* from septicemic and diarrheic pigs. FEMS Microbiol. Lett. *152*, 307–312.

Doye, A., Mettouchi, A., Bossis, G., Clément, R., Buisson-Touati, C., Flatau, G., Gagnoux, L., Piechaczyk, M., Boquet, P., and Lemichez, E. (2002). CNF1 exploits the ubiquitin-proteasome machinery to restrict Rho GTPase activation for bacterial host cell invasion. Cell *111*, 553–564.

Dubois, D., Delmas, J., Cady, A., Robin, F., Sivignon, A., Oswald, E., and Bonnet, R. (2010). Cyclomodulins in urosepsis strains of *Escherichia coli*. J. Clin. Microbiol. *48*, 2122–2129.

Duda, D.M., Borg, L.A., Scott, D.C., Hunt, H.W., Hammel, M., and Schulman, B.A. (2008). Structural insights into NEDD8 activation of cullin-RING ligases: conformational control of conjugation. Cell *134*, 995–1006.

Elwell, C.A., and Dreyfus, L.A. (2000). DNase I homologous residues in CdtB are critical for cytolethal distending toxin-mediated cell cycle arrest. Mol. Microbiol. *37*, 952–963.

Emödy, L., Kerényi, M., and Nagy, G. (2003). Virulence factors of uropathogenic *Escherichia coli*. Int. J. Antimicrob. Agents *22(Suppl. 2)*, 29–33.

Eshraghi, A., Maldonado-Arocho, F.J., Gargi, A., Cardwell, M.M., Prouty, M.G., Blanke, S.R., and Bradley, K.A. (2010). Cytolethal distending toxin family members are differentially affected by alterations in host glycans and membrane cholesterol. J. Biol. Chem. 285, 18199–18207.

Etienne-Manneville, S., and Hall, A. (2002). Rho GTPases in cell biology. Nature 420, 629–635.

Fabbri, A., Gauthier, M., and Boquet, P. (1999). The 5' region of cnf1 harbours a translational regulatory mechanism for CNF1 synthesis and encodes the cell-binding domain of the toxin. Mol. Microbiol. 33, 108–118.

Fabbri, A., Travaglione, S., and Fiorentini, C. (2010). Escherichia coli Cytotoxic Necrotizing Factor 1 (CNF1): Toxin biology, in vivo applications and therapeutic potential. Toxins 2, 283–296.

Falbo, V., Famiglietti, M., and Caprioli, A. (1992). Gene block encoding production of cytotoxic necrotizing factor 1 and hemolysin in Escherichia coli isolates from extraintestinal infections. Infect. Immun. 60, 2182–2187.

Falbo, V., Pace, T., Picci, L., Pizzi, E., and Caprioli, A. (1993). Isolation and nucleotide sequence of the gene encoding cytotoxic necrotizing factor 1 of Escherichia coli. Infect. Immun. 61, 4909–4914.

Falzano, L., Fiorentini, C., Boquet, P., and Donelli, G. (1993). Interaction of Escherichia coli cytotoxic necrotizing factor type 1 (CNF1) with cultured cells. Cytotechnology 11(Suppl. 1), 56–58.

Falzano, L., Quaranta, M.G., Travaglione, S., Filippini, P., Fabbri, A., Viora, M., Donelli, G., and Fiorentini, C. (2003). Cytotoxic necrotizing factor 1 enhances reactive oxygen species-dependent transcription and secretion of proinflammatory cytokines in human uroepithelial cells. Infect. Immun. 71, 4178–4181.

Falzano, L., Filippini, P., Travaglione, S., Miraglia, A.G., Fabbri, A., and Fiorentini, C. (2006). Escherichia coli cytotoxic necrotizing factor 1 blocks cell cycle G2/M transition in uroepithelial cells. Infect. Immun. 74, 3765–3772.

Fiorentini, C., Arancia, G., Caprioli, A., Falbo, V., Ruggeri, F.M., and Donelli, G. (1988). Cytoskeletal changes induced in HEp-2 cells by the cytotoxic necrotizing factor of Escherichia coli. Toxicon 26, 1047–1056.

Fiorentini, C., Donelli, G., Matarrese, P., Fabbri, A., Paradisi, S., and Boquet, P. (1995). Escherichia coli cytotoxic necrotizing factor 1: evidence for induction of actin assembly by constitutive activation of the p21 Rho GTPase. Infect. Immun. 63, 3936–3944.

Fiorentini, C., Fabbri, A., Flatau, G., Donelli, G., Matarrese, P., Lemichez, E., Falzano, L., and Boquet, P. (1997). Escherichia coli cytotoxic necrotizing factor 1 (CNF1), a toxin that activates the Rho GTPase. J. Biol. Chem. 272, 19532–19537.

Fiorentini, C., Matarrese, P., Straface, E., Falzano, L., Fabbri, A., Donelli, G., Cossarizza, A., Boquet, P., and Malorni, W. (1998). Toxin-induced activation of Rho GTP-binding protein increases Bcl-2 expression and influences mitochondrial homeostasis. Exp. Cell. Res. 242, 341–350.

Flatau, G., Lemichez, E., Gauthier, M., Chardin, P., Paris, S., Fiorentini, C., and Boquet, P. (1997). Toxin-induced activation of the G protein p21 Rho by deamidation of glutamine. Nature 387, 729–733.

Fox, J.G., Rogers, A.B., Whary, M.T., Ge, Z., Taylor, N.S., Xu, S., Horwitz, B.H., and Erdman, S.E. (2004). Gastroenteritis in NF-kappaB-deficient mice is produced with wild-type Camplyobacter jejuni but not with C. jejuni lacking cytolethal distending toxin despite persistent colonization with both strains. Infect. Immun. 72, 1116–1125.

Friedrich, A.W., Lu, S., Bielaszewska, M., Prager, R., Bruns, P., Xu, J., Tschäpe, H., and Karch, H. (2006). Cytolethal distending toxin in Escherichia coli O157:H7: spectrum of conservation, structure, and endothelial toxicity. J. Clin. Microbiol. 44, 1844–1846.

Frisan, T., Cortes-Bratti, X., Chaves-Olarte, E., Stenerlöw, B., and Thelestam, M. (2003). The Haemophilus ducreyi cytolethal distending toxin induces DNA double-strand breaks and promotes ATM-dependent activation of RhoA. Cell. Microbiol. 5, 695–707.

Ge, Z., Schauer, D.B., and Fox, J.G. (2008). In vivo virulence properties of bacterial cytolethal-distending toxin. Cell. Microbiol. 10, 1599–1607.

Gelfanova, V., Hansen, E.J., and Spinola, S.M. (1999). Cytolethal distending toxin of Haemophilus ducreyi induces apoptotic death of Jurkat T cells. Infect. Immun. 67, 6394–6402.

Ghanbarpour, R., and Oswald, E. (2009). Characteristics and virulence genes of Escherichia coli isolated from septicemic calves in southeast of Iran. Trop. Anim. Health Prod. 41, 1091–1099.

Guerra, L., Teter, K., Lilley, B.N., Stenerlöw, B., Holmes, R.K., Ploegh, H.L., Sandvig, K., Thelestam, M., and Frisan, T. (2005). Cellular internalization of cytolethal distending toxin: a new end to a known pathway. Cell. Microbiol. 7, 921–934.

Guerra, L., Carr, H.S., Richter-Dahlfors, A., Masucci, M.G., Thelestam, M., Frost, J.A., and Frisan, T. (2008). A bacterial cytotoxin identifies the RhoA exchange factor Net1 as a key effector in the response to DNA damage. PLoS One 3, e2254.

Guerra, L., Nemec, K.N., Massey, S., Tatulian, S.A., Thelestam, M., Frisan, T., and Teter, K. (2009). A novel mode of translocation for cytolethal distending toxin. Biochim. Biophys. Acta 1793, 489–495.

Guerra, L., Albihn, A., Tronnersjö, S., Yan, Q., Guidi, R., Stenerlöw, B., Sterzenbach, T., Josenhans, C., Fox, J.G., Schauer, D.B., Thelestam, M., Larsson, L., Henriksson, M., and Frisan, T. (2010). Myc is required for activation of the ATM-dependent checkpoints in response to DNA damage. PLoS One 5, e8924.

Guerra, L., Cortes-Bratti, X., Guidi, R., and Frisan, T. (2011). The biology of the cytolethal distending toxins. Toxins (Basel) 3, 172–190.

Grimaldi, D., Bonacorsi, S., Roussel, H., Zuber, B., Poupet, H., Chiche, J.D., Poyart, C., and Mira, J.P. (2010). Unusual 'flesh-eating' strain of Escherichia coli. J. Clin. Microbiol. 48, 3794–3796.

Hacker, J., Blum-Oehler, G., Mühldorfer, I., and Tschäpe, H. (1997). Pathogenicity islands of virulent bacteria:

structure, function and impact on microbial evolution. Mol. Microbiol. 23, 1089–1097.

Haghjoo, E.G.J. (2004). *Salmonella typhi* encodes a functional cytolethal distending toxin that is delivered into host cells by a bacterial-internalization pathway. Proc. Natl. Acad. Sci. U.S.A. 101, 4614–4619.

Hassane, D.C., Lee, R.B., and Pickett, C.L. (2003). *Campylobacter jejuni* cytolethal distending toxin promotes DNA repair responses in normal human cells. Infect. Immun. 71, 541–545.

Heywood, W., Henderson, B., and Nair, S.P. (2005). Cytolethal distending toxin: creating a gap in the cell cycle. J. Med. Microbiol. 54, 207–216.

Hinenoya, A., Naigita, A., Ninomiya, K., Asakura, M., Shima, K., Seto, K., Tsukamoto, T., Ramamurthy, T., Faruque, S.M., and Yamasaki, S. (2009). Prevalence and characteristics of cytolethal distending toxin-producing *Escherichia coli* from children with diarrhoea in Japan. Microbiol. Immunol. 53, 206–215.

Hofman, P., Flatau, G., Selva, E., Gauthier, M., Le Negrate, G., Fiorentini, C., Rossi, B., and Boquet, P. (1998). *Escherichia coli* cytotoxic necrotizing factor 1 effaces microvilli and decreases transmigration of polymorphonuclear leukocytes in intestinal T84 epithelial cell monolayers. Infect. Immun. 66, 2494–2500.

Horiguchi, Y. (2001). *Escherichia coli* cytotoxic necrotizing factors and *Bordetella* dermonecrotic toxin: the dermonecrosis-inducing toxins activating Rho small GTPases. Toxicon. 39, 1619–1627.

Houdouin, V., Bonacorsi, S., Brahimi, N., Clermont, O., Nassif, X., and Bingen, E. (2002). A uropathogenicity island contributes to the pathogenicity of *Escherichia coli* strains that cause neonatal meningitis. Infect. Immun. 70, 5865–5869.

Hsu, Y., Jubelin, G., Taieb, F., Nougayrède, J.P., Oswald, E., and Stebbins, C.E. (2008). Structure of the cyclomodulin Cif from pathogenic *Escherichia coli*. J. Mol. Biol. 384, 465–477.

Janka, A., Bielaszewska, M., Dobrindt, U., Greune, L., Schmidt, M.A., and Karch, H. (2003). Cytolethal distending toxin gene cluster in enterohemorrhagic *Escherichia coli* O157:H– and O157:H7: characterization and evolutionary considerations. Infect. Immun. 71, 3634–3638.

Jinadasa, R.N., Bloom, S.E., Weiss, R.S., and Duhamel, G.E. (2011). Cytolethal distending toxin: a conserved bacterial genotoxin that blocks cell cycle progression, leading to apoptosis of a broad range of mammalian cell lineages. Microbiology 157, 1851–1875.

Johnson, J.R., Oswald, E., O'Bryan, T.T., Kuskowski, M.A., and Spanjaard, L. (2002). Phylogenetic distribution of virulence-associated genes among *Escherichia coli* isolates associated with neonatal bacterial meningitis in the Netherlands. J. Infect. Dis. 185, 774–784.

Johnson, J.R., Johnston, B., Kuskowski, M.A., Nougayrède, J., and Oswald, E. (2008). Molecular epidemiology and phylogenetic distribution of the *Escherichia coli* pks genomic island. J. Clin. Microbiol. 46, 3906–3911.

Johnson, T.J., Kariyawasam, S., Wannemuehler, Y., Mangiamele, P., Johnson, S.J., Doetkott, C., Skyberg, J.A., Lynne, A.M., Johnson, J.R., and Nolan, L.K. (2007). The genome sequence of avian pathogenic *Escherichia coli* strain O1:K1:H7 shares strong similarities with human extraintestinal pathogenic *E. coli* genomes. J. Bacteriol. 189, 3228–3236.

Johnson, T.J., DebRoy, C., Belton, S., Williams, M.L., Lawrence, M., Nolan, L.K., and Thorsness, J.L. (2010). Pyrosequencing of the Vir plasmid of necrotoxigenic *Escherichia coli*. Vet. Microbiol. 144, 100–109.

Johnson, W.M., and Lior, H. (1988). A new heat-labile cytolethal distending toxin (CDT) produced by *Escherichia coli* isolates from clinical material. Microb. Pathog. 4, 103–113.

Jubelin, G., Chavez, C.V., Taieb, F., Banfield, M.J., Samba-Louaka, A., Nobe, R., Nougayrède, J.P., Zumbihl, R., Givaudan, A., Escoubas, J.M., and Oswald, E. (2009). Cycle inhibiting factors (CIFs) are a growing family of functional cyclomodulins present in invertebrate and mammal bacterial pathogens. PLoS One 4, e4855.

Kadhum, H.J., Finlay, D., Rowe, M.T., Wilson, I.G., and Ball, H.J. (2008). Occurrence and characteristics of cytotoxic necrotizing factors, Cytolethal distending toxins and other virulence factors in *Escherichia coli* from human blood and faecal samples. Epidemiol. Infect. 136, 752–760.

Kaper, J.B., Nataro, J.P., and Mobley, H.L. (2004). Pathogenic *Escherichia coli*. Nat. Rev. Microbiol. 2, 123–140.

Kim, J., Kim, J., Choo, Y., Jang, H., Choi, Y., Chung, J., Cho, S., Park, M., and Lee, B. (2009). Detection of cytolethal distending toxin and other virulence characteristics of enteropathogenic *Escherichia coli* isolates from diarrheal patients in Republic of Korea. J. Microbiol. Biotechnol. 19, 525–529.

Krieger, J.N., Dobrindt, U., Riley, D.E., and Oswald, E. (2011). Acute *Escherichia coli* prostatitis in previously healthy young men: bacterial virulence factors, antimicrobial resistance, and clinical outcomes. Urology 77, 1420–1425.

Lara-Tejero, M., and Galán, J.E. (2001). CdtA, CdtB, and CdtC form a tripartite complex that is required for cytolethal distending toxin activity. Infect. Immun. 69, 4358–4365.

Lee, R.B., Hassane, D.C., Cottle, D.L., and Pickett, C.L. (2003). Interactions of *Campylobacter jejuni* cytolethal distending toxin subunits CdtA and CdtC with HeLa cells. Infect. Immun. 71, 4883–4890.

Lemichez, E., Flatau, G., Bruzzone, M., Boquet, P., and Gauthier, M. (1997). Molecular localization of the *Escherichia coli* cytotoxic necrotizing factor CNF1 cell-binding and catalytic domains. Mol. Microbiol. 24, 1061–1070.

Lemonnier, M., Landraud, L., and Lemichez, E. (2007). Rho GTPase-activating bacterial toxins: from bacterial virulence regulation to eukaryotic cell biology. FEMS Microbiol. Rev. 31, 515–534.

Lerm, M., Selzer, J., Hoffmeyer, A., Rapp, U.R., Aktories, K., and Schmidt, G. (1999). Deamidation of Cdc42 and Rac by *Escherichia coli* cytotoxic necrotizing factor 1: activation of c-Jun N-terminal kinase in HeLa cells. Infect. Immun. 67, 496–503.

Lewis, D.A., Stevens, M.K., Latimer, J.L., Ward, C.K., Deng, K., Blick, R., Lumbley, S.R., Ison, C.A., and Hansen, E.J. (2001). Characterization of *Haemophilus ducreyi* cdtA, cdtB, and cdtC mutants in *in vitro* and *in vivo* systems. Infect. Immun. *69*, 5626–5634.

Li, L., Shapiro, A., Chaves-Olarte, E., Masucci, M.G., Levitsky, V., Thelestam, M., and Frisan, T. (2002). The *Haemophilus ducreyi* cytolethal distending toxin activates sensors of DNA damage and repair complexes in proliferating and non-proliferating cells. Cell. Microbiol. *4*, 87–99.

Liyanage, N.P.M., Manthey, K.C., Dassanayake, R.P., Kuszynski, C.A., Oakley, G.G., and Duhamel, G.E. (2010). *Helicobacter hepaticus* cytolethal distending toxin causes cell death in intestinal epithelial cells via mitochondrial apoptotic pathway. Helicobacter *15*, 98–107.

Loukiadis, E., Nobe, R., Herold, S., Tramuta, C., Ogura, Y., Ooka, T., Morabito, S., Kérourédan, M., Brugère, H., Schmidt, H., *et al.* (2008). Distribution, functional expression, and genetic organization of Cif, a phage-encoded type III-secreted effector from enteropathogenic and enterohemorrhagic *Escherichia coli*. J. Bacteriol. *190*, 275–285.

McAuley, J.L., Linden, S.K., Png, C.W., King, R.M., Pennington, H.L., Gendler, S.J., Florin, T.H., Hill, G.R., Korolik, V., and McGuckin, M.A. (2007). MUC1 cell surface mucin is a critical element of the mucosal barrier to infection. J. Clin. Invest. *117*, 2313–2324.

McSweeney, L.A., and Dreyfus, L.A. (2004). Nuclear localization of the *Escherichia coli* cytolethal distending toxin CdtB subunit. Cell. Microbiol. *6*, 447–458.

McSweeney, L.A., and Dreyfus, L.A. (2005). Carbohydrate-binding specificity of the *Escherichia coli* cytolethal distending toxin CdtA-II and CdtC-II subunits. Infect. Immun. *73*, 2051–2060.

Mainil, J.G., Jacquemin, E., and Oswald, E. (2003). Prevalence and identity of cdt-related sequences in necrotoxigenic *Escherichia coli*. Vet. Microbiol. *94*, 159–165.

Marchès, O., Ledger, T.N., Boury, M., Ohara, M., Tu, X., Goffaux, F., Mainil, J., Rosenshine, I., Sugai, M., De Rycke, J., and Oswald, E. (2003). Enteropathogenic and enterohaemorrhagic *Escherichia coli* deliver a novel effector called Cif, which blocks cell cycle G2/M transition. Mol. Microbiol. *50*, 1553–1567.

Mayer, M.P., Bueno, L.C., Hansen, E.J., and DiRienzo, J.M. (1999). Identification of a cytolethal distending toxin gene locus and features of a virulence-associated region in *Actinobacillus actinomycetemcomitans*. Infect. Immun. *67*, 1227–1237.

Munro, P., Flatau, G., Anjuère, F., Hofman, V., Czerkinsky, C., and Lemichez, E. (2005). The Rho GTPase activators CNF1 and DNT bacterial toxins have mucosal adjuvant properties. Vaccine *23*, 2551–2556.

Munro, P., Flatau, G., Doye, A., Boyer, L., Oregioni, O., Mege, J.L., Landraud, L., and Lemichez, E. (2004). Activation and proteasomal degradation of rho GTPases by cytotoxic necrotizing factor-1 elicit a controlled inflammatory response. J. Biol. Chem. *279*, 35849–35857.

Nesić, D., Hsu, Y., and Stebbins, C.E. (2004). Assembly and function of a bacterial genotoxin. Nature *429*, 429–433.

Nishikubo, S., Ohara, M., Ueno, Y., Ikura, M., Kurihara, H., Komatsuzawa, H., Oswald, E., and Sugai, M. (2003). An N-terminal segment of the active component of the bacterial genotoxin cytolethal distending toxin B (CDTB) directs CDTB into the nucleus. J. Biol. Chem. *278*, 50671–50681.

Nougayrède, J.P., Marches, O., Boury, M., Mainil, J., Charlier, G., Pohl, P., De Rycke, J., Milon, A., and Oswald, E. (1999). The long-term cytoskeletal rearrangement induced by rabbit enteropathogenic *Escherichia coli* is Esp dependent but intimin independent. Mol. Microbiol. *31*, 19–30.

Nougayrède, J.P., Boury, M., Tasca, C., Marches, O., Milon, A., Oswald, E., and De Rycke, J. (2001). Type III secretion dependent cell cycle block caused in HeLa cells by enteropathogenic *Escherichia coli* O103. Infect. Immun. *69*, 6785–6795.

Nougayrède, J.P., Taieb, F., De Rycke, J., and Oswald, E. (2005). Cyclomodulins: bacterial effectors that modulate the eukaryotic cell cycle. Trends Microbiol. *13*, 103–110.

Nougayrède, J.P., Homburg, S., Taieb, F., Boury, M., Brzuszkiewicz, E., Gottschalk, G., Buchrieser, C., Hacker, J., Dobrindt, U., and Oswald, E. (2006). *Escherichia coli* induces DNA double-strand breaks in eukaryotic cells. Science *313*, 848–851.

Nowrouzian, F.L., and Oswald, E. (2012). *Escherichia coli* strains with the capacity for long-term persistence in the bowel microbiota carry the potentially genotoxic pks island. Microb. Pathog. *53*, 180–182.

Ohara, M., Oswald, E., and Sugai, M. (2004). Cytolethal distending toxin: a bacterial bullet targeted to nucleus. J. Biochem. *136*, 409–413.

Ohara, M., Miyauchi, M., Tsuruda, K., Takata, T., and Sugai, M. (2011). Topical application of *Aggregatibacter actinomycetemcomitans* cytolethal distending toxin induces cell cycle arrest in the rat gingival epithelium *in vivo*. J. Periodont. Res. *46*, 389–395.

Okeke, I.N., Lamikanra, A., Steinrück, H., and Kaper, J.B. (2000). Characterization of *Escherichia coli* strains from cases of childhood diarrhoea in provincial southwestern Nigeria. J. Clin. Microbiol. *38*, 7–12.

Orden, J.A., Ruiz-Santa-Quiteria, J.A., Cid, D., Garcia, S., and de la Fuente, R. (1999). Prevalence and characteristics of necrotoxigenic *Escherichia coli* (NTEC) strains isolated from diarrhoeic dairy calves. Vet. Microbiol. *66*, 265–273.

Orden, J.A., Cid, D., Ruiz-Santa-Quiteria, J.A., García, S., Martínez, S., and de la Fuente R. (2002). Verotoxin-producing *Escherichia coli* (VTEC), enteropathogenic *E. coli* (EPEC) and necrotoxigenic *E. coli* (NTEC) isolated from healthy cattle in Spain. J. Appl. Microbiol. *93*, 29–35.

Orden, J.A., Dominguez-Bernal, G., Martinez-Pulgarin, S., Blanco, M., Blanco, J.E., Mora, A., Blanco, J., Blanco, J., and de la Fuente, R. (2007). Necrotoxigenic *Escherichia coli* from sheep and goats produce a new type of cytotoxic necrotizing factor (CNF3) associated with the *eae* and *ehxA* genes. Int. Microbiol. *10*, 47–55.

Orth, D., Grif, K., Dierich, M.P., and Würzner, R. (2006). Cytolethal distending toxins in Shiga toxin-producing *Escherichia coli*: alleles, serotype distribution and biological effects. J. Med. Microbiol. 55, 1487–1492.

Osek, J. (2001). Characterization of necrotoxigenic *Escherichia coli* (NTEC) strains isolated from healthy calves in Poland. J. Vet. Med. B Infect. Dis. Vet. Public Health 48, 641–646.

Osek, J. (2005). Detection of cytolethal distending toxin genes in Shiga toxin-producing *Escherichia coli* isolated from different sources. Bull. Vet. Inst. Pulawy 49, 153–156.

Oswald, E., De Rycke, J., Guillot, J.F., and Boivin, R. (1989). Cytotoxic effect of multinucleation in HeLa cell cultures associated with the presence of Vir plasmid in *Escherichia coli* strains. FEMS Microbiol. Lett. 49, 95–99.

Oswald, E., De Rycke, J., Lintermans, P., van Muylem, K., Mainil, J., Daube, G., and Pohl, P. (1991). Virulence factors associated with cytotoxic necrotizing factor type two in bovine diarrheic and septicemic strains of *Escherichia coli*. J. Clin. Microbiol. 29, 2522–2527.

Oswald, E., Sugai, M., Labigne, A., Wu, H.C., Fiorentini, C., Boquet, P., and O'Brien, A.D. (1994). Cytotoxic necrotizing factor type 2 produced by virulent *Escherichia coli* modifies the small GTP-binding proteins Rho involved in assembly of actin stress fibers. Proc. Natl. Acad. Sci. U.S.A. 26, 3814–3818.

Oswald, E., Nougayrède, J.P., Taieb, F., and Sugai, M. (2005). Bacterial toxins that modulate host cell-cycle progression. Curr. Opin. Microbiol. 8, 83–91.

Pandey, M., Khan, A., Das, S.C., Sarkar, B., Kahali, S., Chakraborty, S., Chattopadhyay, S., Yamasaki, S., Takeda, Y., Nair, G.B., and Ramamurthy, T. (2003). Association of cytolethal distending toxin locus cdtB with enteropathogenic *Escherichia coli* isolated from patients with acute diarrhoea in Calcutta, India. J. Clin. Microbiol. 41, 5277–5281.

Parkhill, J., Dougan, G., James, K.D., Thomson, N.R., Pickard, D., Wain, J., Churcher, C., Mungall, K.L., Bentley, S.D., Holden, M.T., et al. (2001). Complete genome sequence of a multiple drug resistant *Salmonella enterica* serovar Typhi CT18. Nature 413, 848–852.

Pérès, S.Y., Marchès, O., Daigle, F., Nougayrède, J.P., Herault, F., Tasca, C., De Rycke, J., and Oswald, E. (1997). A new cytolethal distending toxin (CDT) from *Escherichia coli* producing CNF2 blocks HeLa cell division in G2/M phase. Mol. Microbiol. 24, 1095–1107.

Petroski, M.D., and Deshaies, R.J. (2005). Function and regulation of cullin-RING ubiquitin ligases. Nat. Rev. Mol. Cell. Biol. 6, 9–20.

Pfeifer, B.A., Wang, C.C., Walsh, C.T., and Khosla, C. (2003). Biosynthesis of Yersiniabactin, a complex polyketide-nonribosomal peptide, using *Escherichia coli* as a heterologous host. Appl. Environ. Microbiol. 69, 6698–6702.

Pickart, C.M., and Eddins, M.J. (2004). Ubiquitin: structures, functions, mechanisms. Biochim. Biophys. Acta 1695, 55–72.

Pickett, C.L., Cottle, D.L., Pesci, E.C., and Bikah, G. (1994). Cloning, sequencing, and expression of the *Escherichia coli* cytolethal distending toxin genes. Infect. Immun. 62, 1046–1051.

Pizarro-Cerdá, J., and Cossart, P. (2006). Bacterial adhesion and entry into host cells. Cell 124, 715–727.

Purdy, D., Buswell, C.M., Hodgson, A.E., McAlpine, K., Henderson, I., and Leach, S.A. (2000). Characterisation of cytolethal distending toxin (CDT) mutants of *Campylobacter jejuni*. J. Med. Microbiol. 49, 473–479.

Putze, J., Hennequin, C., Nougayrède, J., Zhang, W., Homburg, S., Karch, H., Bringer, M., Fayolle, C., Carniel, E., Rabsch, W., et al. (2009). Genetic structure and distribution of the colibactin genomic island among members of the family Enterobacteriaceae. Infect. Immun. 77, 4696–4703.

Rabin, S.D.P., Flitton, J.G., and Demuth, D.R. (2009). *Aggregatibacter actinomycetemcomitans* cytolethal distending toxin induces apoptosis in nonproliferating macrophages by a phosphatase-independent mechanism. Infect. Immun. 77, 3161–3169.

Rippere-Lampe, K.E., Lang, M., Ceri, H., Olson, M., Lockman, H.A., and O'Brien, A.D. (2001a). Cytotoxic necrotizing factor type 1-positive *Escherichia coli* causes increased inflammation and tissue damage to the prostate in a rat prostatitis model. Infect. Immun. 69, 6515–6519.

Rippere-Lampe, K.E., O'Brien, A.D., Conran, R., and Lockman, H.A. (2001b). Mutation of the gene encoding cytotoxic necrotizing factor type 1 (cnf(1)) attenuates the virulence of uropathogenic *Escherichia coli*. Infect. Immun. 69, 3954–3964.

Saiki, K., Konishi, K., Gomi, T., Nishihara, T., and Yoshikawa, M. (2001). Reconstitution and purification of cytolethal distending toxin of *Actinobacillus actinomycetemcomitans*. Microbiol. Immunol. 45, 497–506.

Salvarani, S., Tramuta, C., Nebbia, P., and Robino, P. (2012). Occurrence and functionality of cycle inhibiting factor, cytotoxic necrotising factors and cytolethal distending toxins in *Escherichia coli* isolated from calves and dogs in Italy. Res. Vet. Sci. 92, 372–377.

Samba-Louaka, A., Nougayrède, J.P., Watrin, C., Jubelin, G., Oswald, E., and Taieb, F. (2008). Bacterial cyclomodulin Cif blocks the host cell cycle by stabilizing the cyclin-dependent kinase inhibitors p21[wafl] and p27[kip1]. Cell. Microbiol. 10, 2496–2508.

Schmidt, G., Sehr, P., Wilm, M., Selzer, J., Mann, M., and Aktories, K. (1997). Gln 63 of Rho is deamidated by *Escherichia coli* cytotoxic necrotizing factor-1. Nature 387, 725–729.

Schmidt, G., Selzer, J., Lerm, M., and Aktories, K. (1998). The Rho-deamidating cytotoxic necrotizing factor 1 from *Escherichia coli* possesses transglutaminase activity. Cysteine 866 and histidine 881 are essential for enzyme activity. J. Biol. Chem. 273, 13669–13674.

Scott, D.A., and Kaper, J.B. (1994). Cloning and sequencing of the genes encoding *Escherichia coli* cytolethal distending toxin. Infect. Immun. 62, 244–251.

Sert, V., Cans, C., Tasca, C., Bret-Bennis, L., Oswald, E., Ducommun, B., and De Rycke, J. (1999). The bacterial cytolethal distending toxin (CDT) triggers a G2 cell cycle checkpoint in mammalian cells without preliminary induction of DNA strand breaks. Oncogene 18, 6296–6304.

Shenker, B.J., Hoffmaster, R.H., Zekavat, A., Yamaguchi, N., Lally, E.T., and Demuth, D.R. (2001). Induction of apoptosis in human T cells by Actinobacillus actinomycetemcomitans cytolethal distending toxin is a consequence of G2 arrest of the cell cycle. J. Immunol. 167, 435–441.

Smith, H.W. (1974). A search for transmissible pathogenic characters in invasive strains of Escherichia coli: the discovery of a plasmid-controlled toxin and a plasmid-controlled lethal character closely associated, or identical, with colicine V. J. Gen. Microbiol. 83, 95–111.

Smith, H.W., and Huggins, M.B. (1976). Further observations on the association of the Colicine V plasmid of Escherichia coli with pathogenicity and with survival in the alimentary tract. J. Gen. Microbiol. 92, 335–350.

Spanò, S., Ugalde, J.E., and Galán, J.E. (2008). Delivery of a Salmonella Typhi exotoxin from a host intracellular compartment. Cell Host Microbe 3, 30–38.

Svensson, L., Tarkowski, A., Thelestam, M., and Lagergård, T. (2001). The impact of Haemophilus ducreyi cytolethal distending toxin on cells involved in immune response. Microb. Pathog. 30, 157–166.

Taieb, F., Nougayrède, J.P., Watrin, C., Samba-Louaka, A., and Oswald, E. (2006). Escherichia coli cyclomodulin Cif induces G2 arrest of the host cell cycle without activation of the DNA-damage checkpoint-signalling pathway. Cell. Microbiol. 8, 1910–1921.

Taieb, F., Nougayrède, J.P., and Oswald, E. (2011). Cycle inhibiting factors (cifs): cyclomodulins that usurp the ubiquitin-dependent degradation pathway of host cells. Toxins 3, 356–368.

Touchon, M., Hoede, C., Tenaillon, O., Barbe, V., Baeriswyl, S., Bidet, P., Bingen, E., Bonacorsi, S., Bouchier, C., Bouvet, O., et al. (2009). Organised genome dynamics in the Escherichia coli species results in highly diverse adaptive paths. PLoS Genet. 5, e1000344.

Tóth, I., Oswald, E., Mainil, J.G., Awad-Masalmeh, M., and Nagy, B. (2000). Characterization of intestinal cnf1 + Escherichia coli from weaned pigs. Int. J. Med. Microbiol. 290, 539–542.

Tóth, I., Hérault, F., Beutin, L., and Oswald, E. (2003). Production of cytolethal distending toxins by pathogenic Escherichia coli strains isolated from human and animal sources: establishment of the existence of a new cdt variant (Type IV). J. Clin. Microbiol. 41, 4285–4291.

Tóth, I., Nougayrède, J., Dobrindt, U., Ledger, T.N., Boury, M., Morabito, S., Fujiwara, T., Sugai, M., Hacker, J., and Oswald, E. (2009a). Cytolethal distending toxin type I and type IV genes are framed with lambdoid prophage genes in extraintestinal pathogenic Escherichia coli. Infect. Immun. 77, 492–500.

Tóth, I., Schmidt, H., Kardos, G., Lancz, Z., Creuzburg, K., Damjanova, I., Pászti, J., Beutin, L., and Nagy, B. (2009b). Virulence genes and molecular typing of different groups of Escherichia coli O157 strains in cattle. Appl. Environ. Microbiol. 75, 6282–6291.

Tramuta, C., Nucera, D., Robino, P., Salvarani, S., and Nebbia, P. (2011). Virulence factors and genetic variability of uropathogenic Escherichia coli isolated from dogs and cats in Italy. J. Vet. Sci. 12, 49–55.

Van Bost, S., Roels, S., Oswald, E., and Mainil, J. (2003). Putative roles of the CNF2 and CDTIII toxins in experimental infections with necrotoxigenic Escherichia coli type 2 (NTEC2) strains in calves. Microbes Infect. 13, 1189–1193.

Wising, C., Azem, J., Zetterberg, M., Svensson, L.A., Ahlman, K., and Lagergård, T. (2005). Induction of apoptosis/necrosis in various human cell lineages by Haemophilus ducreyi cytolethal distending toxin. Toxicon 45, 767–776.

Wray, C., Piercy, D.W., Carroll, P.J., and Cooley WA. 1993. Experimental infection of neonatal pigs with CNF toxin-producing strains of Escherichia coli. Res Vet Sci. 54, 290–298.

Wu, Y., Hinenoya, A., Taguchi, T., Nagita, A., Shima, K., Tsukamoto, T., Sugimoto, N., Asakura, M., and Yamasaki, S. (2010). Distribution of virulence genes related to adhesins and toxins in shiga toxin-producing Escherichia coli strains isolated from healthy cattle and diarrhoeal patients in Japan. J. Vet. Med. Sci. 72, 589–597.

Xu, T., Lundqvist, A., Ahmed, H.J., Eriksson, K., Yang, Y., and Lagergård, T. (2004). Interactions of Haemophilus ducreyi and purified cytolethal distending toxin with human monocyte-derived dendritic cells, macrophages and CD4+ T cells. Microbes Infect. 6, 1171–1181.

Yao, Q., Cui, J., Zhu, Y., Wang, G., Hu, L., Long, C., Cao, R., Liu, X., Huang, N., Chen, S., et al. (2009). A bacterial type III effector family uses the papain-like hydrolytic activity to arrest the host cell cycle. Proc. Natl. Acad. Sci. U.S.A. 106, 3716–3721.

Young, R.S., Fortney, K.R., Gelfanova, V., Phillips, C.L., Katz, B.P., Hood, A.F., Latimer, J.L., Munson, Jr., R.S., Hansen, E.J., and Spinola, S.M. (2001). Expression of cytolethal distending toxin and hemolysin is not required for pustule formation by Haemophilus ducreyi in human volunteers. Infect. Immun. 69, 1938–1942.

The Heat-stable and Heat-labile Enterotoxins Produced by Enterotoxigenic *Escherichia coli*

8

Sascha Kopic, Ahmad Saleh and John P. Geibel

Abstract

Enterotoxigenic *Escherichia coli* (ETEC) produces two toxins: a heat-stable and a heat-labile enterotoxin. Both toxins differ strongly in their structure, their respective receptors and their intracellular mode of action. However, they target common molecular endpoints on the apical membrane of the enterocyte, leading to an intestinal hypersecretion of chloride and a concomitant inhibition of sodium absorption. This chapter will review toxin structure, their receptors and the intracellular processing and signalling involved in toxin exposure. Furthermore, basic principles of intestinal water and ion homeostasis are discussed.

Introduction

Enterotoxigenic *Escherichia coli* (ETEC) infection is a major cause of infantile diarrhoea and travellers' diarrhoea. Given the faecal transmission route, ETEC associated diarrhoea is mostly a problem of developing countries where poor sanitary conditions prevail. The incidence of ETEC associated diarrhoea among children in underdeveloped countries is staggering. In rural Egypt, for example, it has been estimated that ETEC diarrhoea has an incidence of 1.5 episodes per child per year (< 3 years of age) (Rao *et al.*, 2003). The total incidence of diarrhoea attacks was estimated to be 5.5 episodes/year with ETEC infection representing the most common cause of diarrhoea in this cohort (Rao *et al.*, 2003). Young children are especially vulnerable to the loss of electrolytes and water taking place in the course of secretory diarrhoea. It has been estimated that ETEC causes up to 370,000 deaths each year

(Steffen *et al.*, 2005; Wenneras and Erling, 2004). The second group that is vulnerable to ETEC infection is travellers from industrialized countries. 20 to 40% of all cases of travellers' diarrhoea are attributable to ETEC infection (Black, 1990; Jiang *et al.*, 2002; Sack, 1990).

A plethora of ETEC strains exists. They mostly differ in the expression of colonization factors, allowing them to effectively adhere to the intestinal mucosa, their respective O and H serogroups and in the production of one of two (or both) enterotoxins. To date, 78 O– and 34 H– antigen serogroups have been described (Wolf, 1997). For the purpose of this chapter we will closely examine the two enterotoxins and their pathophysiological mode of action. It has been discovered in the late 1960s that ETEC produces two major enterotoxins (Gyles and Barnum, 1969; Smith and Gyles, 1970; Smith and Halls, 1967). In that period, ETEC infection had not yet been linked to diarrhoeal disease in human, but was extensively investigated in the context of post-weaning diarrhoea in piglets. The major factor delineating the two toxins was their resilience to heat (Gyles and Barnum, 1969; Smith and Gyles, 1970; Smith and Halls, 1967). It had been observed that one toxin was inactivated by temperatures over 60°C, whereas the other retained its activity at temperatures above 100°C (Gyles and Barnum, 1969; Smith and Halls, 1967). The toxins were therefore named heat-stable enterotoxin (ST) and heat-labile enterotoxin (LT), respectively (Smith and Gyles, 1970). This nomenclature is in use to the present day. In the 1970s it became apparent that STa is also responsible for acute diarrhoeal disease in humans (Hughes *et al.*, 1980; Ryder *et al.*, 1976;

Sack *et al.*, 1975). It should be noted that not all ETEC strains produce both toxins. A large study conducted in Bangladesh, for example, reports that of all ETEC positive samples 50% produce STa, 25% produce LT and 25% produce both STa and LT (Qadri *et al.*, 2000). However, this distribution is subject to a high variability between different regions (Qadri *et al.*, 2005).

Heat-stable enterotoxin

Two major classes of heat-stable enterotoxins have been isolated: the methanol soluble heat-stable enterotoxin a (STa, also abbreviated STI), which is the causative diarrhoeal agent in humans, and the methanol insoluble heat-stable enterotoxin b (STb, also abbreviated STII), which is mostly associated with diarrhoeal disease in pigs (Alderete and Robertson, 1978; Burgess *et al.*, 1978; Takeda *et al.*, 1979). STa can be further subdivided into two classes STp (STIa) and STh (STIb) based on differences in the amino acid sequence (Moseley *et al.*, 1983; So and McCarthy, 1980).

Structure

STa is a small 18–19 amino acid (AA) peptide encoded by bacterial plasmids (Aimoto *et al.*, 1982; Moseley *et al.*, 1983; So and McCarthy, 1980; Staples *et al.*, 1980). The heat stability is conferred to the molecule by the presence of three characteristic disulfide bonds (Staples *et al.*, 1980). The three disulfide bonds form between 6 cysteine residues, which are part of an essential tridecapeptide sequence, the so-called toxic domain of the peptide, which is sufficient to elicit diarrhoeagenic effects by itself (Aimoto *et al.*, 1983; Gariepy *et al.*, 1986; Gariepy and Schoolnik, 1986; Takao *et al.*, 1985). This sequence is also conserved among heat-stable enterotoxins produced by other bacteria, such as *Yersinia enterocolitica* or *Vibrio cholerae* non-O1 (Takao *et al.*, 1985). Furthermore, mutational analysis has demonstrated that the disulfide bridges are important for toxicity, as disruption of individual disulfide bridges by mutation impedes the toxin's potency (Gariépy *et al.*, 1987). Owing to its short sequence and its solubility the toxin has been crystalized fairly early (Gariepy *et al.*, 1986).

More recent structural reports suggest that the individual toxin molecules arrange in a hexameric ring formation, which may facilitate the clustering of the toxin's receptor, guanylyl cyclase C (GC-C) (see below), on the enterocyte surface (Sato and Shimonishi, 2004).

Receptor

Soon after identifying the structure of STa, attempts were made at elucidating its pathophysiological function. A fundamental observation with regard to the effects of STa on the enterocyte was made in 1978 when it was reported that STa exposure leads to an increase in the intracellular levels of cGMP (Fig. 8.1) (Field *et al.*, 1978; Hughes *et al.*, 1978). STa directly increases the activity of guanylate cyclase, an enzyme catalysing the conversion of GTP to cGMP (Field *et al.*, 1978; Guerrant *et al.*, 1980). Furthermore, investigators demonstrated that the toxin increases cyclase activity in the particulate, rather than in the soluble fraction of intestinal homogenates, which already pointed towards a membrane bound localization (Guerrant *et al.*, 1980).

Although some light had been shed on the intracellular effects of STa, not much was known about its receptor. First insights were gained in the 1980s by radioisotope labelling of STa, which revealed that the toxin binds to a receptor on the surface of the enterocyte (Gariepy and Schoolnik, 1986; Giannella *et al.*, 1983; Kuno *et al.*, 1986). Gianella *et al.* described the binding characteristics as 'rapid, reversible, linear with cell number and saturable' (Giannella *et al.*, 1983). Specificity could be demonstrated by the occurrence of competitive inhibition when non-labelled STa was added (Giannella *et al.*, 1983). However, the exact molecular identity of the receptor remained elusive. It was later independently discovered that membrane bound guanylyl cyclases (GC) could serve as peptide receptors in mammalian tissues (Chang *et al.*, 1989; Shimomura *et al.*, 1986). Furthermore, two isoforms of GC (GC-A and GC–B), both receptors for atrial natriuretic peptide (ANP), were cloned (Chinkers *et al.*, 1989; Lowe *et al.*, 1989). The observations that STa increases cGMP concentration and that guanylyl cyclases can serve as membrane bound receptors eventually culminated in the discovery of GC-C,

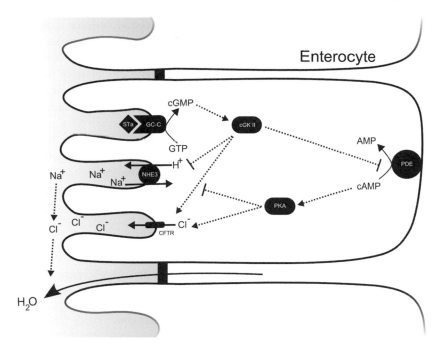

Figure 8.1 Cell model summarizing the intracellular effects of STa binding to GC-C.

the receptor of STa, in 1990 by Schulz *et al.* (1990) (Fig. 8.1). Human GC-C was cloned a year later and it was shown that GC-C overexpressing cells responded with an increase in intracellular cGMP levels upon STa exposure, thereby convincingly unravelling the identity of the STa receptor (de Sauvage *et al.*, 1991; Schulz *et al.*, 1990). The central role of GC-C in the pathophysiology of ETEC infection has been further corroborated by the generation of GC-C (−/−) animals. Mice lacking the cyclase are resistant to the secretory effects of STa (Mann *et al.*, 1997; Schulz *et al.*, 1997). In addition, baseline intestinal activity of guanylyl cyclase is severely reduced in GC-C (−/−) animals, suggesting that GC-C is the predominant intestinal isoform (Schulz *et al.*, 1997).

GC-C is not distributed evenly across the intestine. In the longitudinal axis, GC-C density decreases from proximal to distal small intestine (Krause *et al.*, 1994). With regard to the vertical axis, it has been reported that GC-C expression is most pronounced at the tip of the villus and decreases along the villus–crypt axis (Cohen *et al.*, 1992). Others have reported strongest expression at the base of the villus (Krause *et al.*, 1994). In the colon, expression is more pronounced in

the enterocytes lining the lumen of the crypt as opposed to the surface cells (Krause *et al.*, 1994).

Structurally, the GC-C protein can be divided into multiple domains. The extracellular domain serves as a binding site for STa and other peptide ligands (see below). It is followed by a single alpha-helix transmembrane domain. The cytoplasmic portion of the cyclase contains a kinase homology and a guanylyl cyclase domain.

The extracellular domain is responsible for ligand binding. Besides STa, two physiological ligands can bind to GC-C: guanylin and uroguanylin (Carpick and Gariepy, 1993; Currie *et al.*, 1992; Forte *et al.*, 1996; Kita *et al.*, 1994). Both peptides share significant homology with STa and function as intrinsic secretagogues in the intestine (Currie *et al.*, 1992; Forte *et al.*, 1996). They are produced by intestinal cells and locally modulate ion and water homeostasis. In analogy to STa, they induce intestinal chloride, bicarbonate and fluid secretion (Forte *et al.*, 1993; Guba *et al.*, 1996; Ieda *et al.*, 1999) (see below). Furthermore, guanylin and uroguanylin also exert endocrine effects in the kidney, where they induce natriuresis (Carrithers *et al.*, 1999, 2004; Greenberg *et al.*, 1997). Interestingly, knockout studies indicate that this process is GC-C independent and may function via distinct G-protein coupled receptors (Carrithers *et al.*, 2004; Sindic *et al.*, 2005). Apart from actions on ion transport, uroguanylin was recently shown to affect satiety in the hypothalamus in a GC-C dependent fashion (Valentino *et al.*, 2011).

Several attempts have been made at identifying the exact binding site of STa on the extracellular domain of GC-C. By using a photoaffinity labelled STa analogue, Hasegawa *et al.* (1999a) predicted that binding occurs in a subdomain close to the transmembrane region of the protein. Independent overexpression of this microdomain (Met341-Gln407) showed that it was indeed capable of binding to STa, confirming that it contained the ligand binding site (Hidaka *et al.*, 2002). As of now, we have no exact structural information on the extracellular domain of GC-C. However, homology modelling has been conducted by using the extracellular domain of GC-A as a template (Hasegawa *et al.*, 2005; Lauber *et al.*, 2009). Analyses further show that the extracellular domain contains four disulfide bonds (Hasegawa *et al.*, 2005). The extracellular domain of GC-C is heavily glycosylated (Hasegawa *et al.*, 1999b; Vaandrager *et al.*, 1993a). Expression experiments of GC-C in HEK293 cells demonstrated that cells express two forms of GC-C, a 145 kDa and a 130 kDa form, the former of which is expressed on the plasma membrane of the cells, whereas the latter represents an immature form of the protein localizing to the endoplasmic reticulum (Ghanekar *et al.*, 2004; Scheving *et al.*, 1996). Mutational analysis of the glycosylation sites suggests that one glycosylation site (N379) is of particular importance, as its mutation significantly decreased the STa-binding ability (decreased maximum capacity at normal affinity) and the structural stability (assayed by resistance to detergent) of GC-C (Hasegawa *et al.*, 1999b). This observation was expanded by a more recent investigation in which STa binding was assessed between normally glycosylated and entirely unglycosylated GC-C (Ghanekar *et al.*, 2004). Inhibition of glycosylation did not impair STa binding, however, cGMP production was severely reduced (Ghanekar *et al.*, 2004). The authors suggested that glycosylation may be

pivotal for conformational changes to take place after ligand binding (Ghanekar *et al.*, 2004).

The kinase homology domain of GC-C is located on the cytoplasmic side of the protein. It lacks kinase activity (hence the name kinase homology domain) and its function is not entirely clear. Multiple groups have performed deletion experiments and came to divergent results. It has been reported that deletion of the kinase homology domain leads to a constitutively active enzyme (suggesting an auto inhibitory function of the domain), whereas others report that deletion causes a loss of cyclase activity (Deshmane *et al.*, 1997; Rudner *et al.*, 1995). The discrepancy in the results may be attributable to the extent of the performed deletion. It is known that the activity of GC-C is also modulated by intracellular ATP (Bhandari *et al.*, 1999; Jaleel *et al.*, 2006; Rudner *et al.*, 1995). ATP potentiates cyclase activity in response to STa by stabilizing GC-C in its active conformation (de Sauvage *et al.*, 1991; Rudner *et al.*, 1995; Schulz *et al.*, 1990; Vaandrager *et al.*, 1993b). Mutational analysis has revealed that this modulation is mediated by the kinase homology domain (Bhandari *et al.*, 1999; Rudner *et al.*, 1995). The interaction between ATP and the kinase homology domain has been mapped to K516 (Jaleel *et al.*, 2006).

The guanylyl cyclase domain is responsible for generating cGMP and is followed by the C-terminal domain. The C-terminal domain contains an 11aa motif required for apical membrane targeting of GC-C (Hodson *et al.*, 2006). Furthermore, it attaches to the PDZ protein NHERF4 (aka IKEPP; intestinal and kidney-enriched PDZ protein), which can directly regulate the catalytic activity of GC-C (Scott *et al.*, 2002). It has also been proposed that the C-terminal domain contains a PKC phosphorylation site at S1029. PKC phosphorylation enhances the effects of ligand binding on cGMP production (Crane and Shanks, 1996).

GC-C as a therapeutic target

Besides its physiological expression on enterocytes, GC-C has also been detected on colon carcinoma cells (Carrithers *et al.*, 1996). This expression pattern prompted investigators to explore how GC-C can be used as a tumour marker. Specifically, any expression of GC-C by colorectal tumour cells in tissues outside of the intestine would allow GC-C to be used as a selective marker for the spread and metastasis of the cancer (Carrithers *et al.*, 1996). For example, it has been shown that tumour cells could be detected in the blood of patients with colon cancer by PCR detection of GC-C. GC-C was not found in the blood of patients with non-malignant intestinal pathologies and healthy control subjects, which emphasizes the adequacy of GC-C as a tumour marker (Carrithers *et al.*, 1996). Furthermore, investigators reported that GC-C could also be utilized as a therapeutic target. For example, GC-C activation was found to decrease the spread of colon cancer metastasis by regulating matrix metalloproteinase-9 (MMP-9) (Lubbe *et al.*, 2009). MMP-9, a gelatinase produced by colorectal cancer cells, is a critical determinant of metastatic disease, as it facilitates tumour micro invasion by digesting extracellular matrix components (Lubbe *et al.*, 2009). MMP-9 is regulated by cGMP and it was shown that GC-C activation decreases the accumulation of MMP-9, which in turn reduced metastatic spread by up to 20% (Lubbe *et al.*, 2009). Because intestinal tumours lack effective levels of endogenous GC-C activators, the reactivation of GC-C signalling can be achieved by exogenous administration of STa, which may represent a putative therapeutic strategy for colon cancer (Lubbe *et al.*, 2009).

As previously mentioned, the stimulation of GC-C leads to diarrhoeal disease. This mechanism can also be exploited for treatment of constipation. Linaclotide is a novel GC-C agonist that effectively alleviates symptoms of constipation by increasing intestinal fluid accumulation. Researchers found that 290 µg once daily significantly improved abdominal and bowel symptoms associated with IBS-C over 26 weeks of treatment (Chey *et al.*, 2012).

Furthermore, GC-C has been targeted to alleviate symptoms of STa mediated diarrhoea. Investigators have tried to cut the GC-C signalling cascade utilizing multiple approaches. One of them was to inhibit the binding of STa to GC-C by administering a GC-C inhibitor such as 2-chloroadenosine (2ClAdo), which was successful in disrupting ST-dependent signalling in intestinal

cells (Zhang *et al.*, 1999). This opened the door for a therapy that targets the molecular mechanism of secretory diarrhoea rather than treating symptoms by anti-motility therapy and conservative therapy such as fluid replacement.

Intracellular effects of GC-C activation

As described previously, STa binding to GC-C results in an acute increase in the intracellular levels of cGMP (Fig. 8.1) (Field *et al.*, 1978; Hughes *et al.*, 1978). The elevated cGMP levels initiate two pathways within the enterocyte: (1) a GMP-dependent protein kinase (cGK aka PKG) mediated pathway, (2) a PKA-mediated pathway (Vaandrager *et al.*, 2000). It has been suggested that both pathways are preferentially activated in different parts of the intestine. The cGKII pathway is thought to predominate in the small intestine, whereas the PKA pathway is mostly specific to the colon, although cross-activation between pathways is highly likely (Vaandrager *et al.*, 2000).

Elevations of cGMP activate cGK, ultimately leading to an increase in chloride and water secretion and inhibition of sodium absorption (see below and Fig. 8.1). In the epithelial cells of the intestine the isoform II of cGK (cGKII) predominates (de Jonge, 1981; Lohmann *et al.*, 1997; Markert *et al.*, 1995). Experiments in Ussing chambers utilizing cGKII inhibitors demonstrated that pharmacological inhibition of cGKII decreases the chloride secretory response of the epithelium following STa exposure (Vaandrager *et al.*, 1997a). This working model has been confirmed by the generation of cGKII deficient animals, which exhibit increased baseline water absorption and a severely blunted secretory response to STa (Pfeifer *et al.*, 1996; Vaandrager *et al.*, 2000). The altered baseline conditions in the (−/−) animals suggest that a constitutive secretory tonus is present in the intestinal mucosa, which is most likely mediated by the physiological secretagogues guanylin or uroguanylin (Vaandrager *et al.*, 2000). In the colon, the disruption of cGKII had a comparably small effect on the STa evoked salt and fluid response, illustrating the point that PKA mediated pathways may play a more central role in this intestinal segment (Vaandrager *et al.*, 2000).

There is evidence for the importance of PKA

activation in the process of STa mediated secretion. It has been shown in T84 cells (colon-like cells) that cGKII activators had little effect on chloride secretion, whereas agents elevating cAMP levels caused robust secretion (the latter, however, is an already well-established observation) (Forte *et al.*, 1992). The authors propose that cGMP leads to a cross activation of PKA, which in turn promotes fluid and chloride secretion in the crypt (Forte *et al.*, 1992). This cross-activation may be attributable to a cGMP-mediated inhibition of phosphodiesterases, leading to an accumulation of cAMP and subsequent PKA activation in the colonic enterocyte (Fig. 8.1) (Vaandrager *et al.*, 2000). It has for example been confirmed that STa exposure leads to an activation of the phosphodiesterase type 5 (PDE5) in the intestine (Sopory *et al.*, 2004). The hypothesis of a predominance of the cAMP/PKA pathway in the colon has been contested by a more recent investigation, which reports that elevation of cGMP lead to cGKII dependent secretion in primary human colonic cell cultures and T84 cells and that this effect was reversible by pharmacological inhibition of cGKII (Selvaraj *et al.*, 2000). In summary, the relative degree of involvement of PKA and cGKII dependent pathways in the intestine remains to be clarified, however, a parallel activation of both pathways seems plausible.

Plasma membrane effectors

GC-C activation by STa has three reported effects on enterocyte ion transport: (i) An increase in chloride secretion, (ii) an increase in bicarbonate secretion, and (iii) a reduction in sodium absorption (Fig. 8.1). Before we focus on the exact molecular players involved in these processes it is worthwhile examining basic aspects of intestinal salt homeostasis. Ion transport in the intestine occurs through two distinct routes: a transcellular route and a paracellular route. The cellular permeability of enterocytes to ions (transcellular route) is governed by the number and type of ion transporters/channels on the apical surface of the cell. Contrarily, the paracellular route is gated by so-called tight junctions, which are protein complexes surrounding the enterocytes on the lateral cell surface in an annular form. The exact molecular composition of these tight junctions affects the permeability of the paracellular route to certain ions. Secretory diarrhoea, i.e. fulminant water loss in the intestine, is the result of an increase in the intraluminal osmolarity of the intestine, which in turn is the result of an excessive secretion of ions (chloride and bicarbonate) and a concomitant inhibition of ion absorption (sodium) (Fig. 8.1). The increased luminal osmolarity causes water to be secreted. The exact route of the water flow is under debate. Both transcellular water movement through aquaporins and paracellular water movement have been discussed (Murek *et al.*, 2010).

CFTR

It has been reported very early that STa causes an increase in the secretion of chloride from enterocytes (Field *et al.*, 1978). This hypersecretion of chloride occurs through the cystic fibrosis transmembrane conductance regulator (CFTR) (Fig. 8.1). CFTR mostly localizes to the crypt region of the small intestine, although a subpopulation of cells with extremely high levels of CFTR expression exists in the villi (Ameen *et al.*, 1995). However, the physiological function of these CFTR high expresser cells is still elusive. CFTR is a complex channel protein, belonging to the ABC transporter family. Each channel is composed of five domains: The channel pore is formed by two membrane spanning domains (with six transmembrane helices each). Furthermore CFTR has two intracellular nucleotide binding domains, where ATP hydrolysis occurs, and an intracellular regulatory R-domain, which allows for modulation of channel gating by kinase phosphorylation (Sheppard and Welsh, 1999). CFTR is expressed in a variety of epithelia. Besides the intestine, it is most notably expressed in the pancreas and the airway epithelium, where it is responsible for the hydration of the mucinous secretions of the airways. CFTR has gained tremendous scientific attention, as its mutation causes cystic fibrosis (CF), which is the most common inherited disease in the Caucasian population. To date up to 1900 mutations of CFTR have been reported; however, the F508del is the most common, accounting for up to two-thirds of all alleles of affected patients (Bobadilla *et al.*, 2002; Database). F508del causes CFTR to misprocess and being retained in the ER, thereby effectively decreasing channel numbers on the cell surface (Cheng *et al.*, 1990). In accordance with its physiological role, CF patients are prone to pancreatic dysfunction and recurrent airway infection, due to inadequate mucocilliary clearance because of hyperviscous mucus. Patients eventually develop resistances to antibiotic treatment, resulting in potentially fatal infections of the respiratory tract. In the intestine, CFTR plays a crucial role in intestinal water balance. Apical CFTR channels are a primary route of intestinal chloride secretion and can thereby regulate the water content of the stool. CF and genetic mouse models are an effective way of understanding the physiological function of CFTR in this tissue. For example, one of the first manifestation of CF in patients is meconium ileus, a prenatal obstruction of stool (Eggermont, 1996). Furthermore, mice with mutated or deleted CFTR also frequently succumb to intestinal obstruction, emphasizing the channel's central role in the hydration of the intestinal contents (Colledge *et al.*, 1995; Grubb and Gabriel, 1997).

CFTR is mostly a cAMP-activated channel. Increasing levels in intracellular cAMP lead to PKA activation, which in turn phosphorylates CFTR at its R-domain leading to channel opening (Cheng *et al.*, 1991; Tabcharani *et al.*, 1991). Besides PKA various other kinases have also been reported to phosphorylate CFTR. These include: calcium/calmodulin dependent kinase, protein kinase C (PKC), the AMP-activated protein kinase (AMPK) (inhibitory action) and – in the context of secretory diarrhoea notably – cGKII (Jia *et al.*, 1997; Kongsuphol *et al.*, 2009; Vaandrager *et al.*, 1997b; Wagner *et al.*, 1991). There is, however, some doubt whether PKC phosphorylates CFTR directly or whether it exerts its action mostly indirectly via the regulatory protein NHERF (NHE regulatory factor) (Chappe *et al.*, 2003; Li *et al.*, 2007). As mentioned previously, GC-C activation increases the intracellular levels of cGMP, which activates cGKII. cGKII was shown to directly phosphorylate CFTR at the R-domain thereby functionally coupling the binding of STa to an increase in chloride secretion through CFTR (French *et al.*, 1995; Picciotto *et al.*, 1992; Vaandrager *et al.*, 1997b, 1998). Naturally, the spatial proximity of cGKII is essential for CFTR phosphorylation to take place. This has been

effectively illustrated by experiments employing a cGKII mutant, which lacks its correct membrane targeting sequence. Increases in cGMP in these cells do not lead to a phosphorylation of CFTR and thereby lack an adequate chloride secretory response (Vaandrager *et al.*, 1998). As an alternate route, cGMP may also cross activate PKA, which in turn also leads to CFTR phosphorylation and opening (see above). It should be noted that apart from directly modulating channel gating, an increase in channel trafficking to the apical membrane of the enterocyte represents a second strategy for increasing total chloride flux. In accordance with this hypothesis, it has been observed that both cGKII and PKA activation lead to an increase (up to 4-fold) of CFTR trafficking to the cell surface of the enterocyte (Golin-Bisello *et al.*, 2005).

The central role of CFTR in the process of ETEC-mediated diarrhoea is underlined by observations made in CFTR deficient animals. It has been demonstrated that intestinal tissues from CFTR mutant or (−/−) animals are immune to the secretory effects of STa (Seidler *et al.*, 1997). This resistance to the diarrhoeagenic effects of enterotoxins has led to the hypothesis that the high prevalence of CFTR mutations may be attributable to an innate resistance against ETEC, cholera and other diarrhoea entities, thereby conferring an evolutionary advantage to mutation carriers (Baxter *et al.*, 1988; Guggino, 1994; Hansson, 1988). However, this hypothesis could not be corroborated by investigations in heterozygous CF mice, although only acute and no chronic effects of toxin exposure were examined (Cuthbert *et al.*, 1995). Given that the hypersecretion of chloride through CFTR is the molecular endpoint of ETEC mediated diarrhoea, but also of other diarrhoea entities, such as cholera, the channel has become a therapeutic target for the treatment of acute diarrhoeal illness. In particular, non-absorbable small molecule CFTR inhibitors have been developed, with the aim of locally reducing chloride flux in the intestine (Ma *et al.*, 2002; Muanprasat *et al.*, 2004; Sonawane *et al.*, 2005). These inhibitors were shown to effectively reduce water and chloride secretion in various animal models of STa and cholera toxin (CTX) mediated diarrhoea (Ma *et al.*, 2002; Muanprasat *et al.*, 2004; Sonawane *et al.*, 2005). However, concerns have emerged regarding the specificity of these inhibitors as an impairment of mitochondrial function and increased reactive oxygen species (ROS) production have been observed following exposure (Kelly *et al.*, 2010). As of now, the efficacy of these compounds remains to be tested in a clinical setting.

NHE

STa exposure leads to an inhibition of sodium absorption in the intestine. Intestinal electroneutral salt absorption is mainly mediated by the apical isoforms of the Na, H-Exchanger (NHE) (Fig. 8.1). It has been observed in the mid 1970s that isolated intestinal brush boarder vesicles are capable of acidifying their environment, while simultaneously absorbing sodium (Murer *et al.*, 1976). The authors postulated the existence of a sodium, proton antiporter system (Murer *et al.*, 1976). Subsequently NHE isoform 1 was cloned (Sardet *et al.*, 1989). NHE1 is mostly expressed on the basolateral surface of epithelial cells, where it functions as a housekeeping protein by maintaining intracellular pH and volume homeostasis. Currently nine isoforms of NHE have been characterized in various tissues.

NHEs have 12 transmembrane segments and exchange one sodium ion for one proton (Levine *et al.*, 1993; Orlowski, 1993; Sardet *et al.*, 1989; Wakabayashi *et al.*, 2000; Yu *et al.*, 1993). Functional regulation by other kinases and PDZ proteins occurs by interactions at the intracellular C-terminal domain.

The apical surface of enterocytes expresses NHE2 and NHE3, which are responsible for bulk sodium absorption in the small intestine (Hoogerwerf *et al.*, 1996). Both isoforms have been detected on the protein level in human small and large intestine (Hoogerwerf *et al.*, 1996). Functionally, the importance of apical sodium absorption through NHEs is effectively illustrated by patients suffering from congenital sodium diarrhoea (Booth *et al.*, 1985; Keller *et al.*, 1990). Congenital sodium diarrhoea is characterized by defective sodium absorption through NHEs resulting in metabolic acidosis, dehydration and hyponatremia (Booth *et al.*, 1985; Keller *et al.*, 1990). Although a functional impairment

of sodium/proton exchange was demonstrated in tissue samples from patients, surprisingly, the disorder is not linked to direct mutations of NHE2 or NHE3 (Muller *et al.*, 2000). Physiologically, the relative contribution of each isoform to sodium absorption seems to vary between species, however, at least in the mouse NHE3 seems to be the predominant mediator of small intestinal sodium absorption (Gawenis *et al.*, 2002). This conclusion has been reached with the help of NHE2 and NHE3 deficient animals. While NHE2 (–/–) animals do not show prominent absorptive defects, NHE3 deficient mice suffer from sodium malabsorption and diarrhoea, thereby emulating the phenotype of congenital sodium diarrhoea (Gawenis *et al.*, 2002). NHE3 is strongly expressed in the villus region of the small intestine, with much lower expression occurring in the crypt region (Jakab *et al.*, 2011). This is in line with the current paradigm that the villus has mostly absorptive functions, whereas the crypt region is mostly secretory in its nature. However, it should be noted that crypts were also shown to absorb fluid and that this rigid functional delineation therefore may not be entirely appropriate and may be dependent on environmental cues (Singh *et al.*, 1995).

The activity of NHE is regulated by association to PDZ-adaptor proteins from the NHERF family. Currently four members of the NHERF family have been described (NHERF1–4). NHERF proteins serve as membrane scaffolding proteins that facilitate protein–protein interactions by creating large regulatory complexes. NHERF proteins are required for the inhibitory effects of cAMP, cGMP or calcium on NHE activity to take place. For example, it has been shown that NHERF1 is a prerequisite for cAMP triggered NHE3 inhibition in the kidney and various cell lines (Lamprecht *et al.*, 1998; Weinman *et al.*, 1995, 2001; Zizak *et al.*, 1999). However, in the small intestine NHERF1 is most likely not responsible for the inhibition of NHE3, as the inhibitory response to cAMP on NHE3 activity was still shown to be intact in NHERF1 deficient animals (Broere *et al.*, 2009; Murtazina *et al.*, 2007). It has rather been suggested that NHERF3 (aka PDZK1) functionally links elevations of cAMP and calcium to an inhibition of NHE3 in the small intestine (Broere *et al.*, 2009; Zachos *et al.*, 2009). This has been demonstrated by in vivo knockout studies and by knockdown of NHERF3 in cell culture models (Broere *et al.*, 2009; Zachos *et al.*, 2009). Of interest for STa mediated diarrhoea, cGMP linked NHE3 inhibition is dependent on NHERF2 (aka E3KARP) (Cha *et al.*, 2005; Chen *et al.*, 2010). NHERF2 ablation resulted in higher baseline fluid absorption in the ileum of affected animals (Chen *et al.*, 2010). More importantly, STa exposure did not lead to NHE3 inhibition in NHERF2 (–/–) animals (Chen *et al.*, 2010). In support of these findings it was shown that cGKII can bind to NHERF2 in vitro (Cha *et al.*, 2005). It is therefore highly likely that STa inhibits NHE3 in the small intestine via the cGMP-cGKII pathway and that this interaction is facilitated by NHERF2.

Bicarbonate

The third ion affected by STa is bicarbonate. While it is well known that STa increases the secretion of bicarbonate especially in the duodenum, much less is known about the molecular pathway of the bicarbonate movement. As CFTR can also serve as a bicarbonate conductance pathway, it seemed plausible that the increase in bicarbonate secretion was CFTR mediated. However, increased bicarbonate secretion following STa exposure can also be observed in tissues from CF patients and in CFTR deficient animals, making the channel's involvement in this process highly unlikely (Pratha *et al.*, 2000; Sellers *et al.*, 2005). Even more interestingly, bicarbonate secretion also occurs in the genetic absence of GC-C and CFTR (Sellers *et al.*, 2008). These findings not only further mystify the source of the bicarbonate flux, but also raise the question whether STa has GC-C independent effects, presumably by interacting with another receptor protein (Sellers *et al.*, 2008) (see above). In summary, further studies will be necessary to unravel the exact mechanism of increased bicarbonate secretion.

Heat-stable enterotoxin b (STb)

STb was historically believed to be associated only with ETEC of porcine origin. After thorough investigation, scientists were able to detect STb-producing strains in bovine, chicken and, most importantly, human ETEC genes (Dubreuil,

1997). To assess the relevance for human disease, several investigators gathered information about STb from countries in South-East Asia, such as Thailand and Malaysia. A STb+ strain was isolated from a farmer without recent history of diarrhoea. In a subsequent study a ETEC STb+ strain was found in 1 of 177 villagers with diarrhoea and in 12 of 1307 without diarrhoea (Dubreuil, 1997). Although the existence of STb producing strains in humans was uncovered, a causal relationship between STb and diarrhoeal disease could not be established.

Heat-labile enterotoxin

Structure and secretion

The ETEC heat-labile enterotoxin belongs to the family of AB_5 toxins. AB_5 toxins are large multimeric proteins comprised of a single A subunit and five identical B subunits, forming a ring (Gill *et al.*, 1981). The A subunit is the enzymatically active subunit (see below), whereas the B subunits serve as a scaffold and mediate binding to the enterocyte surface. The A subunit itself consist of two peptides (A1 and A2), which are linked by a disulfide bond (Gill *et al.*, 1981). The available crystal structure of LT shows that the A2 subunit links the A subunit to the pentameric B subunit via interactions in the pore of the B subunit ring (Sixma *et al.*, 1993). Besides LT, cholera toxin (CTX) and Shiga toxins are prominent members of the AB_5 toxin family. Given the close relationship there is strong homology of both the A and B subunits between CTX and LT (approx. 80%) (Dallas and Falkow, 1980; Spicer *et al.*, 1981; Yamamoto and Yokota, 1983). Both toxins are also immunologically related, as is reflected by the ability of CTX antiserum to inactivate LT (Gyles and Barnum, 1969). Despite the close homology between CTX and LT, it has been demonstrated that CTX has a greater toxicity in terms of inducing chloride secretion (short circuit current in Ussing chambers) of T84 monolayers (Rodighiero *et al.*, 1999). It has recently been suggested that this discrepancy may be attributable to differences in a 10aa segment in the A2 subunit of both toxins. By utilizing hybrid toxins investigators postulated that the CTX A2 domain creates a closer linkage to the B subunit ring, thereby increasing holotoxin stability (Rodighiero *et al.*, 1999).

Two forms of LT exist: LT-I and LT-II. While LT-I represents the initially identified LT with close relationship to CTX, LT-II is less common, chromosomally encoded (unlike LT-I which is plasmid encoded) and antigenically unrelated to LT-I (Green *et al.*, 1983). For example it was shown that only 2% of LT-producing *E. coli* strains isolated from humans contained LT-II coding genes (Seriwatana *et al.*, 1988). LT-II is a heterogeneous group of toxins and can further be subdivided into LT-IIa, b and c (Holmes *et al.*, 1986; Nardi *et al.*, 2005; Nawar *et al.*, 2010; Pickett *et al.*, 1986, 1987, 1989). It has been proposed that LTIIb is not associated with human disease, because it preferentially binds to a different ganglioside receptor than LT-I, which cannot associate with lipid rafts (see below) (Fujinaga *et al.*, 2003). In light of the low relevance for human disease, this chapter will exclusively focus on LT-I and refer to it as LT.

Following synthesis, the individual A and B subunits are transported into the periplasm of the bacterium (Hirst *et al.*, 1984). Translocation is encoded by a 18aa (for A subunit) and 21aa (for B subunit) signal sequence, which is rich in hydrophobic aas, allowing the subunits to pass through the bacterial inner membrane (Spicer and Noble, 1982; Yamamoto *et al.*, 1982). The individual subunits accumulate in the periplasm, where they assemble into holotoxin (Hofstra and Witholt, 1985). The holotoxin is then secreted from the periplasm into the extracellular space by the type II secretion system (T2SS), which has also been identified as the secretory pathway for CTX in *Vibrio cholerae* (Sandkvist *et al.*, 1997). The T2SS is a complex multi-protein (12–15 proteins) machinery that allows bacteria to translocate proteins across the bacterial outer membrane into their environment. Besides *V. cholerae* and ETEC also other bacteria, such as enterohaemorrhagic *E. coli* (EHEC), *Yersinia enterocolitica*, *Klebsiella*, *Pseudomonas aeruginosa* and *Legionella pneumophila* employ this system (Bally *et al.*, 1992; D'Enfert and Pugsley, 1989; Iwobi *et al.*, 2003; Possot and Pugsley, 1994). The T2SS consists of multiple parts that span both the inner and outer bacterial membrane. The pore through which LT translocates from the periplasm into the extracellular space is formed by secretins (GspD), which arrange as dodecamers in the bacterial outer membrane (Chami *et al.*, 2005; Reichow *et al.*, 2010). The translocation process is catalysed by ATP hydrolysis by the secretion ATPase (GspE) located in the cytoplasm of the bacterium (Camberg and Sandkvist, 2005; Sandkvist *et al.*, 1995; Turner *et al.*, 1993). It has been suggested that individual LT toxins are pushed through the secretin pore by a pilus-like protein consisting of individual subunits (pseudopilin, GspG). The pseudopilus is being elongated by addition of further GspG subunits following ATP hydrolysis by the secretion ATPase (GspE), thereby pushing the toxin into the extracellular space through the secretin pore (Korotkov *et al.*, 2012). Although many bacteria employ the T2SS, it was unknown for a long time how ETEC secretes its LT. The functionality of the ETEC T2SS has been reported in 2002, by demonstrating that insertional inactivation of gspD rendered the bacterium incapable of secreting LT (Tauschek *et al.*, 2002). It was further discovered that LT is secreted at focal points of the membrane that are in intimate contact with the host cell and that the focally express the secretion apparatus (Dorsey *et al.*, 2006).

Interestingly, it has been reported that up to 95% of secreted LT is associated to outer membrane vesicles shed by ETEC (Horstman and Kuehn, 2000, 2002). These vesicles contain periplasmic components on the inside (including LT) and also have LT bound to their surface. It has been proposed that following its translocation by T2SS, the B subunit of LT binds to the Kdo core of lipopolysaccharide (LPS – the major component of the outer membrane of Gram-negative bacteria) expressed on the bacterial surface and that LT covered vesicles shedding from the bacteria then mediate toxicity (Horstman *et al.*, 2004; Horstman and Kuehn, 2000, 2002). Secretion of LT also seems to promote the colonization of the intestine by ETEC. ETEC strains with mutations in either LT production or LT secretion (T2SS mutations) are far less efficient at colonizing the intestine of host animals than wild-type strains (Allen *et al.*, 2006). The molecular basis of this mechanism is not entirely understood, but the authors speculate that structural changes of the enterocyte induced by LT binding may facilitate colonization by promoting ETEC adherence to the mucosa (Allen *et al.*, 2006). Following secretion of LT an additional activation step of the toxin takes place, which is called nicking. A secreted protease attacks a peptide loop of the A-subunit of CTX, thereby yielding the A1 and A2 subunits, which are only linked by a disulfide bond. The importance of nicking of LT has been discussed somewhat controversially and in conclusion it seems that unnicked LT still has toxic activity, albeit a delayed one (Grant *et al.*, 1994; Lencer *et al.*, 1997; Tsuji *et al.*, 1984). LT mutants, that do not undergo nicking, were shown to still elicit a secretory response that is slightly blunted and delayed compared to *wt* toxin (Grant *et al.*, 1994; Lencer *et al.*, 1997).

Receptor

Indirect evidence for the LT receptor had already been gained in 1971, when van Heyningen *et al.* (1971) demonstrated that CTX can be deactivated by a crude ganglioside mixture. Gangliosides are glycosphingolipids embedded in the cell membrane. The authors postulated that 'fixation to gangliosides may play a role in the

binding of cholera toxin to the cell membrane' (Van Heyningen *et al.*, 1971). Two years later it was discovered that CTX specifically binds to the ganglioside GM1 and that LT was capable of binding to the same ganglioside (Fig. 8.2) (Holmgren, 1973; Holmgren *et al.*, 1973a,b). A crystal structure of CTX bound to GM1 is available to us (Merritt *et al.*, 1994a). As mentioned previously, it is the B-subunit of LT, which interacts with GM1. The GM1 binding site is predominantly located on a single B-subunit, however, a secondary interaction exists with the adjacent B-subunit in the pentameric ring (Merritt *et al.*, 1994a). Since adjacent B-subunits are involved in GM1 interactions, B-subunit monomers are incapable of effectively binding to GM1 (Merritt *et al.*, 1994a). Although LT and CTX differ in 20% of their AA sequence, the GM1 binding sites are entirely conserved between both toxins (except for one AA residue in specific ETEC strains) (Merritt *et al.*, 1994a). Given the pentameric nature of the B-subunit, the holotoxin can bind to five GM1 gangliosides. It should be noted that LT was also shown to bind other gangliosides, albeit with lower affinity. These include GD1b, GT1b, GM2 and asialo-GM1 (Ångström *et al.*, 1994; Fukuta *et al.*, 1988;

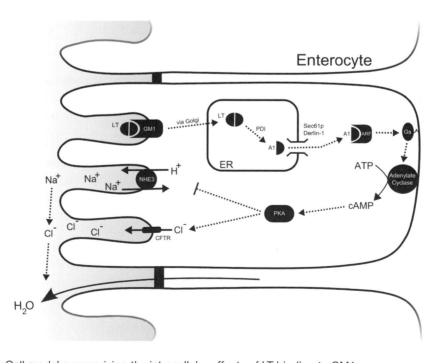

Figure 8.2 Cell model summarizing the intracellular effects of LT binding to GM1.

Holmgren *et al.*, 1985). The pathophysiological relevance of these binding partners has not yet been elucidated. The binding of LT to GM1 has also been exploited as a pharmacological target (Merritt *et al.*, 1994b). Our detailed structural knowledge of the LT/GM1 complex has enabled investigators to develop LT and CTX antagonists that inhibit GM1 binding (Fan *et al.*, 2001; Merritt *et al.*, 1997; Minke *et al.*, 1999; Mitchell *et al.*, 2004; Thompson and Schengrund, 1997, 1998). While these treatment approaches are undoubtedly promising they await clinical testing.

It should be noted that apart from gangliosides and LPS, LT can also bind to blood group antigens. This observation was already made in the 1980s, when investigators demonstrated that labelled LT and CTX was bound to GM1, but that LT could also additionally bind glycoproteins in the intestine (Griffiths *et al.*, 1986; Holmgren *et al.*, 1985). Today, a crystal structure of the B-subunit bound to a blood group A antigen is available to us and reveals that the binding site for glycoprotein binding differs from the site of GM1 binding, but overlaps with site of LPS binding (Holmner *et al.*, 2007). Blood antigen binding could be of importance to determine whether patients with certain blood groups are more prone to pathological ETEC infection than others.

Internalization and intracellular effects

Although GM1 is dispersed over the entire enterocyte membrane it particularly accumulates in caveolae (Parton, 1994). Caveolae, which are a special type of lipid rafts, are small bulb-shaped membrane invaginations characterized by the expression of caveolin. Caveolae can be internalized, but are distinct from clathrin-coated pits, which are involved in the initial stages of classical protein receptor mediated endocytosis. In line with this distinction it has been shown that GM1 is not present in clathrin-coated pits (Nichols, 2003).

Following binding, the holotoxin is internalized into enterocytes, which are mostly located in the villus region of the small intestine (Lindner *et al.*, 1994). This has, for example, been demonstrated with fluorescently labelled B-subunits of LT in mouse small intestine (Lindner *et al.*, 1994).

As mentioned above, the caveolar internalization process is clathrin independent (Singh *et al.*, 2003). At least with CTX, also clathrin-dependent and clathrin and caveolin independent endocytosis have been described; albeit mostly in non-intestinal cell-types (Kirkham *et al.*, 2005; Massol *et al.*, 2004; Singh *et al.*, 2003). Whether LT can utilize similar alternative routes of endocytosis is not entirely clear. Following endocytosis, LT embarks on a complex journey throughout the cell. Given the close homology to CTX, this process has mainly been investigated in the context of CTX mediated diarrhoea. In summary, reverse trafficking takes place before the toxin is released into the cytosol (plasma membrane > early endosomes > Trans Golgi network > endoplasmic reticulum > cytosol). The canonical model of reverse trafficking suggests that the caveolar vesicles bud in a dynamin-dependent fashion and then transiently fuse with early endosomes (Henley *et al.*, 1998; Pelkmans *et al.*, 2004). From the endosomes, the toxin is then delivered to the Trans Golgi network (Pelkmans *et al.*, 2004; Richards *et al.*, 2002). This translocation process is sensitive to Brefeldin A (BFA) treatment, which is known to be an inhibitor of endosome-Golgi transport (Mallard *et al.*, 1998; Richards *et al.*, 2002). From the Golgi apparatus the toxin is transported into the ER (Fig. 8.2). This has been demonstrated by introducing antibodies (anti-betaCOP) or mutations (Arf1 or Sarf1 mutations), which target and inhibit the Golgi-ER trafficking pathway (Aoe *et al.*, 1997; Aridor *et al.*, 1995). Under each condition labelled CTX was shown to be trapped in endosomes or the Golgi apparatus (Majoul *et al.*, 1998; Richards *et al.*, 2002). The enzymatically active A-subunit of LT contains an ER retention signal (RDEL), which was implicated to play a role in Golgi-ER trafficking. Mutation of this sequence was shown to delay the onset of toxin activity (measured by changes in short circuit current in Ussing chambers) (Cieplak Jr. *et al.*, 1995; Lencer *et al.*, 1995). Once the holotoxin reaches the ER, it is disassembled and unfolded, i.e. the A1 subunit is released from the holotoxin. The unfolding and dissociation process of A1 is most likely catalysed by the ER enzyme protein disulfide isomerase (PDI) (Fig. 8.2) (Tsai *et al.*, 2001) Interestingly, it has been argued that CTX needs to be nicked, in order to be recognized by PDI as a substrate, however this would not explain why protease resistant A-subunit mutants still show some toxic activity in other reports (Grant *et al.*, 1994; Lencer *et al.*, 1997; Orlandi, 1997; Tsai *et al.*, 2001). A more recent investigation alternatively proposes that PDI only dissociates the A1 subunit from the holotoxin and that subsequently A1 will unfold spontaneously without the aid of PDI (Taylor *et al.*, 2011). This is supported by the observation that A1 has an unstable tertiary structure at 37°C (Massey *et al.*, 2009; Pande *et al.*, 2007). This thermal instability does not occur when A1 is still bound to A2 (Pande *et al.*, 2007). A1 unfolding is necessary for the toxin subunit to leave the ER (Pande *et al.*, 2007). The exact pathway allowing A1 to pass into the cytosol is not entirely clear, but most likely the LT exploits the ER-associated degradation (ERAD) system, which usually recognizes misfolded endogenous proteins, retro-translocates them from the ER into the cytosol and degrades them in proteasome following ubiquitination. The unfolded nature of the A1 subunit may mimic a misfolded endogenous protein, thereby exploiting the endogenous ERAD quality control system to translocate into the cytosol (Bernardi *et al.*, 2008; Hazes and Read, 1997; Massey *et al.*, 2009; Pande *et al.*, 2007; Taylor *et al.*, 2010). It has been argued that the Sec61p complex, which usually is responsible for protein uptake into the ER, mediates A1 translocation across the ER membrane (Fig. 8.2) (Schmitz *et al.*, 2000). The Sec61 complex forms a pore through the lipid bilayer of the ER allowing peptides to pass through its aqueous environment. Derlin-1, another protein conducting ER channel, may represent an alternate route (Bernardi *et al.*, 2008). It has also been demonstrated that CTX exposure causes up-regulation of Derlin-1, thereby further promoting its release into the cytosol (Dixit *et al.*, 2008). The translocation process through the ER membrane is aided by Hsp90, as loss of Hsp90 resulted in ER trapping of the A1 subunit (Taylor *et al.*, 2010). Usually, proteins that translocate from the ER through the ERAD system are tagged by ubiquitin ligases and thereby directed towards proteasomal degradation in the cytosol (Carvalho *et al.*, 2006). The A1 subunit escapes this fate by its lack of lysine residues, which are necessary for

the attachment of the ubiquitin tag (Hazes and Read, 1997; Rodighiero *et al.*, 2002). Of note and as mentioned previously, most of these studies have been conducted in models employing CTX, however, it is highly likely that LT utilizes the same intracellular mechanisms.

Following release from the ER, the A1 subunit exerts its catalytic activity in the cytosol. In analogy to CTX, the A1 subunit of LT has ADP-ribosyltransferase activity, which transfers an ADP-ribose moiety from NAD+ to the alpha-subunit of a G-protein (Gs) regulating adenylate cyclase activity (Fig. 8.2) (Lee *et al.*, 1991; Moss and Richardson, 1978). To activate its ADP-ribosyltransferase activity, the A1-subunit has to first bind a protein co-factor, called ADP-ribosylation factor (ARF) (Kahn and Gilman, 1984, 1986). ARF is a small GTP binding protein itself (Kahn and Gilman, 1986). Structural insights have recently been gained for the interaction between the A1 subunit of CTX and ARF6 (O'Neal *et al.*, 2005). ARF binding uncovers the binding site of NAD+ on A1, thereby allowing for the transfer of the ADP-ribose moiety onto the G-protein (O'Neal *et al.*, 2005). ADP-ribosylation of Gs then results in G-protein activation and increased adenylate cyclase activity, which raises the intracellular levels of cAMP. As outlined in a previous section, cAMP promotes PKA activation, leading to increased chloride secretion through CFTR and concomitant NHE3 inhibition (Fig. 8.2). The increase in luminal osmolarity causes water retention in the intestine and diarrhoea. In summary, LT relies on a complex retrograde trafficking mechanism to exert its pathophysiological action. The complexity of this process is nicely reflected in short-circuit current measurement of intestinal epithelia (Ussing chambers), during which the chloride secretion slowly ramps up after approximately 40 min of toxin exposure. Conversely, the effects of STa on short circuit current are instantaneous reflecting direct GC-C activation.

Conclusion

The two major toxins secreted by ETEC profoundly differ in their mode of action. However, the molecular endpoints, i.e. a hypersecretion of chloride through CFTR and a concomitant inhibition of NHE3, are identical. Although pharmacological approaches targeting the pathophysiological processes underlying the toxicity of either toxin, e.g. GC-C inhibitors or GM1 binding inhibitors, have been developed, it seems more intuitive to target their common molecular endpoints, give that some ETEC strains produce both toxins. In summary, the story of ETEC nicely exemplifies how only a detailed scientific knowledge about the pathophysiological basis of a disease, allows us to craft targeted pharmacological countermeasures against it.

References

Aimoto, S., Takao, T., Shimonishi, Y., Hara, S., Takeda, T., Takeda, Y., and Miwatani, T. (1982). Amino-acid sequence of a heat-stable enterotoxin produced by human enterotoxigenic *Escherichia coli*. Eur. J. Biochem. *129*, 257–263.

Aimoto, S., Watanabe, H., Ikemura, H., Shimonishi, Y., Takeda, T., Takeda, Y., and Miwatani, T. (1983). Chemical synthesis of a highly potent and heat-stable analog of an enterotoxin produced by a human strain of enterotoxigenic *Escherichia coli*. Biochem. Biophys. Res. Comm. *112*, 320–326.

Alderete, J.F., and Robertson, D.C. (1978). Purification and chemical characterization of the heat-stable enterotoxin produced by porcine strains of enterotoxigenic *Escherichia coli*. Infect. Immun. *19*, 1021–1030.

Allen, K.P., Randolph, M.M., and Fleckenstein, J.M. (2006). Importance of heat-labile enterotoxin in colonization of the adult mouse small intestine by human enterotoxigenic *Escherichia coli* strains. Infect. Immun. 74, 869–875.

Ameen, N.A., Ardito, T., Kashgarian, M., and Marino, C.R. (1995). A unique subset of rat and human intestinal villus cells express the cystic fibrosis transmembrane conductance regulator. Gastroenterology *108*, 1016–1023.

Ångström, J., Teneberg, S., and Karlsson, K.A. (1994). Delineation and comparison of ganglioside-binding epitopes for the toxins of *Vibrio cholerae*, *Escherichia coli*, and *Clostridium tetani*: Evidence for overlapping epitopes. Proc. Natl. Acad. Sci. U.S.A. *91*, 11859–11863.

Aoe, T., Cukierman, E., Lee, A., Cassel, D., Peters, P.J., and Hsu, V.W. (1997). The KDEL receptor, ERD2, regulates intracellular traffic by recruiting a GTPase-activating protein for ARF1. EMBO J. *16*, 7305–7316.

Aridor, M., Bannykh, S.I., Rowe, T., and Balch, W.E. (1995). Sequential coupling between COPII and COPI vesicle coats in endoplasmic reticulum to Golgi transport. J. Cell Biol. *131*, 875–893.

Bally, M., Filloux, A., Akrim, M., Ball, G., Lazdunski, A., and Tommassen, J. (1992). Protein secretion in *Pseudomonas aeruginosa*: characterization of seven xcp genes and processing of secretory apparatus components by prepilin peptidase. Mol. Microbiol. 6, 1121–1131.

Baxter, P.S., Goldhill, J., Hardcastle, J., Hardcastle, P.T., and Taylor, C.J. (1988). Accounting for cystic fibrosis. Nature 335, 211.

Bernardi, K.M., Forster, M.L., Lencer, W.I., and Tsai, B. (2008). Derlin-1 facilitates the retro-translocation of cholera toxin. Mol. Biol. Cell. *19*, 877–884.

Bhandari, R., Suguna, K., and Visweswariah, S.S. (1999). Guanylyl cyclase C receptor: regulation of catalytic activity by ATP. Biosci. Rep. *19*, 179–188.

Black, R.E. (1990). Epidemiology of travelers' diarrhoea and relative importance of various pathogens. Rev. Infect. Dis. *12(Suppl. 1)*, S73–79.

Bobadilla, J.L., Macek, M., Jr., Fine, J.P., and Farrell, P.M. (2002). Cystic fibrosis: a worldwide analysis of CFTR mutations--correlation with incidence data and application to screening. Human Mut. *19*, 575–606.

Booth, I.W., Stange, G., Murer, H., Fenton, T.R., and Milla, P.J. (1985). Defective jejunal brush-border Na+/H+ exchange: a cause of congenital secretory diarrhoea. Lancet 1, 1066–1069.

Broere, N., Chen, M., Cinar, A., Singh, A.K., Hillesheim, J., Riederer, B., Lunnemann, M., Rottinghaus, I., Krabbenhoft, A., Engelhardt, R., *et al.* (2009). Defective jejunal and colonic salt absorption and alteredNa(+)/H (+) exchanger 3 (NHE3) activity in NHE regulatory factor 1 (NHERF1) adaptor protein-deficient mice. Pflugers Arch. *457*, 1079–1091.

Burgess, M.N., Bywater, R.J., Cowley, C.M., Mullan, N.A., and Newsome, P.M. (1978). Biological evaluation of a methanol-soluble, heat-stable *Escherichia coli* enterotoxin in infant mice, pigs, rabbits, and calves. Infect. Immun. *21*, 526–531.

Camberg, J.L., and Sandkvist, M. (2005). Molecular analysis of the *Vibrio cholerae* type II secretion ATPase EpsE. J. Bacteriol. *187*, 249–256.

Carpick, B.W., and Gariepy, J. (1993). The *Escherichia coli* heat-stable enterotoxin is a long-lived superagonist of guanylin. Infect. Immun. *61*, 4710–4715.

Carrithers, S.L., Barber, M.T., Biswas, S., Parkinson, S.J., Park, P.K., Goldstein, S.D., and Waldman, S.A. (1996). Guanylyl cyclase C is a selective marker for metastatic colorectal tumors in human extraintestinal tissues. Proc. Natl. Acad. Sci. U.S.A. *93*, 14827–14832.

Carrithers, S.L., Hill, M.J., Johnson, B.R., O'Hara, S.M., Jackson, B.A., Ott, C.E., Lorenz, J., Mann, E.A., Giannella, R.A., Forte, L.R., *et al.* (1999). Renal effects of uroguanylin and guanylin in vivo. Brazilian J. Med. Biol. *32*, 1337–1344.

Carrithers, S.L., Ott, C.E., Hill, M.J., Johnson, B.R., Cai, W., Chang, J.J., Shah, R.G., Sun, C., Mann, E.A., Fonteles, M.C., *et al.* (2004). Guanylin and uroguanylin induce natriuresis in mice lacking guanylyl cyclase-C receptor. Kidney Int. *65*, 40–53.

Carvalho, P., Goder, V., and Rapoport, T.A. (2006). Distinct ubiquitin–ligase complexes define convergent pathways for the degradation of ER proteins. Cell *126*, 361–373.

Cha, B., Kim, J.H., Hut, H., Hogema, B.M., Nadarja, J., Zizak, M., Cavet, M., Lee-Kwon, W., Lohmann, S.M., Smolenski, A., *et al.* (2005). cGMP inhibition of

Na+/H+ antiporter 3 (NHE3) requires PDZ domain adapter NHERF2, a broad specificity protein kinase G-anchoring protein. J. Biol. Chem. *280*, 16642–16650.

Chami, M., Guilvout, I., Gregorini, M., Remigy, H.W., Muller, S.A., Valerio, M., Engel, A., Pugsley, A.P., and Bayan, N. (2005). Structural insights into the secretin PulD and its trypsin-resistant core. J. Biol. Chem. *280*, 37732–37741.

Chang, M.S., Lowe, D.G., Lewis, M., Hellmiss, R., Chen, E., and Goeddel, D.V. (1989). Differential activation by atrial and brain natriuretic peptides of two different receptor guanylate cyclases. Nature *341*, 68–72.

Chappe, V., Hinkson, D.A., Zhu, T., Chang, X.B., Riordan, J.R., and Hanrahan, J.W. (2003). Phosphorylation of protein kinase C sites in NBD1 and the R domain control CFTR channel activation by PKA. J. Physiol. *548*, 39–52.

Chen, M., Sultan, A., Cinar, A., Yeruva, S., Riederer, B., Singh, A.K., Li, J., Bonhagen, J., Chen, G., Yun, C., *et al.* (2010). Loss of PDZ-adaptor protein NHERF2 affects membrane localization and cGMP- and [Ca2+]- but not cAMP-dependent regulation of Na+/H+ exchanger 3 in murine intestine. J. Physiol. *588*, 5049–5063.

Cheng, S.H., Gregory, R.J., Marshall, J., Paul, S., Souza, D.W., White, G.A., O'Riordan, C.R., and Smith, A.E. (1990). Defective intracellular transport and processing of CFTR is the molecular basis of most cystic fibrosis. Cell *63*, 827–834.

Cheng, S.H., Rich, D.P., Marshall, J., Gregory, R.J., Welsh, M.J., and Smith, A.E. (1991). Phosphorylation of the R domain by cAMP-dependent protein kinase regulates the CFTR chloride channel. Cell *66*, 1027–1036.

Chey, W.D., Lembo, A.J., Lavins, B.J., Shiff, S.J., Kurtz, C.B., Currie, M.G., MacDougall, J.E., Jia, X.D., Shao, J.Z., Fitch, D.A., *et al.* (2012). Linaclotide for irritable bowel syndrome with constipation: a 26-week, randomized, double-blind, placebo-controlled trial to evaluate efficacy and safety. Am. J. Gastroenterol. *107*, 1702–1712.

Chinkers, M., Garbers, D.L., Chang, M.S., Lowe, D.G., Chin, H.M., Goeddel, D.V., and Schulz, S. (1989). A membrane form of guanylate cyclase is an atrial natriuretic peptide receptor. Nature *338*, 78–83.

Cieplak Jr, W., Messer, R.J., Konkel, M.E., and Grant, C.C.R. (1995). Role of a potential endoplasmic reticulum retention sequence (RDEL) and the Golgi complex in the cytotonic activity of *Escherichia coli* heat-labile enterotoxin. Mol. Microbiol. *16*, 789–800.

Cohen, M.B., Mann, E.A., Lau, C., Henning, S.J., and Giannella, R.A. (1992). A gradient in expression of the *Escherichia coli* heat-stable enterotoxin receptor exists along the villus–to-crypt axis of rat small intestine. Biochem. Biophys. Res. Comm. *186*, 483–490.

Colledge, W.H., Abella, B.S., Southern, K.W., Ratcliff, R., Jiang, C., Cheng, S.H., MacVinish, L.J., Anderson, J.R., Cuthbert, A.W., and Evans, M.J. (1995). Generation and characterization of a delta F508 cystic fibrosis mouse model. Nature Genet. *10*, 445–452.

Crane, J.K., and Shanks, K.L. (1996). Phosphorylation and activation of the intestinal guanylyl cyclase receptor for *Escherichia coli* heat-stable toxin by protein kinase C. Mol. Cell. Biochem. *165*, 111–120.

Currie, M.G., Fok, K.F., Kato, J., Moore, R.J., Hamra, F.K., Duffin, K.L., and Smith, C.E. (1992). Guanylin: an endogenous activator of intestinal guanylate cyclase. Proc. Natl. Acad. Sci. U.S.A. *89*, 947–951.

Cuthbert, A.W., Halstead, J., Ratcliff, R., Colledge, W.H., and Evans, M.J. (1995). The genetic advantage hypothesis in cystic fibrosis heterozygotes: A murine study. J. Physiol. *482*, 449–454.

D'Enfert, C., and Pugsley, A.P. (1989). *Klebsiella pneumoniae* pulS gene encodes an outer membrane lipoprotein required for pullulanase secretion. J. Bacteriol. *171*, 3673–3679.

Dallas, W.S., and Falkow, S. (1980). Amino acid sequence homology between cholera toxin and *Escherichia coli* heat-labile toxin. Nature *288*, 499–501.

Database, C.F.M., de Jonge, H.R. (1981). Cyclic GMP-dependent protein kinase in intestinal brushborders. Adv. Cyclic Nucleotide Res. *14*, 315–333.

Deshmane, S.P., Parkinson, S.J., Crupper, S.S., Robertson, D.C., Schulz, S., and Waldman, S.A. (1997). Cytoplasmic domains mediate the ligand-induced affinity shift of guanylyl cyclase C. Biochemistry *36*, 12921–12929.

Dixit, G., Mikoryak, C., Hayslett, T., Bhat, A., and Draper, R.K. (2008). Cholera toxin up-regulates endoplasmic reticulum proteins that correlate with sensitivity to the toxin. Exp. Biol. Med. (Maywood) *233*, 163–175.

Dorsey, F.C., Fischer, J.F., and Fleckenstein, J.M. (2006). Directed delivery of heat-labile enterotoxin by enterotoxigenic *Escherichia coli*. Cell. Microbiol. *8*, 1516–1527.

Dubreuil, J.D. (1997). *Escherichia coli* STb enterotoxin. Microbiology *143*, 1783–1795.

Eggermont, E. (1996). Gastrointestinal manifestations in cystic fibrosis. Eur. J. Gastroenterol. Hepatol. *8*, 731–738.

Fan, E., Merritt, E.A., Zhang, Z., Pickens, J.C., Roach, C., Ahn, M., and Hol, W.G.J. (2001). Exploration of the GM1 receptor-binding site of heat-labile enterotoxin and cholera toxin by phenyl-ring-containing galactose derivatives. Acta Crystallographica Section D: Biol. Crystallog. *57*, 201–212.

Field, M., Graf, L.H., Jr., Laird, W.J., and Smith, P.L. (1978). Heat-stable enterotoxin of *Escherichia coli*: in vitro effects on guanylate cyclase activity, cyclic GMP concentration, and ion transport in small intestine. Proc. Natl. Acad. Sci. U.S.A. *75*, 2800–2804.

Forte, L.R., Thorne, P.K., Eber, S.L., Krause, W.J., Freeman, R.H., Francis, S.H., and Corbin, J.D. (1992). Stimulation of intestinal Cl- transport by heat-stable enterotoxin: activation of cAMP-dependent protein kinase by cGMP. Am. J. physiol. *263*, C607–615.

Forte, L.R., Eber, S.L., Turner, J.T., Freeman, R.H., Fok, K.F., and Currie, M.G. (1993). Guanylin stimulation of Cl- secretion in human intestinal T84 cells via cyclic guanosine monophosphate. J. Clin. Invest. *91*, 2423–2428.

Forte, L.R., Fan, X., and Hamra, F.K. (1996). Salt and water homeostasis: uroguanylin is a circulating peptide hormone with natriuretic activity. Am. J. Kidney Dis. *28*, 296–304.

French, P.J., Bijman, J., Edixhoven, M., Vaandrager, A.B., Scholte, B.J., Lohmann, S.M., Nairn, A.C., and de Jonge, H.R. (1995). Isotype-specific activation of cystic fibrosis transmembrane conductance regulator-chloride channels by cGMP-dependent protein kinase II. J. Biol. Chem. *270*, 26626–26631.

Fujinaga, Y., Wolf, A.A., Rodighiero, C., Wheeler, H., Tsai, B., Allen, L., Jobling, M.G., Rapoport, T., Holmes, R.K., and Lencer, W.I. (2003). Gangliosides that associate with lipid rafts mediate transport of cholera and related toxins from the plasma membrane to endoplasmic reticulm. Mol. Biol. Cell *14*, 4783–4793.

Fukuta, S., Magnani, J.L., Twiddy, E.M., Holmes, R.K., and Ginsburg, V. (1988). Comparison of the carbohydrate-binding specificities of cholera toxin and *Escherichia coli* heat-labile enterotoxins LTh-I, LT-IIa, and LT-IIb. Infect. Immun. *56*, 1748–1753.

Gariepy, J., and Schoolnik, G.K. (1986). Design of a photoreactive analogue of the *Escherichia coli* heat-stable enterotoxin STIb: use in identifying its receptor on rat brush border membranes. Proc. Natl. Acad. Sci. U.S.A. *83*, 483–487.

Gariepy, J., Lane, A., Frayman, F., Wilbur, D., Robien, W., Schoolnik, G.K., and Jardetzky, O. (1986). Structure of the toxic domain of the *Escherichia coli* heat-stable enterotoxin ST I. Biochemistry *25*, 7854–7866.

Gariépy, J., Judd, A.K., and Schoolnik, G.K. (1987). Importance of disulfide bridges in the structure and activity of *Escherichia coli* enterotoxin ST1b. Proc. Natl. Acad. Sci. U.S.A. *84*, 8907–8911.

Gawenis, L.R., Stien, X., Shull, G.E., Schultheis, P.J., Woo, A.L., Walker, N.M., and Clarke, L.L. (2002). Intestinal NaCl transport in NHE2 and NHE3 knockout mice. Am. J. Physiol. Gastrointest. Liver Physiol. *282*, G776–784.

Ghanekar, Y., Chandrashaker, A., Tatu, U., and Visweswariah, S.S. (2004). Glycosylation of the receptor guanylate cyclase C: role in ligand binding and catalytic activity. Biochem. J. *379*, 653–663.

Giannella, R.A., Luttrell, M., and Thompson, M. (1983). Binding of *Escherichia coli* heat-stable enterotoxin to receptors on rat intestinal cells. Am. J. Physiol. Gastrointest. Liver Physiol. *8*, G492-G498.

Gill, D.M., Clements, J.D., Robertson, D.C., and Finkelstein, R.A. (1981). Subunit number and arrangement in *Escherichia coli* heat-labile enterotoxin. Infect. Immun. *33*, 677–682.

Golin-Bisello, F., Bradbury, N., and Ameen, N. (2005). STa and cGMP stimulate CFTR translocation to the surface of villus enterocytes in rat jejunum and is regulated by protein kinase G. Am. J. Physiol. Cell Physiol. *289*, C708–716.

Grant, C.C., Messer, R.J., and Cieplak, W., Jr. (1994). Role of trypsin–like cleavage at arginine 192 in the enzymatic and cytotonic activities of *Escherichia coli* heat-labile enterotoxin. Infect. Immun. *62*, 4270–4278.

Green, B.A., Neill, R.J., Ruyechan, W.T., and Holmes, R.K. (1983). Evidence that a new enterotoxin of *Escherichia coli* which activates adenylate cyclase in eucaryotic

target cells is not plasmid mediated. Infect. Immun. *41*, 383–390.

Greenberg, R.N., Hill, M., Crytzer, J., Krause, W.J., Eber, S.L., Hamra, F.K., and Forte, L.R. (1997). Comparison of effects of uroguanylin, guanylin, and *Escherichia coli* heat-stable enterotoxin STa in mouse intestine and kidney: evidence that uroguanylin is an intestinal natriuretic hormone. J. Invest. Med. *45*, 276–282.

Griffiths, S.L., Finkelstein, R.A., and Critchley, D.R. (1986). Characterization of the receptor for cholera toxin and *Escherichia coli* heat-labile toxin in rabbit intestinal brush borders. Biochem. J. *238*, 313–322.

Grubb, B.R., and Gabriel, S.E. (1997). Intestinal physiology and pathology in gene-targeted mouse models of cystic fibrosis. Am. J. Physiol. *273*, G258–266.

Guba, M., Kuhn, M., Forssmann, W.G., Classen, M., Gregor, M., and Seidler, U. (1996). Guanylin strongly stimulates rat duodenal HCO3- secretion: proposed mechanism and comparison with other secretagogues. Gastroenterology *111*, 1558–1568.

Guerrant, R.L., Hughes, J.M., Chang, B., Robertson, D.C., and Murad, F. (1980). Activation of intestinal guanylate cyclase by heat-stable enterotoxin of *Escherichia coli*: Studies of tissue specificity, potential receptors, and intermediates. J. Infect. Dis. *142*, 220–228.

Guggino, S.E. (1994). Gates of Janus: cystic fibrosis and diarrhoea. Trends Microbiol. *2*, 91–94.

Gyles, C.L., and Barnum, D.A. (1969). A heat-labile enterotoxin from strains of *Eschericha coli* enteropathogenic for pigs. J. Infect. Dis. *120*, 419–426.

Hansson, G.C. (1988). Cystic fibrosis and chloride-secreting diarrhoea. Nature *333*, 711.

Hasegawa, M., Hidaka, Y., Matsumoto, Y., Sanni, T., and Shimonishi, Y. (1999a). Determination of the binding site on the extracellular domain of guanylyl cyclase C to heat-stable enterotoxin. J. Biol. Chem. *274*, 31713–31718.

Hasegawa, M., Hidaka, Y., Wada, A., Hirayama, T., and Shimonishi, Y. (1999b). The relevance of N-linked glycosylation to the binding of a ligand to guanylate cyclase C. Eur. J. Biochem. *263*, 338–345.

Hasegawa, M., Matsumoto-Ishikawa, Y., Hijikata, A., Hidaka, Y., Go, M., and Shimonishi, Y. (2005). Disulfide linkages and a three-dimensional structure model of the extracellular ligand-binding domain of guanylyl cyclase C. Protein J. *24*, 315–325.

Hazes, B., and Read, R.J. (1997). Accumulating evidence suggests that several AB-toxins subvert the endoplasmic reticulum-associated protein degradation pathway to enter target cells. Biochemistry *36*, 11051–11054.

Henley, J.R., Krueger, E.W., Oswald, B.J., and McNiven, M.A. (1998). Dynamin-mediated internalization of caveolae. J. Cell Biol. *141*, 85–99.

Hidaka, Y., Matsumoto, Y., and Shimonishi, Y. (2002). The micro domain responsible for ligand-binding of guanylyl cyclase C. FEBS Lett. *526*, 58–62.

Hirst, T.R., Randall, L.L., and Hardy, S.J.S. (1984). Cellular location of heat-labile enterotoxin in *Escherichia coli*. J. Bacteriol. *157*, 637–642.

Hodson, C.A., Ambrogi, I.G., Scott, R.O., Mohler, P.J., and Milgram, S.L. (2006). Polarized apical sorting of guanylyl cyclase C is specified by a cytosolic signal. Traffic 7, 456–464.

Hofstra, H., and Witholt, B. (1985). Heat-labile enterotoxin in *Escherichia coli*. Kinetics of association of subunits into periplasmic holotoxin. J. Biol. Chem. *260*, 16037–16044.

Holmes, R.K., Twiddy, E.M., and Pickett, C.L. (1986). Purification and characterization of type II heat-labile enterotoxin of *Escherichia coli*. Infect. Immun. *53*, 464–473.

Holmgren, J. (1973). Comparison of the tissue receptors for *Vibrio cholerae* and *Escherichia coli* enterotoxins by means of gangliosides and natural cholera toxoid. Infect. Immun. *8*, 851–859.

Holmgren, J., Lonnroth, I., and Svennerholm, L. (1973a). Fixation and inactivation of cholera toxin by GM1 ganglioside. Scand. J. Infect. Dis. *5*, 77–78.

Holmgren, J., Lonnroth, I., and Svennerholm, L. (1973b). Tissue receptor for cholera exotoxin: postulated structure from studies with GM1 ganglioside and related glycolipids. Infect. Immun. *8*, 208–214.

Holmgren, J., Lindblad, M., Fredman, P., Svennerholm, L., and Myrvold, H. (1985). Comparison of receptors for cholera and *Escherichia coli* enterotoxins in human intestine. Gastroenterology *89*, 27–35.

Holmner, A., Askarieh, G., Okvist, M., and Krengel, U. (2007). Blood group antigen recognition by *Escherichia coli* heat-labile enterotoxin. J. Mol. Biol. *371*, 754–764.

Hoogerwerf, W.A., Tsao, S.C., Devuyst, O., Levine, S.A., Yun, C.H., Yip, J.W., Cohen, M.E., Wilson, P.D., Lazenby, A.J., Tse, C.M., et al. (1996). NHE2 and NHE3 are human and rabbit intestinal brush-border proteins. Am. J. Physiol. *270*, G29–41.

Horstman, A.L., and Kuehn, M.J. (2000). Enterotoxigenic *Escherichia coli* secretes active heat-labile enterotoxin via outer membrane vesicles. J. Biol. Chem. *275*, 12489–12496.

Horstman, A.L., and Kuehn, M.J. (2002). Bacterial surface association of heat-labile enterotoxin through lipopolysaccharide after secretion via the general secretory pathway. J. Biol. Chem. *277*, 32538–32545.

Horstman, A.L., Bauman, S.J., and Kuehn, M.J. (2004). Lipopolysaccharide 3-deoxy-D-manno-octulosonic acid (Kdo) core determines bacterial association of secreted toxins. J. Biol. Chem. *279*, 8070–8075.

Hughes, J.M., Murad, F., Chang, B., and Guerrant, R.L. (1978). Role of cyclic GMP in the action of heat-stable enterotoxin of *Escherichia coli*. Nature *271*, 755–756.

Hughes, J.M., Rouse, J.D., Barada, F.A., and Guerrant, R.L. (1980). Etiology of summer diarrhoea among the Navajo. Am. J. Trop. Med. Hygiene *29*, 613–619.

Ieda, H., Naruse, S., Kitagawa, M., Ishiguro, H., and Hayakawa, T. (1999). Effects of guanylin and uroguanylin on rat jejunal fluid and electrolyte transport: Comparison with heat-stable enterotoxin. Regulat. Peptides *79*, 165–171.

Iwobi, A., Heesemann, J., Garcia, E., Igwe, E., Noelting, C., and Rakin, A. (2003). Novel virulence-associated type II secretion system unique to high-pathogenicity *Yersinia enterocolitica*. Infect. Immun. *71*, 1872–1879.

Jakab, R.L., Collaco, A.M., and Ameen, N.A. (2011). Physiological relevance of cell-specific distribution patterns of CFTR, NKCC1, NBCe1, and NHE3 along the crypt–villus axis in the intestine. Am. J. Physiol. Gastrointest. Liver Physiol. *300*, G82–98.

Jaleel, M., Saha, S., Shenoy, A.R., and Visweswariah, S.S. (2006). The kinase homology domain of receptor guanylyl cyclase C: ATP binding and identification of an adenine nucleotide sensitive site. Biochemistry *45*, 1888–1898.

Jia, Y., Mathews, C.J., and Hanrahan, J.W. (1997). Phosphorylation by protein kinase C is required for acute activation of cystic fibrosis transmembrane conductance regulator by protein kinase A. J. Biol. Chem. *272*, 4978–4984.

Jiang, Z.D., Lowe, B., Verenkar, M.P., Ashley, D., Steffen, R., Tornieporth, N., von Sonnenburg, F., Waiyaki, P., and DuPont, H.L. (2002). Prevalence of enteric pathogens among international travelers with diarrhoea acquired in Kenya (Mombasa), India (Goa), or Jamaica (Montego Bay). J. Infect. Dis. *185*, 497–502.

Kahn, R.A., and Gilman, A.G. (1984). Purification of a protein cofactor required for ADP-ribosylation of the stimulatory regulatory component of adenylate cyclase by cholera toxin. J. Biol. Chem. *259*, 6228–6234.

Kahn, R.A., and Gilman, A.G. (1986). The protein cofactor necessary for ADP-ribosylation of Gs by cholera toxin is itself a GTP binding protein. J. Biol. Chem. *261*, 7906–7911.

Keller, K.M., Wirth, S., Baumann, W., Sule, D., and Booth, I.W. (1990). Defective jejunal brush border membrane sodium/proton exchange in association with lethal familial protracted diarrhoea. Gut *31*, 1156–1158.

Kelly, M., Trudel, S., Brouillard, F., Bouillaud, F., Colas, J., Nguyen-Khoa, T., Ollero, M., Edelman, A., and Fritsch, J. (2010). Cystic fibrosis transmembrane regulator inhibitors CFTR(inh)–172 and GlyH-101 target mitochondrial functions, independently of chloride channel inhibition. J. Pharmacol. Exp. Ther. *333*, 60–69.

Kirkham, M., Fujita, A., Chadda, R., Nixon, S.J., Kurzchalia, T.V., Sharma, D.K., Pagano, R.E., Hancock, J.F., Mayor, S., and Parton, R.G. (2005). Ultrastructural identification of uncoated caveolin-independent early endocytic vehicles. J. Cell Biol. *168*, 465–476.

Kita, T., Smith, C.E., Fok, K.F., Duffin, K.L., Moore, W.M., Karabatsos, P.J., Kachur, J.F., Hamra, F.K., Pidhorodeckyj, N.V., Forte, L.R., et al. (1994). Characterization of human uroguanylin: a member of the guanylin peptide family. Am. J. Physiol. *266*, F342–348.

Kongsuphol, P., Cassidy, D., Hieke, B., Treharne, K.J., Schreiber, R., Mehta, A., and Kunzelmann, K. (2009). Mechanistic insight into control of CFTR by AMPK. J. Biol. Chem. *284*, 5645–5653.

Korotkov, K.V., Sandkvist, M., and Hol, W.G. (2012). The type II secretion system: biogenesis, molecular architecture and mechanism. Nature Rev. Microbiol. *10*, 336–351.

Krause, W.J., Cullingford, G.L., Freeman, R.H., Eber, S.L., Richardson, K.C., Fok, K.F., Currie, M.G., and Forte, L.R. (1994). Distribution of heat-stable enterotoxin/guanylin receptors in the intestinal tract of man and other mammals. J. Anat. *184 (Pt 2)*, 407–417.

Kuno, T., Kamisaki, Y., Waldman, S.A., Gariepy, J., Schoolnik, G., and Murad, F. (1986). Characterization of the receptor for heat-stable enterotoxin from *Escherichia coli* in rat intestine. J. Biol. Chem. *261*, 1470–1476.

Lamprecht, G., Weinman, E.J., and Yun, C.H. (1998). The role of NHERF and E3KARP in the cAMP-mediated inhibition of NHE3. J. Biol. Chem. *273*, 29972–29978.

Lauber, T., Tidten, N., Matecko, I., Zeeb, M., Rosch, P., and Marx, U.C. (2009). Design and characterization of a soluble fragment of the extracellular ligand-binding domain of the peptide hormone receptor guanylyl cyclase-C. PEDS *22*, 1–7.

Lee, C.M., Chang, P.P., Tsai, S.C., Adamik, R., Price, S.R., Kunz, B.C., Moss, J., Twiddy, E.M., and Holmes, R.K. (1991). Activation of *Escherichia coli* heat-labile enterotoxins by native and recombinant adenosine diphosphate-ribosylation factors, 20 kD guanine nucleotide-binding proteins. J. Clin. Invest. *87*, 1780–1786.

Lencer, W.I., Constable, C., Moe, S., Jobling, M.G., Webb, H.M., Ruston, S., Madara, J.L., Hirst, T.R., and Holmes, R.K. (1995). Targeting of cholera toxin and *Escherichia coli* heat labile toxin in polarized epithelia: Role of COOH-terminal KDEL. J. Cell Biol. *131*, 951–962.

Lencer, W.I., Constable, C., Moe, S., Rufo, P.A., Wolf, A., Jobling, M.G., Ruston, S.P., Madara, J.L., Holmes, R.K., and Hirst, T.R. (1997). Proteolytic activation of cholera toxin and *Escherichia coli* labile toxin by entry into host epithelial cells. Signal transduction by a protease-resistant toxin variant. J. Biol. Chem. *272*, 15562–15568.

Levine, S.A., Montrose, M.H., Tse, C.M., and Donowitz, M. (1993). Kinetics and regulation of three cloned mammalian Na+/H+ exchangers stably expressed in a fibroblast cell line. J. Biol. Chem. *268*, 25527–25535.

Li, J., Poulikakos, P.I., Dai, Z., Testa, J.R., Callaway, D.J., and Bu, Z. (2007). Protein kinase C phosphorylation disrupts Na+/H+ exchanger regulatory factor 1 autoinhibition and promotes cystic fibrosis transmembrane conductance regulator macromolecular assembly. J. Biol. Chem. *282*, 27086–27099.

Lindner, J., Geczy, A.F., and Russell-Jones, G.J. (1994). Identification of the site of uptake of the *E. coli* heat-labile enterotoxin, LTB. Scand. J. Immunol. *40*, 564–572.

Lohmann, S.M., Vaandrager, A.B., Smolenski, A., Walter, U., and De Jonge, H.R. (1997). Distinct and specific functions of cGMP-dependent protein kinases. Trends Biochem. Sci. *22*, 307–312.

Lowe, D.G., Chang, M.S., Hellmiss, R., Chen, E., Singh, S., Garbers, D.L., and Goeddel, D.V. (1989). Human atrial natriuretic peptide receptor defines a new paradigm for second messenger signal transduction. EMBO J. *8*, 1377–1384.

Lubbe, W.J., Zuzga, D.S., Zhou, Z., Fu, W., Pelta-Heller, J., Muschel, R.J., Waldman, S.A., and Pitari, G.M. (2009). Guanylyl cyclase C prevents colon cancer metastasis by regulating tumor epithelial cell matrix metalloproteinase-9. Cancer Res. *69*, 3529–3536.

Ma, T., Thiagarajah, J.R., Yang, H., Sonawane, N.D., Folli, C., Galietta, L.J., and Verkman, A.S. (2002). Thiazolidinone CFTR inhibitor identified by high-throughput screening blocks cholera toxin-induced intestinal fluid secretion. J. Clin. Invest. *110*, 1651–1658.

Majoul, I., Sohn, K., Wieland, F.T., Pepperkok, R., Pizza, M., Hillemann, J., and Soling, H.D. (1998). KDEL receptor (Erd2p)-mediated retrograde transport of the cholera toxin A subunit from the Golgi involves COPI, p23, and the COOH terminus of Erd2p. J. Cell Biol. *143*, 601–612.

Mallard, F., Antony, C., Tenza, D., Salamero, J., Goud, B., and Johannes, L. (1998). Direct pathway from early/recycling endosomes to the Golgi apparatus revealed through the study of shiga toxin B-fragment transport. J. Cell Biol. *143*, 973–990.

Mann, E.A., Jump, M.L., Wu, J., Yee, E., and Giannella, R.A. (1997). Mice lacking the guanylyl cyclase C receptor are resistant to STa-induced intestinal secretion. Biochem. Biophys. Res. Comm. *239*, 463–466.

Markert, T., Vaandrager, A.B., Gambaryan, S., Pohler, D., Hausler, C., Walter, U., De Jonge, H.R., Jarchau, T., and Lohmann, S.M. (1995). Endogenous expression of type II cGMP-dependent protein kinase mRNA and protein in rat intestine. Implications for cystic fibrosis transmembrane conductance regulator. J. Clin. Invest. *96*, 822–830.

Massey, S., Banerjee, T., Pande, A.H., Taylor, M., Tatulian, S.A., and Teter, K. (2009). Stabilization of the tertiary structure of the cholera toxin A1 subunit inhibits toxin dislocation and cellular intoxication. J. Mol. Biol. *393*, 1083–1096.

Massol, R.H., Larsen, J.E., Fujinaga, Y., Lencer, W.I., and Kirchhausen, T. (2004). Cholera toxin toxicity does not require functional Arf6- and dynamin-dependent endocytic pathways. Mol. Biol. Cell *15*, 3631–3641.

Merritt, E.A., Sarfaty, S., Van den Akker, F., L'Hoir, C., Martial, J.A., and Hol, W.G.J. (1994a). Crystal structure of cholera toxin B-pentamer bound to receptor G(M1) pentasaccharide. Protein Sci. *3*, 166–175.

Merritt, E.A., Sixma, T.K., Kalk, K.H., Van Zanten, B.A.M., and Hol, W.G.J. (1994b). Galactose-binding site in *Escherichia coli* heat-labile enterotoxin (LT) and cholera toxin (CT). Mol. Microbiol. *13*, 745–753.

Merritt, E.A., Sarfaty, S., Feil, I.K., and Hol, W.G.J. (1997). Structural foundation for the design of receptor antagonists targeting *Escherichia coli* heat-labile enterotoxin. Structure *5*, 1485–1499.

Minke, W.E., Hong, F., Verlinde, C.L.M.J., Hol, W.G.J., and Fan, E. (1999). Using a galactose library for exploration of a novel hydrophobic pocket in the receptor binding site of the *Escherichia coli* heat-labile enterotoxin. J. Biol. Chem. *274*, 33469–33473.

Mitchell, D.D., Pickens, J.C., Korotkov, K., Fan, E., and Hol, W.G.J. (2004). 3,5-Substituted phenyl galactosides as leads in designing effective cholera toxin antagonists: Synthesis and crystallographic studies. Bioorgan. Med. Chem. *12*, 907–920.

Moseley, S.L., Hardy, J.W., Hug, M.I., Echeverria, P., and Falkow, S. (1983). Isolation and nucleotide sequence determination of a gene encoding a heat-stable enterotoxin of *Escherichia coli*. Infect. Immun. *39*, 1167–1174.

Moss, J., and Richardson, S.H. (1978). Activation of adenylate cyclase by heat-labile *Escherichia coli* enterotoxin. Evidence for ADP-ribosyltransferase activity similar to that of choleragen. J. Clin. Invest. *62*, 281–285.

Muanprasat, C., Sonawane, N.D., Salinas, D., Taddei, A., Galietta, L.J., and Verkman, A.S. (2004). Discovery of glycine hydrazide pore-occluding CFTR inhibitors: mechanism, structure–activity analysis, and *in vivo* efficacy. J. Gen. Physiol. *124*, 125–137.

Muller, T., Wijmenga, C., Phillips, A.D., Janecke, A., Houwen, R.H., Fischer, H., Ellemunter, H., Fruhwirth, M., Offner, F., Hofer, S., *et al.* (2000). Congenital sodium diarrhoea is an autosomal recessive disorder of sodium/proton exchange but unrelated to known candidate genes. Gastroenterology *119*, 1506–1513.

Murek, M., Kopic, S., and Geibel, J. (2010). Evidence for intestinal chloride secretion. Exp. Physiol. *95*, 471–478.

Murer, H., Hopfer, U., and Kinne, R. (1976). Sodium/proton antiport in brush-border-membrane vesicles isolated from rat small intestine and kidney. Biochem. J. *154*, 597–604.

Murtazina, R., Kovbasnjuk, O., Zachos, N.C., Li, X., Chen, Y., Hubbard, A., Hogema, B.M., Steplock, D., Seidler, U., Hoque, K.M., *et al.* (2007). Tissue-specific regulation of sodium/proton exchanger isoform 3 activity in Na(+)/H(+) exchanger regulatory factor 1 (NHERF1) null mice. cAMP inhibition is differentially dependent on NHERF1 and exchange protein directly activated by cAMP in ileum versus proximal tubule. J. Biol. Chem. *282*, 25141–25151.

Nardi, A.R.M., Salvadori, M.R., Coswig, L.T., Gatti, M.S.V., Leite, D.S., Valadares, G.F., Neto, M.G., Shocken-Iturrino, R.P., Blanco, J.E., and Yano, T. (2005). Type 2 heat-labile enterotoxin (LT-II)-producing *Escherichia coli* isolated from ostriches with diarrhoea. Vet. Microbiol. *105*, 245–249.

Nawar, H.F., King-Lyons, N.D., Hu, J.C., Pasek, R.C., and Connell, T.D. (2010). LT-IIc, a new member of the type II heat-labile enterotoxin family encoded by an *Escherichia coli* strain obtained from a nonmammalian host. Infect. Immun. *78*, 4705–4713.

Nichols, B.J. (2003). GM1-containing lipid rafts are depleted within clathrin-coated pits. Curr. Biol. *13*, 686–690.

O'Neal, C.J., Jobling, M.G., Holmes, R.K., and Hol, W.G. (2005). Structural basis for the activation of cholera toxin by human ARF6-GTP. Science *309*, 1093–1096.

Orlandi, P.A. (1997). Protein-disulfide isomerase-mediated reduction of the A subunit of cholera toxin in a human intestinal cell line. J. Biol. Chem. *272*, 4591–4599.

Orlowski, J. (1993). Heterologous expression and functional properties of amiloride high affinity (NHE-1) and low affinity (NHE-3) isoforms of the rat Na/H exchanger. J. Biol. Chem. *268*, 16369–16377.

Pande, A.H., Scaglione, P., Taylor, M., Nemec, K.N., Tuthill, S., Moe, D., Holmes, R.K., Tatulian, S.A., and Teter, K. (2007). Conformational instability of the cholera toxin A1 polypeptide. J. Mol. Biol. *374*, 1114–1128.

Parton, R.G. (1994). Ultrastructural localization of gangliosides; GM1 is concentrated in caveolae. J. Histochem. Cytochem. *42*, 155–166.

Pelkmans, L., Burli, T., Zerial, M., and Helenius, A. (2004). Caveolin-stabilized membrane domains as multifunctional transport and sorting devices in endocytic membrane traffic. Cell *118*, 767–780.

Pfeifer, A., Aszodi, A., Seidler, U., Ruth, P., Hofmann, F., and Fassler, R. (1996). Intestinal secretory defects and dwarfism in mice lacking cGMP-dependent protein kinase II. Science *274*, 2082–2086.

Picciotto, M.R., Cohn, J.A., Bertuzzi, G., Greengard, P., and Nairn, A.C. (1992). Phosphorylation of the cystic fibrosis transmembrane conductance regulator. J. Biol. Chem. *267*, 12742–12752.

Pickett, C.L., Twiddy, E.M., Wolfganger Belisle, B., and Holmes, R.K. (1986). Cloning of genes that encode a new heat-labile enterotoxin of *Escherichia coli*. J. Bacteriol. *165*, 348–352.

Pickett, C.L., Weinstein, D.L., and Holmes, R.K. (1987). Genetics of type IIa heat-labile enterotoxin of *Escherichia coli*: operon fusions, nucleotide sequence, and hybridization studies. J. Bacteriol. *169*, 5180–5187.

Pickett, C.L., Twiddy, E.M., Coker, C., and Holmes, R.K. (1989). Cloning, nucleotide sequence, and hybridization studies of the type IIb heat-labile enterotoxin gene of *Escherichia coli*. J. Bacteriol. *171*, 4945–4952.

Possot, O., and Pugsley, A.P. (1994). Molecular characterization of PulE, a protein required for pullulanase secretion. Mol. Microbiol. *12*, 287–299.

Pratha, V.S., Hogan, D.L., Martensson, B.A., Bernard, J., Zhou, R., and Isenberg, J.I. (2000). Identification of transport abnormalities in duodenal mucosa and duodenal enterocytes from patients with cystic fibrosis. Gastroenterology *118*, 1051–1060.

Qadri, F., Das, S.K., Faruque, A.S., Fuchs, G.J., Albert, M.J., Sack, R.B., and Svennerholm, A.M. (2000). Prevalence of toxin types and colonization factors in enterotoxigenic *Escherichia coli* isolated during a 2-year period from diarrhoeal patients in Bangladesh. J. Clin. Microbiol. *38*, 27–31.

Qadri, F., Svennerholm, A.M., Faruque, A.S., and Sack, R.B. (2005). Enterotoxigenic *Escherichia coli* in developing countries: epidemiology, microbiology, clinical features, treatment, and prevention. Clin. Microbiol. Rev. *18*, 465–483.

Rao, M.R., Abu-Elyazeed, R., Savarino, S.J., Naficy, A.B., Wierzba, T.F., Abdel-Messih, I., Shaheen, H., Frenck, R.W., Jr., Svennerholm, A.M., and Clemens, J.D. (2003). High disease burden of diarrhoea due to enterotoxigenic *Escherichia coli* among rural Egyptian infants and young children. J. Clin. Microbiol. *41*, 4862–4864.

Reichow, S.L., Korotkov, K.V., Hol, W.G., and Gonen, T. (2010). Structure of the cholera toxin secretion channel in its closed state. Nat. Struct. Mol. Biol. *17*, 1226–1232.

Richards, A.A., Stang, E., Pepperkok, R., and Parton, R.G. (2002). Inhibitors of COP-mediated transport and cholera toxin action inhibit simian virus 40 infection. Mol. Biol. Cell *13*, 1750–1764.

Rodighiero, C., Aman, A.T., Kenny, M.J., Moss, J., Lencer, W.I., and Hirst, T.R. (1999). Structural basis for the differential toxicity of cholera toxin and *Escherichia coli* heat-labile enterotoxin: Construction of hybrid toxins identifies the A2-domain as the determinant of differential toxicity. J. Biol. Chem. *274*, 3962–3969.

Rodighiero, C., Tsai, B., Rapoport, T.A., and Lencer, W.I. (2002). Role of ubiquitination in retro-translocation of cholera toxin and escape of cytosolic degradation. EMBO Rep. *3*, 1222–1227.

Rudner, X.L., Mandal, K.K., de Sauvage, F.J., Kindman, L.A., and Almenoff, J.S. (1995). Regulation of cell signaling by the cytoplasmic domains of the heat-stable enterotoxin receptor: identification of autoinhibitory and activating motifs. Proc. Natl. Acad. Sci. U.S.A. *92*, 5169–5173.

Ryder, R.W., Wachsmuth, I.K., Buxton, A.E., Evans, D.G., DuPont, H.L., Mason, E., and Barrett, F.F. (1976). Infantile diarrhoea produced by heat-stable enterotoxigenic *Escherichia coli*. N. Engl. J. Med. *295*, 849–853.

Sack, D.A., Merson, M.H., Wells, J.G., Sack, R.B., and Morris, G.K. (1975). Diarrhoea associated with heat-stable enterotoxin-producing strains of *Escherichia coli*. Lancet *2*, 239–241.

Sack, R.B. (1990). Travelers' diarrhoea: microbiologic bases for prevention and treatment. Rev. Infect. Dis. *12(Suppl 1)*, S59–63.

Sandkvist, M., Bagdasarian, M., Howard, S.P., and DiRita, V.J. (1995). Interaction between the autokinase EpsE and EpsL in the cytoplasmic membrane is required for extracellular secretion in *Vibrio cholerae*. EMBO J. *14*, 1664–1673.

Sandkvist, M., Michel, L.O., Hough, L.P., Morales, V.M., Bagdasarian, M., Koomey, M., DiRita, V.J., and Bagdasarian, M. (1997). General secretion pathway (eps) genes required for toxin secretion and outer membrane biogenesis in *Vibrio cholerae*. J. Bacteriol. *179*, 6994–7003.

Sardet, C., Franchi, A., and Pouyssegur, J. (1989). Molecular cloning, primary structure, and expression of the human growth factor-activatable Na+/H+ antiporter. Cell *56*, 271–280.

Sato, T., and Shimonishi, Y. (2004). Structural features of *Escherichia coli* heat-stable enterotoxin that activates membrane-associated guanylyl cyclase. J. Peptide Res. *63*, 200–206.

de Sauvage, F.J., Camerato, T.R., and Goeddel, D.V. (1991). Primary structure and functional expression of the human receptor for *Escherichia coli* heat-stable enterotoxin. J. Biol. Chem. *266*, 17912–17918.

Scheving, L.A., Russell, W.E., and Chong, K.M. (1996). Structure, glycosylation, and localization of rat intestinal guanylyl cyclase C: modulation by fasting. Am. J. Physiol. *271*, G959–968.

Schmitz, A., Herrgen, H., Winkeler, A., and Herzog, V. (2000). Cholera toxin is exported from microsomes by the Sec61p complex. J. Cell. Biol. *148*, 1203–1212.

Schulz, S., Green, C.K., Yuen, P.S., and Garbers, D.L. (1990). Guanylyl cyclase is a heat-stable enterotoxin receptor. Cell *63*, 941–948.

Schulz, S., Lopez, M.J., Kuhn, M., and Garbers, D.L. (1997). Disruption of the guanylyl cyclase-C gene leads to a paradoxical phenotype of viable but heat-stable enterotoxin-resistant mice. J. Clin. Invest. *100*, 1590–1595.

Scott, R.O., Thelin, W.R., and Milgram, S.L. (2002). A novel PDZ protein regulates the activity of guanylyl cyclase C, the heat-stable enterotoxin receptor. J. Biol. Chem. *277*, 22934–22941.

Seidler, U., Blumenstein, I., Kretz, A., Viellard-Baron, D., Rossmann, H., Colledge, W.H., Evans, M., Ratcliff, R., and Gregor, M. (1997). A functional CFTR protein is required for mouse intestinal cAMP-, cGMP- and Ca(2+)-dependent HCO3- secretion. J. Physiol. *505 (Pt 2)*, 411–423.

Sellers, Z.M., Childs, D., Chow, J.Y., Smith, A.J., Hogan, D.L., Isenberg, J.I., Dong, H., Barrett, K.E., and Pratha, V.S. (2005). Heat-stable enterotoxin of *Escherichia coli* stimulates a non-CFTR-mediated duodenal bicarbonate secretory pathway. Am. J. Physiol. Gastrointest. Liver Physiol. *288*, G654–663.

Sellers, Z.M., Mann, E., Smith, A., Ko, K.H., Giannella, R., Cohen, M.B., Barrett, K.E., and Dong, H. (2008). Heat-stable enterotoxin of *Escherichia coli* (STa) can stimulate duodenal HCO3(-) secretion via a novel GC-C - and CFTR-independent pathway. FASEB J. *22*, 1306–1316.

Selvaraj, N.G., Prasad, R., Goldstein, J.L., and Rao, M.C. (2000). Evidence for the presence of cGMP-dependent protein kinase-II in human distal colon and in T84, the colonic cell line. Biochim. Biophys. Acta *1498*, 32–43.

Seriwatana, J., Echeverria, P., Taylor, D.N., Rasrinaul, L., Brown, J.E., Peiris, J.S., and Clayton, C.L. (1988). Type II heat-labile enterotoxin-producing *Escherichia coli* isolated from animals and humans. Infect. Immun. *56*, 1158–1161.

Sheppard, D.N., and Welsh, M.J. (1999). Structure and function of the CFTR chloride channel. Physiol. Rev. *79*, S23–45.

Shimomura, H., Dangott, L.J., and Garbers, D.L. (1986). Covalent coupling of a resact analogue to guanylate cyclase. J. Biol. Chem. *261*, 15778–15782.

Sindic, A., Velic, A., Basoglu, C., Hirsch, J.R., Edemir, B., Kuhn, M., and Schlatter, E. (2005). Uroguanylin and guanylin regulate transport of mouse cortical collecting duct independent of guanylate cyclase C. Kidney Int. *68*, 1008–1017.

Singh, R.D., Puri, V., Valiyaveettil, J.T., Marks, D.L., Bittman, R., and Pagano, R.E. (2003). Selective caveolin-1-dependent endocytosis of glycosphingolipids. Mol. Biol. Cell *14*, 3254–3265.

Singh, S.K., Binder, H.J., Boron, W.F., and Geibel, J.P. (1995). Fluid absorption in isolated perfused colonic crypts. J. Clin. Invest. *96*, 2373–2379.

Sixma, T.K., Kalk, K.H., Van Zanten, B.A.M., Dauter, Z., Kingma, J., Witholt, B., and Hol, W.G.J. (1993). Refined structure of *Escherichia coli* heat-labile enterotoxin, a close relative of cholera toxin. J. Mol. Biol. *230*, 890–918.

Smith, H.W., and Gyles, C.L. (1970). The relationship between two apparently different enterotoxins produced by enteropathogenic strains of *Escherichia coli* of porcine origin. J. Med. Microbiol. *3*, 387–401.

Smith, H.W., and Halls, S. (1967). Studies on *Escherichia coli* enterotoxin. J. Pathol. Bacteriol. *93*, 531–543.

So, M., and McCarthy, B.J. (1980). Nucleotide sequence of the bacterial transposon Tn1681 encoding a heat-stable (ST) toxin and its identification in enterotoxigenic *Escherichia coli* strains. Proc. Natl. Acad. Sci. U.S.A. *77*, 4011–4015.

Sonawane, N.D., Muanprasat, C., Nagatani, R., Jr., Song, Y., and Verkman, A.S. (2005). *In vivo* pharmacology and antidiarrhoeal efficacy of a thiazolidinone CFTR inhibitor in rodents. J. Pharm. Sci. *94*, 134–143.

Sopory, S., Kaur, T., and Visweswariah, S.S. (2004). The cGMP-binding, cGMP-specific phosphodiesterase (PDE5): intestinal cell expression, regulation and role in fluid secretion. Cell. Signal. *16*, 681–692.

Spicer, E.K., and Noble, J.A. (1982). *Escherichia coli* heat-labile enterotoxin. Nucleotide sequence of the A subunit gene. J. Biol. Chem. *257*, 5716–5721.

Spicer, E.K., Kavanaugh, W.M., Dallas, W.S., Falkow, S., Konigsberg, W.H., and Schafer, D.E. (1981). Sequence homologies between A subunits of *Escherichia coli* and *Vibrio cholerae* enterotoxins. Proc. Natl. Acad. Sci. U.S.A. *78*, 50–54.

Staples, S.J., Asher, S.E., and Giannella, R.A. (1980). Purification and characterization of heat-stable enterotoxin produced by a strain of *E. coli* pathogenic for man. J. Biol. Chem. *255*, 4716–4721.

Steffen, R., Castelli, F., Dieter Nothdurft, H., Rombo, L., and Jane Zuckerman, N. (2005). Vaccination against enterotoxigenic *Escherichia coli*, a cause of travelers' diarrhoea. J. Travel Med. *12*, 102–107.

Tabcharani, J.A., Chang, X.B., Riordan, J.R., and Hanrahan, J.W. (1991). Phosphorylation-regulated Cl- channel in CHO cells stably expressing the cystic fibrosis gene. Nature *352*, 628–631.

Takao, T., Tominaga, N., Yoshimura, S., Shimonishi, Y., Hara, S., Inoue, T., and Miyama, A. (1985). Isolation, primary structure and synthesis of heat-stable enterotoxin produced by *Yersinia enterocolitica*. FEBS *152*, 199–206.

Takeda, Y., Takeda, T., Yano, T., Yamamoto, K., and Miwatani, T. (1979). Purification and partial characterization of heat-stable enterotoxin of enterotoxigenic *Escherichia coli*. Infect. Immun. *25*, 978–985.

Tauschek, M., Gorrell, R.J., Strugnell, R.A., and Robins-Browne, R.M. (2002). Identification of a protein secretory pathway for the secretion of heat-labile enterotoxin by an enterotoxigenic strain of *Escherichia coli*. Proc. Natl. Acad. Sci. U.S.A. *99*, 7066–7071.

Taylor, M., Navarro-Garcia, F., Huerta, J., Burress, H., Massey, S., Ireton, K., and Teter, K. (2010). Hsp90 is required for transfer of the cholera toxin A1 subunit from the endoplasmic reticulum to the cytosol. J. Biol. Chem. *285*, 31261–31267.

Taylor, M., Banerjee, T., Ray, S., Tatulian, S.A., and Teter, K. (2011). Protein-disulfide isomerase displaces the cholera toxin A1 subunit from the holotoxin without unfolding the A1 subunit. J. Biol. Chem. *286*, 22090–22100.

Thompson, J.P., and Schengrund, C.L. (1997). Oligosaccharide-derivatized dendrimers: Defined multivalent inhibitors of the adherence of the cholera toxin B subunit and the heat labile enterotoxin of *E. coli* to GM1. Glycoconjugate J. *14*, 837–845.

Thompson, J.P., and Schengrund, C.L. (1998). Inhibition of the adherence of cholera toxin and the heat-labile enterotoxin of *Escherichia coli* to cell-surface GM1 by oligosaccharide-derivatized dendrimers. Biochem. Pharmacol. *56*, 591–597.

Tsai, B., Rodighiero, C., Lencer, W.I., and Rapoport, T.A. (2001). Protein disulfide isomerase acts as a redox-dependent chaperone to unfold cholera toxin. Cell *104*, 937–948.

Tsuji, T., Honda, T., and Miwatani, T. (1984). Comparison of effects of nicked and unnicked *Escherichia coli* heat-labile enterotoxin on Chinese hamster ovary cells. Infect. Immun. *46*, 94–97.

Turner, L.R., Lara, J.C., Nunn, D.N., and Lory, S. (1993). Mutations in the consensus ATP-binding sites of XcpR and PilB eliminate extracellular protein secretion and pilus biogenesis in *Pseudomonas aeruginosa*. J. Bacteriol. *175*, 4962–4969.

Vaandrager, A.B., Schulz, S., De Jonge, H.R., and Garbers, D.L. (1993a). Guanylyl cyclase C is an N-linked glycoprotein receptor that accounts for multiple heat-stable enterotoxin-binding proteins in the intestine. J. Biol. Chem. *268*, 2174–2179.

Vaandrager, A.B., Van der Wiel, E., and De Jonge, H.R. (1993b). Heat-stable enterotoxin activation of immunopurified guanylyl cyclase C. Modulation by adenine nucleotides. J. Biol. Chem. *268*, 19598–19603.

Vaandrager, A.B., Bot, A.G., and De Jonge, H.R. (1997a). Guanosine 3′,5′-cyclic monophosphate-dependent protein kinase II mediates heat-stable enterotoxin-provoked chloride secretion in rat intestine. Gastroenterology *112*, 437–443.

Vaandrager, A.B., Tilly, B.C., Smolenski, A., Schneider-Rasp, S., Bot, A.G., Edixhoven, M., Scholte, B.J., Jarchau, T., Walter, U., Lohmann, S.M., et al. (1997b). cGMP stimulation of cystic fibrosis transmembrane conductance regulator Cl- channels co-expressed with cGMP-dependent protein kinase type II but not type Ibeta. J. Biol. Chem. *272*, 4195–4200.

Vaandrager, A.B., Smolenski, A., Tilly, B.C., Houtsmuller, A.B., Ehlert, E.M., Bot, A.G., Edixhoven, M., Boomaars, W.E., Lohmann, S.M., and de Jonge, H.R. (1998). Membrane targeting of cGMP-dependent protein kinase is required for cystic fibrosis transmembrane conductance regulator Cl– channel activation. Proc. Natl. Acad. Sci. U.S.A. *95*, 1466–1471.

Vaandrager, A.B., Bot, A.G., Ruth, P., Pfeifer, A., Hofmann, F., and De Jonge, H.R. (2000). Differential role of cyclic GMP-dependent protein kinase II in ion transport in murine small intestine and colon. Gastroenterology *118*, 108–114.

Valentino, M.A., Lin, J.E., Snook, A.E., Li, P., Kim, G.W., Marszalowicz, G., Magee, M.S., Hyslop, T., Schulz, S., and Waldman, S.A. (2011). A uroguanylin-GUCY2C endocrine axis regulates feeding in mice. J. Clin. Invest. *121*, 3578–3588.

Van Heyningen, W.E., Carpenter, C.C., Pierce, N.F., and Greenough, W.B., 3rd (1971). Deactivation of cholera toxin by ganglioside. J. Infect. Dis. *124*, 415–418.

Wagner, J.A., Cozens, A.L., Schulman, H., Gruenert, D.C., Stryer, L., and Gardner, P. (1991). Activation of chloride channels in normal and cystic fibrosis airway epithelial cells by multifunctional calcium/calmodulin-dependent protein kinase. Nature *349*, 793–796.

Wakabayashi, S., Pang, T., Su, X., and Shigekawa, M. (2000). A novel topology model of the human Na(+)/H(+) exchanger isoform 1. J. Biol. Chem. *275*, 7942–7949.

Weinman, E.J., Steplock, D., Wang, Y., and Shenolikar, S. (1995). Characterization of a protein cofactor that mediates protein kinase A regulation of the renal brush border membrane Na(+)-H+ exchanger. J. Clin. Invest. *95*, 2143–2149.

Weinman, E.J., Steplock, D., Wade, J.B., and Shenolikar, S. (2001). Ezrin binding domain-deficient NHERF attenuates cAMP-mediated inhibition of Na(+)/H(+) exchange in OK cells. Am. J. Physiol. Renal Physiol. *281*, F374–380.

Wenneras, C., and Erling, V. (2004). Prevalence of enterotoxigenic *Escherichia coli*-associated diarrhoea and carrier state in the developing world. J. Health Population Nutr. *22*, 370–382.

Wolf, M.K. (1997). Occurrence, distribution, and associations of O and H serogroups, colonization factor antigens, and toxins of enterotoxigenic *Escherichia coli*. Clin. Microbiol. Rev. *10*, 569–584.

Yamamoto, T., and Yokota, T. (1983). Sequence of heat-labile enterotoxin of *Escherichia coli* pathogenic for humans. J. Bacteriol. *155*, 728–733.

Yamamoto, T., Tamura, T., Ryoji, M., Kaji, A., Yokota, T., and Takano, T. (1982). Sequence analysis of the heat-labile enterotoxin subunit B gene originating in human enterotoxigenic *Escherichia coli*. J. Bacteriol. *152*, 506–509.

Yu, F.H., Shull, G.E., and Orlowski, J. (1993). Functional properties of the rat Na/H exchanger NHE-2 isoform expressed in Na/H exchanger-deficient Chinese hamster ovary cells. J. Biol. Chem. *268*, 25536–25541.

Zachos, N.C., Li, X., Kovbasnjuk, O., Hogema, B., Sarker, R., Lee, L.J., Li, M., de Jonge, H., and Donowitz, M. (2009). NHERF3 (PDZK1) contributes to basal and calcium inhibition of NHE3 activity in Caco-2BBe cells. J. Biol. Chem. *284*, 23708–23718.

Zhang, W., Mannan, I., Schulz, S., Parkinson, S.J., Alekseev, A.E., Gomez, L.A., Terzic, A., and Waldman, S.A. (1999). Interruption of transmembrane signaling as a novel antisecretory strategy to treat enterotoxigenic diarrhoea. FASEB J. *13*, 913–922.

Zizak, M., Lamprecht, G., Steplock, D., Tariq, N., Shenolikar, S., Donowitz, M., Yun, C.H., and Weinman, E.J. (1999). cAMP-induced phosphorylation and inhibition of Na(+)/H(+) exchanger 3 (NHE3) are dependent on the presence but not the phosphorylation of NHE regulatory factor. J. Biol. Chem. *274*, 24753–24758.

Haemolysins

Bernt E. Uhlin, Jan Oscarsson and Sun N. Wai

9

Abstract

As is the case with several other bacterial species, there are strains of *Escherichia coli* that produce haemolysins. The production may be detected by the use of red blood cells in solid growth media or in other lysis assays. Here we are summarizing findings about properties of three haemolysins/cytolysins: the α-haemolysin (HlyA), the EHEC-Hly and the ClyA (SheA, HlyE) proteins that are found in different *E. coli*.

The HlyA protein

HlyA (α-haemolysin) is produced and disseminated by uropathogenic and many commensal faecal *Escherichia coli* isolates, and its synthesis, activation, and secretion are directed by the *hlyCABD* gene locus (Goebel and Hedgpeth, 1982; Koronakis and Hughes, 1996). The *hlyA* gene product is a ~ 107 kDa polypeptide that consists of 1023 or 1024 amino acid residues, depending on strain. The haemolysin is synthesized as an inactive protoxin, ProHlyA, which is post-translationally activated in two steps, first via HlyC-directed acylation in the cytoplasm, and then by binding Ca^{2+} in the extracellular medium (Stanley *et al.*, 1998). Albeit no crystal structure of HlyA has yet been obtained, the secondary structure reveals the presence of an N-terminal domain, characterized by nine amphipathic α-helices, which is presumed to interact with the hydrophobic matrix of the host membrane (Hyland *et al.*, 2001), and a C-terminal domain containing 11–17 glycine- and aspartate-rich nonapeptide β-strand repeats in tandem, having the consensus sequence GGxGxDxUx (U, large, hydrophobic amino acid; X, any amino acid). The 'GG repeat' region is a hallmark of the 'repeats-in-toxin' (RTX) family of toxins (Welch, 1991), which is widespread among Gram-negative bacteria (Linhartova *et al.*, 2010), and serves as a Ca^{2+} binding domain of HlyA (Boehm *et al.*, 1990; Ludwig *et al.*, 1988; Ostolaza *et al.*, 1995). Crystal structures of two other RTX toxins have indicated that the GG repeats bind one Ca^{2+} ion per nonapeptide with high affinity (Baumann *et al.*, 1993; Hamada *et al.*, 1996). Moreover, a non-cleavable secretion signal is encoded within the last 50–60 C-terminal amino acids of HlyA (Grey *et al.*, 1989; Nicaud *et al.*, 1986; Stanley *et al.*, 1991).

HlyA mechanism of action

HlyA exhibits cytolytic and cytotoxic activity towards a wide spectrum of cells of various mammalian species, including erythrocytes (Bhakdi *et al.*, 1986), granulocytes (Bhakdi *et al.*, 1989; Cavalieri and Snyder, 1982; Gadeberg and Orskov, 1984), monocytes (Bhakdi *et al.*, 1990), endothelial cells (Suttorp *et al.*, 1990), renal epithelial cells (Keane *et al.*, 1987; Mobley *et al.*, 1990), and T lymphocytes (Jonas *et al.*, 1993) derived from mice, primates, and ruminants. Mechanisms governing the target cell specificity of HlyA are not yet fully understood as can be illustrated by the strong resistance of python (*Python regius*, and *Python molurus*) erythrocytes towards HlyA action, which could not be attributed to either increased osmotic resistance, lack of purinergic haemolysis amplification (see below), or differences in HlyA affinity (Larsen *et al.*, 2011). Analogously to several other pore-forming toxins, the activities of HlyA against target cells are highly dose dependent with pore-induced changes in ion balance causing a diverse range of secondary effects at sublytic toxin levels, whereas it is less clear how often HlyA reaches local concentrations high enough to lyse host cells during infection (Bien *et al.*, 2012; Bischofberger *et al.*, 2012; Wiles *et al.*, 2008b).

Membrane binding

Several investigators have explored the mode of binding of HlyA to membranes, and searched for a toxin-specific receptor, albeit the results have proven difficult to accommodate in a single model. The lymphocyte function-associated antigen (LFA-1) (CD11a/CD18; $\alpha_1\beta_2$ integrin) has been reported to serve as a receptor for HlyA on polymorphonuclear neutrophils (Lally *et al.*, 1997), and HlyA appears to recognize and bind to the N-linked oligosaccharides to their β_2-integrin receptors (Morova *et al.*, 2008). Moreover, on erythrocytes glycophorin, a surface sialoglycoprotein acts as an apparent receptor for HlyA (Aulik *et al.*, 2013; Cortajarena *et al.*, 2001), with the amino acids 914–936 near the C-terminus of HlyA representing a glycophorin-binding domain (Cortajarena *et al.*, 2003). On the other hand, it has also been reported that the binding of HlyA to cells occurs in a non-saturable manner and that the toxin does not interact with a specific protein receptor neither on granulocytes nor erythrocytes (Valeva *et al.*, 2005). Similarly, studies using protein-free unilamellar vesicles have demonstrated that the presence of a receptor is not required for HlyA to induce membrane lesions (Ostolaza *et al.*, 1993). These contradictory findings may perhaps be attributed to different amounts of toxin and/or target cell species used in the various studies. At all events, HlyA appears to interact with target cell membranes in two consecutive steps, with an initial reversible adsorption that is sensitive to electrostatic forces, followed by an irreversible insertion of the toxin into the membrane (Bakas *et al.*, 1996; Ostolaza *et al.*, 1997). The insertion of HlyA apparently occurs independently of membrane lysis as non-haemolytic mutant HlyA derivatives can still insert into lipid monolayers (Sanchez-Magraner *et al.*, 2006). Studies with the Ca^{2+}-binding C-terminal domain of

HlyA (amino acids 602–1024) have revealed that this may be the portion of the toxin that is reversibly adsorbed onto the membrane during the early stages of HlyA–membrane interaction (Sanchez-Magraner *et al.*, 2007). Concomitantly, as suggested (Sanchez-Magraner *et al.*, 2007), the N-terminal principal region of HlyA inserting into lipid bilayers (amino acids 177–411; Hyland *et al.*, 2001) may be responsible for the later stages, involving irreversible interaction with the membrane hydrophobic matrix and subsequent loss of the permeability barrier.

The HlyA pore-forming mechanism

Once inserted in target membranes, HlyA induces cell lysis by the formation of transient, non-selective pores with a size of ~ 1 nm or ~ 1.5–3 nm as estimated by conductance (Benz *et al.*, 1989) and permeability (Bhakdi *et al.*, 1986), respectively. However, it is not yet entirely clear how HlyA creates these lesions, and whether oligomerization is an absolute requirement for the pore formation. It has been estimated that only one to three HlyA molecules form the pore (Benz *et al.*, 1989; Menestrina *et al.*, 1987, 1994). Fluorescence resonance energy transfer (FRET) measurements of labelled HlyA bound to sheep ghost erythrocytes supports that the toxin forms oligomers on erythrocyte membranes (Herlax *et al.*, 2009). Cholesterol-rich microdomains within the host cell plasma membrane may act as platforms that concentrate the toxin in this oligomerization process (Dhakal and Mulvey, 2012; Herlax *et al.*, 2009). It has also been suggested that small oligomers are first formed, followed by their assembly into multimeric structures (Herlax *et al.*, 2009), which would be in accordance with the presence of heterogeneous erythrocyte lesions that increase in size over time (Moayeri and Welch, 1994). On the other hand, results from liposome studies using a fluorescence requenching method are not consistent with oligomer formation (Fiser and Konopasek, 2009). Moreover, no HlyA oligomer structure has been isolated up to now, and HlyA recovered from detergent-solubilized erythrocyte membranes is monomeric, indicating either that oligomerization is not required for pore formation or that oligomers dissociates in the detergent (Bhakdi *et al.*, 1986).

Mechanisms of cell damage

The mechanism by which HlyA induces cell damage has been assumed to result from the pore itself (Bhakdi *et al.*, 1986, 1988), however it was recently shown that the toxin takes advantage of a specific cellular amplification system to produce the full haemolytic response on erythrocytes *in vitro*, which would be consistent with earlier reports that a single HlyA molecule is sufficient to lyse an erythrocyte (Jorgensen *et al.*, 1980). More exactly, the haemolytic process requires the activation of purinergic (P2) receptors and pannexin channels, and consists of sequential shrinkage and swelling of the erythrocytes (Skals *et al.*, 2009, 2010). The observed HlyA-induced volume reduction of erythrocytes prior to lysis appears to be a consequence of a rapid rise in the intracellular Ca^{2+} concentration, which triggers K^+ efflux via activation of K^+ channels (Sanchez *et al.*, 2011; Skals *et al.*, 2010). Further experimental evidence suggests that this initial consequence of HlyA exposure targets erythrocytes for recognition and phagocytosis through monocytes, thereby preventing acute lysis of erythrocytes and reducing the risk of intravascular haemolysis (Fagerberg *et al.*, 2013). At sublytic HlyA concentrations, pore-induced changes in ion balance cause a multitude of downstream effects on host cells. For instance, HlyA can trigger multiple signalling cascades by inducing oscillations in the intracellular levels of Ca^{2+} (Koschinski *et al.*, 2006; Laestadius *et al.*, 2002; Uhlen *et al.*, 2000), and by influencing key regulators of host cell survival and inflammation such as Akt (Wiles *et al.*, 2008a), Mitogen-activated protein kinase (MAPK) p38 (Kloft *et al.*, 2009), and NFκB (Dhakal and Mulvey, 2012) to subvert important physiological host cell functions. A mechanism for such HlyA-induced modulation was recently proposed. Upon insertion into epithelial cell and macrophage membranes at sublytic levels, HlyA induces the activation of host cell proteases, which triggers degradation of the cytoskeletal scaffolding protein paxillin and other host regulatory proteins, including components of the proinflammatory NFκB signalling cascade (Dhakal and Mulvey, 2012). Consistent with HlyA-induced apoptosis (Chen *et al.*, 2006; Jonas *et al.*, 1993; Russo *et al.*, 2005), HlyA-intoxication, seemingly independent

of host serine proteases, stimulates caspase activation (Dhakal and Mulvey, 2012), implicating a coordinated mechanism by which HlyA can modulate epithelial cell functions, disable macrophages and suppress inflammatory responses. According to a recent study, HlyA contributes to the activation of the intrinsic apoptosis pathway in platelets by UPEC strains, via degradation of the pro-survival protein Bcl-x$_L$, suggesting the possibility that the toxin could promote the development of thrombocytopenia in patients with bloodstream infections (Kraemer *et al.*, 2012). Moreover, low concentrations of HlyA induce oxidative stress in human whole blood cultures, resulting in oxidative damage to macromolecules such as proteins and lipids (Baronetti *et al.*, 2013). Further determination of the sublytic, secondary effects of HlyA, especially during infection, will likely be of great importance to clarify the roles(s) this toxin may play during pathogenesis.

HlyA role in virulence

Epidemiological studies support that expression of HlyA is linked to pathogenesis in upper urinary tract infections, with prevalence of the *hly* locus of 28–45% among UPEC strains, and 7–20% among faecal isolates of *E. coli* (Johnson *et al.*, 2005; Kerenyi *et al.*, 2005; Mao *et al.*, 2012; Wang *et al.*, 2002), and an association of HlyA with enteric disease has also been reported (Elliott *et al.*, 1998). Although being a virulence determinant in several animal models of ExPEC infection, including peritonitis, pneumonia, pyelonephritis, and septicaemia (May *et al.*, 2000; O'Hanley *et al.*, 1991; Russo *et al.*, 2005; Welch *et al.*, 1981), HlyA has not yet been assigned a specific role in pathogenesis. A major role has generally been attributed to the destruction of host cells, thereby facilitating the release of nutrients and other factors such as iron, which are critical for bacterial growth (Johnson, 1991; Wiles *et al.*, 2008b). Indeed, HlyA-producing UPEC strains cause more pronounced tissue damage within the urinary tract, correlating with more severe clinical outcomes (Bien *et al.*, 2012; Marrs *et al.*, 2005; Smith *et al.*, 2008), and murine and tissue culture model systems have revealed that *hlyA*-positive isolates, in contrast to isogenic mutants induce a more rapid and extensive shedding of bladder epithelial cells, (Dhakal and Mulvey, 2012; Smith *et al.*, 2006, 2008). Hence, HlyA may promote bacterial dissemination in the urinary tract, albeit exfoliation of uroepithelial cells is also functioning as a host clearing mechanism (Bower *et al.*, 2005). Analogously, HlyA may enhance translocation of bacteria in the colon by forming focal leaks within epithelial cells (Troeger *et al.*, 2007). A role of HlyA in augmenting the inflammatory response during the early hours of infection has been suggested based on the delayed pathophysiology of exteriorized kidney tubules in live rats when using *hlyA*-deficient as compared to *hlyA*-expressing UPEC strains (Mansson *et al.*, 2007). Consistent with this idea, production of HlyA by UPEC strains appears to correlate with an enhanced proinflammatory response in murine bladders, as evidenced using cDNA microarrays (Garcia *et al.*, 2013). Studies using a zebrafish embryo model system suggest that the toxin may contribute to immune defence evasion by neutralizing phagocytes (Wiles *et al.*, 2009). On the other hand, there appears to be no clear correlation between HlyA and host cytokine production during *in vivo* infection (Hilbert *et al.*, 2012; Real *et al.*, 2007). Moreover, several investigations support the notion that damage caused by HlyA does not result in enhanced colonization and fitness of the infecting *E. coli* strain (Haugen *et al.*, 2007; Linggood and Ingram, 1982; Mansson *et al.*, 2007; Reigstad *et al.*, 2007; Sheshko *et al.*, 2006; Smith *et al.*, 2008).

Extracellular export of HlyA

HlyA is extracellularly released as a soluble protein in a single step via the type I secretion pathway by means of the inner membrane protein HlyB, which is an ATP-binding cassette transporter, the outer membrane protein TolC, and the membrane fusion protein HlyD, anchored in the inner membrane (Grey *et al.*, 1989; Holland *et al.*, 2005; Koronakis *et al.*, 2000; Pimenta *et al.*, 1999; Wandersman and Delepelaire, 1990; Wang *et al.*, 1991). The functional characteristics of this secretion system enable it to be used in different applications, including the presentation of heterologous antigens in live-attenuated bacterial vaccines (Gentschev *et al.*, 1996, 2002;

Hotz *et al.*, 2009; Zhu *et al.*, 2006), and industrial enzyme production, particularly regarding proteins with slow-folding kinetics (Low *et al.*, 2010; Schwarz *et al.*, 2012; Su *et al.*, 2012). Many events involved in the assembly of the HlyA translocator remain unclear. The mechanism for extracellular export of HlyA includes the formation of an HlyB–HlyD complex in the inner membrane, which interacts with the HlyA C-terminal signal peptide to recruit TolC (Benabdelhak *et al.*, 2003; Thanabalu *et al.*, 1998). It has been suggested that HlyD in this assembly process makes a direct contact with TolC through a structural motif similar to the adaptor proteins in drug efflux pumps (Lee *et al.*, 2012). This functional complex then forms a tunnel across the periplasm to ensure the continuous translocation of HlyA without any periplasmic intermediate (Koronakis *et al.*, 2000; Thanabalu *et al.*, 1998). In contrast to the extracellular export of HlyA, which requires ATP hydrolysis catalysed by HlyB, the assembly of the Hly functional complex is nucleotide-independent (Thanabalu *et al.*, 1998). Recent data support that HlyA exists in an unfolded conformation in the cytoplasm (Bakkes *et al.*, 2010), and is protected inside the cell via an N-terminal domain of HlyB that is essential for secretion of the toxin (Lecher *et al.*, 2012). Translocation of HlyA as an unfolded polypeptide may be a requirement considering the restricted diameter of TolC (Koronakis *et al.*, 2000; Pimenta *et al.*, 2005). It has also been reported that HlyA accumulates on the *E. coli* cell surface (Oropeza-Wekerle *et al.*, 1989; Pimenta *et al.*, 2005), and that a fraction (~ 2–30%) of the total amount of extracellular HlyA protein is released via outer membrane vesicles (OMVs) in different ExPEC isolates (Balsalobre *et al.*, 2006). Hence, as suggested (Balsalobre *et al.*, 2006), OMVs may play an important role in the transport/dissemination of the toxin to host cells during infection, possibly via endocytic uptake of the OMVs, analogously to ETEC vesicles carrying the Heat-labile enterotoxin (Kesty *et al.*, 2004). As the specific haemolytic activity of OMV-associated HlyA appears to be higher than that of purified, free HlyA, OMVs may also function as an HlyA-concentrating mechanism near the target cells (Herlax *et al.*, 2010).

Post-translational modification of HlyA

The inactive protoxin (ProHlyA) undergoes post-translational activation in the cytoplasm via HlyC-directed acylation prior to extracellular release. For this, which is essential for toxin activity, the lysine acyl transferase HlyC uses the fatty acyl residues carried by acyl carrier protein (ACP) to form a covalent acyl-HlyC intermediate, which transfers the fatty acyl residues to the ε-amino groups of two lysine residues (Lys-564 and Lys-690, or Lys-563 and Lys-689, respectively depending on length of the amino acid sequence) of ProHlyA via a covalent amide linkage (Issartel et al., 1991; Stanley et al., 1994, 1998; Trent et al., 1998; Worsham et al., 2001). HlyA acylated in vivo is composed of up to nine dissimilar covalent structures with two acylation sites and three possible modifying groups in each site (Lim et al., 2000). In addition to acylation by HlyC, Ca^{2+} is an essential requirement for HlyA activity, and as cytoplasmic $[Ca^{2+}]$ in E. coli is very low (0.1 μM (Gangola and Rosen, 1987)) HlyA must associate with Ca^{2+} in the extracellular medium (Boehm et al., 1990; Ostolaza et al., 1995). This second stage of HlyA activation appears to be acylation-dependent as the Ca^{2+}-binding capacity of HlyA is low when non-acylated (Soloaga et al., 1996). The exact role(s) of these post-translational modifications of HlyA in the haemolytic process are not entirely clear, and they are neither required for secretion of HlyA nor for its binding to membranes (Bauer and Welch, 1996a; Hyland et al., 2001; Nicaud et al., 1985). Acylation of HlyA has been reported to induce a molten globule conformation of the protein that furthers the irreversibility of its binding to membranes (Herlax and Bakas, 2003, 2007), and to promote HlyA protein–protein interactions, which are required to form functional oligomeric pores (Herlax et al., 2009). Consistent with such observations, recent studies using purified HlyA subjected to brief heat treatment support the notion that a stable molten globule state of the toxin is associated with increased cytotoxicity (Aulik et al., 2013). Analogously to the acylation, the extracellular binding of Ca^{2+} via GG repeats has been proposed to ensure productive insertion of HlyA into target cell membranes (Bakas et al., 1998; Hyland et al., 2001; Sanchez-Magraner et al., 2006), seemingly by stabilizing and compacting both the C- and the N-terminal regions of HlyA to induce a folded, active conformation of the toxin (Chenal et al., 2009, 2010; Sanchez-Magraner et al., 2007, 2010). A link between the Ca^{2+}-dependent activation of HlyA and its translocation across the periplasm has been proposed as mutations in hlyD dramatically decrease the folding of HlyA into a haemolytically active form, suggesting that HlyD either directly or indirectly affects HlyA folding following or during its extracellular release (Pimenta et al., 2005). Consistent with this notion, the HlyA C-terminal secretion signal region may play a role in the post-translational folding of the protein (Jumpertz et al., 2010).

Genetics and regulation of hlyA expression

Hierarchical cluster analysis has revealed a common UPEC gene pool that contains determinants encoding virulence factors such as α-haemolysin, and which is widely distributed between ExPEC and many faecal E. coli isolates of ECOR group B2 and D (Brzuszkiewicz et al., 2006). The hly genetic determinants are located on chromosome-bound pathogenicity islands (PAIs) in E. coli strains isolated from human urinary tract infections (Dobrindt et al., 2000; Guyer et al., 1998; Kao et al., 1997). The two unrelated ExPEC model strains J96 (O4:K6) and 536 (O6:K15) carry each on distinct PAIs (PAI-I and PAI-II) two hlyCABD operons, which may have evolved independently as judged by their dissimilar junctions and adjacent sequences (Dobrindt et al., 2000, 2002a; Swenson et al., 1996). Both hly operons are required for full virulence of strain 536 in mouse models of urinary tract infection and lung toxicity (Nagy et al., 2006). In contrast to UPEC strains, α-haemolysin operons in EPEC O26, ETEC, and in Shiga toxin 2e(Stx2e)-producing STEC strains are found on large (48–157 kb) conjugative plasmids (Burgos and Beutin, 2010; Burgos et al., 2009; Wu et al., 2007). Based on sequence analysis of the hlyCABD genes and their 5' regions it has been suggested that the plasmid-encoded and chromosomal operons have evolved in parallel (Burgos and Beutin, 2010; Burgos et al.,

2009). The polycistronic *hlyCABD* operon is transcribed from a promoter located upstream of *hlyC* (Welch and Pellett, 1988), and its transcription is repressed under conditions of high osmolarity, anaerobiosis and low temperature (Carmona *et al.*, 1993; Mourino *et al.*, 1994, 1996). The *hly* operon on plasmid pHly152 has been extensively studied with regards to its regulation *in vitro*. This has revealed the nucleoid protein H-NS, which interacts with Hha to repress *hlyA* transcription, as major components in the thermo- and osmoregulation (Juarez *et al.*, 2000; Madrid *et al.*, 2002; Nieto *et al.*, 1997, 2000). Concomitantly, H-NS has been demonstrated to bind at two sites in the pHly152 *hly* operon upstream regulatory region, also overlapping promoter sequences (Madrid *et al.*, 2002). Similarly, transcription of chromosomally encoded *hlyA* is negatively regulated by H-NS, albeit the upstream regulatory region is distinct from *hly* operons on plasmids (Knapp *et al.*, 1985; Muller *et al.*, 2006; Scott *et al.*, 2003). On the other hand, studies with the two chromosome-bound *hly* operons of uropathogenic strain 536 indicate that their highly different *hlyA* expression levels *in vitro* is a consequence of dissimilar upstream regulatory regions (Nagy *et al.*, 2006). Additionally, synthesis of full-length *hlyCABD* transcripts from both plasmid and chromosomal operons is governed by transcription antitermination in which the elongation factor RfaH acts at a rho-independent terminator in the *hlyA-hlyB* intergenic region in conjunction with a short 5′ operon polarity suppressor DNA element (Bailey *et al.*, 1992, 2000; Koronakis *et al.*, 1989; Leeds and Welch, 1996, 1997; Nieto *et al.*, 1996). In some UPEC-associated pathogenicity islands this antitermination mechanism also governs transcription of a closely located upstream *cnf1* gene, allowing co-regulated production of the two toxins (Landraud *et al.*, 2003). Indeed, HlyA and CNF1 are frequently co-expressed in clinical UPEC isolates and strains producing both toxins are more often isolated from patients with haemorrhagic urinary tract infections (Real *et al.*, 2007). Moreover, the *leuX*-encoded minor tRNA$_5$^Leu is required for efficient expression of HlyA in strain 536, seemingly acting at different levels in the *hlyA* regulatory cascade, including *hha*, *rfaH*, and yet unidentified regulator(s) (Dobrindt *et*

al., 2002b). Although the *in vitro* regulation of the *hly* gene locus has been extensively studied, much less is known about mechanisms governing the expression of a-haemolysin during infection. In light of the complex UTI pathogenesis cycle involving both intracellular and extracellular niches (Hannan *et al.*, 2012), and the highly dose-dependent behaviour of HlyA, the spatiotemporal regulation of HlyA expression *in vivo* likely represents a key factor influencing the pathogenic success of UPEC strains. Indeed, a temporal role of HlyA has been supported using a murine model revealing *in vivo* expression of *hlyA* in exteriorized kidneys that appeared to coincide with haemolysin-dependent augmentation of the inflammatory response during the first hours of the infection process (Mansson *et al.*, 2007). Moreover, up-regulated *hlyA* transcription in UPEC strains in intracellular bacterial communities (IBCs) in the murine bladder relative to when present in the distal intestine is consistent with differential *hlyA* regulation in the intracellular environment (Berry *et al.*, 2009; Reigstad *et al.*, 2007).

The EHEC haemolysin

Enterohaemorrhagic *E. coli* (EHEC) strains of serotype O157:H7 are food-borne pathogens that cause severe intestinal (haemorrhagic colitis) and extraintestinal (haemolytic uraemic syndrome) disease (Kaper *et al.*, 2004). These strains produce EHEC haemolysin (Schmidt *et al.*, 1995), also termed enterohaemolysin, encoded by the *ehxCABD* operon (Bauer and Welch, 1996b) on the non-transmissible plasmid pO157 (Boerlin *et al.*, 1998; Schmidt *et al.*, 1994).

The term EHEC-haemolysin was used to distinguish it from α-haemolysin of uropathogenic *E. coli*. Although homologous to *E. coli* α-haemolysin, the EHEC haemolysin shares only about 60% sequence identity (Bauer and Welch, 1996b; Schmidt *et al.*, 1995) and exhibits distinct biochemical properties (Bauer and Welch, 1996b; Schmidt and Karch, 1996). The *ehxCABD* operon is present in most EHEC isolates, and is under strong selective pressure, suggesting the involvement of EHEC haemolysin in pathogenesis or colonization (Boerlin *et al.*, 1998). The EHEC-Hly is a calcium dependent pore-forming cytolysin and belongs to the RTX (repeat-in-toxin) family toxin (Schmidt *et al.*, 1995; Schmidt and Karch, 1996). The EHEC-Hly lyses erythrocytes from different species and bovine lymphocytes (Bauer and Welch, 1996b; Schmidt *et al.*, 1995). In contrast to α-haemolysin, EHEC-Hly can cause a large clear zones of haemolysis after 4–6 h incubation at 37°C on a standard blood agar plate and when the bacterial strains are grown in a liquid medium, EHEC haemolysin remains cell-associated (Schmidt *et al.*, 1995), whereas α-haemolysin is exported into the culture supernatant via a specific membrane translocator system consisting of HlyB, HlyD and the outer-membrane protein as described above. It was suggested that the association of EHEC-Hly with the bacterial cell surface or its extremely low export levels may contribute to the biological function of this toxin in the intimate interaction of EHEC O157:H7 with the target host (Schmidt *et al.*, 1996). Outer membrane vesicle-mediated toxin delivery has been increasingly recognized as a bacterial virulence mechanism (Rompikuntal *et al.*, 2012). EHEC and *Shigella* sp. release OMVs that carry protein cargo including Stx1 and Stx2 (Dutta *et al.*, 2004; Kolling and Matthews, 1999; Yokoyama *et al.*, 2000) but the biological activity of the OMV-associated Stxs remains unknown. It was demonstrated that also EHEC-Hly may be released from the bacterial cells in association with outer membrane vesicles shed by EHEC (Aldick *et al.*, 2009). The association of toxin to the OMVs mediates an increased stability and the prolonged activity of the toxin. EHEC-Hly associated with outer membrane vesicles was shown to have an up to 80 times increased stability and thereby prolonged activity of the toxin in comparison with the free, vesicle-unbound form (Aldick *et al.*, 2009). It has been shown that the interaction with bacterial proteases modulates biological activity of other members of the RTX family such as the *Vibrio cholerae* cytolysin (Nagamune *et al.*, 1996). Recently, it was demonstrated that the serine protease EspPα degrades and inactivates EHEC-Hly (Brockmeyer *et al.*, 2011). The findings suggest that the bacterial effector molecule interference between EHEC-Hly and the serine protease EspPα reflects the concerted interplay of virulence factors and that it might be an additional strategy to modulate the virulence functions.

The EHEC-Hly gene locus, its conservation and regulation

The *ehxCABD* operon was first discovered by hybridization experiments using the CVD419 probe, a 3.4 kb HindIII fragment derived from the large plasmid of the *E. coli* O157:H7 strain EDL 933. It has been shown to hybridize with 99% of the tested enterohaemorrhagic *E. coli* (EHEC) O157:H7 strains (Levine *et al.*, 1987; Schmidt *et al.*, 1995). It was shown that the large plasmid pO157 of strain EDL 933 harbours the EHEC *ehx* operon encoding an RTX toxin later designated as EHEC haemolysin (Hly) (Schmidt *et al.*, 1995). The results of plasmid analyses showed that, as in most EHEC Hly-positive strains, the EHEC O111 *exhA* gene is located on an approximately 12 kb BamHI fragment, similar to the situation found in *E. coli* O157. Moreover, the nucleotide sequence of EHEC O111 *exhA* is nearly identical to that of EHEC O157 *hlyA*. These results are consistent with the findings of Campos *et al.* (Campos *et al.*, 1994) who suggested that the rate of acquisition and loss of virulence factors is rapid in comparison with the rate of mutations in so-called housekeeping genes in the evolution of *E. coli* O111 lineages. Moreover, because of the high sequence conservation of EHEC *hlyA* in *E. coli* O157 and *E. coli* O111 strains, one may hypothesize that *E. coli* O111 acquired the EHEC *hly* genes late during development of distinct *E. coli* O111 clones. The genes encoding the EHEC haemolysin (*exhA*), its activator (*exhC*), and part of its secretion machinery (*exhB* and *exhD*) are located on large plasmids having some similarities with the EAF plasmid of EPEC (Burland *et al.*, 1998; Hales *et al.*, 1992; Schmidt *et al.*, 1996). All of these factors are located on potentially mobile elements, and their recent acquisition has been suggested as being responsible for the emergence of STEC as a new major human pathogen (Whittam *et al.*, 1993). The *ehxCABD* operon is highly conserved among EHEC. In the EHEC-Hly protein there are conserved amino acid residues in the region between amino acids 661 and 834, which contains the tandem repeats (Schmidt *et al.*, 1995), typical of all RTX toxins and which are involved in calcium binding (Welch *et al.*, 1992); the lysine residues at positions 550 and 675, which are involved in the activation of the toxin by fatty acylation (Pellett and Welch, 1996); and the 50 C-terminal amino acids, which are involved in secretion (Bauer and Welch, 1996b; Welch et al., 1992).

The exhA gene may serve not only as a marker for detection of clinically important EHEC isolates but also as a tool for differentiation of EHEC clinical isolates. Two major groups of EHEC haemolysin plasmids (type I and type II plasmids) were described and suggested that these may differ in EHEC strains significantly in terms of virulence-associated factors. The presence of a conserved 5.6 kb fragment on the type I EHEC haemolysin plasmid, lacking or significantly different in type II plasmids, strongly supports this suggestion. This is also confirmed by the presence of a conserved sequence within a 4 kb fragment on the type II EHEC haemolysin plasmid, which has only low homology with the type I EHEC haemolysin plasmid. In addition, type II plasmids also seem to be consistently slightly larger than type I plasmids (Boerlin et al., 1998). The transcription of the EHEC haemolysin operon (ehxCABD) in EHEC O157:H7 strain EDL933 has been examined (Li et al., 2008). The study included the influence of H-NS, the σ factor RpoS, and the small RNA DsrA, which is known to inhibit H-NS function and to stimulate RpoS synthesis. During growth at 30°C, DsrA overexpression increased ehxA transcription in the wild-type but not in an hns deletion mutant. During growth at 37°C, DsrA overexpression increased ehxA transcription independent of the hns genotype. It was suggested that DsrA influences ehxCABD operon transcription by two different routes, one (at lower temperature) at least partially dependent on H-NS, and one (at higher temperature) independent of H-NS. An rpoS deletion mutant expressed non-detectable levels of ehxA mRNA regardless of growth temperature or DsrA overexpression, indicating that the RpoS σ factor is essential for ehxCABD operon expression.

EHEC-Hly mechanism of action and role in pathogenesis

EHEC-Hly is a pore-forming cytolysin and it lyses erythrocytes from different species and bovine lymphocytes (Bauer and Welch, 1996b;

Schmidt et al., 1995). In addition, EHEC-Hly can damage the microvascular endothelial cells, suggesting a possible role in the pathogenesis of haemolytic–uraemic syndrome (HUS) (Aldick et al., 2007). Possible involvement of EHEC-Hly in the development of HUS is supported by an epidemiological study which demonstrated that the EHEC-hlyA gene and the expression of EHEC-Hly protein were significantly more frequent in EHEC O111 strains isolated from patients with HUS than in those isolated from patients with uncomplicated diarrhoea (Schmidt and Karch, 1996). Alternatively, these strains may possess additional, as yet unknown, virulence factor(s), which might act in concert with EHEC-Hly to precipitate HUS. Notably, it was described that EHEC-Hly is one of the non-Stx virulence determinants that are associated with the ability of EHEC to cause severe extraintestinal complications including HUS (Schmidt and Karch, 1996). The possible involvement of EHEC-Hly in the pathogenesis of HUS is further supported by the detection of an immune response against EHEC-Hly in sera of patients after HUS (Schmidt et al., 1995) and, in the case of E. coli O157:H7 human infections this gene is expressed in shed organisms in the faeces (Rashid et al., 2006). The earlier studies suggested that EHEC-Hly might reach the bloodstream and could thus cause injury to the microvascular endothelium as it can damage vascular endothelium as demonstrated in vitro. However, whether and how EHEC-Hly reaches the bloodstream, how it is transported to the vascular endothelium and what is the amount required for the endothelial damage in vivo are not well identified yet (Bielaszewska and Karch, 2005). As mentioned above, there is also the possibility of the delivery of EHEC-Hly to the target cells via outer membrane vesicles released by the bacteria, a mechanism, that has been proposed also for α-haemolysin (Aldick et al., 2009; Balsalobre et al., 2006).

Interactions of EHEC-Hly and target cells

When the toxin reaches the blood circulation and stimulates the secretion of reactive oxygen species (ROS) in polymorphonuclear cells (PMNs) it can lead subsequently to apoptosis. The consequences of ROS stress on PMNs in the pathogenesis of HUS, however, remain unclear, but it has been proposed that EHEC-Hly has a direct effect on PMNs, which may contribute to tissue injury early in the disease (King *et al.*, 1999). In addition, experimental and clinical evidence suggested that PMNs contribute to endothelial damage in HUS. Although activation of these cells has been demonstrated in HUS patients, deactivation was also observed during the progress of this syndrome, with degranulation in response to cytokines, and subsequent liberation leading to circulation of the intracellular granule content and ROS (Fernandez *et al.*, 2005). In addition, alterations of plasma healthy donors' PMNs were observed by the direct effects of Stx *in vitro* (Forsyth *et al.*, 1989; te Loo *et al.*, 2000). These previous results suggest that in HUS patients, PMNs are functionally impaired and show features of previous degranulation, indicating an anterior process of activation with release of ROS (Facorro *et al.*, 1997). Subsequently, oxidative injuries that can compromise the survival of PMNs may lead to the pathophysiology of systemic disease and it can be considered that oxidative stress is involved in the action of EHEC-Hly. It was suggested that free radical-related pathways may be considered as targets in the intervention of new drugs against HUS (Aiassa *et al.*, 2011).

The ClyA (SheA, HlyE) protein

Cytolysin A (ClyA) of *Escherichia coli* is a cytotoxic protein encoded by the *clyA* gene. The *clyA* gene locus was originally identified in *E. coli* K-12 that was mutated in the gene encoding the nucleoid protein H-NS. Mutations in the gene (*hns*) encoding the nucleoid-associated protein H-NS have pleiotropic effects in *E. coli*, and it is well established that H-NS may cause silencing of several operons. When *hns* mutant *E. coli* K-12 were streaked onto blood plates they were found to be haemolytic (Gomez-Gomez *et al.*, 1996; Uhlin and Mizunoe, 1994), and apparently the genetic determinant was repressed by H-NS. However, an ability of *E. coli* K-12, lacking the *hly* gene clusters, to lyse erythrocytes was observed a few decades earlier in studies of a nalidixic acid resistant mutant derivative (Walton and Smith,

1969). The novel determinant was referred to as γ-haemolysin but was not further characterized. The gene encoding this latent haemolytic activity was cloned and characterized (del Castillo *et al.*, 1997; Uhlin and Mizunoe, 1994). The 912 bp gene, here referred to as *clyA* for cytolysin A (also denoted *sheA* and *hlyE* (del Castillo *et al.*, 1997; Green and Baldwin, 1997)) is encoding a ~34 kDa protein and was localized to ~26.5 min on the *E. coli* K-12 chromosome. The *clyA* locus is located in close vicinity of the *umuCD* operon, which is transcribed in the opposite direction.

There are several examples of gene products (SlyA, MprA, HlyX, FnrP) that, when overexpressed, have conferred a haemolytic phenotype to *E. coli* by depressing *clyA* (del Castillo *et al.*, 1997; Green and Baldwin, 1997; Oscarsson *et al.*, 1996; Uhlich *et al.*, 1999). Due to the presence of this latent cytolytic determinant even in *E. coli* K-12, the commonly used laboratory strains have given results from screening of recombinant clones for haemolytic activity that has caused some confusion. One example is the cloning of genes encoding the bacteriophage-associated proteins Ehly1 and Ehly2 of non-verotoxigenic EPEC O26:H⁻. These determinants were originally identified as putative haemolysins by screening recombinant *E. coli* K-12 for lytic activity (Beutin *et al.*, 1990, 1993; Stroeher *et al.*, 1993). To test whether Ehly1 and Ehly2 affected the expression of *clyA*, plasmids carrying the genetic determinants for Ehly1 and Ehly2 were introduced into *clyA⁺* and *clyA::kan E. coli* strains (Oscarsson *et al.*, 2002a). Only the *clyA⁺* strain became haemolytic, indicating that the haemolytic phenotype of Ehly1 and Ehly2 was linked to expression of ClyA. To investigate whether transcription of *clyA* was depressed by Ehly1 and Ehly2, the Ehly-encoding plasmids were introduced into a *clyA::luxAB* fusion strain and the luciferase activity was monitored (Oscarsson *et al.*, 2002a). There was no significant effect of Ehly1 or Ehly2 on the luciferase activity compared with the vector control, indicating that Ehly1 and Ehly2 did not confer a haemolytic phenotype by affecting the transcription of the *clyA* gene. However, the colonies of strains carrying Ehly1 and Ehly2 encoding plasmids appeared rough and disrupted, as if there were lysed bacterial cells in the colonies

(Beutin *et al.*, 1993). Thus it was hypothesized that the low amount of ClyA produced in non-haemolytic strains could be enough to confer a haemolytic phenotype if at least part of the cells in a colony was lysed. As a test of this hypothesis, samples of *clyA*+ and *clyA* mutant *E. coli* K-12 were cross-streaked with bacteriophage (P1 or ΦW) lysates on blood agar plates to deliberately lyse the bacteria (Oscarsson *et al.*, 2002a). Lysis of the erythrocytes in the blood agar was evident in and around the centres of the cross streaks with *clyA*+ strains but not with the *clyA* mutant strains. These results demonstrated that the low basal level of ClyA production in *E. coli* K-12 strains in fact is sufficient to cause a haemolytic phenotype on blood agar when the bacteria are disrupted and thereby ClyA protein is released.

Analysis of *clyA* transcription by Northern blot hybridization in an *E. coli* strain overexpressing the SlyA protein suggested that *clyA* is a mono-cistronic operon (Oscarsson *et al.*, 1996). Primer extension analysis performed on RNA isolated from an *hns* mutant revealed that the transcription start site is −72 bp upstream of the *clyA* translational start codon (Westermark *et al.*, 2000). The same start site position was determined when the *clyA* gene was activated by the cloned *slyA* locus (Ludwig *et al.*, 1999).

Mechanism of action of ClyA

E. coli K-12 strains expressing ClyA (H-NS mutants or strains expressing the cloned *clyA* locus on a multicopy plasmid) are able to lyse erythrocytes from several mammalian species in both solid and liquid media (Oscarsson *et al.*, 1999). Macrophages and HeLa cells grown in tissue culture are also lysed by *E. coli* K-12 strains overexpressing ClyA. The 34 kDa ClyA protein has been shown to cause lysis of mammalian cells by a pore-forming mechanism (Oscarsson *et al.*, 1999; Wallace *et al.*, 2000) and to induce apoptosis in cultured macrophages (Lai *et al.*, 2000).

An indication that ClyA is a pore-forming protein was initially reported from studies with extracts of *E. coli* containing the cloned *slyA* locus, although then the actual lytic component was not identified (Ludwig *et al.*, 1995). This finding was supported with a non-native ClyA preparation, which generated 2.5- to 3.0-nm pores in planar lipid bilayers (Ludwig *et al.*, 1999). Osmotic protection represents an alternative means to assess a putative pore-forming activity, with the pore-size being proportional to the molecular diameter of the smallest molecule preventing lysis (Lobo and Welch, 1994). As judged by osmotic protection assays, native ClyA protein formed pores of a size between 2.2 and 3.0–3.5 nm in erythrocytes, thereby confirming previous findings (Ludwig and Goebel, 1999; Oscarsson *et al.*, 1999). Thus, the pores formed by ClyA are smaller than those formed by the thiol-activated cytolysins (Ballard *et al.*, 1995), but larger than those formed by RTX-toxins such as α-haemolysin (Benz *et al.*, 1989).

Structure of ClyA and the pore-forming complexes

Initial studies of the crystal structure of soluble monomeric ClyA showed that it is a rod-shaped, mostly α-helical molecule, and the shaft of the rod is formed by a bundle of four helices, each 70–80 Å long (Wallace *et al.*, 2000). A short β-hairpin, termed as the β-tongue, that is flanked by two short helices is located on one end of the bundle. The β-tongue region forms the major hydrophobic patch on the ClyA surface, and it was suggested that this region interacts with target membranes. At the other end, which is containing the N- and C-terminal regions of the protein, there is a shorter (30 Å) helix that packs against the four long helices and the bundle with five helices constitutes about one-third of the length of the molecule (the tail domain).

Electron microscopy studies and 3D reconstructions indicated that the ClyA pores are longer than the structure seen with the water-soluble form of the protein (Eifler *et al.*, 2006; Tzokov *et al.*, 2006). It was suggested that there must be some conformational changes in the ClyA protein when it is forming a functional pore complex. Different studies suggested different number of ClyA molecules in the pore: mainly 13-meric (Eifler *et al.*, 2006) or mainly octameric (Hunt *et al.*, 2008; Tzokov *et al.*, 2006). More recently, the 3.3 Å crystal structure of the 400 kDa dodecameric transmembrane pore formed by ClyA was reported (Mueller *et al.*, 2009). It confirmed the larger size of the pore complex and that the tertiary structure of ClyA protomers in the pore is substantially different from that in the soluble monomer. The conformational alteration was suggested to involve more than half of all residues and to result in large rearrangements, up to 140 Å, of the monomer. ClyA has also been explored for design of special nanopores and it was shown that the lumen of the dodecameric ClyA pore assembly would be wide enough to fit small to medium-sized proteins (Soskine *et al.*, 2012). ClyA nanopores decorated with covalently attached aptamers, and incorporated into planar lipid bilayers, were shown to selectively capture and internalize cognate protein analytes that were predicted to fit within the lumen of the dodecameric pore structure.

Secretion of ClyA via the 'Type Zero' route

Studies of its cellular localization showed that ClyA accumulates in the periplasmic space of *E. coli* (Ludwig and Goebel, 1999; Oscarsson *et al.*, 1999). However, the ClyA protein is translocated without cleavage of any N-terminal signal sequence, and it has remained unclear how it can reach the periplasm. The ability of ClyA to translocate to the periplasm was abolished in deletion mutants lacking the last 23 or 11 amino acid residues of the C-terminal region (Oscarsson *et al.*, 1999; Wai *et al.*, 2003b). Using TnphoA mutagenesis a PhoA+ non-haemolytic isolate with an in-frame insertion of PhoA after residue 179 was obtained and it was concluded that the C-terminal region of ClyA is not required for translocation to the periplasmic space (Wyborn *et al.*, 2004). Data from a study involving site-directed mutagenesis of different regions of the *E. coli* ClyA protein revealed that ClyA translocation to the periplasm requires several protein segments located closely adjacent to each other in the 'tail' domain of the ClyA monomer, namely, the N- and C-terminal regions and the hydrophobic sequence ranging from residues 89 to 101 (Ludwig *et al.*, 2010). Deletion of most of the 'head' domain of the monomer (residues 181 to 203), on the other hand, did not strongly affect ClyA secretion. It remains to be elucidated if ClyA is translocated as a monomer or, as indicated by the dominant negative effect of some mutants, in the form of a dimeric protein complex (Wai *et al.*, 2003b). It was also concluded that ClyA secretion is independent of the type I, II, III, IV and V secretory systems (Wai *et al.*, 2000).

The actual release of ClyA from the bacterial cells was shown to occur as a consequence of release of outer membrane vesicles (Wai *et al.*, 2003a). Together with some other periplasmic proteins, ClyA appeared readily localized to the OMVs and there the protein assembled into the typical pore complexes (Wai *et al.*, 2003a). These findings, and subsequent studies of OMV-mediated release of different proteins, suggest that the release of membrane vesicles represents a very basic and relevant mode of protein export from bacteria and we suggest that it could be referred to as the 'Type Zero' secretion system.

The features of ClyA export via OMVs have been successfully explored for secretion of chimeric proteins. Similar to native unfused ClyA, chimeric ClyA fusion proteins were found localized in bacterial OMVs and retained activity of the fusion partners, demonstrating for the first time that ClyA can be used to co-localize fully functional heterologous proteins directly in bacterial OMV (Kim *et al.*, 2008). Furthermore, it was demonstrated that such engineered *E. coli* OMVs with chimeric ClyA constructs are an easily purified vaccine-delivery system capable of enhancing the immunogenicity of a low-immunogenicity protein antigen without added adjuvants (Chen *et al.*, 2010).

Role of ClyA in virulence of *E. coli*?

The features of this pore-forming protein have raised questions about a possible role of ClyA in *E. coli* virulence. However, it is the only cytolytic factor found in the *E. coli* K-12 laboratory strains that are considered as non-pathogenic strains. On the other hand, sequences homologous to *clyA* have been identified also in a number of pathogenic isolates of *E. coli* (del Castillo *et al.*, 1997; Oscarsson *et al.*, 2002b; Reingold *et al.*, 1999). Studies of the *clyA* locus in a wide spectrum of *E. coli* isolates revealed that extraintestinal pathogenic isolates carry non-functional *clyA* loci (Kerenyi *et al.*,

2005; Murase *et al.*, 2012; Oscarsson, 1999; von Rhein *et al.*, 2008). All examined extraintestinal pathogenic isolates from urinary tract infections (UPEC) and neonatal meningitis (NBM) *E. coli* strains contained various Δ*clyA* alleles. At least four different variants of Δ*clyA* exist among such strains suggesting that the deletions in *clyA* have arisen at more than one occasion and from recent studies of UPEC with a restored *clyA* locus it appears that the mutations are pathoadaptive alterations due to differential expression of *clyA* (Enow, Oscarsson, Westermark, Duperthuy, Wai, Uhlin, manuscript submitted for publication). Presumably, the clyA transcription in such *E. coli* strains can not be sufficiently controlled to the extent as observed in *E. coli* K-12 where silencing by the H-NS protein is evident (Westermark *et al.*, 2000). It remains unclear if the ClyA protein has a role in virulence of other *E. coli* strains and since the *clyA* gene is present in many non-pathogenic strains we must also consider that there is some role for this pore-forming protein in the bacteria that has not been elucidated.

References

Aiassa, V., Baronetti, J.L., Paez, P.L., Barnes, A.I., Albrecht, C., Pellarin, G., Eraso, A.J., and Albesa, I. (2011). Increased advanced oxidation of protein products and enhanced total antioxidant capacity in plasma by action of toxins of *Escherichia coli* STEC. Toxicol. *in vitro 25*, 426–431.

Aldick, T., Bielaszewska, M., Zhang, W., Brockmeyer, J., Schmidt, H., Friedrich, A.W., Kim, K.S., Schmidt, M.A., and Karch, H. (2007). Hemolysin from Shiga toxin-negative *Escherichia coli* O26 strains injures microvascular endothelium. Microb. Infect. *9*, 282–290.

Aldick, T., Bielaszewska, M., Uhlin, B.E., Humpf, H.U., Wai, S.N., and Karch, H. (2009). Vesicular stabilization and activity augmentation of enterohaemorrhagic *Escherichia coli* haemolysin. Mol. Microbiol. *71*, 1496–1508.

Aulik, N.A., Atapattu, D.N., Czuprynski, C.J., and McCaslin, D.R. (2013). Brief heat treatment causes a structural change and enhances cytotoxicity of the *Escherichia coli* alpha-hemolysin. Immunopharmacol. Immunotoxicol. *35*, 15–27.

Bailey, M.J., Koronakis, V., Schmoll, T., and Hughes, C. (1992). *Escherichia coli* HlyT protein, a transcriptional activator of haemolysin synthesis and secretion, is encoded by the rfaH (sfrB) locus required for expression of sex factor and lipopolysaccharide genes. Mol. Microbiol. *6*, 1003–1012.

Bailey, M.J., Hughes, C., and Koronakis, V. (2000). In vitro recruitment of the RfaH regulatory protein into a specialised transcription complex, directed by the nucleic acid ops element. Mol. Gen. Genet. *262*, 1052–1059.

Bakas, L., Ostolaza, H., Vaz, W.L., and Goni, F.M. (1996). Reversible adsorption and nonreversible insertion of *Escherichia coli* alpha-hemolysin into lipid bilayers. Biophys. J. *71*, 1869–1876.

Bakas, L., Veiga, M.P., Soloaga, A., Ostolaza, H., and Goni, F.M. (1998). Calcium-dependent conformation of *E. coli* alpha-haemolysin. Implications for the mechanism of membrane insertion and lysis. Biochim. Biophys. Acta *1368*, 225–234.

Bakkes, P.J., Jenewein, S., Smits, S.H., Holland, I.B., and Schmitt, L. (2010). The rate of folding dictates substrate secretion by the *Escherichia coli* hemolysin type 1 secretion system. J. Biol. Chem. *285*, 40573–40580.

Ballard, J., Crabtree, J., Roe, B.A., and Tweten, R.K. (1995). The primary structure of *Clostridium septicum* alpha-toxin exhibits similarity with that of *Aeromonas hydrophila* aerolysin. Infect. Immun. *63*, 340–344.

Balsalobre, C., Silvan, J.M., Berglund, S., Mizunoe, Y., Uhlin, B.E., and Wai, S.N. (2006). Release of the type I secreted alpha-haemolysin via outer membrane vesicles from *Escherichia coli*. Mol. Microbiol. *59*, 99–112.

Baronetti, J.L., Villegas, N.A., Aiassa, V., Paraje, M.G., and Albesa, I. (2013). Hemolysin from *Escherichia coli* induces oxidative stress in blood. Toxicon. *70*, 15–20.

Bauer, M.E., and Welch, R.A. (1996a). Association of RTX toxins with erythrocytes. Infect. Immun. *64*, 4665–4672.

Bauer, M.E., and Welch, R.A. (1996b). Characterization of an RTX toxin from enterohemorrhagic *Escherichia coli* O157:H7. Infect. Immun. *64*, 167–175.

Baumann, U., Wu, S., Flaherty, K.M., and McKay, D.B. (1993). Three-dimensional structure of the alkaline protease of *Pseudomonas aeruginosa*: a two-domain protein with a calcium binding parallel beta roll motif. EMBO J. *12*, 3357–3364.

Benabdelhak, H., Kiontke, S., Horn, C., Ernst, R., Blight, M.A., Holland, I.B., and Schmitt, L. (2003). A specific interaction between the NBD of the ABC-transporter HlyB and a C-terminal fragment of its transport substrate haemolysin A. J. Mol. Biol. *327*, 1169–1179.

Benz, R., Schmid, A., Wagner, W., and Goebel, W. (1989). Pore formation by the *Escherichia coli* hemolysin: evidence for an association-dissociation equilibrium of the pore-forming aggregates. Infect. Immun. *57*, 887–895.

Berry, R.E., Klumpp, D.J., and Schaeffer, A.J. (2009). Urothelial cultures support intracellular bacterial community formation by uropathogenic *Escherichia coli*. Infect. Immun. *77*, 2762–2772.

Beutin, L., Bode, L., Ozel, M., and Stephan, R. (1990). Enterohemolysin production is associated with a temperate bacteriophage in *Escherichia coli* serogroup O26 strains. J. Bacteriol. *172*, 6469–6475.

Beutin, L., Stroeher, U.H., and Manning, P.A. (1993). Isolation of enterohemolysin (Ehly2)-associated sequences encoded on temperate phages of *Escherichia coli*. Gene *132*, 95–99.

Bhakdi, S., Mackman, N., Nicaud, J.M., and Holland, I.B. (1986). *Escherichia coli* hemolysin may damage target cell membranes by generating transmembrane pores. Infect. Immun. *52*, 63–69.

Bhakdi, S., Mackman, N., Menestrina, G., Gray, L., Hugo, F., Seeger, W., and Holland, I.B. (1988). The hemolysin of *Escherichia coli*. Eur. J. Epidemiol. *4*, 135–143.

Bhakdi, S., Greulich, S., Muhly, M., Eberspacher, B., Becker, H., Thiele, A., and Hugo, F. (1989). Potent leukocidal action of *Escherichia coli* hemolysin mediated by permeabilization of target cell membranes. J. Exp. Med. *169*, 737–754.

Bhakdi, S., Muhly, M., Korom, S., and Schmidt, G. (1990). Effects of *Escherichia coli* hemolysin on human monocytes. Cytocidal action and stimulation of interleukin 1 release. J. Clin. Invest. *85*, 1746–1753.

Bielaszewska, M., and Karch, H. (2005). Consequences of enterohaemorrhagic *Escherichia coli* infection for the vascular endothelium. Thrombos. Haemostas. *94*, 312–318.

Bien, J., Sokolova, O., and Bozko, P. (2012). Role of Uropathogenic *Escherichia coli* Virulence Factors in Development of Urinary Tract Infection and Kidney Damage. Int. J. Nephrol. *2012*, 681473.

Bischofberger, M., Iacovache, I., and Gisou van der Goot, F. (2012). Pathogenic pore-forming proteins: function and host response. Cell Host Microbe *12*, 266–275.

Boehm, D.F., Welch, R.A., and Snyder, I.S. (1990). Domains of *Escherichia coli* hemolysin (HlyA) involved in binding of calcium and erythrocyte membranes. Infect. Immun. *58*, 1959–1964.

Boerlin, P., Chen, S., Colbourne, J.K., Johnson, R., De Grandis, S., and Gyles, C. (1998). Evolution of enterohemorrhagic *Escherichia coli* hemolysin plasmids and the locus for enterocyte effacement in shiga toxin-producing *E. coli*. Infect. Immun. *66*, 2553–2561.

Bower, J.M., Eto, D.S., and Mulvey, M.A. (2005). Covert operations of uropathogenic *Escherichia coli* within the urinary tract. Traffic *6*, 18–31.

Brockmeyer, J., Aldick, T., Soltwisch, J., Zhang, W., Tarr, P.I., Weiss, A., Dreisewerd, K., Muthing, J., Bielaszewska, M., and Karch, H. (2011). Enterohaemorrhagic *Escherichia coli* haemolysin is cleaved and inactivated by serine protease EspPalpha. Environ. Microbiol. *13*, 1327–1341.

Brzuszkiewicz, E., Bruggemann, H., Liesegang, H., Emmerth, M., Olschlager, T., Nagy, G., Albermann, K., Wagner, C., Buchrieser, C., Emody, L., *et al.* (2006). How to become a uropathogen: comparative genomic analysis of extraintestinal pathogenic *Escherichia coli* strains. Proc. Natl. Acad. Sci. U.S.A. *103*, 12879–12884.

Burgos, Y., and Beutin, L. (2010). Common origin of plasmid encoded alpha-hemolysin genes in *Escherichia coli*. BMC Microbiol. *10*, 193.

Burgos, Y.K., Pries, K., Pestana de Castro, A.F., and Beutin, L. (2009). Characterization of the alpha-haemolysin determinant from the human enteropathogenic *Escherichia coli* O26 plasmid pEO5. FEMS Microbiol. Lett. *292*, 194–202.

Burland, V., Shao, Y., Perna, N.T., Plunkett, G., Sofia, H.J., and Blattner, F.R. (1998). The complete DNA sequence and analysis of the large virulence plasmid of *Escherichia coli* O157:H7. Nucleic Acids Res. *26*, 4196–4204.

Campos, L.C., Whittam, T.S., Gomes, T.A., Andrade, J.R., and Trabulsi, L.R. (1994). *Escherichia coli* serogroup O111 includes several clones of diarrhoeagenic strains with different virulence properties. Infect. Immun. *62*, 3282–3288.

Carmona, M., Balsalobre, C., Munoa, F., Mourino, M., Jubete, Y., De la Cruz, F., and Juarez, A. (1993). *Escherichia coli* hha mutants, DNA supercoiling and expression of the haemolysin genes from the recombinant plasmid pANN202–312. Mol. Microbiol. *9*, 1011–1018.

del Castillo, F.J., Leal, S.C., Moreno, F., and del Castillo, I. (1997). The *Escherichia coli* K-12 sheA gene encodes a 34 kDa secreted haemolysin. Mol. Microbiol. *25*, 107–115.

Cavalieri, S.J., and Snyder, I.S. (1982). Effect of *Escherichia coli* alpha-hemolysin on human peripheral leukocyte function *in vitro*. Infect. Immun. *37*, 966–974.

Chen, D.J., Osterrieder, N., Metzger, S.M., Buckles, E., Doody, A.M., DeLisa, M.P., and Putnam, D. (2010). Delivery of foreign antigens by engineered outer membrane vesicle vaccines. Proc. Natl. Acad. Sci. U.S.A. *107*, 3099–3104.

Chen, M., Tofighi, R., Bao, W., Aspevall, O., Jahnukainen, T., Gustafsson, L.E., Ceccatelli, S., and Celsi, G. (2006). Carbon monoxide prevents apoptosis induced by uropathogenic *Escherichia coli* toxins. Pediatr. Nephrol. *21*, 382–389.

Chenal, A., Guijarro, J.I., Raynal, B., Delepierre, M., and Ladant, D. (2009). RTX calcium binding motifs are intrinsically disordered in the absence of calcium: implication for protein secretion. J. Biol. Chem. *284*, 1781–1789.

Chenal, A., Karst, J.C., Sotomayor Perez, A.C., Wozniak, A.K., Baron, B., England, P., and Ladant, D. (2010). Calcium-induced folding and stabilization of the intrinsically disordered RTX domain of the CyaA toxin. Biophys. J. *99*, 3744–3753.

Cortajarena, A.L., Goni, F.M., and Ostolaza, H. (2001). Glycophorin as a receptor for *Escherichia coli* alpha-hemolysin in erythrocytes. J. Biol. Chem. *276*, 12513–12519.

Cortajarena, A.L., Goni, F.M., and Ostolaza, H. (2003). A receptor-binding region in *Escherichia coli* alpha-haemolysin. J. Biol. Chem. *278*, 19159–19163.

Dhakal, B.K., and Mulvey, M.A. (2012). The UPEC pore-forming toxin alpha-hemolysin triggers proteolysis of host proteins to disrupt cell adhesion, inflammatory, and survival pathways. Cell Host Microbe *11*, 58–69.

Dobrindt, U., Janke, B., Piechaczek, K., Nagy, G., Ziebuhr, W., Fischer, G., Schierhorn, A., Hecker, M., Blum-Oehler, G., and Hacker, J. (2000). Toxin genes on pathogenicity islands: impact for microbial evolution. IJMM *290*, 307–311.

Dobrindt, U., Blum-Oehler, G., Nagy, G., Schneider, G., Johann, A., Gottschalk, G., and Hacker, J. (2002a). Genetic structure and distribution of four pathogenicity islands (PAI I(536) to PAI IV(536)) of uropathogenic *Escherichia coli* strain 536. Infect. Immun. *70*, 6365–6372.

Dobrindt, U., Emody, L., Gentschev, I., Goebel, W., and Hacker, J. (2002b). Efficient expression of the alpha-haemolysin determinant in the uropathogenic *Escherichia coli* strain 536 requires the leuX-encoded tRNA(5)(Leu). Mol. Genet. Genom. *267*, 370–379.

Dutta, S., Iida, K., Takade, A., Meno, Y., Nair, G.B., and Yoshida, S. (2004). Release of Shiga toxin by membrane vesicles in *Shigella dysenteriae* serotype 1 strains and in vitro effects of antimicrobials on toxin production and release. Microbiol. Immunol. *48*, 965–969.

Eifler, N., Vetsch, M., Gregorini, M., Ringler, P., Chami, M., Philippsen, A., Fritz, A., Muller, S.A., Glocksshuber, R., Engel, A., *et al.* (2006). Cytotoxin ClyA from *Escherichia coli* assembles to a 13-meric pore independent of its redox-state. EMBO J. *25*, 2652–2661.

Elliott, S.J., Srinivas, S., Albert, M.J., Alam, K., Robins-Browne, R.M., Gunzburg, S.T., Mee, B.J., and Chang, B.J. (1998). Characterization of the roles of hemolysin and other toxins in enteropathy caused by alpha-hemolytic *Escherichia coli* linked to human diarrhoea. Infect. Immun. *66*, 2040–2051.

Facorro, G., Aguirre, F., Florentin, L., Diaz, M., De Paoli, T., Ihlo, J.E., Hager, A.A., Sanchez Avalos, J.C., Farach, H.A., and Poole, C.P., Jr. (1997). Oxidative stress and membrane fluidity in erythrocytes from patients with hemolytic uremic syndrome. Acta Physiolog. Pharmacolog. Therapeut. Latinoamericana *47*, 137–146.

Fagerberg, S.K., Skals, M., Leipziger, J., and Praetorius, H.A. (2013). P2X Receptor-Dependent Erythrocyte Damage by alpha-Hemolysin from *Escherichia coli* Triggers Phagocytosis by THP-1 Cells. Toxins (Basel) *5*, 472–487.

Fernandez, G.C., Gomez, S.A., Rubel, C.J., Bentancor, L.V., Barrionuevo, P., Alduncin, M., Grimoldi, I., Exeni, R., Isturiz, M.A., and Palermo, M.S. (2005). Impaired neutrophils in children with the typical form of hemolytic uremic syndrome. Pediatr. Nephrol. *20*, 1306–1314.

Fiser, R., and Konopasek, I. (2009). Different modes of membrane permeabilization by two RTX toxins: HlyA from *Escherichia coli* and CyaA from *Bordetella pertussis*. Biochim. Biophys. Acta *1788*, 1249–1254.

Forsyth, K.D., Simpson, A.C., Fitzpatrick, M.M., Barratt, T.M., and Levinsky, R.J. (1989). Neutrophil-mediated endothelial injury in haemolytic uraemic syndrome. Lancet *2*, 411–414.

Gadeberg, O.V., and Orskov, I. (1984). *In vitro* cytotoxic effect of alpha-hemolytic *Escherichia coli* on human blood granulocytes. Infect. Immun. *45*, 255–260.

Gangola, P., and Rosen, B.P. (1987). Maintenance of intracellular calcium in *Escherichia coli*. J. Biol. Chem. *262*, 12570–12574.

Garcia, T.A., Ventura, C.L., Smith, M.A., Merrell, D.S., and O'Brien, A.D. (2013). Cytotoxic necrotizing factor 1 and hemolysin from uropathogenic *Escherichia coli* elicit different host responses in the murine bladder. Infect. Immun. *18*, 99–109.

Gentschev, I., Mollenkopf, H., Sokolovic, Z., Hess, J., Kaufmann, S.H., and Goebel, W. (1996). Development of antigen-delivery systems, based on the *Escherichia coli* hemolysin secretion pathway. Gene *179*, 133–140.

Gentschev, I., Dietrich, G., and Goebel, W. (2002). The *E. coli* alpha-hemolysin secretion system and its use in vaccine development. Trends Microbiol. *10*, 39–45.

Goebel, W., and Hedgpeth, J. (1982). Cloning and functional characterization of the plasmid-encoded hemolysin determinant of *Escherichia coli*. J. Bacteriol. *151*, 1290–1298.

Gomez-Gomez, J.M., Blazquez, J., Baquero, F., and Martinez, J.L. (1996). Hns mutant unveils the presence of a latent haemolytic activity in *Escherichia coli* K-12. Mol. Microbiol. *19*, 909–910.

Gray, L., Baker, K., Kenny, B., Mackman, N., Haigh, R., and Holland, I.B. (1989). A novel C-terminal signal sequence targets *Escherichia coli* haemolysin directly to the medium. J. Cell Sci. Suppl *11*, 45–57.

Green, J., and Baldwin, M.L. (1997). The molecular basis for the differential regulation of the hlyE-encoded haemolysin of *Escherichia coli* by FNR and HlyX lies in the improved activating region 1 contact of HlyX. Microbiology *143*, 3785–3793.

Guyer, D.M., Kao, J.S., and Mobley, H.L. (1998). Genomic analysis of a pathogenicity island in uropathogenic *Escherichia coli* CFT073: distribution of homologous sequences among isolates from patients with pyelonephritis, cystitis, and Catheter-associated bacteriuria and from fecal samples. Infect. Immun. *66*, 4411–4417.

Hales, B.A., Hart, C.A., Batt, R.M., and Saunders, J.R. (1992). The large plasmids found in enterohemorrhagic and enteropathogenic *Escherichia coli* constitute a related series of transfer-defective Inc F-IIA replicons. Plasmid *28*, 183–193.

Hamada, K., Hata, Y., Katsuya, Y., Hiramatsu, H., Fujiwara, T., and Katsube, Y. (1996). Crystal structure of Serratia protease, a zinc-dependent proteinase from Serratia sp. E-15, containing a beta-sheet coil motif at 2.0 A resolution. J. Biochem. *119*, 844–851.

Hannan, T.J., Totsika, M., Mansfield, K.J., Moore, K.H., Schembri, M.A., and Hultgren, S.J. (2012). Host-pathogen checkpoints and population bottlenecks in persistent and intracellular uropathogenic *Escherichia coli* bladder infection. FEMS Microbiol. Rev. *36*, 616–648.

Haugen, B.J., Pellett, S., Redford, P., Hamilton, H.L., Roesch, P.L., and Welch, R.A. (2007). *In vivo* gene expression analysis identifies genes required for enhanced colonization of the mouse urinary tract by uropathogenic *Escherichia coli* strain CFT073 dsdA. Infect. Immun. *75*, 278–289.

Herlax, V., and Bakas, L. (2003). Acyl chains are responsible for the irreversibility in the *Escherichia coli* alpha-hemolysin binding to membranes. Chem. Phys. Lipids *122*, 185–190.

Herlax, V., and Bakas, L. (2007). Fatty acids covalently bound to alpha-hemolysin of *Escherichia coli* are involved in the molten globule conformation: implication of disordered regions in binding promiscuity. Biochemistry *46*, 5177–5184.

Herlax, V., Mate, S., Rimoldi, O., and Bakas, L. (2009). Relevance of fatty acid covalently bound to *Escherichia coli* alpha-hemolysin and membrane microdomains in the oligomerization process. J. Biol. Chem. *284*, 25199–25210.

Herlax, V., Henning, M.F., Bernasconi, A.M., Goni, F.M., and Bakas, L. (2010). The lytic mechanism of *Escherichia coli* α-hemolysin associated to outer membrane vesicles. Health *2*, 484–492.

Hilbert, D.W., Paulish-Miller, T.E., Tan, C.K., Carey, A.J., Ulett, G.C., Mordechai, E., Adelson, M.E., Gygax, S.E., and Trama, J.P. (2012). Clinical *Escherichia coli* isolates utilize alpha-hemolysin to inhibit in vitro epithelial cytokine production. Microb. Infect. *14*, 628–638.

Holland, I.B., Schmitt, L., and Young, J. (2005). Type 1 protein secretion in bacteria, the ABC-transporter dependent pathway (review). Mol. Membr. Biol. *22*, 29–39.

Hotz, C., Fensterle, J., Goebel, W., Meyer, S.R., Kirchgraber, G., Heisig, M., Furer, A., Dietrich, G., Rapp, U.R., and Gentschev, I. (2009). Improvement of the live vaccine strain *Salmonella enterica* serovar Typhi Ty21a for antigen delivery via the hemolysin secretion system of *Escherichia coli*. IJMM *299*, 109–119.

Hunt, S., Moir, A.J., Tzokov, S., Bullough, P.A., Artymiuk, P.J., and Green, J. (2008). The formation and structure of *Escherichia coli* K-12 haemolysin E pores. Microbiology *154*, 633–642.

Hyland, C., Vuillard, L., Hughes, C., and Koronakis, V. (2001). Membrane interaction of *Escherichia coli* hemolysin: flotation and insertion-dependent labeling by phospholipid vesicles. J. Bacteriol. *183*, 5364–5370.

Issartel, J.P., Koronakis, V., and Hughes, C. (1991). Activation of *Escherichia coli* prohaemolysin to the mature toxin by acyl carrier protein-dependent fatty acylation. Nature *351*, 759–761.

Johnson, J.R. (1991). Virulence factors in *Escherichia coli* urinary tract infection. Clin. Microbiol. Rev. *4*, 80–128.

Johnson, J.R., Owens, K., Gajewski, A., and Kuskowski, M.A. (2005). Bacterial characteristics in relation to clinical source of *Escherichia coli* isolates from women with acute cystitis or pyelonephritis and uninfected women. J. Clin. Microbiol. *43*, 6064–6072.

Jonas, D., Schultheis, B., Klas, C., Krammer, P.H., and Bhakdi, S. (1993). Cytocidal effects of *Escherichia coli* hemolysin on human T lymphocytes. Infect. Immun. *61*, 1715–1721.

Jorgensen, S.E., Hammer, R.F., and Wu, G.K. (1980). Effects of a single hit from the alpha hemolysin produced by *Escherichia coli* on the morphology of sheep erythrocytes. Infect. Immun. *27*, 988–994.

Juarez, A., Nieto, J.M., Prenafeta, A., Miquelay, E., Balsalobre, C., Carrascal, M., and Madrid, C. (2000). Interaction of the nucleoid-associated proteins Hha and H-NS to modulate expression of the hemolysin operon in *Escherichia coli*. Adv. Exp. Med. Biol. *485*, 127–131.

Jumpertz, T., Chervaux, C., Racher, K., Zouhair, M., Blight, M.A., Holland, I.B., and Schmitt, L. (2010). Mutations affecting the extreme C terminus of *Escherichia coli* haemolysin A reduce haemolytic activity by altering the folding of the toxin. Microbiology *156*, 2495–2505.

Kao, J.S., Stucker, D.M., Warren, J.W., and Mobley, H.L. (1997). Pathogenicity island sequences of pyelonephritogenic *Escherichia coli* CFT073 are associated with virulent uropathogenic strains. Infect. Immun. *65*, 2812–2820.

Kaper, J.B., Nataro, J.P., and Mobley, H.L. (2004). Pathogenic *Escherichia coli*. Nature Rev. Microbiol. *2*, 123–140.

Keane, W.F., Welch, R., Gekker, G., and Peterson, P.K. (1987). Mechanism of *Escherichia coli* alpha-hemolysin-induced injury to isolated renal tubular cells. Am. J. Pathol. *126*, 350–357.

Kerenyi, M., Allison, H.E., Batai, I., Sonnevend, A., Emody, L., Plaveczky, N., and Pal, T. (2005). Occurrence of hlyA and sheA genes in extraintestinal *Escherichia coli* strains. J. Clin. Microbiol. *43*, 2965–2968.

Kesty, N.C., Mason, K.M., Reedy, M., Miller, S.E., and Kuehn, M.J. (2004). Enterotoxigenic *Escherichia coli* vesicles target toxin delivery into mammalian cells. EMBO J. *23*, 4538–4549.

Kim, J.Y., Doody, A.M., Chen, D.J., Cremona, G.H., Shuler, M.L., Putnam, D., and DeLisa, M.P. (2008). Engineered bacterial outer membrane vesicles with enhanced functionality. J. Mol. Biol. *380*, 51–66.

King, A.J., Sundaram, S., Cendoroglo, M., Acheson, D.W., and Keusch, G.T. (1999). Shiga toxin induces superoxide production in polymorphonuclear cells with subsequent impairment of phagocytosis and responsiveness to phorbol esters. J. Infect. Dis. *179*, 503–507.

Kloft, N., Busch, T., Neukirch, C., Weis, S., Boukhallouk, F., Bobkiewicz, W., Cibis, I., Bhakdi, S., and Husmann, M. (2009). Pore-forming toxins activate MAPK p38 by causing loss of cellular potassium. Biochem. Biophys. Res. Comm. *385*, 503–506.

Knapp, S., Then, I., Wels, W., Michel, G., Tschape, H., Hacker, J., and Goebel, W. (1985). Analysis of the flanking regions from different haemolysin determinants of *Escherichia coli*. MGG *200*, 385–392.

Kolling, G.L., and Matthews, K.R. (1999). Export of virulence genes and Shiga toxin by membrane vesicles of *Escherichia coli* O157:H7. Appl. Environ. Microbiol. *65*, 1843–1848.

Koronakis, V., and Hughes, C. (1996). Synthesis, maturation and export of the *E. coli* hemolysin. Med. Microbiol. Immunol. *185*, 65–71.

Koronakis, V., Cross, M., and Hughes, C. (1989). Transcription antitermination in an *Escherichia coli* haemolysin operon is directed progressively by cis-acting DNA sequences upstream of the promoter region. Mol. Microbiol. *3*, 1397–1404.

Koronakis, V., Sharff, A., Koronakis, E., Luisi, B., and Hughes, C. (2000). Crystal structure of the bacterial membrane protein TolC central to multidrug efflux and protein export. Nature *405*, 914–919.

Koschinski, A., Repp, H., Unver, B., Dreyer, F., Brockmeier, D., Valeva, A., Bhakdi, S., and Walev, I. (2006). Why *Escherichia coli* alpha-hemolysin induces calcium oscillations in mammalian cells – the pore is on its own. FASEB J. *20*, 973–975.

Kraemer, B.F., Campbell, R.A., Schwertz, H., Franks, Z.G., Vieira de Abreu, A., Grundler, K., Kile, B.T., Dhakal, B.K., Rondina, M.T., Kahr, W.H., *et al.*

(2012). Bacteria differentially induce degradation of Bcl-xL, a survival protein, by human platelets. Blood *120*, 5014–5020.

Laestadius, A., Richter-Dahlfors, A., and Aperia, A. (2002). Dual effects of *Escherichia coli* alpha-hemolysin on rat renal proximal tubule cells. Kidney Int. *62*, 2035–2042.

Lai, X.H., Arencibia, I., Johansson, A., Wai, S.N., Oscarsson, J., Kalfas, S., Sundqvist, K.G., Mizunoe, Y., Sjostedt, A., and Uhlin, B.E. (2000). Cytocidal and apoptotic effects of the ClyA protein from *Escherichia coli* on primary and cultured monocytes and macrophages. Infect. Immun. *68*, 4363–4367.

Lally, E.T., Kieba, I.R., Sato, A., Green, C.L., Rosenbloom, J., Korostoff, J., Wang, J.F., Shenker, B.J., Ortlepp, S., Robinson, M.K., *et al.* (1997). RTX toxins recognize a beta2 integrin on the surface of human target cells. J. Biol. Chem. *272*, 30463–30469.

Landraud, L., Gibert, M., Popoff, M.R., Boquet, P., and Gauthier, M. (2003). Expression of cnf1 by *Escherichia coli* J96 involves a large upstream DNA region including the hlyCABD operon, and is regulated by the RfaH protein. Mol. Microbiol. *47*, 1653–1667.

Larsen, C.K., Skals, M., Wang, T., Cheema, M.U., Leipziger, J., and Praetorius, H.A. (2011). Python erythrocytes are resistant to alpha-hemolysin from *Escherichia coli*. J. Membr. Biol. *244*, 131–140.

Lecher, J., Schwarz, C.K., Stoldt, M., Smits, S.H., Willbold, D., and Schmitt, L. (2012). An RTX Transporter Tethers Its Unfolded Substrate during Secretion via a Unique N-Terminal Domain. Structure *20*, 1778–1787.

Lee, M., Jun, S.Y., Yoon, B.Y., Song, S., Lee, K., and Ha, N.C. (2012). Membrane fusion proteins of type I secretion system and tripartite efflux pumps share a binding motif for TolC in Gram-negative bacteria. PLoS One *7*, e40460.

Leeds, J.A., and Welch, R.A. (1996). RfaH enhances elongation of *Escherichia coli* hlyCABD mRNA. J. Bacteriol. *178*, 1850–1857.

Leeds, J.A., and Welch, R.A. (1997). Enhancing transcription through the *Escherichia coli* hemolysin operon, hlyCABD: RfaH and upstream JUMPStart DNA sequences function together via a postinitiation mechanism. J. Bacteriol. *179*, 3519–3527.

Levine, M.M., Xu, J.G., Kaper, J.B., Lior, H., Prado, V., Tall, B., Nataro, J., Karch, H., and Wachsmuth, K. (1987). A DNA probe to identify enterohemorrhagic *Escherichia coli* of O157:H7 and other serotypes that cause hemorrhagic colitis and hemolytic uremic syndrome. J. Infect. Dis. *156*, 175–182.

Li, H., Granat, A., Stewart, V., and Gillespie, J.R. (2008). RpoS, H-NS, and DsrA influence EHEC hemolysin operon (ehxCABD) transcription in *Escherichia coli* O157:H7 strain EDL933. FEMS Microbiol. Lett. *285*, 257–262.

Lim, K.B., Walker, C.R., Guo, L., Pellett, S., Shabanowitz, J., Hunt, D.F., Hewlett, E.L., Ludwig, A., Goebel, W., Welch, R.A., *et al.* (2000). *Escherichia coli* alpha-hemolysin (HlyA) is heterogeneously acylated in vivo with 14-, 15-, and 17-carbon fatty acids. J. Biol. Chem. *275*, 36698–36702.

Linggood, M.A., and Ingram, P.L. (1982). The role of alpha haemolysin in the virulence of *Escherichia coli* for mice. J. Med. Microbiol. *15*, 23–30.

Linhartova, I., Bumba, L., Masin, J., Basler, M., Osicka, R., Kamanova, J., Prochazkova, K., Adkins, I., Hejnova-Holubova, J., Sadilkova, L., *et al.* (2010). RTX proteins: a highly diverse family secreted by a common mechanism. FEMS Microbiol. Rev. *34*, 1076–1112.

Lobo, A.L., and Welch, R.A. (1994). Identification and assay of RTX family of cytolysins. Methods Enzymol. *235*, 667–678.

te Loo, D.M., Monnens, L.A., van Der Velden, T.J., Vermeer, M.A., Preyers, F., Demacker, P.N., van Den Heuvel, L.P., and van Hinsbergh, V.W. (2000). Binding and transfer of verocytotoxin by polymorphonuclear leukocytes in hemolytic uremic syndrome. Blood *95*, 3396–3402.

Low, K.O., Mahadi, N.M., Abdul Rahim, R., Rabu, A., Abu Bakar, F.D., Abdul Murad, A.M., and Illias, R.M. (2010). Enhanced secretory production of hemolysin-mediated cyclodextrin glucanotransferase in *Escherichia coli* by random mutagenesis of the ABC transporter system. J. Biotechnol. *150*, 453–459.

Ludwig, A., and Goebel, W. (1999). The family of the multigenic encoded RTX toxin. In The Comprehensive Sourcebook of Bacterial Protein Toxins, Alouf, J.E., and Freer, J.H., eds. (Academic Press, San Diego, CA), pp. 330–348.

Ludwig, A., Jarchau, T., Benz, R., and Goebel, W. (1988). The repeat domain of *Escherichia coli* haemolysin (HlyA) is responsible for its Ca2+-dependent binding to erythrocytes. MGG *214*, 553–561.

Ludwig, A., Tengel, C., Bauer, S., Bubert, A., Benz, R., Mollenkopf, H.J., and Goebel, W. (1995). SlyA, a regulatory protein from *Salmonella typhimurium*, induces a haemolytic and pore-forming protein in *Escherichia coli*. MGG *249*, 474–486.

Ludwig, A., Bauer, S., Benz, R., Bergmann, B., and Goebel, W. (1999). Analysis of the SlyA-controlled expression, subcellular localization and pore-forming activity of a 34 kDa haemolysin (ClyA) from *Escherichia coli* K-12. Mol. Microbiol. *31*, 557–567.

Ludwig, A., Volkerink, G., von Rhein, C., Bauer, S., Maier, E., Bergmann, B., Goebel, W., and Benz, R. (2010). Mutations affecting export and activity of cytolysin A from *Escherichia coli*. J. Bacteriol. *192*, 4001–4011.

Madrid, C., Nieto, J.M., Paytubi, S., Falconi, M., Gualerzi, C.O., and Juarez, A. (2002). Temperature- and H-NS-dependent regulation of a plasmid-encoded virulence operon expressing *Escherichia coli* hemolysin. J. Bacteriol. *184*, 5058–5066.

Mansson, L.E., Melican, K., Boekel, J., Sandoval, R.M., Hautefort, I., Tanner, G.A., Molitoris, B.A., and Richter-Dahlfors, A. (2007). Real-time studies of the progression of bacterial infections and immediate tissue responses in live animals. Cell. Microbiol. *9*, 413–424.

Mao, B.H., Chang, Y.F., Scaria, J., Chang, C.C., Chou, L.W., Tien, N., Wu, J.J., Tseng, C.C., Wang, M.C., Chang, C.C., *et al.* (2012). Identification of *Escherichia coli* genes associated with urinary tract infections. J. Clin. Microbiol. *50*, 449–456.

Marrs, C.F., Zhang, L., and Foxman, B. (2005). *Escherichia coli* mediated urinary tract infections: are there distinct uropathogenic *E. coli* (UPEC) pathotypes? FEMS Microbiol. Lett. *252*, 183–190.

May, A.K., Gleason, T.G., Sawyer, R.G., and Pruett, T.L. (2000). Contribution of *Escherichia coli* alpha-hemolysin to bacterial virulence and to intraperitoneal alterations in peritonitis. Infect. Immun. *68*, 176–183.

Menestrina, G., Mackman, N., Holland, I.B., and Bhakdi, S. (1987). *Escherichia coli* haemolysin forms voltage-dependent ion channels in lipid membranes. Biochim. Biophys. Acta *905*, 109–117.

Menestrina, G., Moser, C., Pellet, S., and Welch, R. (1994). Pore-formation by *Escherichia coli* hemolysin (HlyA) and other members of the RTX toxins family. Toxicology *87*, 249–267.

Moayeri, M., and Welch, R.A. (1994). Effects of temperature, time, and toxin concentration on lesion formation by the *Escherichia coli* hemolysin. Infect. Immun. *62*, 4124–4134.

Mobley, H.L., Green, D.M., Trifillis, A.L., Johnson, D.E., Chippendale, G.R., Lockatell, C.V., Jones, B.D., and Warren, J.W. (1990). Pyelonephritogenic *Escherichia coli* and killing of cultured human renal proximal tubular epithelial cells: role of hemolysin in some strains. Infect. Immun. *58*, 1281–1289.

Morova, J., Osicka, R., Masin, J., and Sebo, P. (2008). RTX cytotoxins recognize beta2 integrin receptors through N-linked oligosaccharides. Proc. Natl. Acad. Sci. U.S.A. *105*, 5355–5360.

Mourino, M., Munoa, F., Balsalobre, C., Diaz, P., Madrid, C., and Juarez, A. (1994). Environmental regulation of alpha-haemolysin expression in *Escherichia coli*. Microb. Pathog. *16*, 249–259.

Mourino, M., Madrid, C., Balsalobre, C., Prenafeta, A., Munoa, F., Blanco, J., Blanco, M., Blanco, J.E., and Juarez, A. (1996). The Hha protein as a modulator of expression of virulence factors in *Escherichia coli*. Infect. Immun. *64*, 2881–2884.

Mueller, M., Grauschopf, U., Maier, T., Glockshuber, R., and Ban, N. (2009). The structure of a cytolytic alpha-helical toxin pore reveals its assembly mechanism. Nature *459*, 726–730.

Muller, C.M., Dobrindt, U., Nagy, G., Emody, L., Uhlin, B.E., and Hacker, J. (2006). Role of histone-like proteins H-NS and StpA in expression of virulence determinants of uropathogenic *Escherichia coli*. J. Bacteriol. *188*, 5428–5438.

Murase, K., Ooka, T., Iguchi, A., Ogura, Y., Nakayama, K., Asadulghani, M., Islam, M.R., Hiyoshi, H., Kodama, T., Beutin, L., *et al.* (2012). Haemolysin E- and enterohaemolysin-derived haemolytic activity of O55/O157 strains and other *Escherichia coli* lineages. Microbiology *158*, 746–758.

Nagamune, K., Yamamoto, K., Naka, A., Matsuyama, J., Miwatani, T., and Honda, T. (1996). In vitro proteolytic processing and activation of the recombinant precursor of El Tor cytolysin/hemolysin (pro-HlyA) of *Vibrio cholerae* by soluble hemagglutinin/protease of V. cholerae, trypsin, and other proteases. Infect. Immun. *64*, 4655–4658.

Nagy, G., Altenhoefer, A., Knapp, O., Maier, E., Dobrindt, U., Blum-Oehler, G., Benz, R., Emody, L., and Hacker, J. (2006). Both alpha-haemolysin determinants contribute to full virulence of uropathogenic *Escherichia coli* strain 536. Microb. Infect. *8*, 2006–2012.

Nicaud, J.M., Mackman, N., Gray, L., and Holland, I.B. (1985). Characterisation of HlyC and mechanism of activation and secretion of haemolysin from *E. coli* 2001. FEBS Lett. *187*, 339–344.

Nicaud, J.M., Mackman, N., Gray, L., and Holland, I.B. (1986). The C-terminal, 23 kDa peptide of *E. coli* haemolysin 2001 contains all the information necessary for its secretion by the haemolysin (Hly) export machinery. FEBS Lett. *204*, 331–335.

Nieto, J.M., Bailey, M.J., Hughes, C., and Koronakis, V. (1996). Suppression of transcription polarity in the *Escherichia coli* haemolysin operon by a short upstream element shared by polysaccharide and DNA transfer determinants. Mol. Microbiol. *19*, 705–713.

Nieto, J.M., Mourino, M., Balsalobre, C., Madrid, C., Prenafeta, A., Munoa, F.J., and Juarez, A. (1997). Construction of a double hha hns mutant of *Escherichia coli*: effect on DNA supercoiling and alpha-haemolysin production. FEMS Microbiol. Lett. *155*, 39–44.

Nieto, J.M., Madrid, C., Prenafeta, A., Miquelay, E., Balsalobre, C., Carrascal, M., and Juarez, A. (2000). Expression of the hemolysin operon in *Escherichia coli* is modulated by a nucleoid–protein complex that includes the proteins Hha and H-NS. MGG *263*, 349–358.

O'Hanley, P., Lalonde, G., and Ji, G. (1991). Alpha-hemolysin contributes to the pathogenicity of piliated digalactoside-binding *Escherichia coli* in the kidney: efficacy of an alpha-hemolysin vaccine in preventing renal injury in the BALB/c mouse model of pyelonephritis. Infect. Immun. *59*, 1153–1161.

Oropeza-Wekerle, R.L., Muller, E., Kern, P., Meyermann, R., and Goebel, W. (1989). Synthesis, inactivation, and localization of extracellular and intracellular *Escherichia coli* hemolysins. J. Bacteriol. *171*, 2783–2788.

Oscarsson, J. (1999). PhD thesis: Molecular analysis of the cytolysin ClyA in *Escherichia coli* and other Enterobacteria (Umeå, Sweden: Umeå University Medical Dissertations, New series no. 628, ISBN 91-7191-704-7 ISSN 034-6612).

Oscarsson, J., Mizunoe, Y., Uhlin, B.E., and Haydon, D.J. (1996). Induction of haemolytic activity in *Escherichia coli* by the slyA gene product. Mol. Microbiol. *20*, 191–199.

Oscarsson, J., Mizunoe, Y., Li, L., Lai, X.H., Wieslander, A., and Uhlin, B.E. (1999). Molecular analysis of the cytolytic protein ClyA (SheA) from *Escherichia coli*. Mol. Microbiol. *32*, 1226–1238.

Oscarsson, J., Westermark, M., Beutin, L., and Uhlin, B.E. (2002a). The bacteriophage-associated ehly1 and ehly2 determinants from *Escherichia coli* O26:H– strains do not encode enterohemolysins per se but cause release of the ClyA cytolysin. IJMM *291*, 625–631.

Oscarsson, J., Westermark, M., Lofdahl, S., Olsen, B., Palmgren, H., Mizunoe, Y., Wai, S.N., and Uhlin, B.E. (2002b). Characterization of a pore-forming cytotoxin expressed by *Salmonella enterica* serovars typhi and paratyphi A. Infect. Immun. *70*, 5759–5769.

Ostolaza, H., Bartolome, B., Ortiz de Zarate, I., de la Cruz, F., and Goni, F.M. (1993). Release of lipid vesicle contents by the bacterial protein toxin alpha-haemolysin. Biochim. Biophys. Acta *1147*, 81–88.

Ostolaza, H., Soloaga, A., and Goni, F.M. (1995). The binding of divalent cations to *Escherichia coli* alpha-haemolysin. FEBS *228*, 39–44.

Ostolaza, H., Bakas, L., and Goni, F.M. (1997). Balance of electrostatic and hydrophobic interactions in the lysis of model membranes by *E. coli* alpha-haemolysin. J. Membr. Biol. *158*, 137–145.

Pellett, S., and Welch, R.A. (1996). *Escherichia coli* hemolysin mutants with altered target cell specificity. Infect. Immun. *64*, 3081–3087.

Pimenta, A.L., Young, J., Holland, I.B., and Blight, M.A. (1999). Antibody analysis of the localisation, expression and stability of HlyD, the MFP component of the *E. coli* haemolysin translocator. MGG *261*, 122–132.

Pimenta, A.L., Racher, K., Jamieson, L., Blight, M.A., and Holland, I.B. (2005). Mutations in HlyD, part of the type 1 translocator for hemolysin secretion, affect the folding of the secreted toxin. J. Bacteriol. *187*, 7471–7480.

Rashid, R.A., Tabata, T.A., Oatley, M.J., Besser, T.E., Tarr, P.I., and Moseley, S.L. (2006). Expression of putative virulence factors of *Escherichia coli* O157:H7 differs in bovine and human infections. Infect. Immun. *74*, 4142–4148.

Real, J.M., Munro, P., Buisson-Touati, C., Lemichez, E., Boquet, P., and Landraud, L. (2007). Specificity of immunomodulator secretion in urinary samples in response to infection by alpha-hemolysin and CNF1 bearing uropathogenic *Escherichia coli*. Cytokine *37*, 22–25.

Reigstad, C.S., Hultgren, S.J., and Gordon, J.I. (2007). Functional genomic studies of uropathogenic *Escherichia coli* and host urothelial cells when intracellular bacterial communities are assembled. J. Biol. Chem. *282*, 21259–21267.

Reingold, J., Starr, N., Maurer, J., and Lee, M.D. (1999). Identification of a new *Escherichia coli* She haemolysin homolog in avian *E. coli*. Vet. Microbiol. *66*, 125–134.

von Rhein, C., Bauer, S., Simon, V., and Ludwig, A. (2008). Occurrence and characteristics of the cytolysin A gene in *Shigella* strains and other members of the family Enterobacteriaceae. FEMS Microbiol. Lett. *287*, 143–148.

Rompikuntal, P.K., Thay, B., Khan, M.K., Alanko, J., Penttinen, A.M., Asikainen, S., Wai, S.N., and Oscarsson, J. (2012). Perinuclear localization of internalized outer membrane vesicles carrying active cytolethal distending toxin from *Aggregatibacter actinomycetemcomitans*. Infect. Immun. *80*, 31–42.

Russo, T.A., Davidson, B.A., Genagon, S.A., Warholic, N.M., Macdonald, U., Pawlicki, P.D., Beanan, J.M., Olson, R., Holm, B.A., and Knight, P.R., 3rd (2005). *E. coli* virulence factor hemolysin induces neutrophil apoptosis and necrosis/lysis *in vitro* and necrosis/ lysis and lung injury in a rat pneumonia model. Am. J. Physiol. Lung Cell. Mol. Physiol. *289*, L207–216.

Sanchez-Magraner, L., Cortajarena, A.L., Goni, F.M., and Ostolaza, H. (2006). Membrane insertion of *Escherichia coli* alpha-hemolysin is independent from membrane lysis. J. Biol. Chem. *281*, 5461–5467.

Sanchez-Magraner, L., Viguera, A.R., Garcia-Pacios, M., Garcillan, M.P., Arrondo, J.L., de la Cruz, F., Goni, F.M., and Ostolaza, H. (2007). The calcium-binding C-terminal domain of *Escherichia coli* alpha-hemolysin is a major determinant in the surface-active properties of the protein. J. Biol. Chem. *282*, 11827–11835.

Sanchez-Magraner, L., Cortajarena, A.L., Garcia-Pacios, M., Arrondo, J.L., Agirre, J., Guerin, D.M., Goni, F.M., and Ostolaza, H. (2010). Interdomain Ca(2+) effects in *Escherichia coli* alpha-haemolysin: Ca(2+) binding to the C-terminal domain stabilizes both C- and N-terminal domains. Biochim. Biophys. Acta *1798*, 1225–1233.

Sanchez, S., Bakas, L., Gratton, E., and Herlax, V. (2011). Alpha hemolysin induces an increase of erythrocytes calcium: a FLIM 2-photon phasor analysis approach. PLoS One 6, e21127.

Schmidt, H., and Karch, H. (1996). Enterohemolytic phenotypes and genotypes of shiga toxin-producing *Escherichia coli* O111 strains from patients with diarrhoea and hemolytic-uremic syndrome. J. Clin. Microbiol. *34*, 2364–2367.

Schmidt, H., Karch, H., and Beutin, L. (1994). The large-sized plasmids of enterohemorrhagic *Escherichia coli* O157 strains encode hemolysins which are presumably members of the *E. coli* alpha-hemolysin family. FEMS Microbiol. Lett. *117*, 189–196.

Schmidt, H., Beutin, L., and Karch, H. (1995). Molecular analysis of the plasmid-encoded hemolysin of *Escherichia coli* O157:H7 strain EDL 933. Infect. Immun. *63*, 1055–1061.

Schmidt, H., Kernbach, C., and Karch, H. (1996). Analysis of the EHEC hly operon and its location in the physical map of the large plasmid of enterohaemorrhagic *Escherichia coli* O157:h7. Microbiology *142 (Pt 4)*, 907–914.

Schwarz, C.K., Landsberg, C.D., Lenders, M.H., Smits, S.H., and Schmitt, L. (2012). Using an *E. coli* Type 1 secretion system to secrete the mammalian, intracellular protein IFABP in its active form. J. Biotechnol. *159*, 155–161.

Scott, M.E., Melton-Celsa, A.R., and O'Brien, A.D. (2003). Mutations in hns reduce the adherence of Shiga toxin-producing *E. coli* O91:H21 strain B2F1 to human colonic epithelial cells and increase the production of hemolysin. Microb. Pathog. *34*, 155–159.

Sheshko, V., Hejnova, J., Rehakova, Z., Sinkora, J., Faldyna, M., Alexa, P., Felsberg, J., Nemcova, R., Bomba, A., and Sebo, P. (2006). HlyA knock out yields a safer *Escherichia coli* A0 34/86 variant with unaffected colonization capacity in piglets. FEMS Immunol. Med. Microbiol. *48*, 257–266.

Skals, M., Jorgensen, N.R., Leipziger, J., and Praetorius, H.A. (2009). Alpha-hemolysin from *Escherichia coli* uses endogenous amplification through P2X receptor activation to induce hemolysis. Proc. Natl. Acad. Sci. U.S.A. *106*, 4030–4035.

Skals, M., Jensen, U.B., Ousingsawat, J., Kunzelmann, K., Leipziger, J., and Praetorius, H.A. (2010). *Escherichia coli* alpha-hemolysin triggers shrinkage of erythrocytes via K(Ca)3.1 and TMEM16A channels with subsequent phosphatidylserine exposure. J. Biol. Chem. *285*, 15557–15565.

Smith, Y.C., Grande, K.K., Rasmussen, S.B., and O'Brien, A.D. (2006). Novel three-dimensional organoid model for evaluation of the interaction of uropathogenic *Escherichia coli* with terminally differentiated human urothelial cells. Infect. Immun. *74*, 750–757.

Smith, Y.C., Rasmussen, S.B., Grande, K.K., Conran, R.M., and O'Brien, A.D. (2008). Hemolysin of uropathogenic *Escherichia coli* evokes extensive shedding of the uroepithelium and hemorrhage in bladder tissue within the first 24 h after intraurethral inoculation of mice. Infect. Immun. *76*, 2978–2990.

Soloaga, A., Ostolaza, H., Goni, F.M., and de la Cruz, F. (1996). Purification of *Escherichia coli* pro-haemolysin, and a comparison with the properties of mature alpha-haemolysin. FEBS *238*, 418–422.

Soskine, M., Biesemans, A., Moeyaert, B., Cheley, S., Bayley, H., and Maglia, G. (2012). An engineered ClyA nanopore detects folded target proteins by selective external association and pore entry. Nano. Lett. *12*, 4895–4900.

Stanley, P., Koronakis, V., and Hughes, C. (1991). Mutational analysis supports a role for multiple structural features in the C-terminal secretion signal of *Escherichia coli* haemolysin. Mol. Microbiol. *5*, 2391–2403.

Stanley, P., Packman, L.C., Koronakis, V., and Hughes, C. (1994). Fatty acylation of two internal lysine residues required for the toxic activity of *Escherichia coli* hemolysin. Science *266*, 1992–1996.

Stanley, P., Koronakis, V., and Hughes, C. (1998). Acylation of *Escherichia coli* hemolysin: a unique protein lipidation mechanism underlying toxin function. Microbiol. Mol. Biol. Rev. *62*, 309–333.

Stroeher, U.H., Bode, L., Beutin, L., and Manning, P.A. (1993). Characterization and sequence of a 33 kDa enterohemolysin (Ehly 1)-associated protein in *Escherichia coli*. Gene *132*, 89–94.

Su, L., Chen, S., Yi, L., Woodard, R.W., Chen, J., and Wu, J. (2012). Extracellular overexpression of recombinant Thermobifida fusca cutinase by alpha-hemolysin secretion system in *E. coli* BL21(DE3). Microb. Cell Fact. *11*, 8.

Suttorp, N., Floer, B., Schnittler, H., Seeger, W., and Bhakdi, S. (1990). Effects of *Escherichia coli* hemolysin on endothelial cell function. Infect. Immun. *58*, 3796–3801.

Swenson, D.L., Bukanov, N.O., Berg, D.E., and Welch, R.A. (1996). Two pathogenicity islands in uropathogenic *Escherichia coli* J96: cosmid cloning and sample sequencing. Infect. Immun. *64*, 3736–3743.

Thanabalu, T., Koronakis, E., Hughes, C., and Koronakis, V. (1998). Substrate-induced assembly of a contiguous channel for protein export from *E. coli*: reversible bridging of an inner-membrane translocase to an outer membrane exit pore. EMBO J. *17*, 6487–6496.

Trent, M.S., Worsham, L.M., and Ernst-Fonberg, M.L. (1998). The biochemistry of hemolysin toxin activation: characterization of HlyC, an internal protein acyltransferase. Biochemistry *37*, 4644–4652.

Troeger, H., Richter, J.F., Beutin, L., Gunzel, D., Dobrindt, U., Epple, H.J., Gitter, A.H., Zeitz, M., Fromm, M., and Schulzke, J.D. (2007). *Escherichia coli* alpha-haemolysin induces focal leaks in colonic epithelium: a novel mechanism of bacterial translocation. Cell. Microbiol. *9*, 2530–2540.

Tzokov, S.B., Wyborn, N.R., Stillman, T.J., Jamieson, S., Czudnochowski, N., Artymiuk, P.J., Green, J., and Bullough, P.A. (2006). Structure of the hemolysin E (HlyE, ClyA, and SheA) channel in its membrane-bound form. J. Biol. Chem. *281*, 23042–23049.

Uhlen, P., Laestadius, A., Jahnukainen, T., Soderblom, T., Backhed, F., Celsi, G., Brismar, H., Normark, S., Aperia, A., and Richter-Dahlfors, A. (2000). Alpha-haemolysin of uropathogenic *E. coli* induces Ca2+ oscillations in renal epithelial cells. Nature *405*, 694–697.

Uhlich, G.A., McNamara, P.J., Iandolo, J.J., and Mosier, D.A. (1999). Cloning and characterization of the gene encoding Pasteurella haemolytica FnrP, a regulator of the *Escherichia coli* silent hemolysin sheA. J. Bacteriol. *181*, 3845–3848.

Uhlin, B.E., and Mizunoe, Y. (1994). Expression of a novel contacthemolytic activity by *E. coli*. J. Cell Biochem. Suppl. *18A*, 71.

Valeva, A., Walev, I., Kemmer, H., Weis, S., Siegel, I., Boukhallouk, F., Wassenaar, T.M., Chavakis, T., and Bhakdi, S. (2005). Binding of *Escherichia coli* hemolysin and activation of the target cells is not receptor-dependent. J. Biol. Chem. *280*, 36657–36663.

Wai, S.N., Westermark, M., Oscarsson, J., Mizunoe, Y., and Uhlin, B.E. (2000). Localisation and export of the ClyA cytotoxin in *Escherichia coli*. Med. Microbiol. Immunol. *189*, 51.

Wai, S.N., Lindmark, B., Soderblom, T., Takade, A., Westermark, M., Oscarsson, J., Jass, J., Richter-Dahlfors, A., Mizunoe, Y., and Uhlin, B.E. (2003a). Vesicle-mediated export and assembly of pore-forming oligomers of the enterobacterial ClyA cytotoxin. Cell *115*, 25–35.

Wai, S.N., Westermark, M., Oscarsson, J., Jass, J., Maier, E., Benz, R., and Uhlin, B.E. (2003b). Characterization of dominantly negative mutant ClyA cytotoxin proteins in *Escherichia coli*. J. Bacteriol. *185*, 5491–5499.

Wallace, A.J., Stillman, T.J., Atkins, A., Jamieson, S.J., Bullough, P.A., Green, J., and Artymiuk, P.J. (2000). *E. coli* hemolysin E (HlyE, ClyA, SheA): X-ray crystal structure of the toxin and observation of membrane pores by electron microscopy. Cell *100*, 265–276.

Walton, J.R., and Smith, D.H. (1969). New hemolysin (gamma) produced by *Escherichia coli*. J. Bacteriol. *98*, 304–305.

Wandersman, C., and Delepelaire, P. (1990). TolC, an *Escherichia coli* outer membrane protein required for hemolysin secretion. Proc. Natl. Acad. Sci. U.S.A. *87*, 4776–4780.

Wang, M.C., Tseng, C.C., Chen, C.Y., Wu, J.J., and Huang, J.J. (2002). The role of bacterial virulence and host factors in patients with *Escherichia coli* bacteremia who have acute cholangitis or upper urinary tract infection. Clin. Infect. Dis. *35*, 1161–1166.

Wang, R.C., Seror, S.J., Blight, M., Pratt, J.M., Broome-Smith, J.K., and Holland, I.B. (1991). Analysis of the membrane organization of an *Escherichia coli* protein translocator, HlyB, a member of a large family of prokaryote and eukaryote surface transport proteins. J. Mol. Biol. *217*, 441–454.

Welch, R.A. (1991). Pore-forming cytolysins of Gram-negative bacteria. Mol. Microbiol. *5*, 521–528.

Welch, R.A., and Pellett, S. (1988). Transcriptional organization of the *Escherichia coli* hemolysin genes. J. Bacteriol. *170*, 1622–1630.

Welch, R.A., Dellinger, E.P., Minshew, B., and Falkow, S. (1981). Haemolysin contributes to virulence of extra-intestinal *E. coli* infections. Nature *294*, 665–667.

Welch, R.A., Forestier, C., Lobo, A., Pellett, S., Thomas, W., Jr., and Rowe, G. (1992). The synthesis and function of the *Escherichia coli* hemolysin and related RTX exotoxins. FEMS Microbiol. Immunol. *5*, 29–36.

Westermark, M., Oscarsson, J., Mizunoe, Y., Urbonaviciene, J., and Uhlin, B.E. (2000). Silencing and activation of ClyA cytotoxin expression in *Escherichia coli*. J. Bacteriol. *182*, 6347–6357.

Whittam, T.S., Wolfe, M.L., Wachsmuth, I.K., Orskov, F., Orskov, I., and Wilson, R.A. (1993). Clonal relationships among *Escherichia coli* strains that cause hemorrhagic colitis and infantile diarrhoea. Infect. Immun. *61*, 1619–1629.

Wiles, T.J., Dhakal, B.K., Eto, D.S., and Mulvey, M.A. (2008a). Inactivation of host Akt/protein kinase B signaling by bacterial pore-forming toxins. Mol. Biol. Cell *19*, 1427–1438.

Wiles, T.J., Kulesus, R.R., and Mulvey, M.A. (2008b). Origins and virulence mechanisms of uropathogenic *Escherichia coli*. Exp. Mol. Pathol. *85*, 11–19.

Wiles, T.J., Bower, J.M., Redd, M.J., and Mulvey, M.A. (2009). Use of zebrafish to probe the divergent virulence potentials and toxin requirements of extraintestinal pathogenic *Escherichia coli*. PLoS Pathog. *5*, e1000697.

Worsham, L.M., Trent, M.S., Earls, L., Jolly, C., and Ernst-Fonberg, M.L. (2001). Insights into the catalytic mechanism of HlyC, the internal protein acyltransferase that activates *Escherichia coli* hemolysin toxin. Biochemistry *40*, 13607–13616.

Wu, X.Y., Chapman, T., Trott, D.J., Bettelheim, K., Do, T.N., Driesen, S., Walker, M.J., and Chin, J. (2007). Comparative analysis of virulence genes, genetic diversity, and phylogeny of commensal and enterotoxigenic *Escherichia coli* isolates from weaned pigs. Appl. Environ. Microbiol. *73*, 83–91.

Wyborn, N.R., Clark, A., Roberts, R.E., Jamieson, S.J., Tzokov, S., Bullough, P.A., Stillman, T.J., Artymiuk, P.J., Galen, J.E., Zhao, L., *et al.* (2004). Properties of haemolysin E (HlyE) from a pathogenic *Escherichia coli* avian isolate and studies of HlyE export. Microbiology *150*, 1495–1505.

Yokoyama, K., Horii, T., Yamashino, T., Hashikawa, S., Barua, S., Hasegawa, T., Watanabe, H., and Ohta, M. (2000). Production of shiga toxin by *Escherichia coli* measured with reference to the membrane vesicle-associated toxins. FEMS Microbiol. Lett. *192*, 139–144.

Zhu, C., Ruiz-Perez, F., Yang, Z., Mao, Y., Hackethal, V.L., Greco, K.M., Choy, W., Davis, K., Butterton, J.R., and Boedeker, E.C. (2006). Delivery of heterologous protein antigens via hemolysin or autotransporter systems by an attenuated ler mutant of rabbit enteropathogenic *Escherichia coli*. Vaccine *24*, 3821–3831.

Structural, Molecular and Functional Characteristics of Attaching and Effacing Lesions

10

H.T. Law and Julian A. Guttman

Abstract

Pathogenic *Escherichia coli* infections are major causes of illness and mortality throughout the world. Although many different types of *E. coli* exist, a subset of these microbes called the attaching and effacing (A/E) pathogens remain extracellular, colonize the intestine, collapse localized microvilli on enterocytes and generate characteristic 'lesions' that are hallmarks of the infections. These lesions generate morphological membrane protrusions beneath the attached bacteria that are easily recognizable by microscopy. In this chapter we will examine the structural and molecular components of A/E lesions, highlight recent advances in the field, demonstrate similarities of these structures to those generated by other microbes, and will propose functions for these lesions in the disease process.

General characteristics of A/E infections

A multitude of *Escherichia coli* species colonizes the intestinal lumen of humans and other animals, but relatively few cause diarrhoeal diseases. Of those that are intestinal pathogens, the A/E bacteria are some of the most prominent. In humans, two different A/E bacteria, called enteropathogenic *E. coli* (EPEC) and enterohaemorrhagic *E. coli* (EHEC) are the sources of most of the intestinal infections.

Once ingested these microbes colonize different regions of the intestines. The primary site of EPEC attack is in the small intestine whereas EHEC mainly occupies the colon. These bacteria typically remain in the lumen of the gut during

the infections and are thus generally considered as extracellular bacteria. However, despite their extracellular locations, A/E bacteria have devised strategies to control the sub-cellular machinery of their host cells for their benefit. These host cell changes are most clearly exemplified by 2 morphological alterations, (1) the collapse of microvilli on host intestinal epithelial cells (enterocytes), and (2) the formation of 1–5 μm columnar protrusions of the host membrane at sites of bacterial contact, which provide 'pedestal' platforms for the bacteria to sit (Fig. 10.1). It is these physical

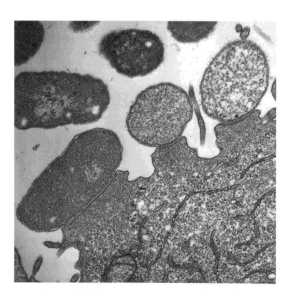

Figure 10.1 Ultrastructural transmission electron microscopic image of EPEC attaching and effacing (A/E) lesions on a HeLa cell. The A/E lesions are identified based on the pedestal-like structures generated beneath the attached bacteria and the localized elimination of microvilli at those sites. Image courtesy of A. Wayne Vogl and J.A. Guttman.

changes to the host cells that are the crux of A/E lesions. These lesions are generated whenever EPEC or EHEC dock onto host cells during the disease process. Thus, they are routinely used as hallmarks of disease both *in vitro* and *in vivo*.

A/E lesion formation

The generation of A/E lesions requires 3 important steps: (1) the initial attachment of the bacteria to the host cells, (2) the collapse of microvilli in the localized area of bacterial docking, and (3) the formation of the pedestal structures themselves. Although research on A/E lesions has increased substantially since their initial discovery in 1969 (Staley *et al.*, 1969), much of the focus remains concentrated on the components that form the pedestal structures.

Initial attachment

A/E bacteria confront a number of obstacles in their quest for pedestal formation. These barriers include a thick mucous layer and microvilli at the surface of epithelial cells. Hence, mucus penetration is a pre-requisite for bacterial adherence to enterocytes (Drumm *et al.*, 1988; Smith *et al.*, 1995; Mack *et al.*, 1999; Larson *et al.*, 2003; Lindén *et al.*, 2008). Unfortunately, few mechanistic data demonstrating how this occurs have been presented in the literature. One piece of evidence involves the zinc metalloprotease StcE, which is expressed by EHEC strains possessing the large virulence plasmid (pO157) (Lathem *et al.*, 2002) and a subset of atypical EPEC strains (Lathem *et al.*, 2003). Studies on EHEC-encoded StcE suggest that this secreted protein can reduce mucus viscosity by cleaving O-linked glycoproteins [mucins] (Lathem *et al.*, 2004; Grys *et al.*, 2005), which are major components of mucus (Reviewed in McGuckin *et al.*, 2011). As such, StcE remains as one of the sole factors implicated in facilitating the penetration of A/E bacteria through the mucus layer (Szabady *et al.*, 2011; Yu *et al.*, 2012).

Once the mucous has been penetrated, 1–2 μm microvilli that protrude from the surface of the enterocytes are likely the first epithelial cell components that interact with the microbes. It is poorly understood how A/E bacteria precisely navigate or associate with the microvilli to trigger the disease process, but a number of hypotheses have been proposed. The first involves bacterial fibrillar structures, called pili. Pili are dynamic structures that can extend out, forming a web-like network to interact with the microvilli (Cleary *et al.*, 2004). As these pili retract, they can pull the bacteria closer to the enterocytes (Humphries *et al.*, 2010). Among the different types of pili expressed by A/E bacteria (Mundy *et al.*, 2003; Xicohtencatl-Cortes *et al.*, 2007; Saldana *et al.*, 2009), bundle-forming pili (BFP) are the best characterized (Girón *et al.*, 1991; Donnenberg *et al.*, 1992). Mechanistically, EPEC BFP can bind to host epithelial cell surface molecules, which include N-acetyllactosamine-containing receptors (Hyland *et al.*, 2008; Humphries *et al.*, 2009) and the phospholipid phosphatidylethanolamine (PE) (Foster *et al.*, 1999). *In vivo* evidence has shown that BFP enhances the disease state and is required to maintain high bacterial loads in infected individuals (Levine *et al.*, 1985; Bieber *et al.*, 1998); however, when A/E bacteria mutated in BFP are found attached to cells in culture A/E lesions are still formed (Nataro *et al.*, 1985; Knutton *et al.*, 1987; Saldana *et al.*, 2009; Law *et al.*, 2012; Lin and Guttman, 2012).

Another hypothesis involves hollow filaments generated by the A/E bacterial protein EspA (*E. coli* secreted protein A). EspA filaments project from the bacterial surface and can reach up to 2 μm in length (Knutton *et al.*, 1998; Sekiya *et al.*, 2001). These structures provide a conduit for bacterial proteins to pass from the microbes directly into the host cells *via* a type III secretion system (Knutton *et al.*, 1998; Shaw *et al.*, 2001), which will be discussed in the next section. In experiments using EPEC mutated in EspA, these microbes were ~ 28% less adherent to the human intestinal epithelium than their parental strain and were unable to accumulate actin beneath surface-bound EPEC (Kenny *et al.*, 1996; Cleary *et al.*, 2004). This is not surprising as certain proteins needed for bacterial adhesion require translocation from the bacterial cytosol into the host through EspA filaments (Knutton *et al.*, 1998; Deng *et al.*, 2004). However, the possibility of EspA itself enhancing adhesion is plausible but requires additional study.

A third hypothesis is that the flagella themselves might serve as anchors to the host cells either by entangling with the microvilli, allowing the bacteria to gain a foothold on the enterocytes, or by embedding themselves within the host cell surface (Girón *et al.*, 2002).

Once A/E bacteria have made their early contact with the epithelial surface, these extracellular pathogens can now develop firm anchors to the host cell. Whether this happens initially on the microvilli themselves or whether this occurs following microvillar effacement remains to be elucidated, but what is known is that this firm docking to the host membrane requires both bacterial surface proteins and proteins rooted in the host plasma membrane.

Intimate bacterial attachment

After the initial attachment of A/E microbes to the epithelial cell surface, these bacteria 'intimately' adhere to the underlying host cell; forming additional connections between the host and bacterial cell membranes. Firm attachment requires two key virulence factors that originate from the bacterium – intimin and translocated intimin receptor (Tir). The bacterial outer membrane protein intimin is the only known bacterial surface adhesion molecule necessary for intestinal colonization of A/E pathogens in animals. Intimin can interact with multiple host cell proteins including $\beta 1$-integrin (Frankel *et al.*, 1996a; Liu *et al.*, 1999a) and nucleolin (Sinclair and O'Brien, 2002, 2004; Sinclair *et al.*, 2006), but it is its strong affinity for Tir that is indispensable for firm bacterial-cell attachment (Kenny *et al.*, 1997; DeVinney *et al.*, 1999b). The ability of intimin to dock with host receptors on the host surface lies in the Tir-binding domain, which is located at the C-terminal 280 amino acids of the protein (Frankel *et al.*, 1994; Adu-Bobie *et al.*, 1998). Intimins are classified into several distinct types based on their divergent Tir-binding domains (Agin and Wolf, 1997; Adu-Bobie *et al.*, 1998). Three common types of intimin molecules are α (e.g. EPEC O127:H7), β (e.g. *C. rodentium* and REPEC), and γ (e.g. O55:H7, O157:H7 EHEC). Although α and γ (expressed by EPEC and EHEC respectively) are the most divergent, they are functionally interchangeable *in vitro* (DeVinney

et al., 1999b) and *in vivo* (Mallick *et al.*, 2012) suggesting that differences in the Tir-binding regions do not significantly affect their ability to attach to host cells. This is likely because despite considerable allelic variation, structural analysis of the Tir-binding domains indicates that a number of key residues responsible for the recognition of Tir remain well conserved.

To control many host cell functions, A/E bacteria use a needle-like protein injection apparatus known as a Type III secretion system (T3SS) to translocate Tir and many other pathogenic bacterial proteins (known as effectors) directly from the bacterial cytosol into the host cytoplasm (Reviewed in Dean and Kenny, 2009). At least partial unfolding of Tir and other effectors is thought to be required for proteins to fit through the narrow type III translocation apparatus (Gauthier and Finlay, 2003; Akeda and Galán, 2005; Zarivach *et al.*, 2007). Tir is one of the first effectors to be delivered through this secretion apparatus and its translocation controls the hierarchical transfer of other effectors through the T3SS (Thomas *et al.*, 2007; Mills *et al.*, 2008). Once in the host cytoplasm, Tir is targeted and becomes embedded in the host plasma membrane (Kenny *et al.*, 1997). Within that membrane, Tir adopts a hairpin loop topology (Fig. 10.2) resulting in its N- and C-terminal domains projecting into the host cytosol, exposing the intimin-binding domain (IBD) extracellularly (DeVinney *et al.*, 1999a; Gauthier *et al.*, 2000). This positions Tir in the ideal orientation for intimin binding and makes Tir an extracellular bacterial receptor on the host cell for the intimin ligand located on the bacterial cell surface (de Grado *et al.*, 1999). The strong interaction between Tir and intimin enables the A/E pathogen to anchor atop the host plasma membrane for pedestal formation.

Microvillar effacement

Microvilli are collapsed in the localized area of A/E bacterial adherence, however details as to how these extracellular pathogens disassemble host microvilli still remain vague. Evidence suggests that the injection of bacterial effectors through the T3SS is a pre-requisite for microvillar collapse (Shaw *et al.*, 2005; Dean *et al.*, 2006) as are host enzymes that are activated by A/E microbes.

EspB, one of three pore forming bacterial effectors that are inserted into the host plasma membrane during A/E bacterial infections is needed for the efficient translocation of other virulence factors into host cells (Kenny and Finlay, 1995; Kenny et al., 1997) and accordingly virulence in humans (Tacket et al., 2000) as well as animal models are significantly attenuated when EspB mutants are used for infections (Abe et al., 1998; Tacket et al., 2000). This protein is multifunctional and can bind to various members of the myosin superfamily (Iizumi et al., 2007), including brush border myosin-1 (Tyska et al., 2005). Experiments studying the effects on myosin-1c demonstrate that the actin-binding domain of myosin-1c has ~ 18 times higher affinity for EspB as compared to filamentous actin (Iizumi et al., 2007; Mattoo et al., 2008). This strong affinity to EspB is thought to inhibit myosin–actin associations, contributing to the effacement of microvilli both in cultured cells and in vivo (Iizumi et al., 2007).

The actin binding protein, ezrin, has also been shown to play an important role in proper microvillar structure and biogenesis (reviewed in Fehon et al., 2010), as intestinal epithelial cells from ezrin knockout mice produce abnormally thick and short microvilli (Saotome et al., 2004). Consequently, the role of ezrin in microvillar collapse during A/E bacterial infections has been examined. Ezrin can be degraded by the enzyme calpain and both ezrin and calpain are targeted by A/E bacteria (Potter et al., 2003; Hardwidge et al., 2004; Lai et al., 2011). However, in CaCo-2 cells that had calpain inhibited, a loss of ezrin from the apical surface did not occur, nor did effacement of brush border microvilli despite these cells showing resistance against A/E bacterial lesions.

EPEC pedestal formation

All A/E pathogens by definition form pedestals; however, the details of the strategies that each A/E bacterium utilizes to form the structures differs. Although it has been known since 1997 that Tir and intimin are critical bacterial proteins needed by all A/E pathogens to elicit pedestal formation (Kenny et al., 1997; Kenny and Finlay, 1997), additional host and bacterial factors have been demonstrated to also be involved in the process.

The discovery of pedestals in 1969 ushered in a wave of research that has focused on identifying the components, functions and mechanisms involved in their formation. Initially it was shown that actin filaments constituted a large proportion of the pedestal structure (Knutton et al., 1989; Fig. 10.3). This led to the identification of N-WASp, the Arp2/3 complex, α-actinin and Nck at pedestals (Kalman et al., 1999; Goosney et al., 2000; Gruenheid et al., 2001) as well as the demonstration that these structures are dynamic and motile, enabling the bacteria to 'surf' on the surface of epithelial cells (Sanger et al., 1996) through the regulation of actin filament polymerization presumably using profilin and cofilin (Goosney et al., 2001) (Fig. 10.2). Subsequent work detailed the interactions of those proteins as well as others. Currently, it is understood that phosphorylation of tyrosine (Y) 474 on the C-terminal arm of Tir is required for > 95% of EPEC pedestals that are formed. The phosphorylation of other Y residues on EPEC Tir including Y454 is responsible for < 5% of the pedestals observed in culture (Campellone and Leong, 2005). Accordingly, the majority of proteins identified at EPEC pedestals have been shown to require the phosphorylation of Y474 (Fig. 10.2). The association of Tir with intimin is crucial for the phosphorylation events. c-Fyn and c-Abl are responsible for the efficient phosphorylation of Tir at Y474 (Phillips et al., 2004; Swimm et al., 2004). However, other tyrosine kinases including Arg, Etk, and Tec have also been implicated in this post-translational modification of Tir (Swimm et al., 2004; Bommarius et al., 2007; Swimm and Kalman, 2008). This phosphorylation recruits the adaptor protein Nck directly to Y474 and ultimately triggers the polymerization of actin filaments beneath the attached bacteria through N-WASp and Arp2/3-based mechanisms (Kalman et al., 1999; Gruenheid et al., 2001). However, in the < 5% of cases that use Y454 instead of Y474 phosphorylation, Nck is dispensable (Campellone and Leong, 2005).

EPEC pedestal proteins

Although Nck is the primary host protein that links EPEC Tir to the actin filament polymerization machinery, there are a number of other

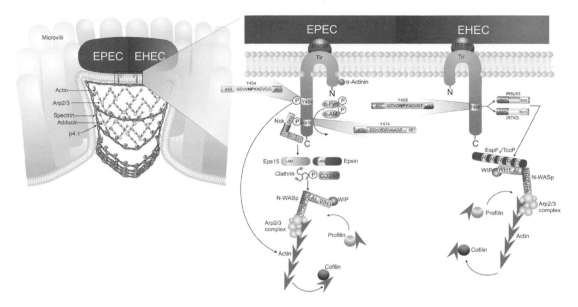

Figure 10.2 EPEC and EHEC recruitment of host proteins at pedestals. An illustration of the organization of crucial components found at EPEC/EHEC pedestals. At sites of A/E lesion formation EPEC and EHEC both generate pedestals and disassemble the localized microvilli. A major structural component of A/E lesions is the highly branched actin network. Here, the Arp2/3 complex is shown at the branch points where new actin filaments are formed; oriented at 70° angles from the pre-existing actin filaments. The pointed (slow-growing) ends of actin filaments are oriented downwards while the barbed (fast-growing) ends points towards the apex of the pedestal. These filaments extend from the sub-apical region of the pedestals to ~2/3 down the pedestal stalk. Spectrin makes-up the basal 2/3 of the pedestal; overlapping with actin in the middle 1/3 of the structures. Spectrin and the associated cytoskeletal elements (adducin and p4.1) also cage the actin along the pedestal membrane. However, adducin does not localize directly beneath the plasma membrane where EPEC/EHEC attach. A closer look at the apical region of the pedestals shows that EPEC and EHEC use the bacterial outer membrane protein intimin and the translocated intimin receptor, Tir, (located in the host plasma membrane) to firmly anchor the bacteria to the surface of host cells. α-actinin binds to the N-terminal arm of EPEC Tir, but it is the C-terminus cytoplasmic domain that is critical for the majority of pedestal formation. Phosphorylation of specific tyrosine (Y) residues by the host tyrosine kinase family members, c-Fyn and c-Abl, on Tir is needed for protein recruitment to ultimately control N-WASp and the Arp2/3 complex for actin polymerization beneath the attached bacteria. EPEC recruits Nck directly to Tir, which is dependent Y474 phosphorylation. Alternatively, phosphorylation of Y454 on Tir can signal actin nucleation in a Nck-independent manner. Downstream of Nck the endocytic proteins Eps15, epsin, CD2AP and clathrin are targeted to the apical region of EPEC pedestals. The ubiquitin interacting motifs (UIM) of Eps15 and epsin are sufficient for localization to pedestals. Unlike EPEC, EHEC does not use Nck or endocytic proteins for pedestal formation. Instead EHEC recruits IRSp53 or IRTKS to Y458 of EHEC Tir. This triggers the docking of a second bacterial protein, EspF$_u$/TccP to the forming apical complex prior to N-WASp and Arp2/3. Actin filaments are polymerized through Arp2/3-dependent nucleation. Additional abbreviations used in this figure to indicate specific regions used for protein binding include: Src homology 2 (SH2) and 3 (SH3) domain, verprolin, cofilin, acidic (VPA) domain, proline-rich domain (PRD), autoinhibitory (AI) domain, WASP homology 1 (WH1) domain, proline-rich repeats (R), IRSp53-MIM homology (I-BAR) domain. A colour version of this figure is available in the plate section at the back of the book.

components involved in EPEC pedestal formation and maintenance. These proteins have been studied to varying degrees and can generally be characterized as localizing to the apical tip, the entire stalk or the distal regions of pedestals (Table 10.1).

Proteins at the apical tip of pedestals

The characteristic formation of a single pedestal beneath EPEC is influenced by phospholipid signalling in the pedestal membrane. Plasma membrane homeostasis is maintained in part by dephosphorylating phosphatidylinositol

Table 10.1 A comparison of identified host proteins in human EPEC versus EHEC pedestals

Host proteins	EPEC Localized at pedestals in general	Apical tip	Stalk	Base	EHEC Localized at pedestals in general	References
14-3-3τ		✓				Patel *et al*. (2006)
Abl	✓					Swimm *et al*. (2004)
Actin		✓*	✓*	✓*	✓*	Knutton *et al*. (1989)
α-actinin		✓	✓	✓	✓	Goosney *et al*. (2000, 2001), Shaner *et al*. (2005)
Adducin			✓*†	✓*	✓*	Ruetz *et al*. (2011), Ruetz *et al*. (2012b,c)
Annexin 2		✓			✓	Zobiac *et al*. (2002), Miyahara *et al*. (2009)
Akt	✓					Sason *et al*. (2009)
Arg	✓					Swimm *et al*. (2004)
Arp2/3		✓*	✓*	✓*	✓*	Kalman *et al*. (1999), Gruenheid *et al*. (2001), Goosney *et al*. (2001)
c-Fyn		✓				Phillips *et al*. (2004), Hayward *et al*. (2009)
Dab2	✓					Bonazzi *et al*. (2011)
Calmodulin		✓	✓	✓		Brown *et al*. (2008)
Calpactin		✓	✓	✓	✓	Goosney *et al*. (2001)
CD2AP		✓*				Guttman *et al*. (2010b)
CD44		✓			✓	Goosney *et al*. (2001)
Clathrin		✓*			✗	Veiga *et al*. (2007), Guttman *et al*. (2010b)
Claudin		✓				Peralta-Ramirez *et al*. (2008)
Cofilin		✓	✓	✓	✓	Goosney *et al*. (2001)
Cortactin		✓*	✓*	✓*	✓*	Cantarelli *et al*. (2002, 2006), Mousnier *et al*. (2008)
CrkII		✓	✓	✓	✗	Goosney *et al*. (2001)
Cytokeratin 8	✓					Batchelor *et al*. (2004)
Cytokeratin 18	✓*				✓	Batchelor *et al*. (2004)
Dynamin 2	✓					Unsworth *et al*. (2007)
Eps15		✓*				Lin *et al*. (2011)
Epsin1		✓*				Lin *et al*. (2011)
Etk	✓					Bommarius *et al*. (2007)
Ezrin		✓	✓	✓	✓	Finlay *et al*. (1992), Goosney *et al*. (2001)
Filamin A			✓	✓		Smith *et al*. (2010)
Fimbrin	✓		✓			Goosney *et al*. (2000)
Gelsolin		✓	✓	✓	✓	Goosney *et al*. (2001)
Grb2			✓	✓	✗	Goosney *et al*. (2001)
Hip1R	✓*					Bonazzi *et al*. (2011)
IQGAP		✓	✓	✓		Brown *et al*. (2008)
IRSp53					✓	Weiss *et al*. (2009)
IRTKS					✓	Vingadassalom *et al*. (2009)

Host proteins	EPEC Localized at pedestals in general	Apical tip	Stalk	Base	EHEC Localized at pedestals in general	References
Lamellipodin		✓			✗	Smith et al. (2010)
Lipoma-preferred partner		✓	✓		✓	Goosney et al. (2001)
Myosin II	✓			✓		Sanger et al. (1996)
N-WASP, WASP	✓*				✓*	Kalman et al. (1999), Lommel et al. (2004)
Nck	✓*				✗	Gruenheid et al. (2001)
Nexilin	✓	✓	✓			Law et al. (2011)
NHERF2	✓					Munera et al. (2012)
Nucleolin					✓	Sinclair et al. (2002), Sinclair et al. (2006)
p4.1		✓*	✓*†	✓*	✓*	Ruetz et al. (2011), Ruetz et al. (2012b)
S100A10 (p11)	✓					Zobiack et al. (2002)
Shank3	✓					Huett et al. (2009)
Shc		✓	✓	✓		Goosney et al. (2001)
SHIP2		✓	✓	✓	✗	Smith et al. (2010)
Spectrin		✓*	✓*†	✓*	✓*	Ruetz et al. (2011), Ruetz et al. (2012b)
Talin		✓	✓	✓	✓	Finlay et al. (1992), Cantarelli et al. (2001), Goosney et al. (2001)
TOCA1					✓	Campellone et al. (2012)
Tropomyosin				✓	✓	Sanger et al. (1996), Vallance and Finlay (2000), Goosney et al. (2001)
VASP		✓			✓	Goosney et al. (2000, 2001)
Villin	✓					Sanger et al. (1996), Vallance and Finlay (2000)
Vinculin		✓	✓	✓	✓	Freeman et al. (2000), Goosney et al. (2001)
WIP		✓	✓	✓	✓	Lommel et al. (2004), Garber et al. (2012)
ZO-1, -2				✓		Hanajima-Ozawa et al. (2007)
Zyxin		✓	✓	✓		Goosney et al. (2001)

✓ protein is present; ✗ protein is absent; * required for efficient pedestal formation; † localized at the periphery.

3,4,5-trisphosphate $(PI(3,4,5)P_3)$ to phosphatidylinositol 4,5-bisphosphate $(PI(3,4)P_2)$, which is involved in the PI3K-dependent signalling pathway (Sason et al., 2009; Smith et al., 2010). Dysregulation of this pathway has been shown to cause multiple pedestals to form underneath a single bacterium (Smith et al., 2010). Several proteins involved in the formation of $PI(3,4)P_2$ are localized at the apex of EPEC pedestals including SHIP2, lamellipodin and Shc (Src homologous and collagen gene) (Smith et al., 2010). Their recruitment beneath adherent EPEC is facilitated mainly by two tyrosine residues (Y483 and Y551) on Tir and both sites are necessary for the recruitment of these proteins to the pedestal tip (Smith et al., 2010).

Upon EPEC attachment, the phospholipid and F-actin-binding protein annexin II accumulates at the plasma membrane where the pathogen makes contact (Zobiack et al., 2002). Redistribution of annexin II during these infections has been shown to be dependent on EPEC Tir, but does not require

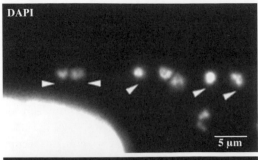

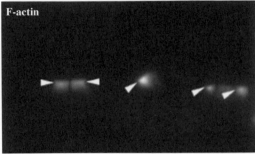

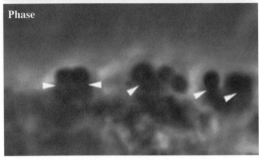

Figure 10.3 Fluorescence and phase contrast microscopy of EPEC pedestals generated *in vitro*. EPEC and the host nucleus were visualized using the DNA stain, DAPI. The actin-rich regions of the pedestals (arrowheads) were stained using fluorescently labelled phalloidin. Phase contrast microscopy image shows extracellular EPEC and pedestals on the surface of a HeLa cell.

Y474 phosphorylation (Munera *et al.*, 2012). The C-terminus of annexin II can bind to both the N- and C-terminal cytoplasmic regions of EPEC Tir (Munera *et al.*, 2012), and it was suggested that its presence at EPEC pedestals could act as a scaffold for other proteins and aid in stabilizing membrane rafts (Zobiack *et al.*, 2002).

WASp-interacting proteins

N-WASp is a crucial protein for pedestal formation and is localized at the apical tip of the structures (Kalman *et al.*, 1999). Accordingly

the WAS/WASL-interacting protein (WIP), a regulator of actin cytoskeletal dynamics and stabilizer of N-WASp (Martinez-Quiles *et al.*, 2001) also accumulates at the tip of pedestals where it directly interacts with the WH1 (WASP homology 1) domain of N-WASp (Lommel *et al.*, 2004). However unlike N-WASp, WIP is dispensable in pedestal formation (Lommel *et al.*, 2004; Garber *et al.*, 2012), suggesting its role may simply be supportive in the actin stability process.

Endocytic proteins

The extracellular location of A/E bacteria inherently suggests that proteins involved in internalization (endocytosis) should not be components of the pedestal structures, as pedestals protrude the bacteria off of the natural surface of epithelial cells rather than internalize the microbes. Surprisingly, clathrin-mediated endocytic (CME) proteins that were thought to be exclusively involved in particle internalization, are present and crucial for pedestal formation (Bonazzi *et al.*, 2011; Unsworth *et al.*, 2007; Veiga *et al.*, 2007; Guttman *et al.*, 2010b; Lin *et al.*, 2011). CME normally requires at least four core proteins to work in concert to enable particle internalization; clathrin, epsin, Eps15 as well as AP-2 and uses the scission protein dynamin-2 to release the internalized particle into the cell cytoplasm. Three of these proteins (clathrin, epsin, and Eps15) as well as the endocytic protein CD2AP are present at the apical tip of EPEC pedestals, require Y474 phosphorylation and Nck in order to be recruited and positioned upstream of the actin polymerization machinery (Veiga *et al.*, 2007; Guttman *et al.*, 2010b). AP-2 is absent from pedestals (Lin *et al.*, 2011) and it is thought that the lack of AP-2 at pedestals likely changes the organization of the CME components beneath the bacteria, which could be a contributing factor involved in blocking EPEC internalization.

EPEC pedestal stalk proteins

Cytoskeletal systems

It is well known that A/E pathogens dramatically alter the cytoskeletal landscape of host cells and this is best exemplified by the dramatic re-organization of actin filaments beneath the bacteria in pedestals, but these microbes also control other cytoskeletal elements during A/E lesion formation. One of these involves the intermediate filament proteins cytokeratins 8 and 18 (CK-8 and CK-18). Both CK-8 and CK-18 have been localized in EPEC pedestals and depletion of CK-18 blocks pedestal formation (Batchelor *et al.*, 2004). Interestingly, despite the presence of CK8/18 at these structures, intermediate filaments themselves have never been seen at pedestals.

The sub-membranous spectrin cytoskeleton is distinct from actin-, microtubule- and intermediate filament-containing systems in function and distribution in mammalian cells. This cytoskeletal network contains four central proteins namely β2-spectrin, adducin, p4.1 and ankyrin. Thus far, all but ankyrin have been identified at pedestals (Ruetz *et al.*, 2011, 2012b,c). Unlike other proteins found in these actin-rich structures, spectrin, adducin, and p4.1 are generally confined to the lateral and distal regions of EPEC pedestals. Spectrin also co-localizes with one-third of actin filaments, occupying the bottom 2/3 of pedestals (Fig. 10.2). With exception to adducin, spectrin and p4.1 localize at the apex directly beneath EPEC thereby forming a continuous system along the pedestal periphery. Y474 phosphorylation-dependent signalling is necessary for the redistribution of spectrin, p4.1, and adducin, though further work is necessary to define the molecular interactions between the actin and spectrin cytoskeletal systems. Depletion of spectrin-associated components by siRNA abolishes pedestal formation and in the case of adducin, completely blocked EPEC attachment to epithelial cells (Ruetz *et al.*, 2011, 2012b) (Table 10.1). Hence, the spectrin cytoskeleton does not only play a key role in pedestal formation, but might also serve as a structural platform needed for EPEC-host cell attachment (Ruetz *et al.*, 2011).

Actin-associated proteins

Because actin filaments are by far the most well studied cytoskeletal component at pedestals, the involvement of a number of actin-associated proteins have been examined to varying degrees and most require the C-terminus of Tir for successful recruitment. Proteins that have been identified at

pedestals in general are found in Table 10.1 and details on actin-related proteins are discussed in depth in the subsequent sections.

Proteins at the N-terminus of Tir

Several proteins associate with the N-terminal domain of EPEC Tir including α-actinin, annexin II, cortactin and talin. As other parts of the chapter will cover annexin II and cortactin, this section will focus on α-actinin and talin.

α-Actinin

α-Actinin, an actin-binding protein, was the first identified host protein to bind directly to the first 200 amino acids of Tir (Goosney *et al.*, 2000). This was substantiated by two additional reports that found α-actinin preferentially bound to the N-terminus of Tir using co-precipitation experiments (Freeman *et al.*, 2000; Huang *et al.*, 2002). However, controversies arose when Cantarelli and co-workers (2001) found no substantial interaction between the two molecules using column affinity and far-Western analysis (Cantarelli *et al.*, 2001). Although the reasons for this disparity have yet to be fully addressed, cultured cells infected with Tir Y474F mutant EPEC indicated that α-actinin recruitment did not require Y474 phosphorylation, which is found at the C-terminus of Tir (Goosney *et al.*, 2000).

Talin

Talin, another actin-related protein, was shown to accumulate beneath adherent EPEC and directly engage the N-terminal domain of Tir when analysed by both *in vitro* binding (Freeman *et al.*, 2000; Cantarelli *et al.*, 2001) and yeast two-hybrid assays (Huang *et al.*, 2002). Further results indicated that talin has a higher affinity for the N-terminus of Tir than α-actinin and the actin-binding protein vinculin (Cantarelli *et al.*, 2001). Microinjection of talin partially impeded actin pedestal formation, suggesting it plays a role in Tir-based actin nucleation (Cantarelli *et al.*, 2001).

Proteins at the C-terminus of Tir

The majority of cytoskeletal proteins found at pedestals require the C-terminus of Tir. Much of this evidence comes from mutants in which EPEC

TirY474 is mutated in one form or another and thus general pedestal formation is for the most part impeded.

Cortactin

Cortactin is a prominent protein in EPEC pedestals, localizing throughout the entire structure (Cantarelli et al., 2002; 2006). Its presence is dependent on clathrin, dynamin, N-WASp and thus also the C-terminus of Tir (Cantarelli et al., 2002; Unsworth et al., 2007). Interestingly, in vitro experiments suggest that cortactin can also interact with the N-terminus of Tir (Nieto-Pelegrin and Martinez-Quiles, 2009). Cortactin contains a variety of domains and it is through those regions that it can interact with many actin-associated proteins. One of its binding partners, CD2AP, requires cortactin for redistribution beneath EPEC (Guttman et al., 2010b). At EPEC pedestals cortactin localizes not only to the apical tip of the pedestals, but also along their entire length (Cantarelli et al., 2006; Guttman et al., 2010b).

The SH3 domain within cortactin can interact with N-WASp and WIP (Kirkbride et al., 2011), two proteins located at the apex of pedestals. When this domain of cortactin is overexpressed in cells, pedestal assembly is inhibited (Cantarelli et al., 2006). In the stalk of EPEC pedestals cortactin can activate the Arp2/3 complex (Uruno et al., 2001; Cantarelli et al., 2006). The overexpression of the N-terminal region of cortactin, which directly interacts with the Arp2/3 complex and actin filaments, inhibits actin accumulation beneath EPEC (Cantarelli et al., 2006).

Distal pedestal proteins

Various proteins have been identified exclusively at the base of EPEC pedestals. These include filamin A (Smith et al., 2010), myosin II (Sanger et al., 1996), tropomyosin (Sanger et al., 1996), ZO-1 and ZO-2 (Hanajima-Ozawa et al., 2007). Filamin A is a large heterodimeric protein that primarily cross-links actin filaments. However, it can also link cellular receptors to the actin cytoskeleton and facilitate signal transduction subjacent to the plasma membrane in other systems (Feng and Walsh, 2004). Its localization at the EPEC pedestal base (Smith et al., 2010) suggests that filamin A might stabilize the highly branched microfilament network at the slow-growing (pointed) end of actin filaments (Stossel et al., 2001; Ruetz et al., 2012c).

The presence of myosin II and tropomyosin at the base of pedestals is reminiscent of their localization beneath brush-border microvilli (Sanger et al., 1996). To understand the role of myosin II at these structures potaroo kidney (Ptk2) fibroblast cells infected with EPEC were treated with the non-muscle myosin II ATPase chemical inhibitor called 2,3-Butanedione monoxime (BDM) (Cramer and Mitchison, 1995). Surprisingly, after applying BDM to EPEC infected cells, the length of actin-rich pedestals increased up to 50 μm (Shaner et al., 2005). A general ATPase inhibitor cocktail containing 2-deoxyglucose and sodium azide had a similar effect, whereas the myosin light chain kinase inhibitor (ML-7) diminished pedestal length. This suggested that inhibition of the myosin II ATPase was involved in regulating pedestal length.

The tight junction zonula occludens (ZO) proteins are cytoplasmic adaptors linking the transmembrane components of the tight junctions to the actin network (Stevenson et al., 1986). During in vitro EPEC infections, ZO-1 and ZO-2 are recruited to the distal regions of pedestals. This localization requires the C-terminal proline-rich domain of ZO-1 (Hanajima-Ozawa et al., 2007) and provides a novel function of the ZO proteins. However, the precise role of tight junction proteins at pedestals remains unknown. Despite the fact that ZO proteins have been studied extensively in vivo during A/E infections (Philpott et al., 1998; Czerucka et al., 2000; Simonovic et al., 2000; Howe et al., 2005), their presence at A/E pedestals during animal infections has remained elusive.

Very few proteins have been shown to change their localization within EPEC pedestal during motility events. This is likely because most studies use fixed samples instead of live preparations. An investigation of the focal adhesion and actin-binding protein nexilin at EPEC pedestals demonstrated a dual location of nexilin; within the entire stalk and at the base of pedestals (Law et al., 2012). Using live cell imaging of fluorescently tagged samples nexilin was located along the entire stalk, co-localizing with actin, in stationary pedestals. When pedestals became motile nexilin segregated from actin and concentrated at the pedestal base (Law *et al.*, 2012). This unique dynamic positioning of nexilin suggested that this protein might act as a launching platform during pedestal movement.

EHEC pedestal formation

In general, EHEC pedestals resemble those of EPEC. Both protrude from directly beneath the attached microbes, are actin and spectrin-rich, and require Tir/intimin interactions to be initiated (Liu *et al.*, 1999b; Ruetz *et al.*, 2012b). However, unlike EPEC, which directly recruits host proteins to Tir, during EHEC infections another bacterial protein [EspFu (*E. coli* secreted protein F-like from prophage U), also referred to as TccP (Tir-cytoskeleton coupling protein)] (Garmendia *et al.*, 2004; Campellone *et al.*, 2004), and two additional host proteins, IRSp53 (insulin receptor tyrosine kinase substrate p53) and IRTKS (insulin receptor tyrosine kinase substrate), are required for pedestal formation (Vingadassalom *et al.*, 2009; Weiss *et al.*, 2009) (Fig. 10.2).

EHEC Tir shares 58% identity with its EPEC homologue, lacks a residue corresponding to Y474 and consequently does not use Nck to induce pedestal formation (DeVinney *et al.*, 1999b). The EHEC Tir C-terminal residues 452–463 and particularly the NPY458 motif found within those 12 residues, is sufficient to recruit IRSp53/IRTKS to the apical tip of the pedestals (Vingadassalom *et al.*, 2009; Weiss *et al.*, 2009), enabling actin polymerization (Campellone *et al.*, 2006) (Fig. 10.2). An NPY motif in EHEC Tir is a critical binding site for the recognition of the N-terminal inverse-Bin-amphiphysin-Rvs (I-BAR) domain of IRSp53/IRTKS (Vingadassalom *et al.*, 2009; Weiss *et al.*, 2009). EPEC Tir, which also possesses an NPY motif within Y454 can also trigger actin assembly through a Nck-independent pathway, but this recruitment is inefficient (Brady *et al.*, 2007). Structural analysis predicted that two I-BAR domains from independent IRSp53/IRTKS molecules can cross-link two adjacent Tir molecules (Millard *et al.*, 2005; de Groot *et al.*, 2011), suggesting a mechanisms for EHEC Tir clustering at the plasma membrane.

The docking of IRSp53/IRTKS to EHEC Tir precedes the recruitment of EspFu/TccP, which is needed for EHEC pedestal formation and was identified almost simultaneously by two independent labs (Garmendia *et al.*, 2004; Campellone *et al.*, 2004). EspFu/TccP shares 35% sequence similarity with the LEE-encoded effector EspF found in both EPEC and EHEC (Campellone *et al.*, 2004). Given that most of the resemblance is situated in the short N-terminal signalling sequence and not at the C-terminal region, which is principally responsible for protein function, EspFu and EspF have noticeably dissimilar functions (Campellone *et al.*, 2004). In EHEC O157:H7, EspFu possesses 5.5 C-terminal proline-rich repeats (Weiss *et al.*, 2009). The number of these nearly identical sequences vary from 2 to 7 amongst genetically classified strains (Garmendia *et al.*, 2005). Each full 47-amino acid repeat interacts directly with IRSp53/IRTKS (Vingadassalom *et al.*, 2009; Weiss *et al.*, 2009) as well as the membrane tubulation protein TOCA1 (Campellone *et al.*, 2012) and N-WASp (Garmendia *et al.*, 2004; Campellone *et al.*, 2004).

EPEC and EHEC subvert the N-WASp functions through different Tir-based mechanisms for pedestal assembly. Rather than exploiting Nck, EHEC uses EspFu/TccP to directly engage N-WASp through its multiple C-terminal repeats, which in turn triggers actin nucleation beneath the site of bacterial attachment (Garmendia *et al.*, 2004; Campellone *et al.*, 2004). Within the C-terminal repeat of EspFu/TccP, a 17-amino acid long peptide derived from the N-terminus was able to sufficiently bind to the autoinhibitory (AI) portion of the N-WASp GTPase-binding domain (GBD) (Sallee *et al.*, 2008). Although a single GBD-binding motif of EspFu can bind to N-WASp, at least two C-terminal repeats were needed to activate N-WASp and trigger actin nucleation beneath EHEC (Garmendia *et al.*, 2006; Campellone *et al.*, 2008; Sallee *et al.*, 2008).

A/E lesions *in vivo*

While *in vitro* studies have provided important insight into the formation of A/E lesions, many of these findings have yet to be recapitulated *in vivo*. A noteworthy reason why *in vivo* investigations have trailed behind is the fact that typical human EPEC and EHEC strains are not naturally virulent in most conventional laboratory animals (Vallance and Finlay, 2000). Thus, to study A/E bacterial infections *in vivo* many of the early animals studies used piglet (Staley *et al.*, 1969; Baldini *et al.*, 1983; Moon *et al.*, 1983), guinea pig (Echeverria *et al.*, 1976), rabbit (Moon *et al.*, 1983; Ritchie *et al.*, 2003), and calf (Dean-Nystrom *et al.*, 1997) infections with EPEC and EHEC. A challenge with using these hosts is that access to the genetic tools afforded through mouse infections are not available, and unfortunately mice are poor hosts for these human pathogens. Additionally, any host/microbe interactions that are species (human) specific might not be identified using non-human hosts with the human targeted pathogens.

To overcome these obstacles human biopsy samples have been attempted, however this material is difficult to obtain and accordingly there are few studies in the literature. Another strategy to study A/E lesions and disease *in vivo* is to use an A/E microbe that is virulent in mice. To do this scientists have exploited *Citrobacter rodentium* as a surrogate model for *in vivo* studies (Frankel *et al.*, 1996b; Higgins *et al.*, 1999; Deng *et al.*, 2001, 2003, 2004; Guttman *et al.*, 2006, 2007, 2010a).

C. rodentium is a rodent-specific A/E bacterium that has homologous effectors as the human A/E pathogens, colonizes the colon at high levels (similar to EHEC), generates small pedestals, and has Tir that most closely resembles EPEC. Y471 is an important tyrosine residue in *C. rodentium* Tir that closely mimics Y474 of EPEC Tir (Deng *et al.*, 2003). Phosphorylation at this residue is mediated by host tyrosine kinases enabling Nck to dock with Tir and the subsequent recruitment of N-WASp, Arp2/3 and actin filaments (Crepin *et al.*, 2010). When an amino acid substitution of Y471F was used, this impeded Tir phosphorylation and blocked Nck recruitment, which in turn inhibited actin nucleation at bacterial/host cell contact sites as well as A/E lesion formation in general (Deng *et al.*, 2003). *C. rodentium* Tir also has a secondary tyrosine residue that mirrors the Y454 of EPEC Tir. This Y451 residue on *C. rodentium* Tir shares a common NPY motif with Y454 of EPEC Tir (Campellone and Leong, 2005; Crepin *et al.*, 2010). *C. rodentium* expressing a non-phosphorylatable Y471 mutant of Tir inefficiently promoted N-WASP-mediated actin nucleation *in vivo* (Deng *et al.*, 2003; Crepin *et al.*, 2010).

Structures generated by other microbes that resemble A/E lesions

Hijacking the host cytoskeleton to generate propulsive actin-rich structures is not unique to EPEC and EHEC, as *Listeria monocytogenes, Shigella flexneri*, the vaccinia virus and other microbes share that property. In a similar manner to the A/E bacteria, when these microbes generate these motile structures, they control the actin-based polymerization machinery only at one end of the microbes. Actin is most highly concentrated at the microbial interface and is polymerized at that interaction point through the mimicry or recruitment of host actin nucleators. The resulting Arp2/3-based polymerized actin network generates 'comet tails' that rocket the microbes within and between host cells enabling microbial spreading to neighbouring cells.

L. monocytogenes uses the bacterial surface protein ActA to induce actin nucleation. ActA mimics the host N-WASp/SCAR family proteins thereby directly activating the Arp2/3 complex (Niebuhr *et al.*, 1997). This tactic allows *L. monocytogenes* to circumvent the recruitment of the actin-nucleating factor N-WASp (Niebuhr *et al.*, 1997; Boujemaa-Paterski *et al.*, 2001). In a similar manner to *L. monocytogenes, S. flexneri* expresses the outer membrane protein IcsA at one pole of the bacterium. IscA mimics the host small GTPase Cdc42, an activator of N-WASp (Egile *et al.*, 1999), thus directly interacting with N-WASp prior to Arp2/3 complex recruitment and activation (Suzuki and Sasakawa, 2001). The docking of IcsA to N-WASp occurs at the N-WASp GBD domain; the same region used for the EHEC effector EspFu/TccP, which shares no sequence homology (Campellone *et al.*, 2004; Hayward *et al.*, 2006). The Vaccinia virus also uses a microbial protein to nucleate actin into comet tails, called A36 (Frischknecht *et al.*, 1999). This protein, like EPEC Tir, exploits receptor-based tyrosine phosphorylation to activate N-WASp and induce

Arp2/3-based actin polymerization (Kenny, 1999; Gruenheid *et al.*, 2001). The phosphorylation of Y112 on A36 permits direct binding of the host adaptor protein Nck (Frischknecht *et al.*, 1999), while phosphorylation of a second tyrosine residue at position 132 allows Grb2 to dock with the viral protein (Scaplehorn *et al.*, 2002). While A36–Nck interaction alone is sufficient to activate N-WASp, the engagement of A36 to both Nck and Grb2 enhances Arp2/3-based comet tail formation (Scaplehorn *et al.*, 2002).

L. monocytogenes and *S. flexneri* do not only exploit the host actin cytoskeleton but like EPEC and EHEC they also commandeer the spectrin cytoskeleton during their infections (Ruetz *et al.*, 2011, 2012a). *L. monocytogenes* uses spectrin, adducin and p4.1 for entry into epithelial cells and recruits spectrin prior to actin at the initial stages of comet tail formation (Ruetz *et al.*, 2011). *S. flexneri* only uses p4.1 during their active invasion of epithelial cells, but then hijacks spectrin to the distal region of the comet tails (Ruetz *et al.*, 2012a).

Proposed functions of A/E lesions

Microvillar effacement

There are two distinct events that underlie A/E lesion formation: (1) microvillar collapse, and (2) pedestal formation. So, why would the bacteria go through extensive effort to form these lesions? This is a question that has plagued the research field for decades. A simple explanation for the microvillar effacement component of lesions could be as a contributing factor in the generation of diarrhoea by decreasing the absorptive surface area in the infected tissue. However, based on the rapid onset of diarrhoea, it was surmised that the malabsorption from the lack of microvilli is not likely to contribute to significant amounts of water loss (reviewed in Hecht, 2001). Additional evidence from Dean and co-workers (2006) found that EPEC substantially impairs the sodium/glucose transporter activity of SGLT-1, a major contributor in water absorption (Meinild *et al.*, 1998), prior to extensive effacement of microvilli (Dean *et al.*, 2006). Other

molecular mechanisms of diarrhoea generation are also at work during A/E bacterial infections which would likely overshadow any need for contributions from microvillar effacement (Spitz *et al.*, 1995; Hecht and Koutsouris, 1999; Higgins *et al.*, 1999; Hecht *et al.*, 2004; Muza-Moons *et al.*, 2004; Borthakur *et al.*, 2006; O'Hara *et al.*, 2006; Gill *et al.*, 2007; Esmaili *et al.*, 2009). Another hypothesis is that microvillar collapse in the localized region of the pedestals is simply a by-product of the pedestal formation process and that pedestals could use the protein-based resources of microvilli for their formation. This is likely as pedestals contain many of the same proteins as those found in microvilli.

Pedestal formation

When pedestals are formed they could be used to enhance microbial maintenance in the intestine during diarrhoeal episodes; when the lumen of the colon is being flushed out the pedestal might increase the likelihood of the bacteria not sheering off of the membrane. However, at the micrometre scales that the bacteria are operating it is questionable how much shearing force occurs at the luminal interface. This is an aspect of the infections that requires further study. Another option is that pedestals are simply used for cell-to cell spreading, as is the case for *L. monocytogenes* and *S. flexneri*. This is plausible as once an A/E microbe docks with the epithelium, transfer to neighbouring cells could be advantageous.

Conclusion

Research on A/E lesions has expanded significantly since their initial discovery. To date over 50 proteins have been identified at these structures. Some have been studied in depth whereas others require follow-up analysis. A/E lesions have proven themselves as not only important structures for the disease process, but also as novel model systems to examine general cell biological principles. The elusive mechanisms A/E bacteria exploit to efface microvilli remains as a challenge to the field, but with the advanced systems currently available there is no doubt that this will be solved. Finally characterizing the molecular interactions occurring between the different types of proteins at pedestals, discovering how and why they are distributed in different pedestal regions and identifying additional proteins at these sites will keep researchers in the field busy for years to come.

Acknowledgements

We would like to thank A. Wayne Vogl for the EM expertise and providing the image in Fig. 10.1. We would also like to thank Fern Ness for rendering the model in Fig. 10.2. JAG is a CIHR New Investigator. Funding was provided by NSERC.

References

Abe, A., Heczko, U., Hegele, R.G., and Brett Finlay, B. (1998). Two enteropathogenic *Escherichia coli* type III secreted proteins, EspA and EspB, are virulence factors. J. Exp. Med. *188*, 1907–1916.

Adu-Bobie, J., Frankel, G., Bain, C., Goncalves, A.G., Trabulsi, L.R., Douce, G., Knutton, S., and Dougan, G. (1998). Detection of intimins alpha, beta, gamma, and delta, four intimin derivatives expressed by attaching and effacing microbial pathogens. J. Clin. Microbiol. *36*, 662–668.

Agin, T.S., and Wolf, M.K. (1997). Identification of a family of intimins common to *Escherichia coli* causing attaching-effacing lesions in rabbits, humans, and swine. J. Clin. Microbiol. *65*, 320–326.

Akeda, Y., and Galán, J.E. (2005). Chaperone release and unfolding of substrates in type III secretion. Nature *437*, 911–915.

Baldini, M.M., Kaper, J.B., Levine, M.M., Candy, D.C., and Moon, H.W. (1983). Plasmid-mediated adhesion in enteropathogenic *Escherichia coli*. J. Pediatr. Gastroenterol. Nutr. *2*, 534–538.

Batchelor, M., Guignot, J., Patel, A., Cummings, N., Cleary, J., Knutton, S., Holden, D.W., Connerton, I., and Frankel, G. (2004). Involvement of the intermediate filament protein cytokeratin-18 in actin pedestal formation during EPEC infection. EMBO Rep. *5*, 104–110.

Bieber, D., Ramer, S.W., Wu, C.Y., Murray, W.J., Tobe, T., Fernandez, R., and Schoolnik, G.K. (1998). Type IV pili, transient bacterial aggregates, and virulence of enteropathogenic *Escherichia coli*. Science *280*, 2114–2118.

Bommarius, B., Maxwell, D., Swimm, A., Leung, S., Corbett, A., Bornmann, W., and Kalman, D. (2007). Enteropathogenic *Escherichia coli* Tir is an SH2/3 ligand that recruits and activates tyrosine kinases required for pedestal formation. Mol. Microbiol. *63*, 1748–1768.

Bonazzi, M., Vasudevan, L., Mallet, A., Sachse, M., Sartori, A., Prevost, M.-C., Roberts, A., Taner, S.B., Wilbur, J.D., Brodsky, F.M., *et al.* (2011). Clathrin phosphorylation is required for actin recruitment at sites of bacterial adhesion and internalization. J. Cell Biol. *195*, 525–536.

Borthakur, A., Gill, R.K., Hodges, K., Ramaswamy, K., Hecht, G., and Dudeja, P.K. (2006). Enteropathogenic *Escherichia coli* inhibits butyrate uptake in Caco-2 cells by altering the apical membrane MCT1 level. Am. J. Physiol. Gastrointest. Liver Physiol. *290*, G30–G35.

Boujemaa-Paterski, R., Gouin, E., Hansen, G., Samarin, S., Clainche, C.L., Didry, D., Dehoux, P., Cossart, P., Kocks, C., Carlier, M., *et al.* (2001). *Listeria* protein ActA mimics WASP family proteins: it activates filament barbed end branching by Arp2/3 complex. Biochemistry *40*, 11390–11404.

Brady, M.J., Campellone, K.G., Ghildiyal, M., and Leong, J.M. (2007). Enterohaemorrhagic and enteropathogenic *Escherichia coli* Tir proteins trigger a common Nck-independent actin assembly pathway. Cell. Microbiol. *9*, 2242–2253.

Brown, M.D., Bry, L., Li, Z., and Sacks, D.B. (2008). Actin pedestal formation by enteropathogenic *Escherichia coli* is regulated by IQGAP1, calcium, and calmodulin. J. Biol. Chem. *283*, 35212.

Campellone, K.G., and Leong, J.M. (2005). Nck-independent actin assembly is mediated by two phosphorylated tyrosines within enteropathogenic *Escherichia coli* Tir. Mol. Microbiol. *56*, 416–432.

Campellone, K.G., Robbins, D., and Leong, J.M. (2004). EspFU is a translocated EHEC effector that interacts with Tir and N-WASP and promotes Nck-independent actin assembly. Develop. Cell *7*, 217–228.

Campellone, K.G., Brady, M.J., Alamares, J.G., Rowe, D.C., Skehan, B.M., Tipper, D.J., and Leong, J.M. (2006). Enterohaemorrhagic *Escherichia coli* Tir requires a C-terminal 12-residue peptide to initiate EspFU-mediated actin assembly and harbours N-terminal sequences that influence pedestal length. Cell. Microbiol. *8*, 1488–1503.

Campellone, K.G., Cheng, H.-C., Robbins, D., Siripala, A.D., McGhie, E.J., Hayward, R.D., Welch, M.D., Rosen, M.K., Koronakis, V., and Leong, J.M. (2008). Repetitive N-WASP-binding elements of the enterohemorrhagic *Escherichia coli* effector EspF(U) synergistically activate actin assembly. PLoS Pathog. *4*, e1000191.

Campellone, K.G., Siripala, A.D., Leong, J.M., and Welch, M.D. (2012). Membrane-deforming proteins play distinct roles in actin pedestal biogenesis by enterohemorrhagic *Escherichia coli*. J. Biol. Chem. *287*, 20613–20624.

Cantarelli, V., Takahashi, A., Yanagihara, I., Akeda, Y., Imura, K., Kodama, T., Kono, G., Sato, Y., and Honda, T. (2001). Talin, a host cell protein, interacts directly with the translocated intimin receptor, Tir, of enteropathogenic *Escherichia coli*, and is essential for pedestal formation. Cell. Microbiol. *3*, 745–751.

Cantarelli, V.V., Takahashi, A., Yanagihara, I., Akeda, Y., Imura, K., Kodama, T., Kono, G., Sato, Y., Iida, T., and Honda, T. (2002). Cortactin is necessary for F-actin accumulation in pedestal structures induced by enteropathogenic *Escherichia coli* infection. Infect. Immun. *70*, 2206–2209.

Cantarelli, V.V., Kodama, T., Nijstad, N., Abolghait, S.K., Iida, T., and Honda, T. (2006). Cortactin is essential for F-actin assembly in enteropathogenic *Escherichia coli* (EPEC)- and enterohaemorrhagic *E. coli* (EHEC)-induced pedestals and the alpha-helical region is involved in the localization of cortactin to bacterial attachment sites. Cell. Microbiol. *8*, 769–780.

Cleary, J., Lai, L.-C., Shaw, R.K., Straatman-Iwanowska, A., Donnenberg, M.S., Frankel, G., and Knutton, S. (2004). Enteropathogenic *Escherichia coli* (EPEC) adhesion to intestinal epithelial cells: role of bundle-forming pili (BFP), EspA filaments and intimin. Microbiology *150*, 527–538.

Cramer, L.P., and Mitchison, T.J. (1995). Myosin is involved in postmitotic cell spreading. J. Cell Biol. *131*, 179–189.

Crepin, V.F., Girard, F., Schuller, S., Phillips, A.D., Mousnier, A., and Frankel, G. (2010). Dissecting the role of the Tir: Nck and Tir: IRTKS/IRSp53 signalling pathways *in vivo*. Mol. Microbiol. *75*, 308–323.

Czerucka, D., Dahan, S., Mograbi, B., Rossi, B., and Rampal, P. (2000). *Saccharomyces boulardii* preserves the barrier function and modulates the signal transduction pathway induced in enteropathogenic *Escherichia coli*-infected T84 cells. Infect. Immun. *68*, 5998–6004.

Dean, P., and Kenny, B. (2009). The effector repertoire of enteropathogenic *E. coli*: ganging up on the host cell. Curr. Opin. Microbiol. *12*, 101–109.

Dean, P., Maresca, M., Schuller, S., Phillips, A.D., and Kenny, B. (2006). Potent diarrhoeagenic mechanism mediated by the cooperative action of three enteropathogenic *Escherichia coli*-injected effector proteins. Proc. Natl. Acad. Sci. U.S.A. *103*, 1876–1881.

Dean-Nystrom, E.A., Bosworth, B.T., Cray, W.C., and Moon, H.W.H. (1997). Pathogenicity of *Escherichia coli* O157:H7 in the intestines of neonatal calves. Infect. Immun. *65*, 1842–1848.

Deng, W., Li, Y., Vallance, B.A., and Finlay, B.B. (2001). Locus of enterocyte effacement from *Citrobacter rodentium*: sequence analysis and evidence for horizontal transfer among attaching and effacing pathogens. Infect. Immun. *69*, 6323–6335.

Deng, W., Vallance, B.A., Li, Y., Puente, J.L., and Finlay, B.B. (2003). *Citrobacter rodentium* translocated intimin receptor (Tir) is an essential virulence factor needed for actin condensation, intestinal colonization and colonic hyperplasia in mice. Mol. Microbiol. *48*, 95–115.

Deng, W., Puente, J.L., Gruenheid, S., Li, Y., Vallance, B.A., Vázquez, A., Barba, J., Ibarra, J.A., O'Donnell, P., and Metalnikov, P. (2004). Dissecting virulence: systematic and functional analyses of a pathogenicity island. Proc. Natl. Acad. Sci. U.S.A. *101*, 3597–3602.

DeVinney, R., Gauthier, A., Abe, A., and Finlay, B.B. (1999a). Enteropathogenic *Escherichia coli*: a pathogen that inserts its own receptor into host cells. CMLS *55*, 961.

DeVinney, R., Stein, M., Reinscheid, D., Abe, A., Ruschkowski, S., and Finlay, B.B. (1999b). Enterohemorrhagic *Escherichia coli* O157:H7 produces Tir, which is translocated to the host cell membrane but is not tyrosine phosphorylated. Infect. Immun. *67*, 2389–2398.

Donnenberg, M.S., Girón, J.A., Nataro, J.P., and Kaper, J.B. (1992). A plasmid-encoded type IV fimbrial gene of enteropathogenic *Escherichia coli* associated with localized adherence. Mol. Microbiol. *6*, 3427–3437.

Donnenberg, M.S., Yu, J., and Kaper, J.B. (1993). A second chromosomal gene necessary for intimate attachment of enteropathogenic *Escherichia coli* to epithelial cells. J. Bacteriol. *175*, 4670–4680.

Drumm, B., Roberton, A.M., and Sherman, P.M. (1988). Inhibition of attachment of *Escherichia coli* RDEC-1 to intestinal microvillus membranes by rabbit ileal mucus and mucin *in vitro*. Infect. Immun. *56*, 2437–2442.

Echeverria, P.D., Chang, C.P., and Smith, D. (1976). Enterotoxigenicity and invasive capacity of 'enteropathogenic' serotypes of *Escherichia coli*. J. Pediatr. *89*, 8–10.

Egile, C., Loisel, T., Laurent, V., Li, R., Pantaloni, D., Sansonetti, P., and Carlier, M. (1999). Activation of the CDC42 effector N-WASP by the *Shigella flexneri* IcsA protein promotes actin nucleation by Arp2/3 complex and bacterial actin-based motility. J. Cell Biol. *146*, 1319–1332.

Esmaili, A., Nazir, S.F., Borthakur, A., Yu, D., Turner, J.R., Saksena, S., Singla, A., Hecht, G.A., Alrefai, W.A., and Gill, R.K. (2009). Enteropathogenic *Escherichia coli* infection inhibits intestinal serotonin transporter function and expression. Gastroenterology *137*, 2074–2083.

Fehon, R.G., McClatchey, A.I., and Bretscher, A. (2010). Organizing the cell cortex: the role of ERM proteins. Nat. Rev. Mol. Cell. Biol. *11*, 276–287.

Feng, Y., and Walsh, C.A. (2004). The many faces of filamin: a versatile molecular scaffold for cell motility and signalling. Nat. Cell. Biol. *6*, 1034–1038.

Finlay, B.B., Rosenshine, I., Donnenberg, M.S., and Kaper, J.B. (1992). Cytoskeletal composition of attaching and effacing lesions associated with enteropathogenic *Escherichia coli* adherence to HeLa cells. Infect. Immun. *60*, 2541–2543.

Foster, D.B., Philpott, D., Abul-Milh, M., Huesca, M., Sherman, P.M., and Lingwood, C.A. (1999). Phosphatidylethanolamine recognition promotes enteropathogenic *E. coli* and enterohemorrhagic *E. coli* host cell attachment. Microb. Pathog. *27*, 289–301.

Frankel, G., Candy, D.C., Everest, P., and Dougan, G. (1994). Characterization of the C-terminal domains of intimin-like proteins of enteropathogenic and enterohemorrhagic *Escherichia coli*, *Citrobacter freundii*, and *hafnia alvei*. Infect. Immun. *62*, 1835–1842.

Frankel, G., Lider, O., Hershkoviz, R., Mould, A.P., Kachalsky, S.G., Candy, D.C., Cahalon, L., Humphries, M.J., and Dougan, G. (1996a). The cell-binding domain of intimin from enteropathogenic *Escherichia coli* binds to beta1 integrins. J. Biol. Chem. *271*, 20359–20364.

Frankel, G., Phillips, A.D., Novakova, M., Field, H., Candy, D.C., Schauer, D.B., Douce, G., and Dougan, G. (1996b). Intimin from enteropathogenic *Escherichia coli* restores murine virulence to a *Citrobacter rodentium* eaeA mutant: induction of an immunoglobulin A response to intimin and EspB. Infect. Immun. *64*, 5315–5325.

Freeman, N.L., Zurawski, D.V., Chowrashi, P., Ayoob, J.C., Huang, L., Mittal, B., Sanger, J.M., and Sanger, J.W. (2000). Interaction of the enteropathogenic *Escherichia coli* protein, translocated intimin receptor (Tir), with focal adhesion proteins. Cell Motil. Cytoskel. 47, 307–318.

Frischknecht, F., Moreau, V., Röttger, S., Gonfloni, S., Reckmann, I., Superti-Furga, G., and Way, M. (1999). Actin-based motility of vaccinia virus mimics receptor tyrosine kinase signalling. Nature 401, 926–929.

Garber, J.J., Takeshima, F., Antón, I.M., Oyoshi, M.K., Lyubimova, A., Kapoor, A., Shibata, T., Chen, F., Alt, F.W., Geha, R.S., *et al.* (2012). Enteropathogenic *Escherichia coli* and vaccinia virus do not require the family of WASP-interacting proteins for pathogen-induced actin assembly. Infect. Immun. 80, 4071–4077.

Garmendia, J., Phillips, A.D., Carlier, M.-F., Chong, Y., Schuller, S., Marchès, O., Dahan, S., Oswald, E., Shaw, R.K., Knutton, S., *et al.* (2004). TccP is an enterohaemorrhagic *Escherichia coli* O157:H7 type III effector protein that couples Tir to the actin-cytoskeleton. Cell. Microbiol. 6, 1167–1183.

Garmendia, J., Ren, Z., Tennant, S., Midolli Viera, M.A., Chong, Y., Whale, A., Azzopardi, K., Dahan, S., Sircili, M.P., Franzolin, M.R., *et al.* (2005). Distribution of *tccP* in clinical enterohemorrhagic and enteropathogenic *Escherichia coli* isolates. J. Clin. Microbiol. 43, 5715–5720.

Garmendia, J., Carlier, M.-F., Egile, C., Didry, D., and Frankel, G. (2006). Characterization of TccP-mediated N-WASP activation during enterohaemorrhagic *Escherichia coli* infection. Cell. Microbiol. 8, 1444–1455.

Gauthier, A., and Finlay, B.B. (2003). Translocated intimin receptor and its chaperone interact with ATPase of the type III secretion apparatus of enteropathogenic *Escherichia coli*. J. Bacteriol. 185, 6747–6755.

Gauthier, A., de Grado, M., and Finlay, B.B. (2000). Mechanical fractionation reveals structural requirements for enteropathogenic *Escherichia coli* Tir insertion into host membranes. Infect. Immun. 68, 4344–4348.

Gill, R.K., Borthakur, A., Hodges, K., Turner, J.R., Clayburgh, D.R., Saksena, S., Zaheer, A., Ramaswamy, K., Hecht, G., and Dudeja, P.K. (2007). Mechanism underlying inhibition of intestinal apical Cl⁻/OH⁻ exchange following infection with enteropathogenic *E. coli*. J. Clin. Invest. 117, 428–437.

Girón, J.A., Ho, A.S., and Schoolnik, G.K. (1991). An inducible bundle-forming pilus of enteropathogenic *Escherichia coli*. Science 254, 710–713.

Girón, J.A., Torres, A.G., Freer, E., and Kaper, J.B. (2002). The flagella of enteropathogenic *Escherichia coli* mediate adherence to epithelial cells. Mol. Microbiol. 44, 361–379.

Goosney, D., DeVinney, R., Pfuetzner, R., Frey, E., Strynadka, N., and Finlay, B. (2000). Enteropathogenic *E. coli* translocated intimin receptor, Tir, interacts directly with alpha-actinin. Curr. Biol. 10, 735–738.

Goosney, D.L., DeVinney, R., and Finlay, B.B. (2001). Recruitment of cytoskeletal and signaling proteins to enteropathogenic and enterohemorrhagic *Escherichia coli* pedestals. Infect. Immun. 69, 3315–3322.

de Grado, M., Abe, A., Gauthier, A., Steele-Mortimer, O., DeVinney, R., and Finlay, B. (1999). Identification of the intimin-binding domain of Tir of enteropathogenic *Escherichia coli*. Cell. Microbiol. 1, 7–17.

de Groot, J.C., Schlüter, K., Carius, Y., Quedenau, C., Vingadassalom, D., Faix, J., Weiss, S.M., Reichelt, J., Standfuß-Gabisch, C., Lesser, C.F., *et al.* (2011). Structural basis for complex formation between human IRSp53 and the translocated intimin receptor Tir of enterohemorrhagic *E. coli*. Structure 19, 1294–1306.

Gruenheid, S., DeVinney, R., Bladt, F., Goosney, D., Gelkop, S., Gish, G.D., Pawson, T., and Finlay, B.B. (2001). Enteropathogenic *E. coli* Tir binds Nck to initiate actin pedestal formation in host cells. Nat. Cell Biol. 3, 856–859.

Grys, T.E., Siegel, M.B., Lathem, W.W., and Welch, R.A. (2005). The StcE protease contributes to intimate adherence of enterohemorrhagic *Escherichia coli* O157: H7 to host cells. Infect. Immun. 73, 1295–1303.

Guttman, J.A., Li, Y., Wickham, M.E., Deng, W., Vogl, A.W., and Finlay, B.B. (2006). Attaching and effacing pathogen-induced tight junction disruption *in vivo*. Cell. Microbiol. 8, 634–645.

Guttman, J.A., Samji, F.N., Li, Y., Deng, W., Lin, A., and Finlay, B.B. (2007). Aquaporins contribute to diarrhoea caused by attaching and effacing bacterial pathogens. Cell. Microbiol. 9, 131–141.

Guttman, J.A., En-Ju Lin, A., Li, Y., Bechberger, J., Naus, C.C., Vogl, A.W., and Finlay, B.B. (2010a). Gap junction hemichannels contribute to the generation of diarrhoea during infectious enteric disease. Gut 59, 218–226.

Guttman, J.A., Lin, A.E., Veiga, E., Cossart, P., and Finlay, B.B. (2010b). Role for CD2AP and other endocytosis-associated proteins in enteropathogenic *Escherichia coli* pedestal formation. Infect. Immun. 78, 3316–3322.

Hanajima-Ozawa, M., Matsuzawa, T., Fukui, A., Kamitani, S., Ohnishi, H., Abe, A., Horiguchi, Y., and Miyake, M. (2007). Enteropathogenic *Escherichia coli*, *Shigella flexneri*, and *Listeria monocytogenes* recruit a junctional protein, zonula occludens-1, to actin tails and pedestals. Infect. Immun. 75, 565–573.

Hardwidge, P.R., Rodriguez-Escudero, I., Goode, D., Donohoe, S., Eng, J., Goodlett, D.R., Aebersold, R., and Finlay, B.B. (2004). Proteomic analysis of the intestinal epithelial cell response to enteropathogenic *Escherichia coli*. J. Biol. Chem. 279, 20127–20136.

Hayward, R.D., Leong, J.M., Koronakis, V., and Campellone, K.G. (2006). Exploiting pathogenic *Escherichia coli* to model transmembrane receptor signalling. Nat. Rev. Micro. 4, 358–370.

Hayward, R.D., Hume, P.J., Humphreys, D., Phillips, N., Smith, K., and Koronakis, V. (2009). Clustering transfers the translocated *Escherichia coli* receptor into lipid rafts to stimulate reversible activation of c-Fyn. Cell. Microbiol. 11, 433–441.

Hecht, G. (2001). Microbes and microbial toxins: paradigms for microbial–mucosal interactions. VII. Enteropathogenic *Escherichia coli*: physiological alterations from an extracellular position. Am. J. Physiol. Gastrointest. Liver Physiol. *281*, G1–G7.

Hecht, G., and Koutsouris, A. (1999). Enteropathogenic *E. coli* attenuates secretagogue-induced net intestinal ion transport but not Cl- secretion. Am. J. Physiol. *276*, G781–G788.

Hecht, G., Hodges, K., Gill, R.K., Kear, F., Tyagi, S., Malakooti, J., Ramaswamy, K., and Dudeja, P.K. (2004). Differential regulation of Na+/H+ exchange isoform activities by enteropathogenic *E. coli* in human intestinal epithelial cells. Am. J. Physiol. Gastrointest. Liver Physiol. *287*, G370–G378.

Higgins, L.M., Frankel, G., Connerton, I., Goncalves, N.S., Dougan, G., and MacDonald, T.T. (1999). Role of bacterial intimin in colonic hyperplasia and inflammation. Science *285*, 588–591.

Howe, K.L., Reardon, C., Wang, A., Nazli, A., and McKay, D.M. (2005). Transforming growth factor-beta regulation of epithelial tight junction proteins enhances barrier function and blocks enterohemorrhagic *Escherichia coli* O157:H7-induced increased permeability. Am. J. Pathol. *167*, 1587–1597.

Huang, L., Mittal, B., Sanger, J.W., and Sanger, J.M. (2002). Host focal adhesion protein domains that bind to the translocated intimin receptor (Tir) of enteropathogenic *Escherichia coli* (EPEC). Cell Motil. Cytoskel. *52*, 255–265.

Huett, A., Leong, J.M., Podolsky, D.K., and Xavier, R.J. (2009). The cytoskeletal scaffold Shank3 is recruited to pathogen-induced actin rearrangements. Exp. Cell Res. *315*, 2001–2011.

Humphries, R.M., Donnenberg, M.S., Strecker, J., Kitova, E., Klassen, J.S., Cui, L., Griener, T.P., Mulvey, G.L., and Armstrong, G.D. (2009). From alpha to beta: identification of amino acids required for the N-acetyllactosamine-specific lectin-like activity of bundlin. Mol. Microbiol. *72*, 859–868.

Humphries, R.M., Griener, T.P., Vogt, S.L., Mulvey, G.L., Raivio, T., Donnenberg, M.S., Kitov, P.I., Surette, M., and Armstrong, G.D. (2010). N-acetyllactosamine-induced retraction of bundle-forming pili regulates virulence-associated gene expression in enteropathogenic *Escherichia coli*. Mol. Microbiol. *76*, 1111–1126.

Hyland, R.M., Sun, J., Griener, T.P., Mulvey, G.L., Klassen, J.S., Donnenberg, M.S., and Armstrong, G.D. (2008). The bundlin pilin protein of enteropathogenic *Escherichia coli* is an N-acetyllactosamine-specific lectin. Cell. Microbiol. *10*, 177–187.

Iizumi, Y., Sagara, H., Kabe, Y., Azuma, M., Kume, K., Ogawa, M., Nagai, T., Gillespie, P.G., Sasakawa, C., and Handa, H. (2007). The enteropathogenic *E. coli* effector EspB facilitates microvillus effacing and antiphagocytosis by inhibiting myosin function. Cell Host Microbe *2*, 383–392.

Kalman, D., Weiner, O.D., Goosney, D.L., Sedat, J.W., Finlay, B.B., Abo, A., and Bishop, J.M. (1999). Enteropathogenic *E. coli* acts through WASP and Arp2/3 complex to form actin pedestals. Nat. Cell Biol. *1*, 389–391.

Kenny, B. (1999). Phosphorylation of tyrosine 474 of the enteropathogenic *Escherichia coli* (EPEC) Tir receptor molecule is essential for actin nucleating activity and is preceded by additional host modifications. Mol. Microbiol. *31*, 1229–1241.

Kenny, B., and Finlay, B.B. (1995). Protein secretion by enteropathogenic *Escherichia coli* is essential for transducing signals to epithelial cells. Proc. Natl. Acad. Sci. U.S.A. *92*, 7991–7995.

Kenny, B., and Finlay, B.B. (1997). Intimin-dependent binding of enteropathogenic *Escherichia coli* to host cells triggers novel signaling events, including tyrosine phosphorylation of phospholipase C-gamma1. Infect. Immun. *65*, 2528–2536.

Kenny, B., Lai, L.C., Finlay, B.B., and Donnenberg, M.S. (1996). EspA, a protein secreted by enteropathogenic *Escherichia coli*, is required to induce signals in epithelial cells. Mol. Microbiol. *20*, 313–323.

Kenny, B., DeVinney, R., Stein, M., Reinscheid, D., Frey, E., and Finlay, B. (1997). Enteropathogenic *E. coli* (EPEC) transfers its receptor for intimate adherence into mammalian cells. Cell *91*, 511–520.

Kirkbride, K.C., Sung, B.H., Sinha, S., and Weaver, A.M. (2011). Cortactin: a multifunctional regulator of cellular invasiveness. Cell. Adh. Migr. *5*, 187–198.

Knutton, S., Lloyd, D.R., and McNeish, A.S. (1987). Adhesion of enteropathogenic *Escherichia coli* to human intestinal enterocytes and cultured human intestinal mucosa. Infect. Immun. *55*, 69–77.

Knutton, S., Baldwin, T., Williams, P.H., and McNeish, A.S. (1989). Actin accumulation at sites of bacterial adhesion to tissue culture cells: basis of a new diagnostic test for enteropathogenic and enterohemorrhagic *Escherichia coli*. Infect. Immun. *57*, 1290–1298.

Knutton, S., Rosenshine, I., Pallen, M.J., Nisan, I., Neves, B.C., Bain, C., Wolff, C., Dougan, G., and Frankel, G. (1998). A novel EspA-associated surface organelle of enteropathogenic *Escherichia coli* involved in protein translocation into epithelial cells. EMBO J. *17*, 2166–2176.

Lai, Y., Riley, K., Cai, A., Leong, J.M., and Herman, I.M. (2011). Calpain mediates epithelial cell microvillar effacement by enterohemorrhagic *Escherichia coli*. Front. Microbiol. *2*, 1–9.

Larson, M.A., Wei, S.H., Weber, A., Mack, D.R., and McDonald, T.L. (2003). Human serum amyloid A3 peptide enhances intestinal MUC3 expression and inhibits EPEC adherence. Biochem. Biophys. Res. Comm. *300*, 531–540.

Lathem, W.W., Grys, T.E., Witowski, S.E., Torres, A.G., Kaper, J.B., Tarr, P.I., and Welch, R.A. (2002). StcE, a metalloprotease secreted by *Escherichia coli* O157:H7, specifically cleaves C1 esterase inhibitor. Mol. Microbiol. *45*, 277–288.

Lathem, W.W., Bergsbaken, T., Witowski, S.E., Perna, N.T., and Welch, R.A. (2003). Acquisition of stcE, a C1 esterase inhibitor-specific metalloprotease, during the evolution of *Escherichia coli* O157:H7. J. Infect. Dis. *187*, 1907–1914.

Lathem, W.W., Bergsbaken, T., and Welch, R.A. (2004). Potentiation of C1 esterase inhibitor by StcE, a metalloprotease secreted by *Escherichia coli* O157:H7. J. Exp. Med. *199*, 1077–1087.

Law, H.T., Bonazzi, M., Jackson, J., Cossart, P., and Guttman, J.A. (2012). Nexilin is a dynamic component of *Listeria monocytogenes* and enteropathogenic *Escherichia coli* actin-rich structures. Cell. Microbiol. *14*, 1097–1108.

Levine, M.M., Nataro, J.P., Karch, H., Baldini, M.M., Kaper, J.B., Black, R.E., Clements, M.L., and O'Brien, A.D. (1985). The diarrhoeal response of humans to some classic serotypes of enteropathogenic *Escherichia coli* is dependent on a plasmid encoding an enteroadhesiveness factor. J. Infect. Dis. *152*, 550–559.

Lin, A.E., and Guttman, J.A. (2012). The *Escherichia coli* adherence factor plasmid of enteropathogenic *Escherichia coli* causes a global decrease in ubiquitylated host cell proteins by decreasing ubiquitin E1 enzyme expression through host aspartyl proteases. Int. J. Biochem. Cell Biol. *44*, 2223–2232.

Lin, A.E., Benmerah, A., and Guttman, J.A. (2011). Eps15 and epsin1 are crucial for enteropathogenic *Escherichia coli* pedestal formation despite the absence of adaptor protein 2. J. Infect. Dis. *204*, 695–703.

Lindén, S.K., Florin, T.H.J., and McGuckin, M.A. (2008). Mucin dynamics in intestinal bacterial infection. PLoS ONE *3*, e3952.

Liu, H., Magoun, L., and Leong, J.M. (1999a). beta1-chain integrins are not essential for intimin-mediated host cell attachment and enteropathogenic *Escherichia coli*-induced actin condensation. Infect. Immun. *67*, 2045–2049.

Liu, H., Magoun, L., Luperchio, S., Schauer, D.B., and Leong, J.M. (1999b). The Tir-binding region of enterohaemorrhagic *Escherichia coli* intimin is sufficient to trigger actin condensation after bacterial-induced host cell signalling. Mol. Microbiol. *34*, 67–81.

Lommel, S., Benesch, S., Rohde, M., Wehland, J., and Rottner, K. (2004). Enterohaemorrhagic and enteropathogenic *Escherichia coli* use different mechanisms for actin pedestal formation that converge on N-WASP. Cell. Microbiol. *6*, 243–254.

McGuckin, M.A.M., Lindén, S.K.S., Sutton, P.P., and Florin, T.H.T. (2011). Mucin dynamics and enteric pathogens. Nat. Rev. Micro. *9*, 265–278.

Mack, D.R., Michail, S., Wei, S., McDougall, L., and Hollingsworth, M.A. (1999). Probiotics inhibit enteropathogenic *E. coli* adherence in vitro by inducing intestinal mucin gene expression. Am. J. Physiol. *276*, G941–G950.

Mallick, E.M., Brady, M.J., Luperchio, S.A., Vanguri, V.K., Magoun, L., Liu, H., Sheppard, B.J., Mukherjee, J., Donohue-Rolfe, A., Tzipori, S., et al. (2012). Allele- and Tir-independent functions of intimin in diverse animal infection models. Front. Microbiol. *3*, 1–16.

Martinez-Quiles, N., Rohatgi, R., Antón, I.M., Medina, M., Saville, S.P., Miki, H., Yamaguchi, H., Takenawa, T., Hartwig, J.H., Geha, R.S., et al. (2001). WIP regulates N-WASP-mediated actin polymerization and filopodium formation. Nat. Cell Biol. *3*, 484–491.

Mattoo, S., Alto, N.M., and Dixon, J.E. (2008). Subversion of myosin function by *E. coli*. Develop. Cell *14*, 8–10.

Meinild, A., Klaerke, D.A., Loo, D.D., Wright, E.M., and Zeuthen, T. (1998). The human Na+-glucose cotransporter is a molecular water pump. J. Physiol. (Lond.) *508(Pt 1)*, 15–21.

Millard, T.H., Bompard, G., Heung, M.Y., Dafforn, T.R., Scott, D.J., Machesky, L.M., and Fütterer, K. (2005). Structural basis of filopodia formation induced by the IRSp53/MIM homology domain of human IRSp53. EMBO J. *24*, 240–250.

Mills, E., Baruch, K., Charpentier, X., Kobi, S., and Rosenshine, I. (2008). Real-Time Analysis of Effector Translocation by the Type III Secretion System of Enteropathogenic *Escherichia coli*. Cell Host Microbe *3*, 104–113.

Miyahara, A., Nakanishi, N., Ooka, T., Hayashi, T., Sugimoto, N., and Tobe, T. (2009). Enterohemorrhagic *Escherichia coli* effector EspL2 induces actin microfilament aggregation through annexin 2 activation. Cell. Microbiol. *11*, 337–350.

Moon, H.W., Whipp, S.C., Argenzio, R.A., Levine, M.M., and Giannella, R.A. (1983). Attaching and effacing activities of rabbit and human enteropathogenic *Escherichia coli* in pig and rabbit intestines. Infect. Immun. *41*, 1340–1351.

Mousnier, A., Whale, A.D., Schuller, S., Leong, J.M., Phillips, A.D., and Frankel, G. (2008). Cortactin recruitment by enterohemorrhagic *Escherichia coli* O157:H7 during infection *in vitro* and *ex vivo*. Infect. Immun. *76*, 4669–4676.

Mundy, R., Pickard, D., Wilson, R.K., Simmons, C.P., Dougan, G., and Frankel, G. (2003). Identification of a novel type IV pilus gene cluster required for gastrointestinal colonization of *Citrobacter rodentium*. Mol. Microbiol. *48*, 795–809.

Munera, D., Martinez, E., Varyukhina, S., Mahajan, A., Ayala-Sanmartin, J., and Frankel, G. (2012). Recruitment and membrane interactions of host cell proteins during attachment of enteropathogenic and enterohaemorrhagic *Escherichia coli*. Biochem. J. *445*, 383–392.

Muza-Moons, M.M., Schneeberger, E.E., and Hecht, G.A. (2004). Enteropathogenic *Escherichia coli* infection leads to appearance of aberrant tight junctions strands in the lateral membrane of intestinal epithelial cells. Cell. Microbiol. *6*, 783–793.

Nataro, J.P., Scaletsky, I.C., Kaper, J.B., Levine, M.M., and Trabulsi, L.R. (1985). Plasmid-mediated factors conferring diffuse and localized adherence of enteropathogenic *Escherichia coli*. Infect. Immun. *48*, 378–383.

Niebuhr, K., Ebel, F., Frank, R., Reinhard, M., Domann, E., Carl, U.D., Walter, U., Gertler, F.B., Wehland, J., and Chakraborty, T. (1997). A novel proline-rich motif present in ActA of *Listeria monocytogenes* and cytoskeletal proteins is the ligand for the EVH1 domain, a protein module present in the Ena/VASP family. EMBO J. *16*, 5433–5444.

Nieto-Pelegrin, E., and Martinez-Quiles, N. (2009). Distinct phosphorylation requirements regulate cortactin activation by TirEPEC and its binding to N-WASP. Cell Commun. Signal. *7*, 11.

O'Hara, J.R., Skinn, A.C., MacNaughton, W.K., Sherman, P.M., and Sharkey, K.A. (2006). Consequences of *Citrobacter rodentium* infection on enteroendocrine cells and the enteric nervous system in the mouse colon. Cell. Microbiol. *8*, 646–660.

Patel, A., Cummings, N., Batchelor, M., Hill, P.J., Dubois, T., Mellits, K.H., Frankel, G., Ian Connerton (2006). Host protein interactions with enteropathogenic *Escherichia coli* (EPEC): 14–3–3tau binds Tir and has a role in EPEC-induced actin polymerization. Cell. Microbiol. *8*, 55–71.

Peralta-Ramirez, J., Hernandez, J.M., Manning-Cela, R., Luna-Munoz, J., Garcia-Tovar, C., Nougayrede, J.P., Oswald, E., and Navarro-Garcia, F. (2008). EspF Interacts with Nucleation-Promoting Factors To Recruit Junctional Proteins into Pedestals for Pedestal Maturation and Disruption of Paracellular Permeability. Infect. Immun. *76*, 3854–3868.

Phillips, N., Hayward, R.D., and Koronakis, V. (2004). Phosphorylation of the enteropathogenic *E. coli* receptor by the Src-family kinase c-Fyn triggers actin pedestal formation. Nat. Cell. Biol. *6*, 618–625.

Philpott, D.J., McKay, D.M., Mak, W., Perdue, M.H., and Sherman, P.M. (1998). Signal transduction pathways involved in enterohemorrhagic *Escherichia coli*-induced alterations in T84 epithelial permeability. Infect. Immun. *66*, 1680–1687.

Potter, D.A., Srirangam, A., Fiacco, K.A., Brocks, D., Hawes, J., Herndon, C., Maki, M., Acheson, D., and Herman, I.M. (2003). Calpain regulates enterocyte brush border actin assembly and pathogenic *Escherichia coli*-mediated effacement. J. Biol. Chem. *278*, 30403–30412.

Ritchie, J., Thorpe, C., Rogers, A.B., and Waldor, M. (2003). Critical roles for stx(2), eae, and tir in enterohemorrhagic *Escherichia coli*-induced diarrhoea and intestinal inflammation in infant rabbits. Infect. Immun. *71*, 7129–7139.

Ruetz, T., Cornick, S., and Guttman, J.A. (2011). The spectrin cytoskeleton is crucial for adherent and invasive bacterial pathogenesis. PLoS ONE *6*, e19940.

Ruetz, T.J., Lin, A.E., and Guttman, J.A. (2012a). *Shigella flexneri* utilize the spectrin cytoskeleton during invasion and comet tail generation. BMC Microbiol. *12*, 36.

Ruetz, T.J., Lin, A.E.-J., and Guttman, J.A. (2012b). Enterohaemorrhagic *Escherichia coli* requires the spectrin cytoskeleton for efficient attachment and pedestal formation on host cells. Microb. Pathog. *52*, 149–156.

Ruetz, T.J., Vogl, A.W., and Guttman, J.A. (2012c). Detailed examination of cytoskeletal networks within enteropathogenic *Escherichia coli* pedestals. Anat. Rec. (Hoboken) *295*, 201–207.

Saldana, Z., Erdem, A.L., Schuller, S., Okeke, I.N., Lucas, M., Sivananthan, A., Phillips, A.D., Kaper, J.B., Puente, J.L., and Girón, J.A. (2009). The *Escherichia coli* Common Pilus and the Bundle-Forming Pilus Act in Concert during the Formation of Localized Adherence by Enteropathogenic *E. coli*. J. Bacteriol. *191*, 3451–3461.

Sallee, N.A., Rivera, G.M., Dueber, J.E., Vasilescu, D., Mullins, R.D., Mayer, B.J., and Lim, W.A. (2008). The pathogen protein EspFU hijacks actin polymerization using mimicry and multivalency. Nature *454*, 1005–1008.

Sanger, J.M., Chang, R., Ashton, F., Kaper, J.B., and Sanger, J.W. (1996). Novel form of actin-based motility transports bacteria on the surfaces of infected cells. Cell Motil. Cytoskel. *34*, 279–287.

Saotome, I., Curto, M., and McClatchey, A.I. (2004). Ezrin is essential for epithelial organization and villus morphogenesis in the developing intestine. Develop. Cell *6*, 855–864.

Sason, H., Milgrom, M., Weiss, A.M., Melamed-Book, N., Balla, T., Grinstein, S., Backert, S., Rosenshine, I., and Aroeti, B. (2009). Enteropathogenic *Escherichia coli* subverts phosphatidylinositol 4,5-bisphosphate and phosphatidylinositol 3,4,5-trisphosphate upon epithelial cell infection. Mol. Biol. Cell *20*, 544–555.

Scaplehorn, N., Holmström, A., Moreau, V., Frischknecht, F., Reckmann, I., and Way, M. (2002). Grb2 and Nck Act Cooperatively to Promote Actin-Based Motility of Vaccinia Virus. Curr. Biol. *12*, 740–745.

Sekiya, K., Ohishi, M., Ogino, T., Tamano, K., Sasakawa, C., and Abe, A. (2001). Supermolecular structure of the enteropathogenic *Escherichia coli* type III secretion system and its direct interaction with the EspA-sheath-like structure. Proc. Natl. Acad. Sci. U.S.A. *98*, 11638–11643.

Shaner, N.C., Sanger, J.W., and Sanger, J.M. (2005). Actin and alpha-actinin dynamics in the adhesion and motility of EPEC and EHEC on host cells. Cell Motil. Cytoskel. *60*, 104–120.

Shaw, R.K., Daniell, S., Ebel, F., Frankel, G., and Knutton, S. (2001). EspA filament-mediated protein translocation into red blood cells. Cell. Microbiol. *3*, 213–222.

Shaw, R.K., Cleary, J., Murphy, M.S., Frankel, G., and Knutton, S. (2005). Interaction of enteropathogenic *Escherichia coli* with human intestinal mucosa: role of effector proteins in brush border remodeling and formation of attaching and effacing lesions. Infect. Immun. *73*, 1243–1251.

Simonovic, I., Rosenberg, J., Koutsouris, A., and Hecht, G. (2000). Enteropathogenic *Escherichia coli* dephosphorylates and dissociates occludin from intestinal epithelial tight junctions. Cell. Microbiol. *2*, 305–315.

Sinclair, J.F., and O'Brien, A.D. (2002). Cell surface-localized nucleolin is a eukaryotic receptor for the adhesin intimin-gamma of enterohemorrhagic *Escherichia coli* O157:H7. J. Biol. Chem. *277*, 2876–2885.

Sinclair, J.F., and O'Brien, A.D. (2004). Intimin types alpha, beta, and gamma bind to nucleolin with equivalent affinity but lower avidity than to the translocated intimin receptor. J. Biol. Chem. *279*, 33751–33758.

Sinclair, J.F., Dean-Nystrom, E.A., and O'Brien, A.D. (2006). The established intimin receptor Tir and the putative eucaryotic intimin receptors nucleolin and beta1 integrin localize at or near the site of enterohemorrhagic *Escherichia coli* O157:H7 adherence to enterocytes *in vivo*. Infect. Immun. *74*, 1255–1265.

Smith, C.J., Kaper, J.B., and Mack, D.R. (1995). Intestinal mucin inhibits adhesion of human enteropathogenic *Escherichia coli* to HEp-2 cells. J. Pediatr. Gastroenterol. Nutr. *21*, 269–276.

Smith, K., Humphreys, D., Hume, P.J., and Koronakis, V. (2010). Enteropathogenic *Escherichia coli* recruits the cellular inositol phosphatase SHIP2 to regulate actin-pedestal formation. Cell Host Microbe 7, 13–24.

Spitz, J., Yuhan, R., Koutsouris, A., Blatt, C., Alverdy, J., and Hecht, G. (1995). Enteropathogenic *Escherichia coli* adherence to intestinal epithelial monolayers diminishes barrier function. Am. J. Physiol. *268*, G374–G379.

Staley, T.E., Jones, E.W., and Corley, L.D. (1969). Attachment and penetration of *Escherichia coli* into intestinal epithelium of the ileum in newborn pigs. Am. J. Pathol. *56*, 371–392.

Stevenson, B.R., Siliciano, J.D., Mooseker, M.S., and Goodenough, D.A. (1986). Identification of ZO-1: a high molecular weight polypeptide associated with the tight junction (zonula occludens) in a variety of epithelia. J. Cell Biol. *103*, 755–766.

Stossel, T.P., Condeelis, J., Cooley, L., Hartwig, J.H., Noegel, A., Schleicher, M., and Shapiro, S.S. (2001). Filamins as integrators of cell mechanics and signalling. Nat. Rev. Mol. Cell Biol. *2*, 138–145.

Suzuki, T., and Sasakawa, C. (2001). Molecular basis of the intracellular spreading of *Shigella*. Infect. Immun. *69*, 5959–5966.

Swimm, A.I., and Kalman, D. (2008). Cytosolic extract induces Tir translocation and pedestals in EPEC-infected red blood cells. PLoS Pathog. *4*, e4.

Swimm, A., Bommarius, B., Li, Y., Cheng, D., Reeves, P., Sherman, M., Veach, D., Bornmann, W., and Kalman, D. (2004). Enteropathogenic *Escherichia coli* use redundant tyrosine kinases to form actin pedestals. Mol. Biol. Cell *15*, 3520–3529.

Szabady, R.L., Yanta, J.H., Halladin, D.K., Schofield, M.J., and Welch, R.A. (2011). TagA is a secreted protease of *Vibrio cholerae* that specifically cleaves mucin glycoproteins. Microbiology *157*, 516–525.

Tacket, C.O., Sztein, M.B., Losonsky, G., Abe, A., Finlay, B.B., McNamara, B.P., Fantry, G.T., James, S.P., Nataro, J.P., Levine, M.M., *et al.* (2000). Role of EspB in experimental human enteropathogenic *Escherichia coli* infection. Infect. Immun. *68*, 3689–3695.

Thomas, N.A., Deng, W., Baker, N., Puente, J., and Finlay, B.B. (2007). Hierarchical delivery of an essential host colonization factor in enteropathogenic *Escherichia coli*. J. Biol. Chem. *282*, 29634–29645.

Tyska, M.J., Mackey, A.T., Huang, J.-D., Copeland, N.G., Jenkins, N.A., and Mooseker, M.S. (2005). Myosin-1a

is critical for normal brush border structure and composition. Mol. Biol. Cell *16*, 2443–2457.

Unsworth, K.E., Mazurkiewicz, P., Senf, F., Zettl, M., McNiven, M., Way, M., and Holden, D.W. (2007). Dynamin is required for F-actin assembly and pedestal formation by enteropathogenic *Escherichia coli* (EPEC). Cell. Microbiol. 9, 438–449.

Uruno, T., Liu, J., Zhang, P., Fan Yx, Egile, C., Li, R., Mueller, S.C., and Zhan, X. (2001). Activation of Arp2/3 complex-mediated actin polymerization by cortactin. Nat. Cell Biol. 3, 259–266.

Vallance, B.A., and Finlay, B.B. (2000). Exploitation of host cells by enteropathogenic *Escherichia coli*. Proc. Natl. Acad. Sci. U.S.A. 97, 8799–8806.

Veiga, E., Guttman, J.A., Bonazzi, M., Boucrot, E., Toledo-Arana, A., Lin, A.E., Enninga, J., Pizarro-Cerda, J., Finlay, B.B., Kirchhausen, T., *et al.* (2007). Invasive and adherent bacterial pathogens co-opt host clathrin for infection. Cell Host Microbe *2*, 340–351.

Vingadassalom, D., Kazlauskas, A., Skehan, B., Cheng, H.-C., Magoun, L., Robbins, D., Rosen, M.K., Saksela, K., and Leong, J.M. (2009). Insulin receptor tyrosine kinase substrate links the *E. coli* O157:H7 actin assembly effectors Tir and EspF(U) during pedestal formation. Proc. Natl. Acad. Sci. *106*, 6754–6759.

Weiss, S.M., Ladwein, M., Schmidt, D., Ehinger, J., Lommel, S., Städing, K., Beutling, U., Disanza, A., Frank, R., Jänsch, L., *et al.* (2009). IRSp53 links the enterohemorrhagic *E. coli* effectors Tir and EspFU for actin pedestal formation. Cell Host Microbe *5*, 244–258.

Xicohtencatl-Cortes, J., Monteiro-Neto, V., Ledesma, M.A., Jordan, D.M., Francetic, O., Kaper, J.B., Puente, J.L., and Girón, J.A. (2007). Intestinal adherence associated with type IV pili of enterohemorrhagic *Escherichia coli* O157:H7. J. Clin. Invest. *117*, 3519–3529.

Yu, A.C.Y., Worrall, L.J., and Strynadka, N.C.J. (2012). Structural insight into the bacterial mucinase StcE essential to adhesion and immune evasion during enterohemorrhagic *E. coli* infection. Structure *20*, 707–717.

Zarivach, R., Vuckovic, M., Deng, W., Finlay, B.B., and Strynadka, N.C.J. (2007). Structural analysis of a prototypical ATPase from the type III secretion system. Nat. Struct. Mol. Biol. *14*, 131–137.

Zobiack, N., Rescher, U., Laarmann, S., Michgehl, S., Schmidt, M.A., and Gerke, V. (2002). Cell-surface attachment of pedestal-forming enteropathogenic *E. coli* induces a clustering of raft components and a recruitment of annexin 2. J. Cell Sci. *115*, 91–98.

Colonization Factor Antigens of Enterotoxigenic *Escherichia coli*

11

Felipe Del Canto and Alfredo G. Torres

Abstract

Enterotoxigenic *Escherichia coli* (ETEC) cause toxin-mediated diarrhoea, which requires that ETEC attaches to epithelial cells using a diverse repertoire of adhesins generically named as colonization factor antigens (CFs). Currently, 22 CFs variants have been identified, displaying different types of structures (fimbriae, fibres and non-fimbrial), which are assembled and displayed at the bacterial surface by three different mechanisms. To date, there is no an effective therapy to prevent ETEC-caused diarrhoea and investigators have considered CFs as suitable potential basis of vaccine formulations. In this chapter, we describe the CFs carried by human ETEC strains, addressing structural and functional features as well as their distribution in clinical isolates as determined by epidemiological studies. Furthermore, three non-classical/non-fimbrial ETEC adhesins are described. The information presented here is an overview about human ETEC adherence mechanisms and about how basic and epidemiological research has advanced to obtain crucial data for the design of a therapy to effectively prevent ETEC-caused diarrhoea.

Introduction

Enterotoxigenic *Escherichia coli* (ETEC) strains are one of the leading causes of diarrhoea in developing countries, mainly in children less than 5 years of age (Sanchez and Holmgren, 2005). ETEC strains are also the most frequent cause of travellers' diarrhoea, illness that affects people travelling from developed countries to endemic regions (Steffen *et al.*, 2005). It has been estimated that ETEC causes about 1 billion diarrhoeal episodes and approximately 300,000–400,000 deaths annually, due to serious dehydration in children (Sanchez and Holmgren, 2005).

ETEC is of one of six categories of classical diarrhoeagenic *E. coli* (Nataro and Kaper, 1998) and overall, this category is the most frequently associated with diarrhoea worldwide (Qadri *et al.*, 2005). An ETEC strain is defined as an *E. coli* strain able to produce heat-labile enterotoxin (LT) and/or heat-stable enterotoxin (ST) (Nataro and Kaper, 1998). According to this criterion, a highly diverse set of *E. coli* strains typed according to their variability in structural antigenic determinants (the somatic O antigen, the flagellar H antigen and the capsular K antigen), are included in this category (Wolf, 1997).

ETEC strains colonize the small bowel intestinal epithelium and secrete LT and/or ST, causing an increase of intracellular levels of the second messengers' cyclic nucleotides, cAMP and cGMP, respectively (Turner *et al.*, 2006a). Subsequently, these molecules induce opening of integral membrane channel proteins allowing massive efflux of water and electrolytes, resulting in watery diarrhoea (Guttman and Finlay, 2008). ETEC adherence to the gut epithelium has been established as a key step to develop full illness (Dorsey *et al.*, 2006) and; therefore, ETEC strains carry a diverse repertoire of adhesins to attach to cells, including more than twenty generically named structures, known as colonization factors (CFs) [also named colonization factor antigens (CFAs) or coli surface-associated antigens (CS)] (Gaastra and Svennerholm, 1996).

Colonization factor antigens

Classification

CFs are the classical adhesins carried by human ETECs and whose classification started with the discovery of the first representative, the colonization factor antigen I (CFA/I) carried by ETEC strain H10407 (Evans et al., 1975). This strain was isolated from an adult suffering from severe cholera like-diarrhoea in Dhaka, Bangladesh (Evans and Evans, 1973), and it has been the prototype for studies deciphering the molecular mechanisms of ETEC pathogenesis. ETEC H10407 displayed the capacity to colonize and proliferate within the rabbit gut whereas a mutant strain derived from successive laboratory passages (ETEC H10407-P) was not. However, cell free extracts obtained from ETEC H10407-P cultures retained the same toxic effect as the wild-type on rabbit ligated loop, which made the investigators concluded that the mutant had lost a key colonization determinant. A serum raised against ETEC H10407 and adsorbed with ETEC H10407-P was able to significantly inhibit colonization of the rabbit gut. Further, pilus-like structures were observed on the ETEC H10407 bacterial surface but absent in ETEC H10407-P. Therefore, it was concluded that these pili were associated with the colonization ability of ETEC H10407 (Evans et al., 1975).

Another CF in LT-producing E. coli strains belonging to serogroups O6 and O8 was reported three years later (Evans and Evans, 1978), and it was designated as CFA/II because it showed some functional similarities to CFA/I. When evaluated, CFA/II also showed the capacity to colonize the rabbit gut and agglutinate bovine erythrocytes, but unlike CFA/I, it was unable to agglutinate human group A erythrocytes (Evans and Evans, 1978). However, years later, CFA/II was found to actually include more than one structure, and the first clues were obtained by performing immuno-diffusion assays with two different sera (Cravioto et al., 1982). These observations were confirmed demonstrating that three structures named as coli surface-associated antigens (CS): CS1, CS2 and CS3, were part of the formerly identified CFA/II (Smyth, 1982; Mullany et al., 1983).

CFA/III was discovered as the third adherence determinant of ETEC, particularly ETEC strain 260-1 (Honda et al., 1984). This strain was examined for its surface hydrophobicity, a feature that had been previously related to the presence of pilus-like structures. This strain did not agglutinate human nor bovine red blood cells, but it adhered to rabbit gut epithelium (Honda et al., 1984). Electron microscopy analysis visualized pilus-like structures at the bacterial surface, which were not produced by an ETEC 260-1 mutant strain that was found later to lack this adherence determinant.

A similar situation to that mentioned for CFA/II occurred with CFA/IV. It was first described as a single adherence determinant (Thomas et al., 1982), but three different structures were identified later within the formerly described CFA/IV. Therefore, the fimbrial-type CSs (CS4 and CS5), and the afimbrial adhesin CS6 were identified (Thomas et al., 1985). Given that the classification was becoming complicated, as novel adhesins were further identified, a uniform nomenclature was proposed, naming each variant as 'CS' followed by a number that indicates order of discovery (Gaastra and Svennerholm, 1996). This classification is currently used for all variants except for CFA/I. Up to date, there are 22 different colonization factors described: CFA/I, CS1-CS8, CS10–15, and CS17-CS23. Strangely, no experimental evidence is available for CS9 and CS16. A list of currently known CFs and some of their main characteristics can be found in Table 11.1.

Structure

In general, CFs are composed of one or more structural protein subunits arranged in diverse conformations and giving rise to fimbriae, fibres and/or afimbrial structures (Gaastra and Svennerholm, 1996). Usually, there is one main subunit which is forming the body of the CF and that is more abundant than the rest. The protein is known as the 'major structural subunit' and therefore, the rest are considered as 'minor subunits' (Jansson et al., 2006). Some minor structural subunits are exposed at the end of the structure, so they are designated as the 'tip subunit' (Baker et al., 2009). Adhesin activity has been attributed to different

Table 11.1 ETEC adhesins

Colonization factors

Original nomenclature	Current nomenclature	Type	Diameter	Assays evaluating activity	References
CFA/I	CFA/I	Fimbrial	7 nm	Haemagglutination (BE, HE) – Cell adhesion assays (HEN, HCL) – Colonization of small intestine in rabbits	Evans et al. (1975), Knutton et al. (1985), Sakerallis et al. (1999), Baker et al. (2009)
CFA/II	CS1	Fimbrial	7 nm	Haemagglutination (BE)	Sakerallis et al. (1999)
CFA/II	CS2	Fimbrial	7 nm	Cell adhesion assays (HEN)	Honda et al. (1989)
CFA/II	CS3	Fibrillae	2–3 nm	Cell adhesion assays (REN – HEN – HCL)	Helander et al. (1997)
CFA/IV	CS4	Fimbrial	6 nm	Haemagglutination (BE)	Thomas et al. (1985)
CFA/IV	CS5	Helical	5 nm	Haemagglutination (BE)	Thomas et al. (1985)
CFA/IV	CS6	Non fimbrial	NR	Cell adhesion assays (REN – HEN – HCL)	Helander et al. (1997)
CS7	CS7	Helical	3–6 nm	Colonization of small intestine in rabbits	Hibberd et al. (1990), Helander et al. (1997)
CFA/III	CS8	Fimbrial	7 nm	Cell adhesion assays (HEN – HCL)	Darfeuille-Michaud et al. (1990)
2230	CS10	Non fimbrial	NR	Cell adhesion assays (HEN)	Darfeuille-Michaud et al. (1986)
PCFO148	CS11	Fibrillae	3 nm	NR	Knutton et al. (1987)
PCFO159	CS12	Fimbrial	7 nm	Cell adhesion assays (REN – HEN – HCL)	Helander et al. (1997)
PCFO9	CS13	Fibrillae	NR	NR	Heuzenroeder et al. (1990)
PCFO166	CS14	Fimbrial	7 nm	Haemagglutination (BE, HE)	McConnel et al. (1989)
8786	CS15	Non fimbrial	NR	Cell adhesion assays (HCL)	Aubel et al. (1991)
CS17	CS17	Fimbrial	7 nm	Haemagglutination (BE)	McConnel et al. (1990)
CS18	CS18	Fimbrial	7 nm	Cell adhesion assays (HCL)	Viboud et al. (1993)
CS19	CS19	Fimbrial	7 nm	Cell adhesion assays (HCL)	Grewal et al. (1997)
CS10	CS20	Fimbrial	7 nm	Cell adhesion assays (HCL)	Valvatne et al. (1996)
Longus	CS21	Fimbrial	7 nm	Cell adhesion assays (HCL)	Mazariego-Espinoza et al. (2010)
CS22	CS22	Fibrillae	7 nm	Cell adhesion assays (HCL)	Pichel et al. (2000)
CS23	CS23	Non fimbrial	NR	Cell adhesion assays (HCL)	Del Canto et al. (2012)

Non Classical Adhesins

Tia	Afimbrial		–	Cell adhesion assays (HCL)	Mammarappallil and Elsinghorst (2000)
TibA	Afimbrial autotransporter		–	Cell adhesion assays (HCL)	Elsinghorst and Weitz (1994)
EtpA	Afimbrial		–	Cell adhesion assays (HCL)	Fleckenstein et al. (2006)

BE, bovine erythrocytes; HE, human erythrocytes; REN, rabbit enterocytes (primary cultures); HEN, human enterocytes (primary cultures); HCL, human cell lines; NR, not reported.

components within CFs, even when it has not been studied in all the cases. For example, CfaE, the tip subunit in CFA/I exerts the adhesin activity responsible for haemagglutination and binding to epithelial cells (Sakellaris *et al.*, 1999; Baker *et al.*, 2009). Specifically, the arginine residue in position 181 (R181) within the amino terminal domain in CfaE is determinant for this property and which is dispensable for fimbrial assembly (Sakellaris *et al.*, 1999; Baker *et al.*, 2009). On the other hand, CfaB is able to bind glycosphingolipids (Jansson *et al.*, 2006), which suggest that the role of CFA/I as adhesin is determined by these two components. The same R181 residue is conserved in CooD, the tip subunit of CS1, and it is also required for the agglutination of red blood cells (Sakellaris *et al.*, 1999). A unique case is the afimbrial CS6, because it is composed by the same proportion of two different structural subunits, CssA and CssB (Wolf *et al.*, 1997). By blocking the adherence capacity of an ETEC strain carrying CS6 with purified afimbrial subunits or with specific antibodies directed to each one of these proteins, the adhesin activity was found to be contained only within CssA (Ghosal *et al.*, 1999).

Assembly and secretion

The majority of CFs of human ETEC strains are encoded in plasmids, except for CS2, which it's encoded within the chromosome (Froehlich *et al.*, 1995). In addition, the ETEC plasmids also bear genes encoding proteins involved in virulence functions, such as enterotoxins, proteases and iron-uptake systems (Nuccio and Bäumler, 2007; Johnson and Nolan, 2009). As CFs are extracellular organelles, they must be secreted and assembled on the outer membrane to be fully functional. The structural CFs subunits are synthetized in the cytoplasm and secreted by the Sec machinery to the periplasm space (Nuccio and Bäumler, 2007). The CFs are synthetized as pre-proteins, carrying a signal sequence in their amino terminal end, which is removed by proteolysis before secretion (Mori and Ito, 2001). Once the proteins have reached the periplasmic space, three pathways have been found involved in the translocation through the outer membrane and for subsequent assembly of the CF structures: the usher/chaperone pathway (in the case of CS3,

CS6, CS12–13, CS18, CS20 and CS23), the alternative usher/chaperone pathway (CFA/I, CS1–2, CS4–5, CS7, CS14, CS17 and CS19) (Korea *et al.*, 2011) and the general secretion pathway for the type IV pilus (CS8 and CS21) (Clavijo *et al.*, 2010). The assignment for the secretion mechanism used is based upon assumptions derived from sequence similarities of the CFs genetic clusters and their protein products, with others structures whose assembly has been experimentally demonstrated. A schematic representation of the three mechanisms involved in CFs assembly is shown in Fig. 11.1.

The usher/chaperone pathway involves the participation of two proteins (the usher and the chaperone), to secrete the structural subunits (Waskman and Hultgren, 2009). Adhesins assembled by this mechanism include thick-rigid structures, as the type 1 fimbria and the P pili produced by other *E. coli* strains; and fibres as the K88 pili, produced by porcine ETEC strains (Soto and Hultgren, 1999). The chaperone is a periplasmic protein that binds structural subunits and assists in their export through the usher, an integral pore-forming membrane protein (Waskman and Hultgren, 2009) (Fig. 11.1). Periplasmic disulfide isomerases, as DsbA, catalyse formation of disulfide bonds within the structural subunits in order to stabilize them (Jacob-Dubuisson *et al.*, 1994) and chaperone binds to them fitting a β-strand in a hydrophobic cleft, in a process known as 'donor strand complementation' (Sauer *et al.*, 2004). The complex is able to bind the usher and then be exported (Li *et al.*, 2011). The first subunits exported are those that constitute the tip and the structure begins to growth from the base. During assembly, the hydrophobic cleft initially occupied by chaperones is replaced by donor strands of other structural subunits, in a process known as 'donor strand exchange' (Sauer *et al.*, 2004). As subunits are assembled, the structure can make turns forming flexible thin fibres if 2 subunits per turn are joined (2–3 nm in diameter), or thicker rigid fimbriae if 3 subunits (6–7 nm) are assembled (Nuccio and Bäumler, 2007). The alternate usher-chaperone pathway, by which a set of CFs is assembled, proceeds in a similar mechanism regarding functionality of their components, but amino acid sequences of the

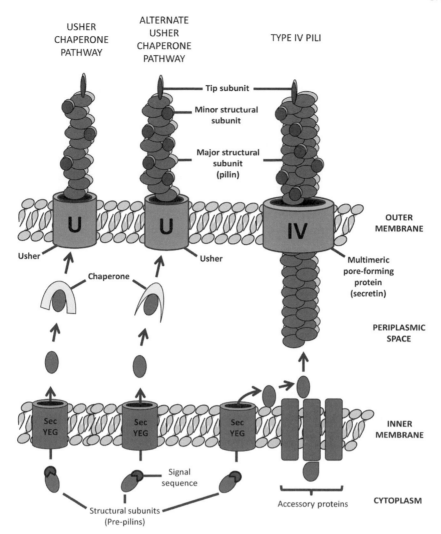

Figure 11.1 Simplified schematic representation of the three mechanisms involved in CFs assembly. Structural subunits containing signal peptides (pre-pilins) are processed and secreted into the periplasm by the Sec machinery. Depending on the CF, the structures may be assembled by the usher/chaperone pathway (in the case of CS3, CS6, CS12, CS13, CS18, CS20 and CS23), the alternate usher/chaperone pathway (CFA/I, CS1, CS2, CS4, CS5, CS7, CS14, CS17 and CS19) or the type IV pilus assembly mechanism (CS8 and CS21).

involved proteins are dissimilar to those involved in the classic usher–chaperone pathway (Soto and Hultgren, 1999). This observation has suggested that the two pathways emerged independently by convergent evolution (Soto and Hultgren, 1999).

Type IV pili structures are also found within CFs. This kind of pili is broadly distributed among bacteria and with diverse roles in motility, electron transfer, DNA uptake and adherence to different surfaces (Pelicic, 2008). Particularly, type IV structures described in human ETEC strains are

similar to the bundle-forming pilus, an adhesin produced by enteropathogenic *E. coli* (Nougay-rède *et al.*, 2003). The assembly of pilins, including major and minor subunits produced as pre-pilins, is assisted by a larger set of proteins compared to the usher/chaperone systems. There is an essential 'core' of proteins involved in assembly of most type IV structures, but additional components can be found in particular cases. This core includes pre-pilin peptidases that release mature subunits, ATPases for obtaining energy required for assembly, integral inner membrane proteins of unknown function and secretins to translocate the pilins at the bacterial surface (Pelicic, 2008) (Fig. 11.1).

Induction and expression

CFs genetic clusters contain between 5 and 16 genes and are organized as operons (Kaper and Nataro, 1998). CFs expression can be modulated by some chemical compounds and by temperature. The culture medium most commonly used for standard detection of CFs in the laboratory is the CFA medium, which it is assumed to include the basic formulation to promote CFs expression (Evans et al., 1979). The CFA medium is composed of 0.15% yeast extract, 1% casamino acids, 0.05% $MgSO_4$ and 0.05% $MnSO_4$, and it is usually supplemented with commercially available bile salts. Experimental evidence indicating that bile salts are needed to induce consistent in vitro expression of CS5, CS7, CS8, CS14 and CS17 has been reported (Sjöling et al., 2007). Similar findings have been reported for fimbrial/afimbrial adhesins produced by other pathogenic E. coli strains and several bacterial species, reinforcing the idea that bile salts are general inducers of adhesins expression and their presence is exploited by enteric pathogens at the site of infection (Torres et al., 2007; Faherty et al., 2012). Further, several CFs have been identified because it is known that adhesins important for human gut colonization are better expressed at 37°C but not at 25°C or at lower temperatures (Göransson et al., 1989). This characteristic was first reported for the K99 adhesin expressed by an animal ETEC strain (Burrows et al., 1976) and was also displayed by CFA/I (Evans et al., 1977, 1978a). Thus, later, electrophoretic separation of surface protein extracts obtained from ETEC cultures grown in both conditions has allowed identification of proteins that have become part of the CFs repertoire (Grewal et al., 1997; Valvatne et al., 1996; Del Canto et al., 2012). In some cases, as in the case of CS8, functionality has been also evaluated at these different temperatures, along with the visualization of pili by electron microscopy (Honda et al., 1984). These evidences are in agreement with the environmental conditions experienced within the gut, to which an ETEC strain needs to adapt

in order to express its virulence factors and display full pathogenicity. Studies addressing the molecular mechanisms involved in the regulation of CFs expression have been carried out for some of these adhesins. For example, repression ('silencing') of CFA/I synthesis occurs at the transcriptional level by the histone-like protein H-NS when bacteria are growth at 20°C. This effect is reversed by an AraC-like activator, CfaD (otherwise known as CfaR), which in turns anti-silence expression when temperature is raised to 37°C (Jordi et al., 1992). A protein from the same family of positive regulators, named Rns, is involved in activation of expression of CS1, CS3, CS17 and CS19 (Caron et al., 1989; Bodero et al., 2008).

CFs as virulence factors

Although CFs are considered virulence factors, several of them may not fulfil the traditional Koch's molecular postulates (Falkow, 2004), and therefore, they cannot be considered as such. The best examples are CS11 and CS13, for which no experimental evidences suggesting a role in ETEC adherence have been found (Table 11.1). Evidence suggesting a role of CFs in adherence of different ETEC strains have emerged mainly from assays performed in vitro using three main experimental models. Capacity of bacteria to agglutinate human or animal erythrocytes has been considered as an indicator of the presence of functional adhesins (Wilson, 1979). So, the haemagglutination assay was performed to demonstrate adherence ability in several CFs, including CFA/I, CS1, CS4, CS5, CS14 and CS17 (Table 11.1). To demonstrate that ETEC infects the small bowel, adherence to human enterocytes obtained after jejunostomy or duodenal biopsies, has been a good indicator of colonization activity mediated by CFs (Helander et al., 1997; Knutton et al., 1985). These types of experiments were performed to demonstrate the role of CFA/I, CS2, CS3, CS6, CS8 and CS12 (Table 11.1). Human cell lines derived from colonic epithelium carcinoma, mainly Caco-2 cells, have also been used as experimental model for ETEC adherence (Darfeuille-Michaud et al., 1990). Caco-2 cells have been used to assay adhesive activity of CFA/I, CS6, CS8, CS15, CS18, CS19, CS20, CS21, CS22 and CS23 (Table 11.1). Adherence levels to the human duodenum-derived cell line HuTu80 has been very low, so they have not being consistently used, even though they originated from the tissue infected by ETEC during diarrhoea (Darfeuille-Michaud *et al.*, 1990). This suggests that expression of adhesin receptors may be not optimal *in vitro* or that mutations occurred during transformation of normal to carcinogenic cells, which may affect receptor recognition by bacterial surface structures.

Non-classical adhesins

Although the CFs are the main group of ETEC adhesins and therefore, a nomenclature system has been proposed and implemented, there are other proteins which have been identified in the prototype ETEC strain H10407 as adherence determinants, based on *in vitro* assays performed in Caco-2 and HCT-8 cell lines (Fleckenstein *et al.*, 2010). Tia, TibA and EtpA are non-fimbrial, non-multimeric surface structures, which are not translocated to the outer membrane by the pathways used by CFs, and therefore, they are not included in the CF nomenclature (Fleckenstein *et al.*, 2010) (Table 11.1). In the next paragraphs, we describe the main features of these proteins.

Tia

The Tia protein was found within a cosmid library as one of two determinants for ETEC H10407 strain to adhere/invade HCT-8 cells (Elsinghorst and Kopecko, 1992). Even when ETEC has always been considered as a non-invasive diarrhoeagenic category of pathogenic *E. coli*, invasion assays and electron microscopy analysis demonstrated that this strain was able to enter intestinal epithelial cells (Elsinghorst, 1992). The gene encoding the Tia (enterotoxigenic invasion locus A) protein is located within a pathogenicity island inserted into the chromosome (Fleckenstein *et al.*, 2000). Tia is a 25 kDa protein that shares limited homology with Ail, and adhesin produced by *Yersinia* species and according to bioinformatic analysis, it is anchored to the outer surface by eight transmembrane domains (Mammarappallil *et al.*, 2000). Expression of Tia in non-pathogenic *E. coli* HB101 and *E. coli* DH5α strains confers the adhesive/invasive phenotype displayed by the wild-type strain, suggesting that this protein would not require additional factors to exert its function. Additionally, inactivation of the *tia* gene by allelic replacement significantly affected association between ETEC H10407 and the epithelial cells (Mammarappallil *et al.*, 2000).

TibA

The *tib* (enterotoxigenic invasion locus B) locus was found as another adhesion/invasion-mediated chromosomal region of ETEC H10407 (Elsinghorst and Kopecko, 1992). This locus encodes the 104 kDa glycoprotein TibA, which is the adhesin/invasin, as well as the glycosyltransferase TibC, presumably required for the post-translational modification of TibA (Lindenthal and Elsinghorst, 1999). Based on sequence homology, TibA is an autotransporter protein, which is produced as a pre-protein and secreted into periplasm, being able to catalyse its own insertion into the outer membrane and subsequent exposure to the extracellular milieu (Wells *et al.*, 2007). This self-assisted mechanism is also known as type V secretion system (Henderson *et al.*, 2004). Bioinformatic analysis indicated that TibA shared similarity with previously described autotransporter proteins AIDA-I and pertactin, two adhesins produced by diffusely adhering *E. coli* and *Bordetella pertussis*, respectively. Furthermore, this analysis suggested that TibA is inserted in the outer phospholipid layer using 10–14 transmembrane domains. As occurred with Tia, expression of TibA in *E. coli* HB101 conferred this strain with the ability to adhere/invade. Inactivation of the *tib* gene in ETEC H10407 affected significantly the associated phenotype in the wild-type background (Elsinghorst and Weist, 1994).

EtpA

After the discovery of CFA/I, Tia and TibA, a fourth adherence determinant protein was found in ETEC H10407. Using a random transposon insertion library, a clone exhibiting a lower adherence compared to the wild-type strain was identified. The insertion was mapped to a locus encoding a two-partner secretion system, subsequently referred to as *etpBAC* (enterotoxigenic two partner), which encodes the EtpB, EtpA and EtpC proteins (Fleckenstein *et al.*, 2006). The *etpBAC* locus in ETEC H10407 is located in the same plasmid that encodes ST, CFA/I and the serine-protease autotransporter EatA protein, which is able to cleave EtpA (Crossman *et al.*, 2010; Roy *et al.*, 2011). Evidence indicates that this proteolytic processing moderate adherence to epithelial cells and also allowing for optimal delivery of LT (Roy *et al.*, 2011). The EtpBAC comprises a type of secretion machinery that is classified as Vb, including EtpB, a pore-forming outer membrane protein by which EtpA is secreted. According to sequence homology, EtpC is a glycosyltransferase required for glycosylation of the exoprotein EtpA (Fleckenstein *et al.*, 2006). The adhesion mechanism mediated by EtpA involves interaction with the tip of the flagellum with subsequent binding to epithelial receptors (Roy *et al.*, 2009). This interaction was found not to involve the flagellin variable domain, so it has been presumed that EtpA could drive adherence of ETEC strains belonging to diverse flagellar (H) serogroups (Roy *et al.*, 2009).

ETEC adhesins' cell receptors

Given that adherence has been established as a key step for ETEC pathogenesis, recognition of cellular receptors by the bacterial adhesins is the molecular event that governs this stage within the infectious mechanism. Certainly, the fact that a pathogen could infect a specific host depends on matching between their virulence factors and host receptors. In that sense, a few host molecules that could function as receptors for human ETEC adhesins during natural infection, particularly in the case of CFs, have been identified. In general, these receptors are composed by sugars found in glycoproteins or glycolipids (De Greve *et al.*, 2007; Korea *et al.*, 2011). The major structural subunit of CFA/I, CfaB, has been found to interact with glycosphingolipids and sialylated proteins (Jansson *et al.*, 2006; Li *et al.*, 2006). Purified CS2, CS5 and CS6 bind to components of rabbit intestinal mucus and this interaction is prevented by treatment with *meta*-periodate, suggesting recognition of specific carbohydrates (Helander *et al.*, 1997). Further, binding of CS6 to fibronectin (a glycoprotein) has been reported, indicating that extracellular matrix proteins could also serve

as a focal contact point prior to reach epithelial cells (Chatterjee *et al.*, 2011). If the information about CFs receptors is considered limited, the data regarding non-classical adhesins targets is even scarcer. The only evidence available indicates that purified EtpA locates preferably in close proximity to mucin-producing cells, when this adhesin is added to frozen sections of murine intestine (Roy *et al.*, 2009). Because adherence of ETEC is a crucial step during pathogenesis, more investigation about the adhesins' receptors is crucial and needed. The identification of target molecules and interactions with the epithelia required for the toxigenic effect in the intestine would potentiate the development of preventative therapies.

Distribution of ETEC adhesins

ETEC strains are considered a group of pathogenic *E. coli* strains producing LT and/or ST but which possess a large diversity of somatic O-antigens, flagella types and surface-exposed adhesins (Wolf, 1997), which is considered one of the main obstacles to develop an effective preventative therapy. Because there is a significant interest to develop ETEC vaccines, substantial efforts have been made to design different strategies to identify novel ETEC adhesins, characterize the ETEC isolates and to define which the isolates are more frequently associated with diarrhoea. Serological identification has been a gold standard method to characterize ETEC strains according to their CFs repertoire (Vidal *et al.*, 2009). However, as more genetic sequences are identified and become available, detection of CF-encoding genes, either using DNA probes or multiplex-PCR (mPCR), has progressively been routinely incorporated for characterization studies (Steinsland *et al.*, 2003; Vidal *et al.*, 2009; Rodas *et al.*, 2009).

Regardless of the methodology used to identify CFs, some of these variants are the most frequently detected and their distribution is quite similar along different geographical locations. CFA/I, CS1–7, CS14 and CS21 are most common CFs carried by ETEC strains causing diarrhoea and they are prime candidates for their incorporation into effective vaccines (Svennerholm and Tobias, 2008). A recent review that collected results obtained from 136 independent

ETEC studies worldwide, which is equivalent to compile data from 17,205 isolates, found that the three most frequent CFs variants are CFA/I, CS6 and CS21 (Isidean *et al.*, 2011). Results obtained in five different studies, which included detection of at least 18 variants, are shown in Fig. 11.2.

Because one ETEC strain may carry between one to four adhesins simultaneously, different repertoires of CFs are found for individual strains. For example, CS21 seems to be the most prevalent and it has been found associated to several CFs including CFA/I, CS1, CS2, CS3 and CS6 (Girón

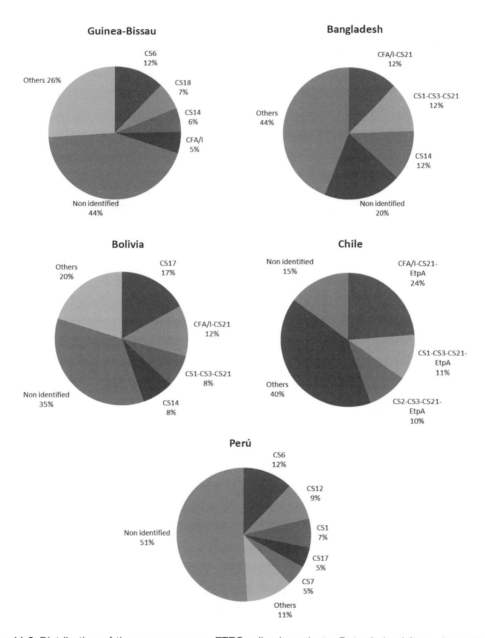

Figure 11.2 Distribution of the more common ETEC adhesin variants. Data derived from characterization of ETEC strains including at least 18 different CFs variants, are shown. Strains were obtained from Guinea Bissau (Steinsland *et al.*, 2002), Bangladesh (Rodas *et al.*, 2009), Bolivia (Rodas *et al.*, 2009), Chile (Del Canto *et al.*, 2011) and Peru (Rivera *et al.*, 2010). Graphics show distribution of individual variants or adhesin profiles, depending on how they were presented in the original reports. Percentages were rounded to facilitate understanding.

et al., 1995; Del Canto *et al.*, 2011). The rest of the CF variants are found in significantly lower frequencies, even when some of them, as CS17 and CS18, have been found as the most frequent in some independent reports (Rodas *et al.*, 2009; Steinsland *et al.*, 2002). This overall low frequency of the CFs is probably due to the fact that this group of adhesins has not been included in the panel screen of virulence factors in several characterization studies (Regua-Mangia *et al.*, 2004; Harris *et al.*, 2008).

Some CFs repertoires have been also associated to the presence of particular enterotoxin profiles and O serogroups. Typically, these virulence factors are also associated to specific patterns, i.e. when strains are genotyped by methodologies involving analysis of polymorphisms within the genome, such as those identified by restriction fragment length polymorphism and pulsed field electrophoresis (PFGE), random amplification of polymorphic DNA (RAPD), multilocus enzyme electrophoresis (MLEE) and even in the multilocus variable number of tandem repeats analysis (MLVA) (Regua-Mangia *et al.*, 2004). Therefore, strains producing CFA/I are also commonly producing ST toxin and they belong to the O78 and O153 serogroups (Wolf *et al.*, 1997). On the other hand, strains carrying CS1/CS3 or CS2/CS3 repertoires produce LT and ST and are frequently members of the O6 serogroup (Wolf *et al.*, 1997). Finally, CS21 has been found widely distributed within all this variable repertoires (Girón *et al.*, 1995; Del Canto *et al.*, 2011).

Distribution of non-classical adhesins has only been reported by a small number of studies and those are mainly in the manuscripts that describe the adhesins for their first time. Tia and TibA has been detected at very low frequencies in clinical isolates (Turner *et al.*, 2006b; Del Canto *et al.*, 2011), but to the contrary, EtpA seems to be produced by a wide range of ETEC strains carrying diverse enterotoxin profiles, CFs repertoires, and belonging to diverse serogroups (Del Canto *et al.*, 2011).

Adhesin-negative ETEC strains

Even when more than twenty different fimbrial/afimbrial adhesins have been described, about a third of the globally characterized ETEC strains are reported to be adhesin-negative isolates, which adds another level of complexity to find suitable antigens for developing broad-protective therapies against ETEC infections. Overall, results of studies carried out worldwide indicated that ETEC strains lacking known adhesins represent about 30% of the total number of isolates (Isidean *et al.*, 2011). Results may be different depending on the geographical location and the number of adhesin variants included in the screening analysis, since the non-classical adhesins are not usually included and, therefore, the strains are reported just as CF-negative. In the case of studies including the detection of 18 CF variants, the number of CF-negative isolates may fluctuate from 15% to 51%. For example, in a study including 1,200 ETEC isolates recovered in Guinea-Bissau, 44% of the strains were negative for CFs (Steinsland *et al.*, 2002). Similarly, negative results in CFs detection have been reported in 35.5% of 65 isolates from Bolivia (Rodas *et al.*, 2009) and 19.5% of 41 isolates from Bangladesh (Rodas *et al.*, 2009) (Fig. 11.2). In a recent study which included the detection of non-classical adhesins in addition to 19 CF variants, 15% of 103 ETEC isolated in Chile were adhesin-negative (Del Canto *et al.*, 2011) (Fig. 11.2). Finally, in a study including ETEC strains isolated in Peru, the detection of CFs was found to be more frequent between isolates obtained from diarrhoeal cases (64%), than in those samples obtained from patients without diarrhoea (37%) (Rivera *et al.*, 2010). This study suggests that ETEC strains associated with diarrhoeal cases within this geographical region are preferentially carrying known CFs.

Negative results in detection of adhesin-encoding genes can be explained by two different ways:

1 Adhesin-negative ETEC strains might carry known adhesins but the methodology used has not allowed their detection.
2 Adhesin-negative ETEC strains might contain adhesins-encoding genes that have not been identified yet.

Regarding the first possibility, at least three aspects are worth discussing. First, ETEC strains may not express the adhesins under laboratory conditions to a level sufficient to be detected by standard serological methods. Second, negative results can be obtained due to the presence of mutations affecting antigenic residues located in the target protein or mutations that might affect the framework in the genetic sequence. The reason this limitation is important, it has to do with the current typing methods that detect just one component of sometimes complex multimeric structures or one gene within a genetic locus. A third reason has to do with the loss of mobile genetic elements or inactivation of virulence genes by insertion sequences that can occur at low frequency during laboratory passages of the clinical strains (Sommerfelt et al., 1989; Kusumoto et al., 2000). As mentioned above, most ETEC adhesins are carried by plasmids or are found within genomic islands (Fleckenstein et al., 2010) and frequently, they are usually flanked by insertions sequences associated with recombination events (Johnson and Nolan, 2009).

In the second option, clinical ETEC strains may carry uncharacterized adhesins and the currently known variants represent a small portion of a larger set of adhesion factors. Fortunately, advances in whole genome sequencing technologies promises to be a key tool in discovering novel adherence determinants. Further, proteomic analysis is also being useful to identify known adhesins previously characterized in other bacterial pathogens but which have not been previously detected in ETEC strains.

Evaluation of ETEC adhesins as vaccine candidates

Advances in vaccine development to prevent *E. coli*-caused diarrhoea will be described in depth in Chapter 15. Therefore, this chapter will include just a brief overview of this topic. As it has been mentioned before, an ideal ETEC vaccine formulation should be able to induce production of antibodies directed to adhesins or other virulence factors and be effective in preventing illness caused by a wide range of ETEC strains. This premise is the basis for vaccine studies including the most frequent CFs and EtpA, and experimental evidence indicates that immune responses

against these surface-exposed structures occur during diarrhoeal episodes. Both immunoglobulins A and G directed against CFA/I and CS6 have been detected in volunteers infected with ETEC strains carrying these adhesins (Evans et al., 1978b; Coster et al., 2007). Additionally, it has been reported that sera obtained from convalescent patients, previously infected with ETEC H10407, recognize CFA/I, TibA and EtpA (Roy et al., 2010).

Given that some ETEC CFs share similarities at the DNA sequence and structural levels, cross-reactivity of anti-CF antibodies has been observed. For example, in the case of a child suffering from diarrhoea, serum antibodies produced in response to a strain carrying CFA/I recognized this CF as well as CS14, whereas CS7 was recognized by serum obtained from a child who suffered from diarrhoea caused by a CS5-expressing strain (Qadri et al., 2006). In a scenario where protection against a pathogen expressing a wide range of antigenic determinants, the cross-reactivity obtained with the antiserum is undoubtedly, a desirable outcome. As such, investigations are looking at conserved regions shared between the ETEC adhesins in order to obtain a response against the widest range of CFs, but with the idea of administered in a formulation containing the minimal number of antigens.

Vaccine preparations containing the majority of the prevalent CFs (CFA/I, CS1 to 6) have been tested in humans, but they have been only able to stimulate immune response against two (Savarino et al., 1999) or three CFs (Turner et al., 2011; Harro et al., 2011), when they included seven or six variants, respectively. In the other hand, EtpA, the only non-classical adhesin evaluated as potential vaccine candidate, has shown to confer protection to subsequent infection with ETEC H10407 in a murine model (Roy et al., 2008) but no evidence is available about its immunogenic potential in humans.

References

Aubel, D., Darfeuille-Michaud, A., and Joly, B. (1991). New adhesive factor (antigen 8787) on a human enterotoxigenic *Escherichia coli* 0117:H4 strain isolated in Africa. Infect. Immun. 59, 1280–1299.

Baker, K.K., Levine, M.M., Morison, J., Phillips, A., and Barry, E.M. (2009). CfaE tip mutations in enterotoxigenic *Escherichia coli* CFA/I fimbriae define critical human intestinal binding sites. Cell. Microbiol. *11*, 742–754.

Bodero, M.D., Harden, E.A., and Munson, G.P. (2008). Transcriptional regulation of subclass 5b fimbriae. BMC Microbiol. *8*, 180.

Burrows, M.R., Sellwood, R., and Gibbons, R.A. (1976). Haemagglutinating and adhesive properties associated with the K99 antigen of bovine strains of *Escherichia coli*. J. Gen. Microbiol. *96*, 269–275.

Caron, J., Coffield, L.M., and Scott, J.R. (1989). A plasmid-encoded regulatory gene, *rns*, required for expression of the CS1 and CS2 adhesins of enterotoxigenic *Escherichia coli*. Proc. Natl. Acad. Sci. U.S.A. *86*, 963–967.

Chatterjee, R., Ghosal, A., Sabui, S., and Chatterjee, N.S. (2011). Three dimensional modeling of C-terminal loop of CssA subunit in CS6 of enterotoxigenic *Escherichia coli* and its interaction with the 70 KDa domain of Fibronectin. Bioinformation *6*, 307–310.

Clavijo, A.P., Bai, J., and Gómez-Duarte, O.G. (2010). The Longus type IV pilus of enterotoxigenic *Escherichia coli* (ETEC) mediates bacterial self-aggregation and protection from antimicrobial agents. Microb. Pathog. *48*, 230–238.

Coster, T.S., Wolf, M.K., Hall, E.R., Cassels, F.J., Taylor, D.N., Liu, C.T., Trespalacios, F.C., Delorimier, A., Angleberger, D.R., and Mcqueen, C.E. (2007). Immune response, ciprofloxacin activity, and gender differences after human experimental challenge by two strains of enterotoxigenic *Escherichia coli*. Infect. Immun. *75*, 252–259.

Cravioto, A., Scotland, S.M., and Rowe, B. (1982). Hemagglutination activity and colonization factor antigens I and II in enterotoxigenic and non-enterotoxigenic strains of *Escherichia coli* isolated from humans. Infect. Immun. *36*, 187–197.

Crossman, L.C., Chaudhuri, R.R., Beatson, S.A., Wells, T.J., Desvaux, M., Cunningham, A.F., Petty, N.K., Mahon, V., Brinkley, C., Hobman, J.L., *et al.* (2010). A commensal gone bad: complete genome sequence of the prototypical enterotoxigenic *Escherichia coli* strain H10407. J. Bacteriol. *192*, 5822–5831.

Darfeuille-Michaud, A., Forestier, C., Joly, B., and Cluzel, R. (1986). Identification of a nonfimbrial adhesive factor of an enterotoxigenic *Escherichia coli* strain. Infect. Immun. *52*, 468–475.

Darfeuille-Michaud, A., Aubel, D., Chauviere, G., Rich, C., Bourges, M., Servin, A., and Joly, B. (1990). Adhesion of enterotoxigenic *Escherichia coli* to the human colon carcinoma cell line Caco-2 in culture. Infect. Immun. *58*, 893–902.

De Greve, H., Wyns, L., and Bouckaert, J. (2007). Combining sites of bacterial fimbriae. Curr. Opin. Struct. Biol. *17*, 506–512.

Del Canto, F.D., Valenzuela, P., Cantero, L., Bronstein, J., Blanco, E., Blanco, J.J.E.J.J.E., Prado, V., Levine, M., Nataro, J., Sommerfelt, H., *et al.* (2011). Distribution of classical and nonclassical virulence genes in enterotoxigenic *Escherichia coli* isolates from Chilean children and tRNA gene screening for putative insertion sites for genomic islands. J. Clin. Microbiol. *49*, 3198–3203.

Del Canto, F., Botkin, D.J., Valenzuela, P., Popov, V., Ruiz-Perez, F., Nataro, J.P., Levine, M.M., Stine, O.C., Pop, M., Torres, A.G., *et al.* (2012). Identification of coli surface antigen 23, a novel adhesin of enterotoxigenic *Escherichia coli*. Infect. Immun. *80*, 2791–2801.

Dorsey, F.C., Fischer, J.F., and Fleckenstein, J.M. (2006). Directed delivery of heat-labile enterotoxin by enterotoxigenic *Escherichia coli*. Cell. Microbiol. *8*, 1516–1527.

Elsinghorst, E.A., and Kopecko, D.J. (1992). Molecular cloning of epithelial cell invasion determinants from enterotoxigenic *Escherichia coli*. Infect. Immun. *60*, 2409–2417.

Elsinghorst, E.A., and Weitz, J.A. (1994). Epithelial cell invasion and adherence directed by the enterotoxigenic *Escherichia coli* tib locus is associated with a 104-kilodalton outer membrane protein. Infect. Immun. *62*, 3463–3471.

Evans, D.J., and Evans, D.G. (1973). Three characteristics associated with enterotoxigenic *Escherichia coli* isolated from man. Infect. Immun. *8*, 322–328.

Evans, D.G., and Evans, D.J. (1978). New surface-associated heat-labile colonization factor antigen (CFA/II) produced by enterotoxigenic *Escherichia coli* of serogroups 06 and 08. Infect. Immun. *21*, 638–647.

Evans, D.G., Silver, R.P., Evans, D.J., Jr., D.G.C., and Gorbach, S.L. (1975). Plasmid-controlled colonization factor associated with virulence in *Escherichia coli* enterotoxigenic for humans. Infect. Immun. *12*, 656–667.

Evans, D.G., Evans, D.J. Jr, and Tjoa, W. (1977). Hemagglutination of human group A erythrocytes by enterotoxigenic *Escherichia coli* isolated from adults with diarrhoea: correlation with colonization factor. Infect. Immun. *18*, 330–337.

Evans, D.G., Evans, D.J. Jr., Tjoa, W.S., and Dupont, H.L. (1978a). Detection and characterization of colonization factor of enterotoxigenic *Escherichia coli* isolated from adults with with Diarrhea. Infect. Immun. *19*, 727–736.

Evans, D.G., Satterwhite, T.K., Evans, D.J., and DuPont, H.L. (1978b). Differences in serological responses and excretion patterns of volunteers challenged with enterotoxigenic *Escherichia coli* with and without the colonization factor antigen. Infect. Immun. *19*, 883–888.

Evans, D.G., Evans, D.J., Clegg, S., and Pauley, J.A. (1979). Purification and characterization of the CFA/I antigen of enterotoxigenic *Escherichia coli*. Infect. Immun. *25*, 738–748.

Faherty, C.S., Redman, J.C., Rasko, D.A., Barry, E.M., and Nataro, J.P. (2012). *Shigella flexneri* effectors OspE1 and OspE2 mediate induced adherence to the colonic epithelium following bile salts exposure. Mol. Microbiol. *85*, 107–121.

Falkow, S. (2004). Molecular Koch's postulates applied to bacterial pathogenicity – a personal recollection 15 years later. Nat. Rev. Microbiol. *2*, 67–72.

Fleckenstein, J.M., Lindler, L.E., Elsinghorst, E.A., and Dale, J.B. (2000). Identification of a gene within a pathogenicity island of enterotoxigenic *Escherichia coli* H10407 required for maximal secretion of the heat-labile enterotoxin. Infect. Immun. *68*, 2766–2774.

Fleckenstein, J.M., Roy, K., Fischer, J.F., and Burkitt, M. (2006). Identification of a two-partner secretion locus of enterotoxigenic *Escherichia coli*. Infect. Immun. *74*, 2245–2258.

Fleckenstein, J.M., Hardwidge, P.R., Munson, G.P., Rasko, D.A., Sommerfelt, H., and Steinsland, H. (2010). Molecular mechanisms of enterotoxigenic *Escherichia coli* infection. Microbes Infect. *12*, 89–98.

Froehlich, B.J., Karakashian, a, Sakellaris, H., and Scott, J.R. (1995). Genes for CS2 pili of enterotoxigenic *Escherichia coli* and their interchangeability with those for CS1 pili. Infect. Immun. *63*, 4849–4856.

Gaastra, W., and Svennerholm, A.M. (1996). Colonization factors of human enterotoxigenic *Escherichia coli* (ETEC). Trends Microbiol. *4*, 444–452.

Ghosal, A., Bhowmick, R., Banerjee, R., Ganguly, S., Yamasaki, S., Ramamurthy, T., Hamabata, T., and Chatterjee, N.S. (2009). Characterization and studies of the cellular interaction of native colonization factor CS6 purified from a clinical isolate of enterotoxigenic *Escherichia coli*. Infect. Immun. *77*, 2125–2135.

Girón, J.A., Viboud, G.I., Sperandio, V., Go, O.G., Maneval, D.R., Albert, M.J., Levine, M.M., and Kaper, J.B. (1995). Prevalence and association of the Longus pilus structural gene (*lngA*) with colonization factor antigens, enterotoxin types, and serotypes of enterotoxigenic *Escherichia coli*. Infect. Immun. *63*, 4195–4198.

Göransson, M., Forsman, K., and Uhlin, B.E. (1989). Regulatory genes in the thermoregulation of *Escherichia coli* pili gene transcription. Genes Develop. *3*, 123–130.

Grewal, H.M., Valvatne, H., Bhan, M.K., Dijk, V., L., Gaastra, W., and Sommerfelt, H. (1997). A new putative fimbrial colonization factor, CS19, of human enterotoxigenic *Escherichia coli*. Infect. Immun. *65*, 507–513.

Guttman, J.A., and Finlay, B.B. (2008). Subcellular alterations that lead to diarrhoea during bacterial pathogenesis. Trends Microbiol. *16*, 535–542.

Harris, A.M., Chowdhury, F., Begum, Y.A., Khan, A.I., Faruque, A.S.G., Svennerholm, A., Harris, J.B., Ryan, E.T., Cravioto, A., and Calderwood, S.B. (2008). Shifting prevalence of major diarrhoeal pathogens in patients seeking hospital care during floods in 1998, 2004, and 2007 in Dhaka, Bangladesh. Trop. Med. *79*, 708–714.

Harro, C., Sack, D., Bourgeois, a L., Walker, R., DeNearing, B., Feller, A., Chakraborty, S., Buchwaldt, C., and Darsley, M.J. (2011). A combination vaccine consisting of three live attenuated enterotoxigenic *Escherichia coli* strains expressing a range of colonization factors and heat-labile toxin subunit B is well tolerated and immunogenic in a placebo-controlled double-blind phase I trial in healthy adults. Clin. Vaccine Immunol. *18*, 2118–2127.

Helander, A., Hansson, G.C., and Svennerholm, A.M. (1997). Binding of enterotoxigenic *Escherichia coli* to isolated enterocytes and intestinal mucus. Microb. Pathog. *23*, 335–346.

Henderson, I.R., and Navarro-garcia, F. (2004). Type V protein secretion pathway : the autotransporter story. Microbiol. Mol. Biol. Rev. *68*, 692–744.

Heuzenroeder, M.W., Elliot, T.R., Thomas, C.J., Halter, R., and Manning, P.A. (1990). A new fimbrial type (PCFO9) on enterotoxigenic *Escherichia coli* O9:H-LT+ isolated from a case of infant diarrhea in central Australia. FEMS Microbiol. Lett. *66*, 55–60.

Hibberd, M.L., McConnell, M.M., Field, A.M., and Rowe, B. (1990). The fimbriae of human enterotoxigenic *Escherichia coli* strain 334 are related to CS5 fimbriae. J. Gen. Microbiol. *136*, 2449–2456.

Honda, T., Arita, M., and Miwatani, T. (1984). Characterization of new hydrophobic pili of human enterotoxigenic *Escherichia coli* : a possible new colonization factor. Infect. Immun. *43*, 959–965.

Isidean, S.D., Riddle, M.S., Savarino, S.J., and Porter, C.K. (2011). A systematic review of ETEC epidemiology focusing on colonization factor and toxin expression. Vaccine *29*, 6167–6178.

Jacob-Dubuisson, F., Pinkner, J., Xu, Z., Striker, R., Padmanhaban, A., and Hultgren, S.J. (1994). PapD chaperone function in pilus biogenesis depends on oxidant and chaperone-like activities of DsbA. Proc. Natl. Acad. Sci. U.S.A. *91*, 11552–11556.

Jansson, L., Tobias, J., Lebens, M., Svennerholm, A.M., and Teneberg, S. (2006). The major subunit, CfaB, of colonization factor antigen I from enterotoxigenic *Escherichia coli* is a glycosphingolipid binding protein. Infect. Immun. *74*, 3488–3497.

Johnson, T.J., and Nolan, L.K. (2009). Pathogenomics of the virulence plasmids of *Escherichia coli*. Microbiol. Mol. Biol. Rev. *73*, 750–774.

Jordi, B.J., Dagberg, B., de Haan, L.A., Hamers, A.M., van der Zeijst, B.A., Gaastra, W., and Uhlin, B.E. (1992). The positive regulator CfaD overcomes the repression mediated by histone-like protein H-NS (H1) in the CFA/I fimbrial operon of *Escherichia coli*. EMBO *11*, 2627–2632.

Knutton, S., Lloyd, D.R., Candy, D.C., and McNeish, A.S. (1985). Adhesion of enterotoxigenic *Escherichia coli* to human small intestinal enterocytes. Infect. Immun. *48*, 824–831.

Knutton, S., Lloyd, D.R., and McNeish, A.S. (1987). Identification of a new fimbrial structure in enterotoxigenic *Escherichia coli* (ETEC) serotype 0148:H28 which adheres to human intestinal mucosa: a potentially new human ETEC colonization factor. Infect. Immun. *55*, 86–92.

Korea, C.G., Ghigo, J.M., and Beloin, C. (2011). The sweet connection: Solving the riddle of multiple sugar-binding fimbrial adhesins in *Escherichia coli* : Multiple *E. coli* fimbriae form a versatile arsenal of sugar-binding lectins potentially involved in surface-colonisation and tissue tropism. BioEssays *33*, 300–311.

Kusumoto, M., Nishiya, Y., and Kawamura, Y. (2000). Reactivation of insertionally inactivated Shiga toxin

2 genes of *Escherichia coli* O157:H7 caused by nonreplicative transposition of the insertion sequence. Appl. Environ. Microbiol. *66*, 1133–1138.

Li, Q., Ng, T.W., Dodson, K.W., Shu, S., So, K., Pinkner, J.S., Scarlata, S., Hultgren, S.J., and David, G. (2011). The differential affinity of the usher for chaperone–subunit complexes is required for assembly of complete Pili. Mol. Microbiol. *76*, 159–172.

Li, Y.-F., Poole, S., Rasulova, F., Esser, L., Savarino, S.J., and Xia, D. (2006). Crystallization and preliminary X-ray diffraction analysis of CfaE, the adhesive subunit of the CFA/I fimbriae from human enterotoxigenic *Escherichia coli*. Acta Crystallogr. Sect. F. Struct. Biol. Cryst. Commun. *62*, 121–124.

Lindenthal, C., and Elsinghorst, E.A. (1999). Identification of a glycoprotein produced by enterotoxigenic *Escherichia coli*. Infect. Immun. *67*, 4084–4091.

McConnel, M.M., Chart, H., Field, A.M., Hibberd, M., and Rowe, B. (1989). Characterization of a putative colonization factor (PCFO166) of enterotoxigenic *Escherichia coli* of serogroup O166. J. Gen. Microbiol. *135*, 1135–1144.

McConnel, M.M., Hibberd, H., Field, A.M., Chart, H., and Rowe, B. (1990). Characterization of a new putative colonization factor (CS17) from a human enterotoxigenic *Escherichia coli* of serotype O114:H21 which produces only heat-labile enterotoxin. J. Infect. Dis. *161*, 343–347.

Mammarappallil, J.G., and Elsinghorst, E.A. (2000). Epithelial cell adherence mediated by the enterotoxigenic *Escherichia coli* Tia protein. Infect. Immun. *68*, 6595–6601.

Mazariego-Espinosa, K., Cruz, A., Ledesma, M.A., Ochoa, S.A., and Xicohtencatl-Cortes, J. (2010). Longus, a type IV pilus of enterotoxigenic *Escherichia coli*, is involved in adherence to intestinal epithelial cells. J. Bacteriol. *192*, 2791–2800.

Mori, H., and Ito, K. (2001). The Sec protein-translocation pathway. Trends Microbiol. *9*, 494–500.

Mullany, P., Field, a M., McConnell, M.M., Scotland, S.M., Smith, H.R., and Rowe, B. (1983). Expression of plasmids coding for colonization factor antigen II (CFA/II) and enterotoxin production in *Escherichia coli*. J. Gen. Microbiol. *129*, 3591–3601.

Nataro, J.P., and Kaper, J.B. (1998). Diarrheagenic *Escherichia coli*. Clin. Microbiol. Rev. *11*, 142–201.

Nougayrède, J., Fernandes, P.J., and Donnenberg, M.S. (2003). Microreview: adhesion of enteropathogenic *Escherichia coli* to host cells. Cell. Microbiol. *5*, 359–372.

Nuccio, S.P., and Bäumler, A.J. (2007). Evolution of the chaperone/usher assembly pathway: fimbrial classification goes Greek. Microbiol. Mol. Biol. Rev. *71*, 551–575.

Pelicic, V. (2008). Type IV pili: e pluribus unum? Mol. Microbiol. *68*, 827–837.

Pichel, M., Binsztein, N., and Viboud, G. (2000). CS22, a novel human enterotoxigenic *escherichia coli* adhesin, is related to CS15. Infect. Immun. *68*, 3280–3285.

Qadri, F., Svennerholm, A., Faruque, A.S.G., and Sack, R.B. (2005). Enterotoxigenic *Escherichia coli* in Developing Countries: Epidemiology, Microbiology, Clinical Features, Treatment, and Prevention. Clin. Microbiol. Rev. *18*, 465–483.

Qadri, F., Ahmed, F., Ahmed, T., and Svennerholm, A.M. (2006). Homologous and cross-reactive immune responses to enterotoxigenic *Escherichia coli* colonization factors in Bangladeshi children. Infect. Immun. *74*, 4512–4518.

Regua-Mangia, A.H., Cabilio, B., Irino, K., Beatriz, A., Pacheco, F., Ferreira, S., and Zahner, V. (2004). Genotypic and phenotypic characterization of enterotoxigenic *Escherichia coli* (ETEC) strains isolated in Rio de Janeiro city, Brazil. FEMS Immunol. Med. Microbiol. *40*, 155–162.

Rivera, F.P., Ochoa, T.J., Maves, R.C., Bernal, M., Medina, a M., Meza, R., Barletta, F., Mercado, E., Ecker, L., Gil, a I., et al. (2010). Genotypic and phenotypic characterization of enterotoxigenic *Escherichia coli* strains isolated from Peruvian children. J. Clin. Microbiol. *48*, 3198–3203.

Rodas, C., Iniguez, V., Qadri, F., Wiklund, G., Svennerholm, A.M., and Sjöling, A. (2009). Development of multiplex PCR assays for detection of enterotoxigenic *Escherichia coli* colonization factors and toxins. J. Clin. Microbiol. *47*, 1218–1220.

Roy, K., Hamilton, D., Allen, K.P., Randolph, M.P., and Fleckenstein, J.M. (2008). The EtpA exoprotein of enterotoxigenic *Escherichia coli* promotes intestinal colonization and is a protective antigen in an experimental model of murine infection. Infect. Immun. *76*, 2106–2112.

Roy, K., Hilliard, G.M., Hamilton, D.J., Luo, J., Ostmann, M.M., and Fleckenstein, J.M. (2009). Enterotoxigenic *Escherichia coli* EtpA mediates adhesion between flagella and host cells. Nature *457*, 594–598.

Roy, K., Bartels, S., Qadri, F., and Fleckenstein, J.M. (2010). Enterotoxigenic *Escherichia coli* elicits immune responses to multiple surface proteins. Infect. Immun. *78*, 3027–3035.

Roy, K., Kansal, R., Bartels, S.R., Hamilton, D.J., Shaaban, S., and Fleckenstein, J.M. (2011). Adhesin degradation accelerates delivery of heat-labile toxin by enterotoxigenic *Escherichia coli*. J. Biol. Chem. *286*, 29771–29779.

Sakellaris, H., Munson, G.P., and Scott, J.R. (1999). A conserved residue in the tip proteins of CS1 and CFA/I pili of enterotoxigenic *Escherichia coli* that is essential for adherence. Proc. Natl. Acad. Sci. U.S.A. *96*, 12828–12832.

Sánchez, J., and Holmgren, J. (2005). Virulence factors, pathogenesis and vaccine protection in cholera and ETEC diarrhoea. Curr. Opin. Immunol. *17*, 388–398.

Sauer, F.G., Remaut, H., Hultgren, S.J., and Waksman, G. (2004). Fiber assembly by the chaperone-usher pathway. Biochim. Biophys. Acta *1694*, 259–267.

Savarino, S.J., Hall, E.R., Bassily, S., Brown, F.M., Youssef, F., Wierzba, T.F., Peruski, L., El-Masry, N.A., Safwat, M., Rao, M., et al. (1999). Oral, inactivated, whole cell enterotoxigenic *Escherichia coli* plus cholera toxin B subunit vaccine: results of the initial evaluation in children. J. Infect. Dis. *179*, 107–114.

Sjöling, A., Wiklund, G., Savarino, S.J., Cohen, D.I., and Svennerholm, A.M. (2007). Comparative analyses of phenotypic and genotypic methods for detection of enterotoxigenic *Escherichia coli* toxins and colonization factors. J. Clin. Microbiol. *45*, 3295–3301.

Smyth, C.J. (1982). Two mannose-resistant haemagglutinins on enterotoxigenic *Escherichia coli* of serotype O6:K15:H16 or H-isolated from travellers' and infantile diarrhoea. J. Gen. Microbiol. *128*, 2081–2096.

Sommerfelt, H., Haukanes, B.I., Kalland, K.H., Svennerholm, A.-M., Sanchéz, J., and Bjorvatn, B. (1989). Mechanism of spontaneous loss of heat-stable toxin (STa) production in enterotoxigenic *Escherichia coli*. Apmis *97*, 436–440.

Soto, G.E., and Hultgren, S.J. (1999). Bacterial Adhesins : Common Themes and Variations in Architecture and Assembly. J. Bacteriol. *181*, 1059–1071.

Steffen, R., Castelli, F., Dieter Nothdurft, H., Rombo, L., and Zuckerman, J.N. (2005). Vaccination against enterotoxigenic *Escherichia coli*, a cause of travelers' diarrhoea. J. Travel Med. *12*, 102–107.

Steinsland, H., Valentiner-Branth, P., Perch, M., Dias, F., Fischer, T.K., Aaby, P., Mølbak, K., and Sommerfelt, H. (2002). Enterotoxigenic *Escherichia coli* infections and diarrhoea in a cohort of young children in Guinea-Bissau. J. Infect. Dis. *186*, 1740–1747.

Steinsland, H., Valentiner-Branth, P., Grewal, H.M.S., Gaastra, W., Mølbak, K., K., and Sommerfelt, H. (2003). Development and evaluation of genotypic assays for the detection and characterization of enterotoxigenic *Escherichia coli*. Diag. Microbiol. Infect. Dis. *45*, 97–105.

Svennerholm, A.M., and Tobias, J. (2008). Vaccines against enterotoxigenic *Escherichia coli*. Expert Rev. Vaccines 7, 795–804.

Thomas, L.V., Cravioto, A., Scotland, S.M., and Rowe, B. (1982). New Fimbrial Antigenic Type (E8775) That May Represent a Colonization Factor in Enterotoxigenic *Escherichia coli* in Humans. Infect. Immun. *35*, 1119–1124.

Thomas, L.V., McConnell, M.M., Rowe, B., and Field, a M. (1985). The possession of three novel coli surface antigens by enterotoxigenic *Escherichia coli* strains positive for the putative colonization factor PCF8775. J. Gen. Microbiol. *131*, 2319–2326.

Torres, A.G., Zhou, X., and Kaper, J.B. (2005). Adherence of Diarrheagenic *Escherichia coli* Strains to Epithelial Cells. Infect. Immun. 73, 18–29.

Torres, A.G., Tutt, C.B., Duval, L., Popov, V., Nasr, A.B., Michalski, J., and Scaletsky, I.C.A. (2007). Bile salts induce expression of the afimbrial LDA adhesin of atypical enteropathogenic *Escherichia coli*. Cell. Microbiol. *9*, 1039–1049.

Turner, S.M., Scott-Tucker, A., Cooper, L.M., and Henderson, I.R. (2006a). Weapons of mass destruction: virulence factors of the global killer enterotoxigenic *Escherichia coli*. FEMS Microbiol. Lett. *263*, 10–20.

Turner, S.M., Chaudhuri, R.R., Jiang, Z.-D., DuPont, H., Gyles, C., Penn, C.W., Pallen, M.J., and Henderson, I.R. (2006b). Phylogenetic comparisons reveal multiple acquisitions of the toxin genes by enterotoxigenic *Escherichia coli* strains of different evolutionary lineages. J. Clin. Microbiol. *44*, 4528–4536.

Turner, A.K., Stephens, J.C., Beavis, J.C., Greenwood, J., Gewert, C., Randall, R., Freeman, D., and Darsley, M.J. (2011). Generation and characterization of a live attenuated enterotoxigenic *Escherichia coli* combination vaccine expressing six colonization factors and heat-labile toxin subunit B. Clin. Vaccine Immunol. *18*, 2128–2135.

Valvatne, H., Sommerfelt, H., Gaastra, W., Bhan, M.K., and Grewal, H.M. (1996). Identification and characterization of CS20, a new putative colonization factor of enterotoxigenic *Escherichia coli*. Infect. Immun. *64*, 2635–2642.

Viboud, G.I., Binsztein, N., and Svennerholm, A.M. (1993). A new fimbrial putative colonization factor, PCFO20, in human enterotoxigenic *Escherichia coli*. Infect. Immun. *61*, 5190–5197.

Vidal, R.M., Valenzuela, P., Baker, K., Lagos, R., Esparza, M., Livio, S., Farfán, M., Nataro, J.P., Levine, M.M., and Prado, V. (2009). Characterization of the most prevalent colonization factor antigens present in Chilean clinical enterotoxigenic *Escherichia coli* strains using a new multiplex polymerase chain reaction. Diag. Microbiol. Infect. Dis. *65*, 217–223.

Waksman, G., and Hultgren, S.J. (2009). Structural biology of the chaperone-usher pathway of pilus biogenesis. Nat. Rev. Microbiol. *7*, 765–774.

Wells, T.J., Tree, J.J., Ulett, G.C., and Schembri, M.A. (2007). Autotransporter proteins : novel targets at the bacterial cell surface. FEMS Microbiol. Lett. *274*, 163–172.

Wilson, I. (1979). The fimbrial and non-fimbrial haemagglutinins of *Escherichia coli*. J. Med. Microbiol. *12*, 213–227.

Wolf, M.K. (1997). Occurrence, distribution, and associations of O and H serogroups, colonization factor antigens, and toxins of enterotoxigenic *Escherichia coli*. Clin. Microbiol. Rev. *10*, 569–584.

Wolf, M.K., de Haan, L.A., Cassels, F.J., Willshaw, G.A., Warren, R., Boedeker, E.C., and Gaastra, W. (1997). The CS6 colonization factor of human enterotoxigenic *Escherichia coli* contains two heterologous major subunits. FEMS Microbiol. Lett. *148*, 35–42.

Enteroaggregative *Escherichia coli* and Disease

12

Nadia Boisen, James P. Nataro and Karen A. Krogfelt

Abstract

Enteroaggregative *Escherichia coli* (EAEC) are generally known as causing diarrhoea, especially in developing countries and in travellers. Yet, recently it was shown that EAEC is also able to cause urinary tract infections. A recent European outbreak of Shiga toxin-producing EAEC drew the attention to and increased the interest on this pathotype. Although the pathogenesis of EAEC is not fully understood, a number of putative virulence factors have been described. The characteristic Aggregative Adherence Fimbriae (AAF) are mucosal adhesins that also elicit inflammatory responses from infected surfaces. EAEC are seen to produce a number of virulence factors including – serine protease autotransporter toxins that induce apoptosis of enterocytes. Epidemiological evidence supports a model of EAEC pathogenesis comprising the concerted action of multiple virulence factors along with induction of inflammation.

Introduction

Diarrhoea causes morbidity and mortality worldwide, predominantly in children under five years of age. Additionally, to almost one million directly attributable deaths per year (Black *et al.*, 2010), morbidity associated with repeated episodes during childhood diarrhoea can be long lasting (Petri *et al.*, 2008). Repeated rounds of diarrhoea during infancy result in malabsorption of nutrients leading to developmental disabilities, including both impaired growth and cognition.

Collectively, the diarrhoeagenic *Escherichia coli* (DEC) represent the most common bacterial pathogen worldwide (Farthing, 2000; Ina *et al.*, 2003; Wanke, 2001).

Of the diarrhoeagenic *Escherichia coli* (DEC), six groups – pathotypes – are currently defined: enterotoxigenic *E. coli* (ETEC), verocytotoxin producing *E. coli* (VTEC), enteroinvasive *E. coli* (EIEC), enteropathogenic *E. coli* (EPEC), diffusely adherent *E. coli* (DAEC) and enteroaggregative *E. coli* (EAEC) (Bryce *et al.*, 2005).

The enteroaggregative *Escherichia coli* (EAEC) is a pathogen of significance. Yet the real burden of disease caused by EAEC remains unknown, because most clinical laboratories do not seek for this pathotype. Further, the identification and isolation of pathogenic strains remain elusive, given an as yet imperfect definition of this pathotype. Thus, the global burden of diarrhoeal diseases resulting from EAEC may be vastly underestimated (Bryce *et al.*, 2005). This underestimation is also true of the sequelae of persistent diarrhoea for which EAEC is a prevalent cause (Fang *et al.*, 1995; Lima *et al.*, 2000)

EAEC pathogenesis

The first association of EAEC with diarrhoeal disease was published in 1987, as part of a prospective study of paediatric diarrhoea in Chile (Nataro *et al.*, 1987b) Subsequently, EAEC was associated with persistent diarrhoea among children in several studies (Bhan *et al.*, 1989a,b; Cravioto *et al.*, 1991; Valentiner-Branth *et al.*, 2003).

When EAEC is cultured on HEp-2 cells it forms a typical 'stack brick- layer' pattern, which is the gold standard in the definition of EAEC. Furthermore, EAEC is also defined as a pathotype of

DEC that does not secrete the heat -stable (ST) or heat-labile (LT) toxins of enterotoxigenic *E. coli* (ETEC).

Typical EAEC applies to EAEC strains possessing the AggR master regulon. Typical EAEC strains have been linked to acute diarrhoea (Sarantuya *et al.*, 2004). Importantly, atypical EAEC lack the AggR regulon and are not reliably associated with diarrhoea (Nataro, 2005).

EAEC as presently defined most likely encompasses both pathogenic and non-pathogenic *E. coli* strains (Kaper *et al.*, 2004). EAEC has unmistakably been associated with diarrhoea in some individuals (e.g. in volunteers and outbreak patients), and subclinical colonization in endemic areas is common.

The EAEC genome is immensely mosaic, with several putative virulence genes flanked by insertion sequences, and predictably, many strains harbour a diversity of the identified genes in multitude combinations (Boisen *et al.*, 2012; Lima *et al.*, 2013).

The essential differences between pathogenic and non-pathogenic strains are mostly unknown, but pathogenesis studies suggest three overall stages of infection: (1) adherence to the intestinal mucosa by virtue of aggregative adherence fimbriae (AAF) and possibly other adherence factors (Hicks *et al.*, 1996; Tzipori *et al.*, 1992); (2) stimulation of the production of mucus following biofilm formation on the surface of the mucosa (Hicks *et al.*, 1996); and (3) inflammation of the mucosa, manifested by cytokine release, cell exfoliation, intestinal secretion (Bouckenooghe *et al.*, 2000; Grewal *et al.*, 1997; Harrington *et al.*, 2005; Jiang *et al.*, 2002; Steiner *et al.*, 1998).

Adhesion, which accounts for the typical pattern of EAEC called aggregative adherence – AA – is associated with a 60 mDa plasmid – the pAA (Nataro, 2005). The plasmid carries several virulence factors including the AAF fimbriae and toxins. AAFs have been found to include five major variants AAF/I – AAF/V (Bernier *et al.*, 2002; Boisen *et al.*, 2008; Czeczulin *et al.*, 1997; Dallman *et al.*, 2012; Nataro *et al.*, 1992). The fimbriae bind to components of the extracellular matrix such as laminin, collagen IV and fibronectin (Farfan *et al.*, 2008). By binding to these components as well as

interacting with each other the aggregative pattern is thought to emerge.

Pathogenesis of EAEC involves production of a mucus layer probably stemming from contributions of both bacteria and intestinal mucosa. EAEC survives within the mucus layer on the surface of enterocytes (Hicks *et al.*, 1996). The precise mechanisms for mucus stimulation are yet unknown

Mucosal inflammation and cytokine response has been demonstrated both in clinical (Greenberg *et al.*, 2002) and laboratory reports (Steiner *et al.*, 1998). Clinical studies have shown that lactoferrin, IL-8 and IL-β can be detected in faeces from cases of EAEC diarrhoea at a higher level than in stools of patients infected with non-EAEC diarrhoea (Jiang *et al.*, 2002)

In vitro, the EAEC flagellin induced IL-8 from intestinal epithelial cells (IECs) in culture (Steiner *et al.*, 1997, 1998). It is also seen that fimbriae mediate release of IL-8 and drop in TEER as well as AAF/II fimbriae are sufficient to induce transmigration of neutrophils across an epithelial layer *in vitro* (Saldana *et al.*, 2009). Since a suitable animal model for EAEC-infection is not yet available, a xenotransplant model was used and it was shown that the aggregative fimbriae are important in the development of inflammation in the human intestine. These data suggest that the AAF adhesins may be not only colonization factors, but may also be both necessary and sufficient for the induction of mucosal inflammation (Boll *et al.*, 2012). Recently, Zangari *et al.* (2013) developed a mouse model in order to study pathogenesis and treatment of the O104:H4 Shigatoxin producing EAEC strain (Zangari *et al.*, 2013). C57BL/6 mice infected with the outbreak strain exhibited both morbidity and mortality.

Clinical manifestations of infection

The clinical manifestations of EAEC infections involve complex host–pathogen interaction. At play are the heterogeneity of EAEC strains, amount of ingested EAEC, and specific host susceptibility (Huang *et al.*, 2004b). Jiang *et al.* (2003) identified a specific host genetic determinant that influences the clinical manifestations of EAEC infection (Jiang *et al.*, 2003). These investigators found that a single nucleotide polymorphism in the promoter of the −251 site of IL-8 conferred an increased risk of developing EAEC diarrhoea as well as presence of elevated faecal IL-8 (Jiang *et al.*, 2003).

The clinical features of EAEC disease have been ascertained from outbreaks, volunteer studies, or sporadic cases (Adachi *et al.*, 2002a; Bhan *et al.*, 1989b; Nataro *et al.*, 1995). The symptoms described to be associated with EAEC infection include nausea, anorexia, low grade fever, borborygmi and tenesmus (Cobeljic *et al.*, 1996; Gascon *et al.*, 1998; Gonzalez *et al.*, 1997). Watery diarrhoea and sometimes profound mucoid diarrhoea have been described (Ina *et al.*, 2003; Nataro, 2005; Nataro *et al.*, 1995). Bloody diarrhoea is not a distinctive feature of EAEC illness and has only rarely been reported. Studies have described one-third of > 2-year-old infant patients with EAEC diarrhoea had grossly bloody stools (Cravioto *et al.*, 1991; Itoh *et al.*, 1997). Further, EAEC diarrhoea has frequently been associated with the presence of faecal leucocytes and lactoferrin (Cennimo *et al.*, 2009; Greenberg *et al.*, 2002; Mercado *et al.*, 2011; Opintan *et al.*, 2010). As reviewed by Huang *et al.* (2006), the incubation period spans from 8 to 18 h, however incubation periods up to 52 h have been described. (Adachi *et al.*, 2002a; Frank *et al.*, 2011; Morabito *et al.*, 1998; Smith *et al.*, 1997) Persistent diarrhoea (> 14 days) is observed in malnourished hosts, especially in children living in developing countries (Huang *et al.*, 2006a).

As discussed later, it was seen that in the recent outbreak in northern Europe the disease manifestation was profound and severe since the EAEC strain acquired a Shiga toxin producing phage (Frank *et al.*, 2011; Scheutz *et al.*, 2011).

The site of colonization has been described to include the terminal ileum and the colon (Grad *et al.*, 2012; Wanke, 2001).

A well conducted study by Steiner *et al.* in 1998 reported how children residing in developing countries with EAEC positive stool samples regardless of presence of diarrhoea suffered from growth retardation (Steiner *et al.*, 1998). In a study by Roche *et al.* (2010) growth retardation and worsening of EAEC infection due to malnutrition

was observed in a mouse model. The growth impairment was seen to be dependent on both the microorganism burden and the challenging dose. It was observed that malnutrition intensified the EAEC infection by decreasing growth and increasing stool shedding of EAEC (Roche *et al.*, 2010).

Investigators working in Fortaleza, Brazil, have repeatedly implicated EAEC as the predominant agent of persistent diarrhoea (Fang *et al.*, 1995; Wanke *et al.*, 1991), which is associated with growth retardation. Interestingly, in this study population even asymptomatic patients infected with EAEC exhibit growth retardation compared with uninfected controls. In longitudinal studies of an infant cohort in Guinea-Bissau, EAEC infection was highly prevalent and was accompanied by growth retardation, although a direct link could not be established (Valentiner-Branth *et al.*, 2001). Given the high rate of asymptomatic excretion of EAEC in much of the developing world understanding its potential role in malnutrition and growth retardation is a high priority.

EAEC have been isolated in vast numbers from asymptomatic carriers (Black *et al.*, 2010; Boisen *et al.*, 2012; Hien *et al.*, 2008; Lima *et al.*, 2013; Okeke *et al.*, 2000b, 2003, 2005; Petri *et al.*, 2008; Scaletsky *et al.*, 2002) and the manifestation of gastrointestinal disease, as mentioned has been suggested to be dependent on host factors (Jiang *et al.*, 2003). An American study reported how young students going to Mexico were more likely to develop EAEC-induced diarrhoea when single nucleotide polymorphisms – SNPs – in the IL-8 gene was present.

In stool samples from EAEC-infected patients increased levels of faecal IL-8, IL-1β, leucocytes and lactoferrin were reported (Jiang *et al.*, 2002; Steiner *et al.*, 1998, 2000). Furthermore increased levels of inflammation mediators are seen even in asymptomatic EAEC-positive carriers when compared to healthy EAEC-negative controls. Dispersin, one of the virulence factors of EAEC encoded by the *aap* gene, have shown to cause an immune response (IgG against a recombinant dispersin protein) in travellers going to Mexico (Huang *et al.*, 2008) measured by ELISA. Another study showed increased levels of IgA antibodies binding to crude EAEC antigens measured in stool samples from 5 out of 10 US students returning from Guadalajara, Mexico (Estrada-Garcia *et al.*, 2002).

Diagnosis, identification and characterization of EAEC

The challenge for clinical microbiologists lays in rapid identification of the *E. coli* pathotype in question.

The golden standard for identification of EAEC remains the HEp-2 cell assay (Nataro and Kaper, 1998; Nataro *et al.*, 1987a). Nevertheless, this test is not designed for examining many samples at a time and it is mainly performed in reference laboratories, since the test requires cell culture facilities and is time consuming. Molecular techniques have been developed to detect EAEC of which PCR has great importance. Many primers for detecting EAEC using PCR have been proposed (Cerna *et al.*, 2003; Kahali *et al.*, 2004; Opintan *et al.*, 2010; Pereira *et al.*, 2007; Regua-Mangia *et al.*, 2009; Sarantuya *et al.*, 2004). Because of the heterogeneity of EAEC strains, there is not a single gene or a subset of genes, which will distinguish either pathogenic or non pathogenic EAEC strains. The gold standard for identifying EAEC is the HEp2-cell adherence assay (Nataro and Kaper, 1998) in as much as the pathogen was initially defined by the presence of a characteristic stacked brick pattern, designated aggregative adherence (AA) in this assay (Nataro *et al.*, 1987a). As reviewed, the sensitivity variation was between 20% and 89% when compared with the HEp2-cell adherence assay (Okeke, 2009). The CVD432 probe targeting the *aatA* gene (putative outer membrane protein) is widely used for the detection of EAEC by hybridization (Nishi *et al.*, 2003). The CVD432 probe has been shown to correspond to the *aatA* gene, which encodes a transporter for the dispersin protein (Aap), also regulated by AggR (Nishi *et al.*, 2003). Aap of EAEC is secreted by many EAEC strains and was suggested as a possible target for diagnosis (Sheikh *et al.*, 2002). However, a recent report demonstrates that Aap is also produced by non-EAEC strains.

A multiplex PCR assay was developed that detects the three AA plasmid-borne genes (*aatA*, *aggR* and *aap*) and studies showed that these loci are commonly but not invariably linked (Cerna *et al.*, 2003). The authors of this study found that 82% of the EAEC strains isolated from patients with diarrhoea were positive for the three loci and that use of a multiplex assay increases both the sensitivity and the specificity of EAEC detection (Cerna *et al.*, 2003). Several studies have applied PCR in detecting EAEC targeting genes *aggR* and/or *aatA* (Gomez-Duarte *et al.*, 2010; Kahali *et al.*, 2004; Opintan *et al.*, 2010; Pereira *et al.*, 2007; Regua-Mangia *et al.*, 2009; Rugeles *et al.*, 2010; Sarantuya *et al.*, 2004). The variable results obtained using molecular diagnostics support the heterogeneity of EAEC pathogenic mechanisms (Okeke and Nataro, 2001). PCR targeting both virulence factors on the pAA plasmid and the chromosomal EAEC loci, such as *aaiC* (Dudley *et al.*, 2006), might prove to be the most advantageous approach in detecting EAEC.

EAEC as endemic pathogen

Since the original description, EAEC has emerged as an important pathogen in several clinical scenarios, including travellers' diarrhoea (Adachi *et al.*, 1999, 2001, 2002a; Glandt *et al.*, 1999), endemic paediatric diarrhoea among children in developed countries (Tompkins *et al.*, 1999b), and developing countries (Okeke *et al.*, 2000a), as well as persistent diarrhoea amongst HIV-infected patients (Durrer *et al.*, 2000; Gassama-Sow *et al.*, 2004; Mossoro *et al.*, 2002; Wanke *et al.*, 1998a,b). A meta-analysis showed that EAEC is a cause of acute diarrhoeal illness in both developing and industrialized regions (Huang *et al.*, 2006b).

EAEC is best known for its role in persistent diarrhoea (> 14 days) in infants and children in developing countries. Studies in Mongolia (Sarantuya *et al.*, 2004), India (Dutta *et al.*, 2003), Brazil (Piva *et al.*, 2003; Zamboni *et al.*, 2004), Nigeria (Okeke *et al.*, 2000a,b), Israel (Shazberg *et al.*, 2003), Venezuela (Gonzalez *et al.*, 1997), Congo (Jalaluddin *et al.*, 1998) and many other countries have identified EAEC as a highly prevalent (often the most prevalent) *E. coli* pathotype in infants. The role of EAEC as a cause of persistent diarrhoea and malnutrition in Brazil has been demonstrated repeatedly (Fang *et al.*, 1995; Huang *et al.*, 2006b;

Lima and Guerrant, 1992). In one study, 68% of those with persistent diarrhoea shed EAEC in their stools (Fang *et al.*, 1995). In Guinea –Bissau the most common bacteria isolated from faeces of children < 2 years of age with diarrhoea was EAEC (Valentiner-Branth *et al.*, 2003).

A large prospective study in England (Tompkins *et al.*, 1999a), in which over 3600 cases of diarrhoea and controls were studied, found that typical EAEC was the second most common bacterial cause of gastroenteritis, following *Campylobacter*. EAEC was significantly associated with diarrhoea in both prospective cohorts and in patients presenting to physicians' attention. Typical EAEC was found to be a major cause of bacterial diarrhoea among infants in Cincinnati OH (Cohen *et al.*, 2005). A large prospective study on diarrhoea across all age groups was conducted in Baltimore, MD, USA and New Haven, CT, USA and EAEC was significantly associated with diarrhoea, being the most common bacterial cause of diarrhoea at both states (Nataro *et al.*, 2006), similarly confirmed by a study performed in New Jersey (Cennimo *et al.*, 2009). Moreover, a Scandinavian case–control study (Bhatnagar *et al.*, 1993), found EAEC in significantly more diarrhoeal cases than controls. This is supported by studies from Germany (Huppertz *et al.*, 1997) and Austria (Presterl *et al.*, 1999), which emphasizes the role of EAEC in developed countries. In a recent case control study of children from resource poor urban communities in northeastern Brazil demonstrated a high prevalence of EAEC, although the overall prevalence of EAEC positive samples in the diarrhoea group was similar to non-diarrhoea control group. Further, the frequency of different virulence related genes among these children confirmed the genetic heterogeneity of EAEC (Lima *et al.*, 2013).

The role of EAEC as an important pathogen in AIDS patients continues to develop, and EAEC now ranks among the most important enteric pathogens in this subpopulation (Wanke *et al.*, 1998a,b). EAEC was reported as the predominant cause of diarrhoea among HIV-infected patients in the Central African Republic (Germani *et al.*, 1998). The importance of EAEC in persistent diarrhoea among African AIDS patients was re-emphasized during a study in Senegal

(Gassama-Sow *et al.*, 2004). The finding that HIV replication is enhanced by inflammation and activation of NF-κB raises the concern that EAEC-induced inflammation may add to the struggles of millions of African AIDS patients

EAEC as a cause of travellers' diarrhoea

Over 100 million individuals travel from industrialized countries to developing regions of the tropical or semitropical world each year, which places them at risk of developing travellers' diarrhoea (TD). TD can be characterized by three factors: susceptibility to enteric infectious agents, residence in an industrialized country, and travel to a region of the tropical or semitropical world with lower levels of hygiene (DuPont, 2005, 2008).

In a recent review of all published studies of travellers' diarrhoea, EAEC was in aggregate second only to ETEC as the most common pathogen among patients with travellers' diarrhoea (Paredes-Paredes *et al.*, 2011; Shah *et al.*, 2009; Taylor *et al.*, 2006). EAEC was a major cause of diarrhoea among Spanish travellers going to the developing world, with an incidence identical to that of ETEC (Gascon *et al.*, 1998). Studies in Mexico have shown that in contrast with ETEC, in which travellers are most susceptible during the first weeks of exposure, travellers remain susceptible to EAEC infection throughout their stay (Adachi *et al.*, 2002a), likely reflecting the particular ability of EAEC to evade the immune system and cause persistent diarrhoea (Okhuysen and Dupont, 2010). Travellers in Mexico showed that the rate of EAEC colonization increased proportionally with the length of the stay (Adachi *et al.*, 2002a).

Travellers' diarrhoea in Korean patients who had visited South-East Asian countries were investigated in a prospective study taking place from February 2009 to April 2009 (Ahn *et al.*, 2011). Although ETEC was the most prevalently isolated pathogen by 36.0%, it was followed by EAEC found in 27.0% of the cases. Other pathogens found were *Vibrio parahaemolyticus*, *Vibrio cholerae*, *Salmonella* spp., *Shigella* and norovirus. Of the travellers with EAEC isolated in stool samples, 10 had been to Thailand, nine the Philippines, nine Vietnam, three Cambodia and two Indonesia. Co-infections are mentioned, but are only described as being scattered among pathogens isolated.

Additionally, illness due to EAEC related TD has been associated with irritable bowel syndrome (IBS) as well as chronic gastrointestinal syndromes (Okhuysen *et al.*, 2004; Sobieszczanska *et al.*, 2007) although the exact role of EAEC herein remains unclear.

EAEC causing diarrhoeal outbreaks

The first reported EAEC outbreak occurred in a Serbian nursery in 1995 (Cobeljic *et al.*, 1996); of the 19 afflicted infants, three developed persistent diarrhoea. EAEC outbreaks in Mexico City were presented in an abstract (Eslava C et al., 1993). Itoh *et al.* (1997) described a massive outbreak of EAEC diarrhoea among Japanese children, affecting nearly 2700 patients (Itoh *et al.*, 1997). Outbreaks have also been reported among adults in the UK (Smith *et al.*, 1997).

Another EAEC outbreak took place in 2008 in Italy on a farm holiday resort, where the guest experienced gastroenteritis after consuming unpasteurized cheese (Scavia *et al.*, 2008). EAEC was isolated in stool samples from six restaurant visitors and one staff member.

Shigatoxin producing EAEC and haemolytic–uremic syndrome

From May through June 2011, Europe was struck by an outbreak of massive proportions, infecting 4137 individuals and resulting in 54 deaths (Bielaszewska *et al.*, 2011; Frank *et al.*, 2011; Scheutz *et al.*, 2011). The strain was highly efficient in causing infection since its toxic abilities were enhanced by the aggregative fimbriae. There are major differences between the German outbreak and previous large outbreaks: (i) HUS represented a 22% of the cases (Frank *et al.*, 2011), which is a much higher proportion than in other outbreaks (6–10%); (ii) it was predominantly seen in healthy adults, traditionally diarrhoea-associated HUS occurs primarily in children; and (iii) the *E. coli* strain was a non-O157 Stx-producing EAEC

strain of serotype O104:H4 (Scheutz *et al.*, 2011) which was lysogenized with a Shiga-toxin encoding phage. This unusual combination of virulence factors of Stx and EAEC has rarely been described in humans, but has been identified before associated with limited numbers of cases (Dallman *et al.*, 2012; Morabito *et al.*, 1998; Vashakidze *et al.*, 2010). Limited knowledge exists on EAEC of the O104:H4 serotype. In a recent study of children's diarrhoea in Mali, Stx-negative EAEC O104:H4 was identified in six children with and without diarrhoea (Boisen *et al.*, 2012). In addition, genome analysis demonstrated that the outbreak strain contains virulence factors, originally described in EAEC and/or *Shigella* spp., including AAF, adhesins, Aap/dispersin, the Aat translocator, the Aai type VI secretion system, the AggR regulator, and three SPATE proteases (Pic, SigA and SepA) (Rasko *et al.*, 2011) (Fig. 12.1).

SPATE proteases are present in pathogenic *E. coli* strains and *Shigella* spp. (Henderson *et al.*, 1998); however, the confluence of Pic, SigA, and SepA is a combination almost exclusively among *Shigella flexneri* 2a (Boisen *et al.*, 2009), suggesting that the German outbreak strain has characteristics of a highly invasive and inflammatory diarrhoeagenic pathogen.

The severity of disease seen in this outbreak may be explained by EAEC constituting a unique and successful background reservoir of dangerous clones, and further genetic examination of O104:H4 isolates will not only elucidate virulence of the outbreak strain but also the origin. This combination of EAEC virulence factors and Stx possibly accounts for the virulence of this outbreak.

The possible virulence of the outbreak strain can be shown using explants of viable colonic tissue harvested from the cynomolgus monkey *Macaca fascicularis* (Fig. 12.2) The O104:H4 outbreak strain (C227–11) adheres in an aggregative manner and form heavy biofilm with thick mucus layer to the colonic tissues. Adherence is followed by mucosal toxicity, including crypt dilatation, microvillus vesiculation, and epithelial cell extrusion.

Morabito *et al.* (1998) described an outbreak of HUS in France caused by a Shiga toxin-producing EAEC strain. Sporadic small outbreaks

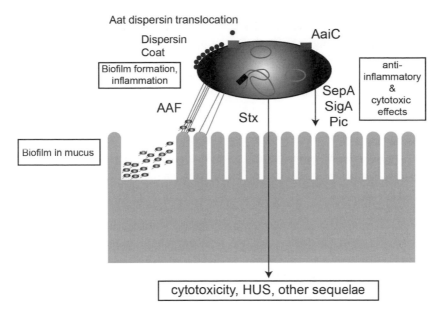

Figure 12.1 Proposed pathogenesis model for the increased virulence associated with the EAEC O104:H4 German outbreak strain. The O104:H4 outbreak strain harbours a Stx2-phage (black box) and a plasmid containing the CTX-M beta lactamase gene (blue), as well as EAEC virulence plasmid pAA (red). The master regulator AggR controls the expression of AAF/I, a dispersin surface coat protein, a dispersin translocator (encoded by *aatPABCD*) as well as a type VI secretion system within a chromosomal island called AAI (encoded by *aaiA-P*). The first stage of infection includes adherence to the intestinal mucosa by virtue of aggregative adherence fimbriae (AAF) followed by stimulation of mucus production, forming a biofilm on the surface of the mucosa; and finally, toxicity (SepA, SigA and Pic) to the mucosa, manifested by cytokine release, cell exfoliation, intestinal secretion and induction of mucosal inflammation. Collectively these factors possibly enable the outbreak strain to bind to and remain closely associated with the intestinal epithelium, which may promote increased uptake of Stx2 into the bloodstream. HUS, haemolytic–uraemic syndrome. A colour version of this figure is available in the plate section at the back of the book.

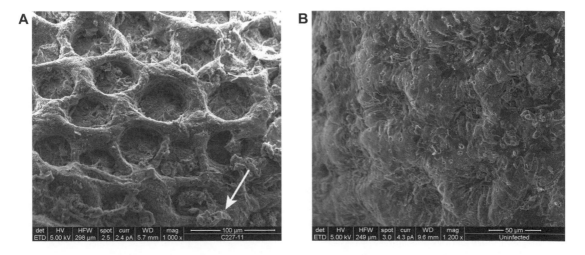

Figure 12.2 Scanning electron microscopy of *Macaca fascicularis* colon infected with O104:H4 EAEC strain C227–11 for 3 h (A) showing biofilm formation (white arrow) and mucosal intestinal toxicity which results in dilatation of crypt openings, rounding and exfoliation of colonic enterocytes, widening of inter-crypt crevices and loss of apical mucus from goblet cells. (B) Uninfected. Bar 50 μm.

of haemolytic uraemic syndrome (HUS) have been attributed to Stx-producing EAEC of the O111:H2, O111:H21 and O104:H4 serotypes over the last 15 years (Dallman *et al.*, 2012; Morabito *et al.*, 1998; Vashakidze *et al.*, 2010); however, they were localized to small populations in France, Ireland and the Republic of Georgia and were not disseminated throughout Europe.

The outbreak of HUS which occurred in the Republic of Georgia from July to October 2009 involving twenty-five people was caused by an EAEC of serotype O104:H4 Stx2+. In this outbreak, fifteen patients with HUS required haemodialysis and five died (Vashakidze *et al.*, 2010). The genome sequence(s) of the Georgian isolates are not available so it is unclear whether the German and Georgian isolates are the same. A recently identified EAEC O111:H21 strain that produces Stx2c caused illness in a family of cattle farmers in Ireland in 2011 (Dallman *et al.*, 2012). Despite the aforementioned reports, EAEC strains that produce Stxs have not been commonly isolated from patients with haemorrhagic colitis and HUS. However, Stx-producing *E. coli* (STEC) that produce intimin and form attach and effacing lesions have been the principal causative agents of post-diarrhoeal HUS. Whether Stx-EAEC will emerge as significant causes of post-diarrhoeal HUS remains to be seen.

EAEC in urinary tract infections (UTI)

In 1991, multi-resistant *Escherichia coli* strains caused an outbreak of urinary tract infections in Copenhagen, Denmark. A cluster of 18 patients in the Copenhagen area experienced predominantly community-acquired urinary tract infection (UTI) caused by multi-resistant *Escherichia coli* O78:H10, the only reported outbreak to date involving this serotype (Olesen *et al.*, 1994, 2012). Unexpectedly, all outbreak isolates fulfilled the molecular criteria for EAEC, but not for ExPEC. Additionally, a representative EAEC outbreak isolate exhibited 'stacked brick' adherence to cultured epithelial cells, the classic phenotype of EAEC. This marks the first time EAEC has been associated with an extraintestinal disease outbreak (Boll *et al.*, 2013; Olesen *et al.*, 2012).

Interestingly, few recent studies have demonstrated the presence of EAEC genes in collections of uropathogenic *E. coli* (UPEC) isolates (Nazemi *et al.*, 2011). The distribution of the pathogenic genes *aatA*, *aap*, *aggR*, among uropathogenic *Escherichia coli* (UPEC) and their linkage with *StbA* gene suggests the need for further characterization of UPEC strains and assessment of EAEC in pathogenesis of urinary tract infections (Nazemi *et al.*, 2011; Wallace-Gadsden *et al.*, 2007).

Epidemiology and transmission

The epidemiology of EAEC is poorly understood. To this day the reservoir for EAEC has still not been determined; there is no evidence for an animal reservoir of the bacterium (Huang *et al.*, 2004a), and it is commonly understood that the reservoir is human (Huang *et al.*, 2006a; Scheutz *et al.*, 2011; Lima *et al.*, 2013). The transmission of EAEC is often described as foodborne or via contaminated water, and as such believed to be transmitted via the faecal–oral route (Itoh *et al.*, 1997). In the recent O104:H4 German outbreak investigators found that fenugreek sprouts made of seeds from Egypt was the only sprout type with an independent association to illness in a multivariable analysis (King *et al.*, 2012).

A study showed how EAEC strains were able to survive in bottled water for up to 60 days at normal storage temperatures (Vasudevan *et al.*, 2003). The bacteria's growth was found to be more extensive in bottled water compared to spring water. Mexican sauces from restaurants in Guadalajara, Mexico, were examined as a possible way of transmission of EAEC (Adachi *et al.*, 2002b). They found that from 32 sauces 14 tested positive for EAEC in comparison to Mexican sauces from restaurants in Texas where only other types of *E. coli* were found.

Risk factors for EAEC infection include travel to developing countries, ingestion of contaminated food and water, poor hygiene, host susceptibility, and possibly immunosuppression (Huang and Dupont, 2004; Huang *et al.*, 2004b, 2006a; Nataro *et al.*, 1985; Okeke and Nataro, 2001).

Some studies suggest that children are more likely to be affected in the first month of life

(Gascon *et al.*, 1998; Gonzalez *et al.*, 1997), while others have stressed that most cases arise in older children (Okeke *et al.*, 2000a). These disparate epidemiological observations could be reconciled by diverse virulence characteristics of the strains circulating at the respective sites, in combination with a variety of host immunity and host resistance.

References

Adachi, J., Glandt, M., Jiang, Z.-D., Mathewson, J., Steffen, R., Ericsson, C., and DuPont, H. (1999). Enteroaggregative *Escherichia coli* as a major etiologic agent in travelers' diarrhoea in three regions of the world. Paper presented at: 37th Annual Meeting of the Infectious Diseases Society of America (Philadelphia: Infectious Diseases Society of America).

Adachi, J.A., Jiang, Z.D., Mathewson, J.J., Verenkar, M.P., Thompson, S., Martinez-Sandoval, F., Steffen, R., Ericsson, C.D., and DuPont, H.L. (2001). Enteroaggregative *Escherichia coli* as a major etiologic agent in traveler's diarrhoea in 3 regions of the world. Clin. Infect. Dis. *32*, 1706–1709.

Adachi, J.A., Ericsson, C.D., Jiang, Z.D., DuPont, M.W., Pallegar, S.R., and DuPont, H.L. (2002a). Natural history of enteroaggregative and enterotoxigenic *Escherichia coli* infection among US travelers to Guadalajara, Mexico. J. Infect. Dis. *185*, 1681–1683.

Adachi, J.A., Mathewson, J.J., Jiang, Z.D., Ericsson, C.D., and DuPont, H.L. (2002b). Enteric pathogens in Mexican sauces of popular restaurants in Guadalajara, Mexico, and Houston, Texas. Ann. Intern. Med. *136*, 884–887.

Ahn, J.Y., Chung, J.W., Chang, K.J., You, M.H., Chai, J.S., Kang, Y.A., Kim, S.H., Jeoung, H., Cheon, D., Jeoung, A., *et al.* (2011). Clinical characteristics and etiology of travelers' diarrhoea among Korean travelers visiting South-East Asia. J. Korean Med. Sci. *26*, 196–200.

Bernier, C., Gounon, P., and Le Bouguenec, C. (2002). Identification of an aggregative adhesion fimbria (AAF) type III-encoding operon in enteroaggregative *Escherichia coli* as a sensitive probe for detecting the AAF-encoding operon family. Infect. Immun. *70*, 4302–4311.

Bhan, M.K., Khoshoo, V., Sommerfelt, H., Raj, P., Sazawal, S., and Srivastava, R. (1989a). Enteroaggregative *Escherichia coli* and *Salmonella* associated with nondysenteric persistent diarrhoea. Pediatr. Infect. Dis. J. *8*, 499–502.

Bhan, M.K., Raj, P., Levine, M.M., Kaper, J.B., Bhandari, N., Srivastava, R., Kumar, R., and Sazawal, S. (1989b). Enteroaggregative *Escherichia coli* associated with persistent diarrhoea in a cohort of rural children in India. J. Infect. Dis. *159*, 1061–1064.

Bhatnagar, S., Bhan, M.K., Sommerfelt, H., Sazawal, S., Kumar, R., and Saini, S. (1993). Enteroaggregative *Escherichia coli* may be a new pathogen causing acute and persistent diarrhoea. Scand. J. Infect. Dis. *25*, 579–583.

Bielaszewska, M., Mellmann, A., Zhang, W., Kock, R., Fruth, A., Bauwens, A., Peters, G., and Karch, H. (2011). Characterisation of the *Escherichia coli* strain associated with an outbreak of haemolytic uraemic syndrome in Germany, 2011: a microbiological study. Lancet Infect. Dis. *11*, 671–676.

Black, R.E., Cousens, S., Johnson, H.L., Lawn, J.E., Rudan, I., Bassani, D.G., Jha, P., Campbell, H., Walker, C.F., Cibulskis, R., *et al.* (2010). Global, regional, and national causes of child mortality in 2008: a systematic analysis. Lancet *375*, 1969–1987.

Boisen, N., Struve, C., Scheutz, F., Krogfelt, K.A., and Nataro, J.P. (2008). New adhesin of enteroaggregative *Escherichia coli* related to the Afa/Dr/AAF family. Infect. Immun. *76*, 3281–3292.

Boisen, N., Ruiz-Perez, F., Scheutz, F., Krogfelt, K.A., and Nataro, J.P. (2009). Short report: high prevalence of serine protease autotransporter cytotoxins among strains of enteroaggregative *Escherichia coli*. Am. J. Trop. Med. Hyg. *80*, 294–301.

Boisen, N., Scheutz, F., Rasko, D.A., Redman, J.C., Persson, S., Simon, J., Kotloff, K.L., Levine, M.M., Sow, S., Tamboura, B., *et al.* (2012). Genomic characterization of enteroaggregative *Escherichia coli* from children in Mali. J. Infect. Dis. *205*, 431–444.

Boll, E.J., Struve, C., Sander, A., Demma, Z., Nataro, J.P., McCormick, B.A., and Krogfelt, K.A. (2012). The fimbriae of enteroaggregative *Escherichia coli* induce epithelial inflammation *in vitro* and in a human intestinal xenograft model. J. Infect. Dis. *206*, 714–722.

Boll, E.J., Struve, C., Boisen, N., Olesen, B., Stahlhut, S.G., and Krogfelt, K.A. (2013). Role of Enteroaggregative *Escherichia coli* Virulence Factors in Uropathogenesis. Infect Immun. *81*, 1164–1171.

Bouckenooghe, A.R., Dupont, H.L., Jiang, Z.D., Adachi, J., Mathewson, J.J., Verenkar, M.P., Rodrigues, S., and Steffen, R. (2000). Markers of enteric inflammation in enteroaggregative *Escherichia coli* diarrhoea in travelers. Am. J. Trop. Med. Hyg. *62*, 711–713.

Bryce, J., Boschi-Pinto, C., Shibuya, K., and Black, R.E. (2005). WHO estimates of the causes of death in children. Lancet *365*, 1147–1152.

Cennimo, D., Abbas, A., Huang, D.B., and Chiang, T. (2009). The prevalence and virulence characteristics of enteroaggregative *Escherichia coli* at an urgent-care clinic in the USA: a case–control study. J. Med. Microbiol. *58*, 403–407.

Cerna, J.F., Nataro, J.P., and Estrada-Garcia, T. (2003). Multiplex PCR for detection of three plasmid-borne genes of enteroaggregative *Escherichia coli* strains. J. Clin. Microbiol. *41*, 2138–2140.

Cobeljic, M., Miljkovic-Selimovic, B., Paunovic-Todosijevic, D., Velickovic, Z., Lepsanovic, Z., Zec, N., Savic, D., Ilic, R., Konstantinovic, S., Jovanovic, B., *et al.* (1996). Enteroaggregative *Escherichia coli* associated with an outbreak of diarrhoea in a neonatal nursery ward. Epidemiol. Infect. *117*, 11–16.

Cohen, M.B., Nataro, J.P., Bernstein, D.I., Hawkins, J., Roberts, N., and Staat, M.A. (2005). Prevalence of diarrhoeagenic *Escherichia coli* in acute childhood enteritis: a prospective controlled study. J. Pediatr. *146*, 54–61.

Cravioto, A., Tello, A., Navarro, A., Ruiz, J., Villafan, H., Uribe, F., and Eslava, C. (1991). Association of *Escherichia coli* HEp-2 adherence patterns with type and duration of diarrhoea. Lancet *337*, 262–264.

Czeczulin, J.R., Balepur, S., Hicks, S., Phillips, A., Hall, R., Kothary, M.H., Navarro-Garcia, F., and Nataro, J.P. (1997). Aggregative adherence fimbria II, a second fimbrial antigen mediating aggregative adherence in enteroaggregative *Escherichia coli*. Infect. Immun. *65*, 4135–4145.

Dallman, T., Smith, G.P., O'Brien, B., Chattaway, M.A., Finlay, D., Grant, K.A., and Jenkins, C. (2012). Characterization of a verocytotoxin-producing enteroaggregative *Escherichia coli* serogroup O111:H21 strain associated with a household outbreak in Northern Ireland. J. Clin. Microbiol. *50*, 4116–4119.

Dudley, E.G., Thomson, N.R., Parkhill, J., Morin, N.P., and Nataro, J.P. (2006). Proteomic and microarray characterization of the AggR regulon identifies a pheU pathogenicity island in enteroaggregative *Escherichia coli*. Mol. Microbiol. *61*, 1267–1282.

DuPont, H.L. (2005). What's new in enteric infectious diseases at home and abroad. Curr. Opin. Infect. Dis. *18*, 407–412.

DuPont, H.L. (2008). Systematic review: prevention of travellers' diarrhoea. Aliment. Pharmacol. Therapeut. *27*, 741–751.

Durrer, P., Zbinden, R., Fleisch, F., Altwegg, M., Ledergerber, B., Karch, H., and Weber, R. (2000). Intestinal infection due to enteroaggregative *escherichia coli* among human immunodeficiency virus-infected persons [In Process Citation]. J. Infect. Dis. *182*, 1540–1544.

Dutta, P.R., Sui, B.Q., and Nataro, J.P. (2003). Structure-function analysis of the enteroaggregative *Escherichia coli* plasmid-encoded toxin autotransporter using scanning linker mutagenesis. J. Biol. Chem. *278*, 39912–39920.

Eslava C, V.J., Morales R, Navarro A, Cravioto A. (1993). Abstracts of the 93rd General Meeting of the American Society for Microbiology 1993. Washington, DC: American Society for Microbiology. Identification of a protein with toxigenic activity produced by enteroaggregative *Escherichia coli*, abstr. B105; p. 44.

Estrada-Garcia, M.T., Jiang, Z.D., Adachi, J., Mathewson, J.J., and DuPont, H.L. (2002). Intestinal immunoglobulin a response to naturally acquired enterotoxigenic *Escherichia coli* in US travelers to an endemic area of Mexico. J. Travel Med. *9*, 247–250.

Fang, G.D., Lima, A.A., Martins, C.V., Nataro, J.P., and Guerrant, R.L. (1995). Etiology and epidemiology of persistent diarrhoea in northeastern Brazil: a hospital-based, prospective, case–control study. J. Pediatr. Gastroenterol. Nutr. *21*, 137–144.

Farfan, M.J., Inman, K.G., and Nataro, J.P. (2008). The major pilin subunit of the AAF/II fimbriae from enteroaggregative *Escherichia coli* mediates binding to extracellular matrix proteins. Infect. Immun. *76*, 4378–4384.

Farthing, M.J. (2000). Diarrhoea: a significant worldwide problem. Int. J. Antimicrob. Agents *14*, 65–69.

Frank, C., Werber, D., Cramer, J.P., Askar, M., Faber, M., an der Heiden, M., Bernard, H., Fruth, A., Prager, R., Spode, A., et al. (2011). Epidemic profile of Shiga-toxin-producing *Escherichia coli* O104:H4 outbreak in Germany. N. Engl. J. Med. *365*, 1771–1780.

Gascon, J., Vargas, M., Quinto, L., Corachan, M., Jimenez de Anta, M.T., and Vila, J. (1998). Enteroaggregative *Escherichia coli* strains as a cause of traveler's diarrhoea: a case–control study. J. Infect. Dis. *177*, 1409–1412.

Gassama-Sow, A., Sow, P.S., Gueye, M., Gueye-N'diaye, A., Perret, J.L., M'Boup, S., and Aidara-Kane, A. (2004). Characterization of pathogenic *Escherichia coli* in human immunodeficiency virus-related diarrhoea in Senegal. J. Infect. Dis. *189*, 75–78.

Germani, Y., Minssart, P., Vohito, M., Yassibanda, S., Glaziou, P., Hocquet, D., Berthelemy, P., and Morvan, J. (1998). Etiologies of acute, persistent, and dysenteric diarrhoeas in adults in Bangui, Central African Republic, in relation to human immunodeficiency virus serostatus. Am. J. Trop. Med. Hyg. *59*, 1008–1014.

Glandt, M., Adachi, J.A., Mathewson, J.J., Jiang, Z.D., DiCesare, D., Ashley, D., Ericsson, C.D., and DuPont, H.L. (1999). Enteroaggregative *Escherichia coli* as a cause of traveler's diarrhoea: clinical response to ciprofloxacin [In Process Citation]. Clin. Infect. Dis. *29*, 335–338.

Gomez-Duarte, O.G., Arzuza, O., Urbina, D., Bai, J., Guerra, J., Montes, O., Puello, M., Mendoza, K., and Castro, G.Y. (2010). Detection of *Escherichia coli* enteropathogens by multiplex polymerase chain reaction from children's diarrhoeal stools in two Caribbean-Colombian cities. Food. Pathog. Dis. *7*, 199–206.

Gonzalez, R., Diaz, C., Marino, M., Cloralt, R., Pequeneze, M., and Perez-Schael, I. (1997). Age-specific prevalence of *Escherichia coli* with localized and aggregative adherence in Venezuelan infants with acute diarrhoea. J. Clin. Microbiol. *35*, 1103–1107.

Grad, Y.H., Lipsitch, M., Feldgarden, M., Arachchi, H.M., Cerqueira, G.C., Fitzgerald, M., Godfrey, P., Haas, B.J., Murphy, C.I., Russ, C., et al. (2012). Genomic epidemiology of the *Escherichia coli* O104:H4 outbreaks in Europe, 2011. Proc. Natl. Acad. Sci. U.S.A. *109*, 3065–3070.

Greenberg, D.E., Jiang, Z.D., Steffen, R., Verenker, M.P., and DuPont, H.L. (2002). Markers of inflammation in bacterial diarrhoea among travelers, with a focus on enteroaggregative *Escherichia coli* pathogenicity. J. Infect. Dis. *185*, 944–949.

Grewal, H.M., Valvatne, H., Bhan, M.K., van Dijk, L., Gaastra, W., and Sommerfelt, H. (1997). A new putative fimbrial colonization factor, CS19, of human enterotoxigenic *Escherichia coli*. Infect. Immun. *65*, 507–513.

Harrington, S.M., Strauman, M.C., Abe, C.M., and Nataro, J.P. (2005). Aggregative adherence fimbriae contribute to the inflammatory response of epithelial cells infected with enteroaggregative *Escherichia coli*. Cell Microbiol. *7*, 1565–1578.

Henderson, I.R., Navarro-Garcia, F., and Nataro, J.P. (1998). The great escape: structure and function of the autotransporter proteins. Trends Microbiol. *6*, 370–378.

Hicks, S., Candy, D.C., and Phillips, A.D. (1996). Adhesion of enteroaggregative *Escherichia coli* to pediatric intestinal mucosa *in vitro*. Infect. Immun. *64*, 4751–4760.

Hien, B.T., Scheutz, F., Cam, P.D., Serichantalergs, O., Huong, T.T., Thu, T.M., and Dalsgaard, A. (2008). Diarrheagenic *Escherichia coli* and *Shigella* strains isolated from children in a hospital case–control study in Hanoi, Vietnam. J. Clin. Microbiol. *46*, 996–1004.

Huang, D.B., and Dupont, H.L. (2004). Enteroaggregative *Escherichia coli*: an emerging pathogen in children. Semin. Pediatr. Infect. Dis. *15*, 266–271.

Huang, D.B., Koo, H., and DuPont, H.L. (2004a). Enteroaggregative *Escherichia coli*: An Emerging Pathogen. Curr. Infect. Dis. Rep. *6*, 83–86.

Huang, D.B., Okhuysen, P.C., Jiang, Z.D., and DuPont, H.L. (2004b). Enteroaggregative *Escherichia coli*: an emerging enteric pathogen. Am. J. Gastroenterol. *99*, 383–389.

Huang, D.B., Mohanty, A., DuPont, H.L., Okhuysen, P.C., and Chiang, T. (2006a). A review of an emerging enteric pathogen: enteroaggregative *Escherichia coli*. J. Med. Microbiol. *55*, 1303–1311.

Huang, D.B., Nataro, J.P., DuPont, H.L., Kamat, P.P., Mhatre, A.D., Okhuysen, P.C., and Chiang, T. (2006b). Enteroaggregative *Escherichia coli* is a cause of acute diarrhoeal illness: a meta-analysis. Clin. Infect. Dis. *43*, 556–563.

Huang, D.B., Brown, E.L., DuPont, H.L., Cerf, J., Carlin, L., Flores, J., Belkind-Gerson, J., Nataro, J.P., and Okhuysen, P.C. (2008). Seroprevalence of the enteroaggregative *Escherichia coli* virulence factor dispersin among USA travellers to Cuernavaca, Mexico: a pilot study. J. Med. Microbiol. *57*, 476–479.

Huppertz, H.I., Rutkowski, S., Aleksic, S., and Karch, H. (1997). Acute and chronic diarrhoea and abdominal colic associated with enteroaggregative *Escherichia coli* in young children living in western Europe. Lancet *349*, 1660–1662.

Ina, K., Kusugami, K., and Ohta, M. (2003). Bacterial hemorrhagic enterocolitis. J. Gastroenterol. *38*, 111–120.

Itoh, Y., Nagano, I., Kunishima, M., and Ezaki, T. (1997). Laboratory investigation of enteroaggregative *Escherichia coli* O untypeable:H10 associated with a massive outbreak of gastrointestinal illness. J. Clin. Microbiol. *35*, 2546–2550.

Jalaluddin, S., de Mol, P., Hemelhof, W., Bauma, N., Brasseur, D., Hennart, P., Lomoyo, R.E., Rowe, B., and Butzler, J.P. (1998). Isolation and characterization of enteroaggregative *Escherichia coli* (EAggEC) by genotypic and phenotypic markers, isolated from diarrhoeal children in Congo. Clin. Microbiol. Infect. *4*, 213–219.

Jiang, Z.D., Greenberg, D., Nataro, J.P., Steffen, R., and DuPont, H.L. (2002). Rate of occurrence and pathogenic effect of enteroaggregative *Escherichia coli* virulence factors in international travelers. J. Clin. Microbiol. *40*, 4185–4190.

Jiang, Z.D., Okhuysen, P.C., Guo, D.C., He, R., King, T.M., DuPont, H.L., and Milewicz, D.M. (2003). Genetic susceptibility to enteroaggregative *Escherichia coli* diarrhoea: polymorphism in the interleukin-8 promotor region. J. Infect. Dis. *188*, 506–511.

Kahali, S., Sarkar, B., Rajendran, K., Khanam, J., Yamasaki, S., Nandy, R.K., Bhattacharya, S.K., and Ramamurthy, T. (2004). Virulence characteristics and molecular epidemiology of enteroaggregative *Escherichia coli* isolates from hospitalized diarrhoeal patients in Kolkata, India. J. Clin. Microbiol. *42*, 4111–4120.

Kaper, J.B., Nataro, J.P., and Mobley, H.L. (2004). Pathogenic *Escherichia coli*. Nat. Rev. Microbiol. *2*, 123–140.

King, L.A., Nogareda, F., Weill, F.X., Mariani-Kurkdjian, P., Loukiadis, E., Gault, G., Jourdan-DaSilva, N., Bingen, E., Mace, M., Thevenot, D., *et al.* (2012). Outbreak of Shiga toxin-producing *Escherichia coli* O104:H4 associated with organic fenugreek sprouts, France, June 2011. Clin. Infect. Dis. *54*, 1588–1594.

Lima, A.A., and Guerrant, R.L. (1992). Persistent diarrhoea in children: epidemiology, risk factors, pathophysiology, nutritional impact, and management. Epidemiol. Rev. *14*, 222–242.

Lima, A.A., Moore, S.R., Barboza, M.S., Jr., Soares, A.M., Schleupner, M.A., Newman, R.D., Sears, C.L., Nataro, J.P., Fedorko, D.P., Wuhib, T., *et al.* (2000). Persistent diarrhoea signals a critical period of increased diarrhoea burdens and nutritional shortfalls: a prospective cohort study among children in northeastern Brazil. J. Infect. Dis. *181*, 1643–1651.

Lima, I.F., Boisen, N., Quetz, J.D., Havt, A., Carvalho, E.B., Soares, A.M., Lima, N.L., Mota, R.M., Nataro, J.P., Guerrant, R.L., *et al.* (2013). Prevalence of enteroaggregative *Escherichia coli* and its virulence-related genes in a case–control study among children from Northeastern Brazil. J. Med. Microbiol. *62*, 683–693.

Mercado, E.H., Ochoa, T.J., Ecker, L., Cabello, M., Durand, D., Barletta, F., Molina, M., Gil, A.I., Huicho, L., Lanata, C.F., *et al.* (2011). Fecal leukocytes in children infected with diarrhoeagenic *Escherichia coli*. J. Clin. Microbiol. *49*, 1376–1381.

Morabito, S., Karch, H., Mariani-Kurkdjian, P., Schmidt, H., Minelli, F., Bingen, E., and Caprioli, A. (1998). Enteroaggregative, Shiga toxin-producing *Escherichia coli* O111:H2 associated with an outbreak of hemolytic-uremic syndrome. J. Clin. Microbiol. *36*, 840–842.

Mossoro, C., Glaziou, P., Yassibanda, S., Lan, N.T., Bekondi, C., Minssart, P., Bernier, C., Le Bouguenec, C., and Germani, Y. (2002). Chronic diarrhoea, hemorrhagic colitis, and hemolytic-uremic syndrome associated with HEp-2 adherent *Escherichia coli* in adults infected with human immunodeficiency virus in Bangui, Central African Republic. J. Clin. Microbiol. *40*, 3086–3088.

Nataro, J.P. (2005). Enteroaggregative *Escherichia coli* pathogenesis. Curr. Opin. Gastroenterol. *21*, 4–8.

Nataro, J.P., and Kaper, J.B. (1998). Diarrheagenic *Escherichia coli*. Clin. Microbiol. Rev. *11*, 142–201.

Nataro, J.P., Baldini, M.M., Kaper, J.B., Black, R.E., Bravo, N., and Levine, M.M. (1985). Detection of an adherence factor of enteropathogenic *Escherichia coli* with a DNA probe. J. Infect. Dis. *152*, 560–565.

Nataro, J.P., Kaper, J.B., Robins-Browne, R., Prado, V., Vial, P., and Levine, M.M. (1987a). Patterns of adherence of diarrhoeagenic *Escherichia coli* to HEp-2 cells. Pediatr. Infect. Dis. J. *6*, 829–831.

Nataro, J.P., Maher, K.O., Mackie, P., and Kaper, J.B. (1987b). Characterization of plasmids encoding the adherence factor of enteropathogenic *Escherichia coli*. Infect. Immun. *55*, 2370–2377.

Nataro, J.P., Deng, Y., Maneval, D.R., German, A.L., Martin, W.C., and Levine, M.M. (1992). Aggregative adherence fimbriae I of enteroaggregative *Escherichia coli* mediate adherence to HEp-2 cells and hemagglutination of human erythrocytes. Infect. Immun. *60*, 2297–2304.

Nataro, J.P., Deng, Y., Cookson, S., Cravioto, A., Savarino, S.J., Guers, L.D., Levine, M.M., and Tacket, C.O. (1995). Heterogeneity of enteroaggregative *Escherichia coli* virulence demonstrated in volunteers. J. Infect. Dis. *171*, 465–468.

Nataro, J.P., Mai, V., Johnson, J., Blackwelder, W.C., Heimer, R., Tirrell, S., Edberg, S.C., Braden, C.R., Glenn Morris, J., Jr., and Hirshon, J.M. (2006). Diarrheagenic *Escherichia coli* infection in Baltimore, Maryland, and New Haven, Connecticut. Clin. Infect. Dis. *43*, 402–407.

Nazemi, A., Mirinargasi, M., Merikhi, N., and Sharifi, S.H. (2011). Distribution of Pathogenic Genes aatA, aap, aggR, among Uropathogenic *Escherichia coli* (UPEC) and Their Linkage with StbA Gene. Indian J. Microbiol. *51*, 355–358.

Nishi, J., Sheikh, J., Mizuguchi, K., Luisi, B., Burland, V., Boutin, A., Rose, D.J., Blattner, F.R., and Nataro, J.P. (2003). The export of coat protein from enteroaggregative *Escherichia coli* by a specific ATP-binding cassette transporter system. J. Biol. Chem. *278*, 45680–45689.

Okeke, I.N. (2009). Diarrheagenic *Escherichia coli* in sub-Saharan Africa: status, uncertainties and necessities. J. Infect. Dev. Ctries. *3*, 817–842.

Okeke, I.N., and Nataro, J.P. (2001). Enteroaggregative *Escherichia coli*. Lancet Infect. Dis. *1*, 304–313.

Okeke, I.N., Lamikanra, A., Czeczulin, J., Dubovsky, F., Kaper, J.B., and Nataro, J.P. (2000a). Heterogeneous virulence of enteroaggregative *Escherichia coli* strains isolated from children in Southwest Nigeria. J. Infect. Dis. *181*, 252–260.

Okeke, I.N., Lamikanra, A., Steinruck, H., and Kaper, J.B. (2000b). Characterization of *Escherichia coli* strains from cases of childhood diarrhoea in provincial southwestern Nigeria. J. Clin. Microbiol. *38*, 7–12.

Okeke, I.N., Ojo, O., Lamikanra, A., and Kaper, J.B. (2003). Etiology of acute diarrhoea in adults in southwestern Nigeria. J. Clin. Microbiol. *41*, 4525–4530.

Okhuysen, P.C., and Dupont, H.L. (2010). Enteroaggregative *Escherichia coli* (EAEC): A Cause of Acute and Persistent Diarrhea of Worldwide Importance. J. Infect. Dis. *202*, 503–505.

Okhuysen, P.C., Jiang, Z.D., Carlin, L., Forbes, C., and DuPont, H.L. (2004). Post-diarrhoea chronic intestinal symptoms and irritable bowel syndrome in North American travelers to Mexico. Am. J. Gastroenterol. *99*, 1774–1778.

Olesen, B., Kolmos, H.J., Orskov, F., and Orskov, I. (1994). Cluster of multiresistant *Escherichia coli* O78:H10 in Greater Copenhagen. Scand. J. Infect. Dis. *26*, 406–410.

Olesen, B., Neimann, J., Bottiger, B., Ethelberg, S., Schiellerup, P., Jensen, C., Helms, M., Scheutz, F., Olsen, K.E., Krogfelt, K., et al. (2005). Etiology of diarrhoea in young children in Denmark: a case–control study. J. Clin. Microbiol. *43*, 3636–3641.

Olesen, B., Scheutz, F., Andersen, R.L., Menard, M., Boisen, N., Johnston, B., Hansen, D.S., Krogfelt, K.A., Nataro, J.P., and Johnson, J.R. (2012). Enteroaggregative *Escherichia coli* O78:H10, the cause of an outbreak of urinary tract infection. J. Clin. Microbiol. *50*, 3703–3711.

Opintan, J.A., Bishar, R.A., Newman, M.J., and Okeke, I.N. (2010). Carriage of diarrhoeagenic *Escherichia coli* by older children and adults in Accra, Ghana. Trans. R. Soc. Trop. Med. Hyg. *104*, 504–506.

Paredes-Paredes, M., Okhuysen, P.C., Flores, J., Mohamed, J.A., Padda, R.S., Gonzalez-Estrada, A., Haley, C.A., Carlin, L.G., Nair, P., and DuPont, H.L. (2011). Seasonality of diarrhoeagenic *Escherichia coli* pathotypes in the US students acquiring diarrhoea in Mexico. J. Travel Med. *18*, 121–125.

Pereira, A.L., Ferraz, L.R., Silva, R.S., and Giugliano, L.G. (2007). Enteroaggregative *Escherichia coli* virulence markers: positive association with distinct clinical characteristics and segregation into 3 enteropathogenic *E. coli* serogroups. J. Infect. Dis. *195*, 366–374.

Petri, W.A., Jr., Miller, M., Binder, H.J., Levine, M.M., Dillingham, R., and Guerrant, R.L. (2008). Enteric infections, diarrhoea, and their impact on function and development. J. Clin. Invest. *118*, 1277–1290.

Piva, I.C., Pereira, A.L., Ferraz, L.R., Silva, R.S., Vieira, A.C., Blanco, J.E., Blanco, M., Blanco, J., and Giugliano, L.G. (2003). Virulence markers of enteroaggregative *Escherichia coli* isolated from children and adults with diarrhoea in Brasilia, Brazil. J. Clin. Microbiol. *41*, 1827–1832.

Presterl, E., Nadrchal, R., Wolf, D., Rotter, M., and Hirschl, A.M. (1999). Enteroaggregative and enterotoxigenic *Escherichia coli* among isolates from patients with diarrhoea in Austria. Eur. J. Clin. Microbiol. Infect. Dis. *18*, 209–212.

Rasko, D.A., Webster, D.R., Sahl, J.W., Bashir, A., Boisen, N., Scheutz, F., Paxinos, E.E., Sebra, R., Chin, C.S., Iliopoulos, D., et al. (2011). Origins of the E. coli Strain Causing an Outbreak of Hemolytic-Uremic Syndrome in Germany. N. Engl. J. Med. *365*, 709–717.

Regua-Mangia, A.H., Gomes, T.A., Vieira, M.A., Irino, K., and Teixeira, L.M. (2009). Molecular typing and virulence of enteroaggregative *Escherichia coli* strains isolated from children with and without diarrhoea in Rio de Janeiro city, Brazil. J. Med. Microbiol. *58*, 414–422.

Roche, J.K., Cabel, A., Sevilleja, J., Nataro, J., and Guerrant, R.L. (2010). Enteroaggregative *Escherichia coli* (EAEC) impairs growth while malnutrition worsens EAEC infection: a novel murine model of the infection malnutrition cycle. J. Infect. Dis. *202*, 506–514.

Rugeles, L.C., Bai, J., Martinez, A.J., Vanegas, M.C., and Gomez-Duarte, O.G. (2010). Molecular characterization of diarrhoeagenic *Escherichia coli* strains from stools samples and food products in Colombia. Int. J. Food Microbiol. *138*, 282–286.

Saldana, Z., Erdem, A.L., Schuller, S., Okeke, I.N., Lucas, M., Sivananthan, A., Phillips, A.D., Kaper, J.B., Puente, J.L., and Giron, J.A. (2009). The *Escherichia coli* common pilus and the bundle-forming pilus act in concert during the formation of localized adherence by enteropathogenic *E. coli*. J. Bacteriol. *191*, 3451–3461.

Sarantuya, J., Nishi, J., Wakimoto, N., Erdene, S., Nataro, J.P., Sheikh, J., Iwashita, M., Manago, K., Tokuda, K., Yoshinaga, M., *et al.* (2004). Typical enteroaggregative *Escherichia coli* is the most prevalent pathotype among *E. coli* strains causing diarrhoea in Mongolian children. J. Clin. Microbiol. *42*, 133–139.

Scaletsky, I.C., Fabbricotti, S.H., Carvalho, R.L., Nunes, C.R., Maranhao, H.S., Morais, M.B., and Fagundes-Neto, U. (2002). Diffusely adherent *Escherichia coli* as a cause of acute diarrhoea in young children in Northeast Brazil: a case–control study. J. Clin. Microbiol. *40*, 645–648.

Scavia, G., Staffolani, M., Fisichella, S., Striano, G., Colletta, S., Ferri, G., Escher, M., Minelli, F., and Caprioli, A. (2008). Enteroaggregative *Escherichia coli* associated with a foodborne outbreak of gastroenteritis. J. Med. Microbiol. *57*, 1141–1146.

Scheutz, F., Moller Nielsen, E., Frimodt-Moller, J., Boisen, N., Morabito, S., Tozzoli, R., Nataro, J., and Caprioli, A. (2011). Characteristics of the enteroaggregative Shiga toxin/verotoxin-producing *Escherichia coli* O104:H4 strain causing the outbreak of haemolytic uraemic syndrome in Germany, May to June 2011. Euro Surveill. *16*.

Shah, N., DuPont, H.L., and Ramsey, D.J. (2009). Global etiology of travelers' diarrhoea: systematic review from 1973 to the present. Am. J. Trop. Med. Hyg. *80*, 609–614.

Shazberg, G., Wolk, M., Schmidt, H., Sechter, I., Gottesman, G., and Miron, D. (2003). Enteroaggregative *Escherichia coli* serotype O126:H27, Israel. Emerg. Infect. Dis. *9*, 1170–1173.

Sheikh, J., Czeczulin, J.R., Harrington, S., Hicks, S., Henderson, I.R., Le Bouguenec, C., Gounon, P., Phillips, A., and Nataro, J.P. (2002). A novel dispersin protein in enteroaggregative *Escherichia coli*. J. Clin. Invest. *110*, 1329–1337.

Smith, H.R., Cheasty, T., and Rowe, B. (1997). Enteroaggregative *Escherichia coll* and outbreaks of gastroenteritis in UK. Lancet *350*, 814–815.

Sobieszczanska, B.M., Osek, J., Wasko-Czopnik, D., Dworniczek, E., and Jermakow, K. (2007). Association of enteroaggregative *Escherichia coli* with irritable bowel syndrome. Clin. Microbiol. Infect. *13*, 404–407.

Steiner, T., Flores, C., Pizarro, T., and Guerrant, R. (1997). Fecal lactoferrin, interleukin-1ß, and interleukin-8 are elevated in patients with severe *Clostridium difficile* colitis. Clin. Diag. Lab. Immunol. *4*, 179–722.

Steiner, T.S., Lima, A.A., Nataro, J.P., and Guerrant, R.L. (1998). Enteroaggregative *Escherichia coli* produce intestinal inflammation and growth impairment and cause interleukin-8 release from intestinal epithelial cells. J. Infect. Dis. *177*, 88–96.

Steiner, T.S., Nataro, J.P., Poteet-Smith, C.E., Smith, J.A., and Guerrant, R.L. (2000). Enteroaggregative *Escherichia coli* expresses a novel flagellin that causes IL-8 release from intestinal epithelial cells. J. Clin. Invest. *105*, 1769–1777.

Taylor, D.N., Bourgeois, A.L., Ericsson, C.D., Steffen, R., Jiang, Z.D., Halpern, J., Haake, R., and Dupont, H.L. (2006). A randomized, double-blind, multicenter study of rifaximin compared with placebo and with ciprofloxacin in the treatment of travelers' diarrhoea. Am. J. Trop. Med. Hyg. *74*, 1060–1066.

Tompkins, D.S., Hudson, M.J., Smith, H.R., Eglin, R.P., Wheeler, J.G., Brett, M.M., Owen, R.J., Brazier, J.S., Cumberland, P., King, V., *et al.* (1999a). A study of infectious intestinal disease in England: microbiological findings in cases and controls. Commun. Dis. Public Health 2, 108–113.

Tompkins, D.S., Hudson, M.J., Smith, H.R., Eglin, R.P., Wheeler, J.G., Brett, M.M., Owen, R.J., Brazier, J.S., Cumberland, P., King, V., *et al.* (1999b). A study of infectious intestinal disease in England: microbiological findings in cases and controls [see comments]. Commun. Dis. Public Health 2, 108–113.

Tzipori, S., Montanaro, J., Robinsbrowne, R.M., Vial, P., Gibson, R., and Levine, M.M. (1992). Studies with Enteroaggregative *Escherichia-Coli* in the Gnotobiotic Piglet Gastroenteritis Model. Infect. Immun. *60*, 5302–5306.

Valentiner-Branth, P., Steinsland, H., Santos, G., Perch, M., Begtrup, K., Bhan, M.K., Dias, F., Aaby, P., Sommerfelt, H., and Molbak, K. (2001). Community-based controlled trial of dietary management of children with persistent diarrhoea: sustained beneficial effect on ponderal and linear growth. Am. J. Clin. Nutr. *73*, 968–974.

Valentiner-Branth, P., Steinsland, H., Fischer, T.K., Perch, M., Scheutz, F., Dias, F., Aaby, P., Molbak, K., and Sommerfelt, H. (2003). Cohort study of guinean children: Incidence, pathogenicity, conferred protection, and attributable risk for enteropathogens during the first 2 years of life. J. Clin. Microbiol. *41*, 4238–4245.

Vashakidze, E., Megrelishvili, T., Pachkoria, E., Tevzadze, L., and Lashkarashvili, M. (2010). Enterohemorrhagic *E. coli* and hemolytic uremic syndrome in Georgia. Georgian Med. News, 38–41.

Vasudevan, P., Annamalai, T., Sartori, L., Hoagland, T., and Venkitanarayanan, K. (2003). Behavior of enteroaggregative *Escherichia coli* in bottled spring and mineral water. J. Food Prot. *66*, 497–500.

Wallace-Gadsden, F., Johnson, J.R., Wain, J., and Okeke, I.N. (2007). Enteroaggregative *Escherichia coli* related to uropathogenic clonal group A. Emerg. Infect. Dis. *13*, 757–760.

Wanke, C.A. (2001). To know *Escherichia coli* is to know bacterial diarrhoeal disease. Clin. Infect. Dis. *32*, 1710–1712.

Wanke, C.A., Schorling, J.B., Barrett, L.J., Desouza, M.A., and Guerrant, R.L. (1991). Potential role of adherence traits of *Escherichia coli* in persistent diarrhoea in an urban Brazilian slum. Pediatr. Infect. Dis. J. *10*, 746–751.

Wanke, C.A., Gerrior, J., Blais, V., Mayer, H., and Acheson, D. (1998a). Successful treatment of diarrhoeal disease associated with enteroaggregative *Escherichia coli* in adults infected with human immunodeficiency virus. J. Infect. Dis. *178*, 1369–1372.

Wanke, C.A., Mayer, H., Weber, R., Zbinden, R., Watson, D.A., and Acheson, D. (1998b). Enteroaggregative *Escherichia coli* as a potential cause of diarrhoeal disease in adults infected with human immunodeficiency virus. J. Infect. Dis. *178*, 185–190.

Zamboni, A., Fabbricotti, S.H., Fagundes-Neto, U., and Scaletsky, I.C. (2004). Enteroaggregative *Escherichia coli* virulence factors are found to be associated with infantile diarrhoea in Brazil. J. Clin. Microbiol. *42*, 1058–1063.

Zangari, T., Melton-Celsa, A.R., Panda, A., Boisen, N., Smith, M.A., Taratov, I., De Tolla, L.J., Nataro, J.P., and O'Brien, A.D. (2013). Virulence of the Shiga toxin type 2-expressing *Escherichia coli* O104:H4 German outbreak isolate in two animal models. Infect. Immun. *81*, 1562–1574.

Host Cell Invasion by Pathogenic *Escherichia coli*

13

Adam J. Lewis, Elizabeth M. Ott, Travis J. Wiles and Matthew A. Mulvey

Abstract

Pathogenic strains of *Escherichia coli* are likely not the first bacteria that come to mind when most scientists and physicians contemplate invasive microorganisms. Yet, select members of this genetically diverse group of pathogens are adept at invading host cells and taking advantage of the numerous benefits afforded to bacteria within intracellular niches. These benefits include access to alternate nutrient sources and protection from a variety of hazards including shear flow of bodily fluids, host phagocytes, complement, antimicrobial peptides, antibodies, and antibiotics. To reap these rewards, bacterial pathogens must manipulate host cell machinery to promote their uptake, subsequently avoid destruction within lysosomal compartments, and ultimately return to the extracellular environment to perpetuate their genetic lineage. Here we cover four invasive pathotypes of *E. coli*: uropathogenic *E. coli* (UPEC), neonatal meningitis *E. coli* (NMEC), enteroinvasive *E. coli* (EIEC), and adherent-invasive *E. coli* (AIEC) (Fig. 13.1). The molecular mechanisms and consequences of host cell invasion by these important pathogens are discussed, with an emphasis on UPEC.

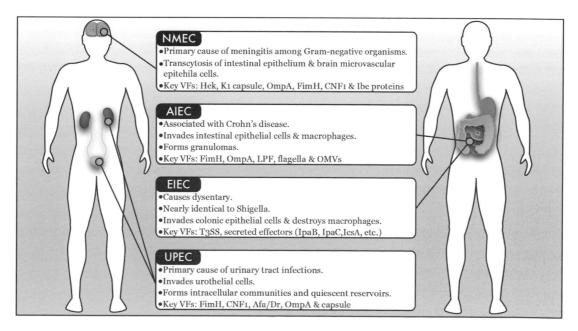

NMEC
- Primary cause of meningitis among Gram-negative organisms.
- Transcytosis of intestinal epithelium & brain microvascular epitehila cells.
- Key VFs: Hek, K1 capsule, OmpA, FimH, CNF1 & Ibe proteins

AIEC
- Associated with Crohn's disease.
- Invades intestinal epithelial cells & macrophages.
- Forms granulomas.
- Key VFs: FimH, OmpA, LPF, flagella & OMVs

EIEC
- Causes dysentery.
- Nearly identical to Shigella.
- Invades colonic epithelial cells & destroys macrophages.
- Key VFs: T3SS, secreted effectors (IpaB, IpaC,IcsA, etc.)

UPEC
- Primary cause of urinary tract infections.
- Invades urothelial cells.
- Forms intracellular communities and quiescent reservoirs.
- Key VFs: FimH, CNF1, Afa/Dr, OmpA & capsule

Figure 13.1 Types of invasive *E. coli* strains with associated virulence factors (VFs) used to enter target host cells. A colour version of this figure is available in the plate section at the back of the book.

UPEC: close encounters with the urothelium

Strains of uropathogenic *Escherichia coli* (UPEC) are the most common causes of urinary tract infection (UTI), capable of colonizing the kidneys as well as the uroepithelial cells (urothelium) that line the urethra, ureters, and bladder (Dielubanza and Schaeffer, 2011; Foxman and Brown, 2003). UTIs are especially problematic due to their propensity to recur and develop into lingering chronic infections (Blango and Mulvey, 2010; Dielubanza and Schaeffer, 2011). The recent emergence of antibiotic-resistant UPEC strains is thwarting the successful treatment of UTIs (Brumbaugh and Mobley, 2012), spurring efforts to better understand and combat these infections.

Historically, UPEC were thought to be strictly extracellular pathogens, despite the fact that early microscopic observations showed that bladder epithelial cells in rodents could internalize UPEC during experimental UTI (Fukushi *et al.*, 1979; McTaggart *et al.*, 1990). Intracellular UPEC were observed both within membrane-bound compartments and free within the host cytosol, and were assumed to be destined for destruction. Subsequent work using mouse models of cystitis (bladder infection) revealed that intracellular pools of UPEC account for greater than 50% of all bacteria present in the bladder within 12 h post inoculation, suggesting that entry into host cells may benefit the pathogen (Mulvey *et al.*, 1998; Schwartz *et al.*, 2011). In fact, intracellular UPEC have a quantifiable advantage over their extracellular counterparts, such as being able to persist longer within the host even in the presence of antibiotics that effectively sterilize the urine (Blango and Mulvey, 2010; Dhakal *et al.*, 2008). In light of these findings, significant effort over the past decade has been invested into understanding how UPEC invades and behaves within the epithelial cells that comprise the urothelium.

UPEC usually enter the urinary tract via an ascending route, moving from the external urethra opening and into the urethra before taking up residence within the bladder. The urothelium that lines the lumen of the bladder is made up of 2–3 layers of immature, partially differentiated epithelial cells underlying a single layer of exceptionally large, often binucleate, terminally differentiated superficial umbrella cells (Apodaca, 2004; Hicks, 1975). The stratified layers of the urothelium comprise a permeability barrier on par in strength with the blood–brain barrier. The barrier function of the urothelium is dependent upon specialized tight junctions and elaboration of the so-called asymmetrical unit membrane (AUM) on the apical surface of the umbrella cells (Apodaca, 2004). The AUM consists of a quasi-crystalline hexagonal array of at least four integral membrane proteins collectively known as uroplakins (Wu *et al.*, 2009). The uroplakins are assembled into plaques that move from the trans Golgi network to the cell surface, adding membrane to increase lumenal surface area as the bladder distends. All but one of the uroplakins have small cytosolic tails and relatively large extracellular domains, giving the AUM its characteristic asymmetric appearance when viewed in cross-section by transmission electron microscopy. As urine stored in the bladder lumen is released, the uroplakin plaques are endocytosed and are, for the most part, eventually degraded within lysosomes (Kreft *et al.*, 2009).

UPEC is able to invade all layers of the bladder urothelium, with superficial umbrella cells being primary targets (Fig. 13.2). Within both umbrella and immature bladder epithelial cells, UPEC is trafficked into acidic, membrane-bound compartments that resemble late endosomes or lysosomes (Eto *et al.*, 2006; Mysorekar and Hultgren, 2006). Specifically, these compartments contain the late endosome/lysosome-associated markers LAMP-1, CD63, and the lipid lysobisphosphatidic acid (LBPA), but lack cathepsin-D, a marker of degradative lysosomes. Although UPEC can survive within these acidic compartments, their growth appears to be restricted until they break out into the host cytosol.

Intracellular bacterial communities – UPEC in party mode

In mouse models of UTI, in which UPEC is inoculated into the bladder via transurethral catheterization, rampant intracellular growth of UPEC is typically limited to the terminally differentiated superficial umbrella cells (Eto *et al.*, 2006; Mulvey *et al.*, 2001; Mysorekar and Hultgren, 2006). Upon entering the cytosol of these host cells, UPEC can

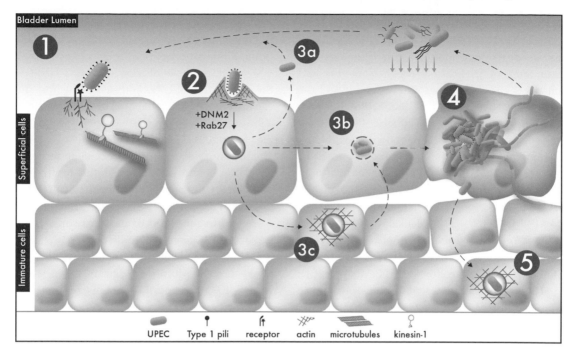

Figure 13.2 UPEC invasion of the urothelium. (**1**) UPEC bind receptors on bladder cells via type 1 pili, triggering actin cytoskeletal rearrangements that are modulated by microtubules, kinesin-1 and many other host factors. (**2**) The host plasma membrane zippers around adherent bacteria, eventually delivering UPEC into a late endosome-like compartment via a DNM2- and Rab27-dependent pathway. Internalized UPEC can be (**3a**) returned back to the extracellular environment via a cAMP-dependent mechanism, (**3b**) exit the endosomal compartment and enter the host cytosol, or (**3c**) remain within actin-bound endosomes in a quiescent state. These reservoirs can persist within either immature urothelial cells or within actin-rich regions of the superficial cells. (**4**) Upon entering the host cytosol UPEC can replicate profusely, forming IBCs. Bacteria are released from IBCs, often exuded as long filamentous forms, and can disseminate further within the urinary tract. Exfoliation of urothelial cells limits the lifespan of IBCs and (**5**) provides UPEC with access to deeper tissue layers. A colour version of this figure is available in the plate section at the back of the book.

rapidly multiply, forming large aggregates known as pods, or intracellular bacterial communities (IBCs) (Anderson *et al.*, 2003; Justice *et al.*, 2004; Mulvey *et al.*, 2001). Within just a few hours after inoculation of the bladder with UPEC, IBCs that contain many hundreds to several thousand bacteria can develop, often swelling the umbrella cells in which they reside. IBCs form in close association with host cytokeratin intermediate filaments (Eto *et al.*, 2006), and display numerous biofilm-like characteristics (Anderson *et al.*, 2003, 2010). Typically, IBCs begin as loose associations of rod-shaped bacteria, but can transition into more densely packed assemblies in which UPEC assumes an overall coccoid morphology (Justice *et al.*, 2004). Bacteria present in IBCs are enmeshed within a matrix of capsular polysaccharide made

up of polymers of α2-8-linked polysialic acid (Anderson *et al.*, 2010). Capsule is needed for normal IBC development, and seems to protect UPEC against infiltrating polymorphonucelar monocytes (PMNs or neutrophils).

Initiation of normal IBC development requires that UPEC express filamentous adhesive organelles known as type 1 pili (or fimbriae) (Wright *et al.*, 2007). These hair-like fibres enable UPEC to bind and invade host cells (see below), but can also promote inter–bacterial interactions and biofilm formation *in vitro* (Pratt and Kolter, 1998; Schembri and Klemm, 2001). Within umbrella cells, type 1 pili help nucleate IBCs, while capsule facilitates condensation of the bacterial communities (Goller and Seed, 2010). Other adhesive molecules like antigen 43 (Ag43) may act in concert with type 1 pili and capsule within IBCs (Anderson et al., 2003). Normal IBC formation is also dependent upon iron acquisition systems and SurA, a periplasmic prolyl isomerase and chaperone that facilitates pilus assembly and the biosynthesis of outer membrane proteins such as OmpA (Justice et al., 2006b). OmpA itself can also promote IBC development, possibly by dampening host proinflammatory responses or by protecting UPEC against antimicrobial peptides (Nicholson et al., 2009). Global regulators of gene transcription, like integration host factor (IHF) (Justice et al., 2012), likely help coordinate expression of type 1 pili and other known and currently undefined gene products that control the initiation and progression of IBC development.

IBCs have been detected in numerous mouse strains infected with a wide variety of UPEC isolates, and evidence that host cell invasion and IBC development contribute to UTIs in human patients is mounting (Garofalo et al., 2007; Rosen et al., 2007). The impact of IBCs on the progression and outcome of a UTI can be substantial. For example, experiments in mice indicate that the more IBCs generated during the first 24 h of a UTI, the more likely that a chronic long-lasting infection would develop (Schwartz et al., 2011). This is in line with current thinking that IBCs provide UPEC with a means to quickly expand bacterial numbers within a protected niche, hidden from many host defences and the shear flow of urine. The release of bacteria from within IBCs can then seed other sites within the urinary tract. Rounds of IBC formation followed by dispersal may help explain the prevalence of chronic and ostensibly recurrent UTIs within the human population.

Bladder cell exfoliation – the beginning or the end?

While IBCs are important for the establishment of UPEC within the bladder, they are probably not long-term UPEC repositories within the host. With time, host umbrella cells that contain IBCs will succumb to the infection, often ripped open from the inside or shed from the urothelial surface and rinsed away with the flow of urine (Mulvey et al., 1998, 2001). By ridding the host of large numbers of bacteria, this shedding process, or exfoliation, can be viewed as an effective host

defence. Indeed, shed umbrella cells containing the remnants of IBCs have been detected in the urine of human patients with UTI (Rosen et al., 2007). However, exfoliation may also facilitate the distribution of UPEC outside of the host and can disrupt the barrier function of the urothelium, allowing UPEC to penetrate deeper layers of the bladder mucosa (Mulvey et al., 2001). The massive influx of neutrophils that occurs in response to UTI can also compromise the urothelium and thereby potentially enhance the dispersal of UPEC (Mulvey et al., 2000; Schilling et al., 2001).

By both live and fixed-tissue microscopy, UPEC can be seen exiting IBCs and moving on to infect neighbouring and underlying epithelial cells (Justice et al., 2004; Mulvey et al., 2001). Often, at this stage, UPEC becomes filamentous, with incomplete septae spaced out along their length. These filamentous bacteria can extend beyond 100 μm and, in some cases, can be seen exiting one host cell while invading another (Mulvey et al., 2001). Interestingly, the long forms of UPEC are highly resistant to killing by host phagocytes (Justice et al., 2006a). The formation of these filamentous bacteria is dependent upon genes like sulA that are components of the SOS response, a system that is activated in E. coli and other bacteria as a means of dealing with oxygen radicals and other potentially mutagenic stresses (Li et al., 2010). SulA-driven filamentation of UPEC promotes the dispersal of UPEC throughout the urothelium and appears to be important to bacterial persistence following the initial establishment of IBCs.

The specific factors that trigger bladder cell exfoliation during a UTI are not entirely clear. Mouse models suggest that exfoliation proceeds via an apoptosis-like mechanism that can be accentuated by indirect type 1 pili-mediated signalling through the AUM component uroplakin III (Mulvey et al., 1998; Thumbikat et al., 2009). In vivo, type 1 pili are necessary to induce high levels of exfoliation, but this likely reflects in part the fact that without type 1 pili bacteria do not adhere well to the urothelium and are rapidly cleared from the host (Mulvey et al., 1998). In ex vivo experiments that use bladder tissue explants, the exfoliation of umbrella cells is stimulated by E. coli even in the absence of type 1 pili (unpublished observations).

Bladder cell exfoliation requires disruption of host tight junctions and probably other intercellular connections (Wood *et al.*, 2012). UPEC strains can enhance the exfoliation process by secreting toxins like cytotoxic necrotizing factor 1 (CNF1, discussed later) and the pore-forming toxin α-haemolysin (HlyA) (Dhakal and Mulvey, 2012; Mills *et al.*, 2000; Smith *et al.*, 2006). HlyA stimulates the activation of pro-apoptotic caspases, as well as host serine proteases like mesotrypsin (Dhakal and Mulvey, 2012; Wood *et al.*, 2012). These proteases in turn can degrade key cytoskeletal elements within target host cells, including factors required for the assembly and maintenance of focal adhesions and other structures that mediate cellular attachment (Dhakal and Mulvey, 2012). Interestingly, HlyA can also trigger the inactivation or degradation of pro-survival and pro-inflammatory factors like the host kinase Akt and the NFκB subunit RelA (Dhakal and Mulvey, 2012; Wiles *et al.*, 2008a). These processes may expedite host cell detachment and death, while simultaneously dampening host inflammatory responses. HlyA and other bacterial factors can have similar anti-inflammatory effects on host phagocytes (Dhakal and Mulvey, 2012; Loughman and Hunstad, 2012), and may thereby potentiate UPEC survival within neutrophils and macrophages (Baorto *et al.*, 1997; Bokil *et al.*, 2011; Nazareth *et al.*, 2007; Shin *et al.*, 2000). In total, these findings indicate that although disruption of the urothelium is potentially detrimental to the establishment of UPEC within the host at the start of a UTI, the processes that mediate exfoliation may actually facilitate the establishment of long-term reservoirs of UPEC within the host.

Quiescent intracellular bacterial reservoirs – swank accommodations for UPEC

Invasion of the immature cells of the urothelium exposes UPEC to an environment that is distinct from that encountered within terminally differentiated umbrella cells. Most notably, immature bladder cells are substantially smaller than umbrella cells, generally lack uroplakin plaques, and have a much more extensive actin cytoskeleton (Eto *et al.*, 2006). Within immature bladder epithelial cells, UPEC-containing vacuoles are often enmeshed within a matrix of actin filaments that crisscross the entire cytosol, whereas in umbrella cells the actin filaments are localized primarily along basolateral surfaces (Eto *et al.*, 2006; Romih *et al.*, 1999). UPEC does not usually replicate to high levels within immature bladder cells, and instead generally remain in a seemingly quiescent state bound within late endosome-like vacuoles (Eto *et al.*, 2006). Interestingly, in cell culture-based assays, the disruption of actin filaments using pharmacological agents like cytochalasin D promotes intravacuolar growth of UPEC and eventual release of the bacteria into the host cell cytosol where they can rapidly multiply, forming large IBC-like aggregates. These develop in close association with host cytokeratin intermediate filaments, as seen *in vivo* within umbrella cells. This process can be expedited by depleting host cholesterol or by disrupting host vacuoles via osmotic shock (Berry *et al.*, 2009; Eto *et al.*, 2006). Such observations suggest that intracellular growth of UPEC is restricted within late endosome-like compartments in an actin- and cholesterol-dependent fashion. Moreover, these results indicate that non-replicating reservoirs of UPEC may be incited to grow as infected immature bladder cells differentiate into umbrella cells, perhaps sensing the realignment of actin filaments during the terminal stages of the differentiation process (Romih *et al.*, 1999).

The quiescent nature of UPEC bound within the immature cells of the bladder, and perhaps also within actin-rich regions of infected umbrella cells, presents huge challenges for treating and eradicating UTIs. First, the quiescent nature of UPEC reservoirs within the urothelium renders them much less susceptible to most antibiotics, the majority of which target only replicating microbes (Blango and Mulvey, 2010). Second, the urothelium is long-lived and able to rapidly repair itself following damage elicited by UPEC or other agents (Jost, 1989; Mulvey *et al.*, 1998; Mysorekar *et al.*, 2002). Consequently, UPEC that manage to invade deeper layers of the urothelium during the acute phase of a UTI may become sequestered behind an exceptionally strong host barrier that is impermeant to most antibiotics. In a mouse UTI model, UPEC residing within the immature cells of the bladder are unfazed by high levels of antibiotics that sterilize the urine and effectively kill intracellular bacteria in cell culture-based assays (Blango and Mulvey, 2010). The normally slow turnover rate of urothelial cells within the bladder mucosa means that UPEC may persist unperturbed for many months to, perhaps, years before re-emerging to cause a fulminant relapsing UTI. Mouse models support this possibility, showing that UPEC can persist at fairly steady levels for weeks to months within the urothelium despite antibiotic treatments (Blango and Mulvey, 2010).

The long-term persistence of UPEC within immature urothelial cells may also be perpetuated by the formation of so-called persister cells (Blango and Mulvey, 2010; Norton and Mulvey, 2012). These cells are in effect dormant bacteria that are resistant to many environmental stresses, including nutrient deprivation and antibiotics. The development of persister cells is mediated in part by genes that encode toxin–antitoxin (TA) systems (Lewis, 2012). UPEC strains carry a set of TA systems that is distinct from most other *E. coli* pathotypes (Norton and Mulvey, 2012). Three of these TA systems have been shown to promote the survival of a reference UPEC isolate within the urinary tract of mice, and one of these also enhances UPEC persistence in the face of antibiotic treatments.

The ability of UPEC to invade and persist within the immature epithelial cells of the urothelium has had a notable impact on how recurrent and relapsing UTIs are viewed. UTIs have an exceptionally high recurrence rate of about 25% (Blango and Mulvey, 2010; Brumbaugh and Mobley, 2012). Oftentimes, these recurrent infections are attributed to re-inoculation of the urinary tract with microbes that reside in the host gastrointestinal tract or elsewhere within the environment. The recognition of UPEC as an opportunistic intracellular pathogen with the capacity to persist within the bladder urothelium even in the presence of antibiotics, suggest that many seemingly recurrent UTIs may actually be relapsing infections caused by the resurgence of UPEC reservoirs and the re-initiation of IBC development. Disruption of this infection cycle by targeting IBC or persister cell development, or by inhibiting the invasion process itself may be of

enormous benefit to the treatment and prevention of chronic and recurrent UTIs.

Invasion mechanisms employed by UPEC – hijacking the host

Prior to invading host cells, bacteria must first make contact via one or more surface adhesins. UPEC strains can encode dozens of adhesins, including type 1, P (Pap), and S/F1C pili and Afa/Dr adhesins (Wiles *et al.*, 2008b; Wright and Hultgren, 2006). These and other adhesins may work in concert with each other and with secreted toxins like HlyA and CNF1, allowing UPEC to better infiltrate tissues within the urinary tract. However, type 1 pili and Afa/Dr adhesins appear to be the primary facilitators of UPEC entry into host cells, with contributions also made by CNF1.

Type 1 pili-mediated invasion of host cells

Type 1 pili are encoded by the *fim* gene cluster and are expressed by most UPEC isolates as well as many other members of the Enterobacteriaceae family (Mulvey, 2002). These filamentous adhesive organelles are assembled at the bacterial surface in a peritrichous manner via a canonical chaperone-usher pathway (Thanassi *et al.*, 2012). Each pilus fibre consists of a 7 nm wide helical rod composed of thousands of FimA pilin subunits that can extend more than 5 microns from the bacterial surface (Jones *et al.*, 1995). Attached to the distal end of each FimA rod is a short 3 nm wide tip fibrillum structure comprised of the adaptor proteins FimF and FimG along with the adhesin protein FimH (Jones *et al.*, 1995; Russell and Orndorff, 1992). The expression of type 1 pili is phase variable, controlled by the invertible genetic element *fimS* and regulated, in part, by environmental conditions. Within the bladder, changes in acidity, osmolarity, and static growth conditions can affect type 1 pili expression (Muller *et al.*, 2009; Schwan, 2011; Schwan *et al.*, 2002). The expression of type 1 pili is critical to the ability of most UPEC strains to effectively colonize the urinary tract (Bahrani-Mougeot *et al.*, 2002; Connell *et al.*, 1996; Langermann *et al.*, 1997; Martinez *et al.*, 2000; Mulvey *et al.*, 1998).

The FimH adhesin situated within the distal tip of each type 1 pilus is a mannose-binding lectin that can interact with a wide assortment of host cell associated receptors, including leucocyte adhesin molecules CD11b and CD18 (Gbarah *et al.*, 1991), the glycophosphoatidylinositol (GPI)-anchored protein CD48 (Baorto *et al.*, 1997; Malaviya *et al.*, 1999), carcinoembryonic antigen-related cell adhesion molecule (CEACAM) family members (Carvalho *et al.*, 2009a; Sauter *et al.*, 1991, 1993), the pattern recognition receptor Toll-like receptor 4 (TLR4) (Mossman *et al.*, 2008), glycoprotein 2 (Hase *et al.*, 2009; Yu and Lowe, 2009), the AUM protein uroplakin 1a (UP1a) (Wu *et al.*, 1996; Zhou *et al.*, 2001), and α3 and β1 integrin subunits (Eto *et al.*, 2007). FimH comprises two domains connected by a short linker peptide (Choudhury, 1999). The C-terminal pilin domain facilitates the incorporation of FimH into the tip fibrillum, while the N-terminal domain contains a mannose-binding pocket. The affinity of FimH for either terminal mono- or trimannose residues present on host glycoprotein receptors is affected by point mutations within FimH (Sokurenko *et al.*, 1997, 1998; Weissman *et al.*, 2006). Within the urinary tract, *fimH* alleles that have high affinity for monomannose residues appear to be selected by enabling UPEC to better bind urothelial receptors and thereby resist elimination with the bulk flow of urine (Sokurenko *et al.*, 1997, 1998). FimH interactions with host glycoprotein receptors are further regulated by an a allosteric catch-bond mechanism, whereby FimH assumes an extended, higher affinity binding conformation in the presence of shear forces as generated by urine flow (Le Trong *et al.*, 2010; Yakovenko *et al.*, 2008).

FimH promotes UPEC entry into bladder cells by triggering actin cytoskeletal rearrangements that cause the host plasma membrane to zipper around and engulf adherent bacteria (Martinez *et al.*, 2000; Wang *et al.*, 2008). By high-resolution and standard transmission electron microscopy, the AUM of superficial bladder cells can be seen wrapping around UPEC, making multiple contacts with the FimH-containing tips of type 1 pili (Mulvey *et al.*, 1998). In assays that use undifferentiated bladder epithelial cells lines, latex beads that are coated with purified FimH, but not other control proteins, are efficiently internalized, while bacteria that express type 1 pili without the

FimH adhesin are unable to invade (Martinez *et al.*, 2000).

The molecular events that mediate FimH-dependent bacterial invasion of host cells have been examined primarily in cell culture-based assays, with *in vivo* experiments that employ mouse UTI models used to corroborate a few of the findings (Fig. 13.2). Pharmacological disruption of actin filaments using cytochalasin D demonstrated that host cell invasion by UPEC is dependent upon F-actin, which accumulates at sites of bacterial entry (Martinez *et al.*, 2000). Entry also requires cholesterol, a key component of host plasma membrane microdomains referred to as lipid rafts that may serve as organizing platforms for the assembly of signalling complexes. Depletion or sequestration of cholesterol inhibits FimH-mediated bacterial entry into bladder cells, suggesting a role for lipid rafts in the invasion process (Duncan *et al.*, 2004; Eto *et al.*, 2008). However, cholesterol depletion may also interfere with the production and activity of phosphoinositides (Grimmer *et al.*, 2002; Kwik *et al.*, 2003), a group of important second messengers that can modulate actin cytoskeletal dynamics and actin regulatory factors such as the Rho GTPase Rac1. The synthesis of phosphoinositides by phosphoinositide 3-kinases (PI 3-kinase) is essential for UPEC entry into bladder cells, as is Rac1 and other Rho GTPase family members like Cdc42 (Martinez and Hultgren, 2002; Martinez *et al.*, 2000). These GTPases can alter actin dynamics at sites of bacterial entry via effects on regulators such as WAVE2 and actin-nucleating Arp2/3 complexes.

Actin and membrane dynamics resulting in UPEC internalization appear to be controlled by many additional host factors. These include focal adhesin kinase (FAK) and Src kinase, the actin stabilizing factors vinculin and α-actinin, and even clathrin, a protein that is typically associated with the endocytosis of cargo that is much smaller than a bacterium (Eto *et al.*, 2007, 2008; Martinez and Hultgren, 2002; Martinez *et al.*, 2000). Rather than forming traditional clathrin-coated pits around invading bacteria, clathrin seems to instead assemble in punctae and transient sheet-like assemblies beneath adherent bacteria (Eto *et al.*, 2008). These, in turn, may help drive the recruitment and localized rearrangement of actin filaments, and perhaps affect membrane curvature during the invasion process. The clathrin adaptor AP-2 and alternate adaptors termed Dab2, Numb and ARH may facilitate this process.

Microtubules also appear important, as the destabilization, disassembly, or aggregation of microtubules by use of pharmacological agents has potent inhibitory effects on the invasion process (Dhakal and Mulvey, 2009). Microtubule outgrowth and stability is influenced by α-tubulin acetylation, which is regulated in part by the enzyme histone deacetylase 6 (HDAC6). HDAC6-mediated deacetylation of α-tubulin also affects the recruitment of kinesin motor complexes that carry cargo along microtubules towards the cell periphery (Dompierre et al., 2007; Reed et al., 2006). Current data suggest that microtubules and kinesins can promote UPEC entry into host cells by delivering as-yet-undefined 'relaxing factors' that increase the fluidity of actin dynamics at the host cell surface (Dhakal and Mulvey, 2009).

Among the known FimH receptors, UP1a has received the most attention as a potential mediator of UPEC entry into bladder cells. FimH binds mannose residues on UP1a, and high-resolution electron microscopy revealed that the FimH-containing tips of type 1 pili could interact with UP1a-containing complexes within the AUM of bladder umbrella cells (Mulvey et al., 1998; Zhou et al., 2001). UP1a has no significant cytosolic tail that might promote signalling cascades leading to bacterial uptake, but this receptor could potentially signal indirectly via interactions with neighbouring proteins like UPIII (Wang et al., 2009). Since uroplakin-containing plaques within the AUM are internalized as the bladder empties (Apodaca, 2004), it is plausible that bound UPEC could also be simultaneously taken in as part of the process. However, uroplakin plaques are significantly smaller than UPEC, suggesting that multiple plaques would be needed to envelope and internalize a single bacterium. Again, microscopy supports this possibility (Mulvey et al., 2000), but so far genetic and biochemical evidence that UP1a can mediate UPEC invasion is missing. In addition, the lack of UP1a and uroplakin plaques on immature bladder cells indicates that alternate receptors for UPEC entry likely exist. Cell culture-based assays implicate α3 and β1 integrin subunits as primary candidates.

Integrins are heterodimeric integral membrane proteins comprised of α and β subunits that normally act as linkers between the host actin cytoskeleton and extracellular matrix components (Arnaout et al., 2005). FimH interacts with N-linked high-mannose type glycan residues present on both α3 and β1 integrin subunits, independent of the canonical ligand-binding pocket formed by α3β1 heterodimers (Eto et al., 2007). Both differentiated and immature cells within the urothelium and elsewhere within the urinary tract express α3β1 integrin dimers (Southgate et al., 1995), making them of potential use to UPEC during all stages of a UTI. Signalling events downstream of integrin activation are regulated by phosphorylation of conserved tyrosine and serine residues within the cytosolic tail of β integrin subunits (Arnaout et al., 2005). Mutation of these residues has variable effects, either boosting or attenuating UPEC invasion frequencies (Eto et al., 2007). Of note, many of the host regulatory and signalling factors that are required for UPEC entry into bladder cells, as described above, can act downstream of integrin receptors. Cumulatively, available data indicate that type 1-piliated UPEC can induce clustering of α3β1 integrins, which in turn can stimulate activation of Src kinase and FAK, leading to the formation of transient complexes between FAK and PI-3-K that activate GTPases like Rac1.

Activation of FAK downstream of integrin receptors can also recruit the guanosine triphosphatase dynamin2 (DNM2) (Wang et al., 2011). This enzyme can mediate membrane scission events and is critical for UPEC entry into bladder cells in both in vivo and cell culture-based experiments (Eto et al., 2008; Wang et al., 2011). Interestingly, DNM2 activation is dependent upon covalent modification by S-nitrosylation, a reaction driven by the production of reactive nitrogen species generated by host nitric oxide synthases like eNOS (Wang et al., 2006, 2009). The infection of host cells with UPEC promotes the phosphorylation and activation of eNOS, thereby stimulating UPEC internalization.

The specific events leading to the uptake and trafficking of type 1-piliated UPEC into late endosome-like compartments are still far from understood. It is feasible that multiple receptors bound by FimH can promote UPEC internalization, possibly via distinct pathways. These may include the involvement of secretory lysosomes, as used by the protozoan parasite *Trypanosoma cruzi* (Bishop *et al.*, 2007), or uroplakin plaques, depending on the host cell type analysed. Cooperation among varied host receptors may also come into play, as suggested by work showing that host proteins like CD46, which recognizes opsonized bacteria, can synergize with type 1 pili to promote UPEC entry into urothelial cells (Li *et al.*, 2009). Complicating matters further is the observation that many bacteria that are internalized into bladder epithelial cells are quickly shuttled back out to the extracellular milieu (Bishop *et al.*, 2007). This process is facilitated by generation of the second messenger cyclic adenosine monophosphate (cAMP) within the target host cells, downstream of TLR4 (Bishop *et al.*, 2007; Song *et al.*, 2009). The ability of FimH to directly engage TLR4 may therefore counter, to some extent, the entry process triggered by engaging other receptors like α3β1 integrins. A more precise understanding of FimH-mediated invasion will require higher resolution analyses, taking advantage of current and emerging technologies and model systems.

CNF1 toxin-mediated invasion of host cells

Rho-family GTPases like Rac1, which are critical mediators of host cell invasion by UPEC and other pathogens, are also substrates for cytotoxic necrotizing factor 1 (CNF1) (Bower *et al.*, 2005). This 113 kDa single-chain toxin is secreted by about one-third of UPEC isolates (Wiles *et al.*, 2008b). CNF1 enters host cells via a low pH-mediated endocytic mechanism employing the host laminin receptor precursor (Chung *et al.*, 2003; Kim *et al.*, 2005). After processing within acidic endosomal compartments, mature CNF1 is released into the host cytosol where it causes constitutive activation of Rho GTPases like Rac1 (Lemonnier *et al.*, 2007). Activated GTP-bound Rac1 is eventually polyubiquitylated by the host E3 ligase HACE1 and subsequently degraded by the ubiquitin-proteasome system (Doye *et al.*, 2002; Lerm *et al.*, 2002; Torrino *et al.*, 2011).

Depletion of Rac1 and other Rho GTPases downstream of CNF1 intoxication can stimulate bladder cell migration, potentially enhancing the dissemination of bound and internalized bacteria within the urothelium (Bower *et al.*, 2005; Doye *et al.*, 2002).

CNF1-modified Rho GTPases are also recognized by other host cell components. Innate immune adapters known as Rip proteins can sense CNF1-modified, constitutively active Rho GTPases, eliciting inflammatory and anti-bacterial responses that can limit pathogen growth and survival (Boyer *et al.*, 2011). On the other hand, the recognition of CNF1-activated and polyubiquitylated Rho GTPases by another adapter known as Tollip can potentially enhance bacterial survival by facilitating host cell invasion (Visvikis *et al.*, 2011). Tollip recruits Tom1 (target of the oncogene v-Myb protein 1) to host membranes, along with Tom1-associated clathrin molecules. These factors can then act cooperatively with Rac1 to promote the internalization of bound UPEC via a process that is also dependent upon α3β1 integrin receptors and the clathrin adapter protein AP2. CNF1-induced internalization of bacteria is thus reminiscent of the type 1 pilus-mediated invasion process described in the preceding section. Together, these observations suggest that the constitutive activation of Rho GTPases by CNF1 drives bacterial entry primarily via stimulatory effects on host components that normally mediate FimH-dependent host cell invasion by UPEC.

Afa/Dr adhesin-mediated invasion of host cells

The Afa/Dr family of bacterial adhesins includes Afa-I, Afa-II, Afa-III, Afa-V, Dr, Dr-II, F1845, and Nfa-I fimbriae (Servin, 2005). These structures are each encoded in operons usually containing five genes, labelled *A* through *E* (e.g. *dra-A* through *dra-E* for Dr fimbriae) (Guignot *et al.*, 2009). The E subunits are assembled by canonical chaperone-usher systems into fibres that are capped by D subunits (Garcia *et al.*, 2000; Jouve *et al.*, 1997; Servin, 2005; Van Loy *et al.*, 2002; Zalewska *et al.*, 2001). Afa/Dr fimbriae are often expressed by diffusely adhering *E. coli* (DAEC), an important cause of diarrhoea in children, but are also associated with some UPEC isolates (Kaper *et al.*, 2004;

Le Bouguenec, 1999). About 25–50% of children with cystitis, and 30% of pregnant women with gestational UTI carry *E. coli* strains that encode Afa/Dr fimbriae (Guignot *et al.*, 2009; Hart *et al.*, 1996; Nowicki *et al.*, 1994). These adhesins also have an epidemiological link with recurrent UTIs and UPEC isolates that carry resistance to multiple antibiotics (Hart *et al.*, 2001; Servin, 2005). Afa/Dr adhesins bind domains called complement control protein repeats 2 and 3 (CCP2 and CCP3) within the broadly expressed glycophosphatidylinositol (GPI)-anchored complement regulatory protein CD55 (a.k.a. decay-accelerating factor, DAF) (Nowicki *et al.*, 1993; Servin, 2005). A subgroup of Afa/Dr adhesins, including Dr, Afa-III, and F1845 fimbriae, also binds members of the CEACAM family (Berger *et al.*, 2004; Guignot *et al.*, 2009).

Interactions between E subunits within Dr or Afa-III fimbriae and either CD55 receptors or CEACAM family receptors like CEA, CEACAM1, and CEACAM6 promote bacterial entry into host cells (Guignot *et al.*, 2009). Entry proceeds via a zipper-like mechanism requiring cholesterol and functional microtubules, reminiscent of host cell invasion by type 1-piliated bacteria (Guignot *et al.*, 2009; Kansau *et al.*, 2004). Interestingly, Afa/Dr-adhesins stimulate the clustering of $\alpha5\beta1$ integrins at sites of bacterial attachment, but these host receptors are not required for bacterial internalization. Actin cytoskeletal rearrangements are partially dispensable. Once internalized, Afa/Dr-expressing bacteria are trafficked into acidic, late endosome-like compartments, similar to those occupied by invasive type 1-piliated bacteria. The signalling events and molecular machinery that promote the microtubule-dependent, actin-independent internalization of Afa/Dr-expressing bacteria remain obscure. Likewise, the functional relevance of Afa/Dr-mediated host cell invasion to the establishment and recurrence of UTIs and other infections has not been addressed. However, it is tempting to speculate that these adhesive organelles may act in a complementary fashion with type 1 pili and toxins like CNF1 to promote long-term bacterial persistence within host cells, keeping bacterial pathogens like UPEC sequestered from hostile elements lurking within the extracellular milieu.

NMEC: something on your mind?

Inflammation of the protective membranes that surround the brain and spinal cord can result in a serious, life-threatening condition known as meningitis (Tunkel *et al.*, 2004). In infants, bacterial strains classified as neonatal meningitis *E. coli* (NMEC) are the primary cause of meningitis by Gram-negative organisms (Houdouin *et al.*, 2008). Individuals with NMEC-induced meningitis have fatality rates approaching 40%, and survivors often have severe neurological disorders (Das *et al.*, 2001). NMEC can be transferred from mother to infant via an oral route during passage through the birth canal (Nizet and Klein, 2010). Once NMEC enters the infant's gastrointestinal tract, it is able to translocate into the circulatory system and establish bacteraemia (Burns *et al.*, 2001). Eventually, NMEC penetrates the central nervous system (CNS) by crossing the blood brain barrier formed by microvascular endothelial cells within the cerebral vasculature (Kim, 2003, 2012).

Although the translocation of NMEC across the intestinal mucosa is seemingly critical for dissemination of NMEC within the host, very little is known about this early stage of the infection process. Recent work indicates that an outer membrane protein known as Hek may enable NMEC to bind and invade intestinal epithelial cells, potentially facilitating bacterial transfer into the bloodstream (Fagan and Smith, 2007). A 25 amino acid loop formed by Hek on the bacterial surface mediates interactions with heparin and other glycosoaminoglycan molecules that may serve as receptors for NMEC within the gastrointestinal tract (Fagan *et al.*, 2008). Hek also promotes the agglutination of red blood cells and bacterial auto-aggregation. The formation of aggregates is known to facilitate bacterial survival within the circulatory system (Fagan *et al.*, 2008), suggesting that Hek may also function outside of the gastrointestinal tract by promoting auto-aggregation of NMEC upon entry into the bloodstream.

After entering the bloodstream, high levels of bacteraemia ($> 10^3$ CFU/ml of blood) must be reached before NMEC can successfully cross the blood brain barrier (Kim, 2003). In addition to Hek, a few other factors have been identified that can aid the persistence and growth of NMEC in the blood. These include the K1 capsule, a homopolymer of sialic acid that is expressed by ~80% of NMEC isolates (Croxen and Finlay, 2010; Silver *et al.*, 1988). Together with O-chain lipopolysaccharides, the K1 capsule provides NMEC with enhanced resistance to both circulating phagocytes and antibacterial factors within serum (Allen *et al.*, 1987). Modification of bacterial surface structures by acetylation via an integrated bacteriophage-encoded enzyme may increase the antigenic diversity of NMEC, enabling the pathogens to better avoid detection by immunosurveillance mechanisms (Ørskov *et al.*, 1979). The outer membrane protein OmpA, which promotes UPEC growth and survival in the urinary tract (Nicholson *et al.*, 2009), may also facilitate the survival of NMEC in blood. Specifically, with help from the outer membrane lipoprotein NlpI, OmpA can bind the complement regulatory factor C4bp present in serum and thereby short-circuit the anti-bacterial effects of the classical complement pathway (Tseng *et al.*, 2012; Wooster *et al.*, 2006).

High levels of bacteraemia increase the likelihood that NMEC make it into the cerebral vasculature where they subsequently bind and invade brain microvascular endothelial cells (BMEC) within the blood brain barrier (Kim, 2003). Within BMEC cells, NMEC is trafficked into late endosome-like compartments before being delivered basally into the CNS (Kim *et al.*, 2003). The K1 capsule facilitates the transcytosis of NMEC, evident as mutants that lack the K1 capsule are destroyed within lytic lysosomal compartments (Hoffman *et al.*, 1999). Once across the blood brain barrier, NMEC can stimulate robust and prolonged inflammatory responses, including the influx of immune effector cells and oedema, triggering tissue damage and overt clinical symptoms that can include fever, chills, headache, and nausea (Kim, 2010).

Crossing the blood brain barrier is an onerous feat from the bacterial perspective. BMECs and other components render the blood brain barrier impenetrable to most microbes and toxic molecules (Kim, 2006). NMEC circumvents the barrier function by employing a small arsenal of secreted and cell surface-associated factors that bind and manipulate BMECs (Fig. 13.3). These factors promote bacterial adherence and subsequent internalization via an actin-dependent zipper-like mechanism that is similar in some respects to that employed by UPEC to gain entry into bladder epithelial cells (Prasadarao *et al.*, 1999). For example, as shown with UPEC, the type 1 pilus-associated adhesin FimH and CNF1 both enhance the ability of NMEC to colonize BMECs. FimH binds the GPI-anchored receptor CD48 on BMECs, initiating calcium fluxes and other signalling events that cause localized actin cytoskeletal rearrangements and the uptake of NMEC (Khan *et al.*, 2007; Moran and Miceli, 1998). Modulation of host actin dynamics during the invasion process is also dependent upon FAK, Src kinases, and paxillin (Kim *et al.*, 2005; Kim, 2006). As with UPEC, host cell invasion by NMEC is augmented by secretion of the bacterial toxin CNF1, which activates Rho GTPases after being processed and delivered into the host cell cytosol via the laminin receptor protein (Khan *et al.*, 2002).

Additional bacterial factors that promote NMEC interactions with BMECs include flagella, OmpA, NlpI, and surface-localized Ibe proteins (Parthasarathy *et al.*, 2007). The functions of OmpA and NlpI in association with BMECs are distinct from those observed in the bloodstream. Specifically, OmpA binds N-acetylglucosamine (GlcNAc) residues present on the host protein Ecgp96 (a.k.a. endoplasmin), a plasma membrane-localized homologue of the chaperone Hsp90B that is typically situated within the endoplasmic reticulum (Kim, 2010; Krishnan *et al.*, 2013). Chaperone proteins like Hsp90B aid in the maturation and delivery of Toll-like receptors to the plasma membrane (Krishnan *et al.*, 2013). Interestingly, Ecgp96 pairs with TLR2 on the surface of BMECs and promotes internalization of bound NMEC (Krishnan *et al.*, 2013). The engagement of Ecg96–TLR2 complexes by OmpA results in the recruitment of the regulatory protein STAT3 (Signal Transducer and Activator of Transcription 3) and subsequent activation of PI 3-kinase and protein kinase C-α (PKC-α) (Krishnan *et al.*, 2012; Maruvada *et al.*, 2008). These signalling factors can modulate

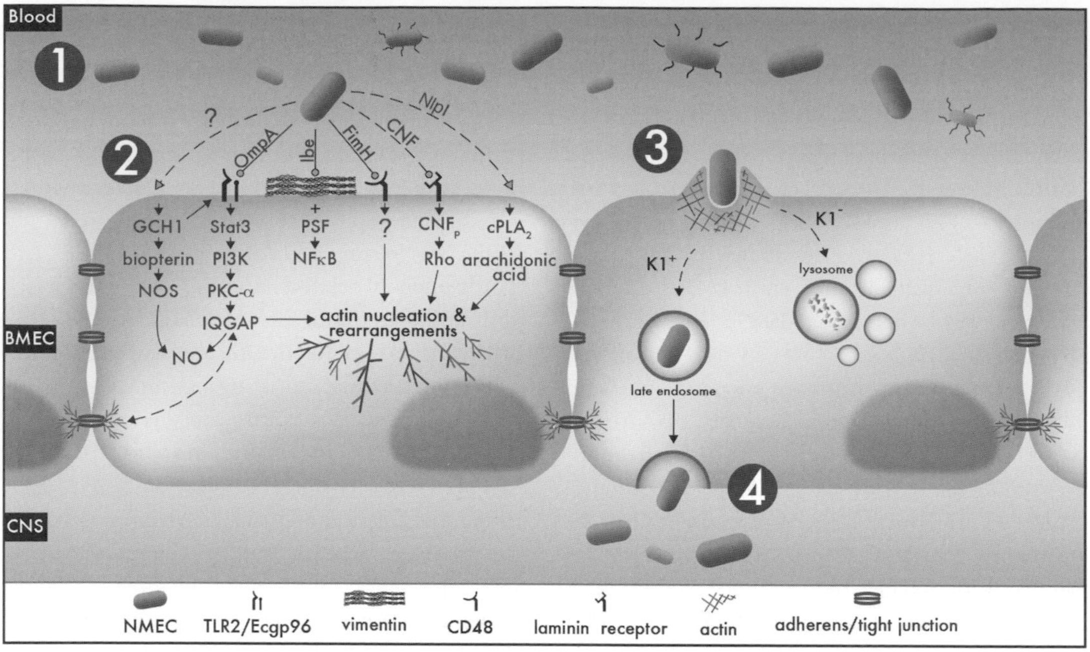

Figure 13.3 Transcytosis of NMEC across the blood brain barrier. NMEC enters the bloodstream via the gastrointestinal tract and (**1**) causes high levels of bacteria. (**2**) Multiple factors expressed by NMEC interact with host receptors and regulatory proteins to promote bacterial attachment and internalization by BMECs, triggering in the process NFκB activation and (**3**) actin cytoskeletal rearrangements. Internalized NMEC are trafficked into late endosomal compartments and (**4**) eventually delivered into the CNS. Bacteria lacking K1 capsule are degraded. A colour version of this figure is available in the plate section at the back of the book.

actin dynamics by multiple pathways, including manipulation of the host protein IQGAP. This multifunctional protein can interact with and regulate a variety proteins, including Rho family GTPases, F-actin, calmodulin, and β-catenin (White *et al.*, 2009). OmpA-mediated activation of PKC-α causes IQGAP to dislodge β-catenin from adherens junctions that connect BMECs, thereby stimulating actin rearrangements that facilitate NMEC internalization (Krishnan *et al.*, 2012).

The disruption of adherens junctions downstream of OmpA-mediated PKC-α activation may also increase the permeability of the blood–brain barrier, possibly allowing NMEC to enter the central nervous system via a paracellular route (Sukumaran and Prasadarao, 2003). Whether or not paracellular entry of NMEC into the CNS impacts the progression of meningitis *in vivo* requires additional work. The analysis of OmpA-mediated effects on NMEC pathogenesis must also take into account observations showing that

the deletion of OmpA notably inhibits the expression of type 1 pili (Teng *et al.*, 2006).

NlpI can promote NMEC entry into BMECs via an alternate pathway involving activation of the host factor cytoplasmic phospholipase A2 ($cPLA_2$) (Teng *et al.*, 2010). This enzyme has been implicated in receptor recycling, vesicle trafficking, endosome fusion, and the generation of arachidonic acid (Brown *et al.*, 2003; Das *et al.*, 2001; de Figueiredo *et al.*, 2001). Cyclooxygenase (COX) and lipooxygenase (LOX) can convert arachidonic acid into eicosanoids, which in turn can affect a variety of cellular functions, including actin remodelling (Glenn and Jacobson, 2002). Pharmacological inhibition of $cPLA_2$, COX, or LOX leads to a reduction in BMEC invasion by NMEC (Das *et al.*, 2001).

The effects of NlpI on arachidonic acid synthesis may also enhance host inflammatory responses that are mediated by eicosanoids (Zhu *et al.*, 2010). Ibe proteins encoded by NMEC and many other *E. coli* pathogens can further escalate inflammatory responses by stimulating activation of the pro-inflammatory, pro-survival transcription factor NFκB (Chi *et al.*, 2012). Interactions between Ibe proteins and vimentin intermediate filaments localized on the surface of BMECs stimulate the recruitment of an Ibe co-receptor, polypyrimidine tract-binding protein associated splicing factor (PSF), which in turn activates NF-κB (Chi *et al.*, 2012). Ibe-induced signalling through vimentin and PSF can also stimulate actin cytoskeletal rearrangements leading to NMEC internalization by BMECs, though the mechanism is not entirely clear (He *et al.*, 2012; Zou *et al.*, 2006).

There are likely many additional factors expressed by NMEC that modulate its ability to invade BMECs and penetrate the blood brain barrier. The growing complexity of the host and bacterial networks that lead to NMEC internalization by BMECs is exemplified by recent work implicating guanosine triphosphate cyclohydrolase (GCH1) in the invasion process. GCH1 associates with the OmpA receptor Ecgp96 and is necessary for both surface expression of Ecgp96 and host cell invasion by NMEC (Shanmuganathan *et al.*, 2013). GCH1 is involved in the biosynthesis of molecules known as biopterins, which can serve as co-factors for the activation of nitric oxide synthases (NOS) and the production of nitric oxide. Interestingly, inhibition of GCH1 activity enhances survival of newborn mice infected with NMEC, possibly by limiting the availability of Ecgp96 on the surface of BMECs. Decreased NOS activity as a consequence of GCH1 inhibition may also attenuate NMEC entry into host cells by preventing full activation of DNM2 via nitrosylation by NO and other reactive nitrogen species, as suggested by observations with UPEC (Wang *et al.*, 2011). IQGAP may feed into this pathway via effects on the Rho GTPases Rac1 and 2 and their association with NOS (Kuncewicz *et al.*, 2001; Kuroda *et al.*, 1996). Antibiotic resistance is increasing among NMEC isolates (Kim, 2010), spurring the need to identify new approaches to combat these often-lethal pathogens. Data from the study of NMEC entry into host cells, including results showing the protective effect of treating NMEC-infected mice with a GCH1 inhibitor, highlight promising therapeutic avenues for the treatment and prevention of bacterial meningitis.

EIEC: colonic catastrophe

Enteroinvasive *E. coli* (EIEC) are almost indistinguishable from *Shigella* species, a major cause of bacillary dysentery that kills hundreds of thousands of people in developing countries each year (Niyogi, 2005; Pupo *et al.*, 1997; van den Beld and Reubsaet, 2012). EIEC and *Shigella* both employ similar virulence factors and are difficult to discern by biochemical or genetic means (Lan *et al.*, 2001, 2004). Indeed, EIEC is more closely related to *Shigella* in evolutionary terms than to other *E. coli* strains (Lan *et al.*, 2001, 2004). Both pathogens carry a 14-megadalton virulence plasmid that encodes a type 3 secretion system (T3SS) and multiple effector molecules that enable the bacteria to invade and replicate within colonic epithelial cells (Fig. 13.4) (Niyogi, 2005; Rolland *et al.*, 1998; Sansonetti *et al.*, 1983). Because of their overall similarity, EIEC and *Shigella* are generally considered to be of the same pathovar and, for the purposes of this chapter, both will be referred to as *Shigella* (Lan *et al.*, 2004).

Shigella is unable to enter colonic epithelial cells via their apical surfaces and must first translocate across the colonic mucosa through microfold M cells (Jensen *et al.*, 1998; Parsot, 2005). These specialized epithelial cells continually sample lumenal antigens for presentation to underlying phagocytes (Kraehenbuhl, 2000; Sansonetti and Phalipon, 2007). *Shigella* is able to transmigrate through M cells into the submucosa where it can be internalized by waiting macrophages (Lafont *et al.*, 2002). Secretion of T3SS effector molecules (including IpaB, IpaC, IpaD, and IpaH) enables *Shigella* to disrupt vacuolar membranes, facilitating bacterial release into the host cytosol (High *et al.*, 1992). Interactions between IpaB and caspase-1 trigger macrophage death via a highly inflammatory process known as pyroptosis, resulting in the release of bacteria and production of the proinflammatory cytokines IL-1β, IL-18, and IL-8 (Sansonetti *et al.*, 2000; Guichon *et al.*, 2001). In response to these inflammatory stimuli, neutrophils are recruited into the tissue, further heightening inflammation and disrupting tight

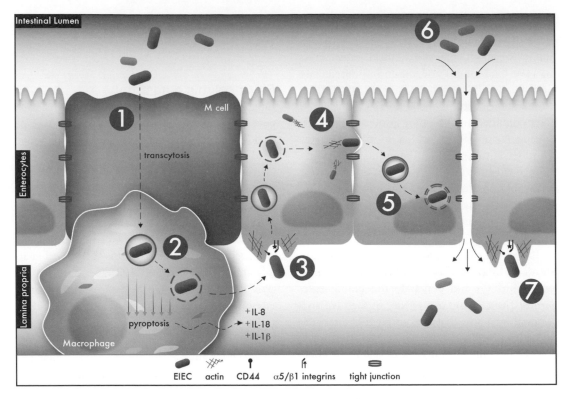

Figure 13.4 EIEC/*Shigella* invasion of the colonic epithelium. (**1**) Bacteria initially cross the epithelium via M cells. (**2**) Within the submucosa, *Shigella* is internalized by macrophages. The secretion of T3SS effector molecules enables *Shigella* to rupture endosomal membranes and kill macrophages via pyroptosis. Inflammatory cytokines released by infected macrophages recruit neutrophils that damage the epithelial barrier, allowing more bacteria to enter the submucosa where they can (**3**) invade enterocytes and (**4**) enter the host cell cytosol via secretion of T3SS effectors. Within enterocytes, *Shigella* can multiply and disseminate, propelling itself within and (**5**) between host cells by polymerizing actin tails. This can disrupt tight junctions, further compromising barrier function and (**6**) providing lumenal bacteria with greater access to the submucosa where they (**7**) can perpetuate the invasion process. A colour version of this figure is available in the plate section at the back of the book.

junctions between colonic epithelial cells (Singer and Sansonetti, 2004). This in turn allows *Shigella* increased access to the submucosa where they can bind and invade the colonic epithelial cells through basolateral surfaces (Perdomo *et al.*, 1994).

The mechanisms by which *Shigella* makes initial contact with target host cells are incompletely understood. Interestingly, binding may be facilitated by the release of granule proteins from infiltrating neutrophils (Björn *et al.*, 2010). Rather than destroying the bacteria, the granule proteins may instead neutralize surface charges on *Shigella* and thereby increase adhesion rates. The T3SS effectors IpaB and IpaC also appear to promote adherence. IpaB binds the hyaluronic

acid receptor CD44, while IpaC binds α5β1 integrin complexes (Lafont *et al.*, 2002; Watarai *et al.*, 1996). With help from IpaD, IpaB and IpaC form a pore within the host plasma membrane that serves as a conduit between the T3SS needle apparatus and the host cytosol (Espina *et al.*, 2006; Hayward *et al.*, 2005; Lafont *et al.*, 2002; Watarai *et al.*, 1996).

Subsequent delivery of multiple T3SS effectors, including IpaB and IpaC, elicits actin cytoskeletal rearrangements and the internalization of bound bacteria (for review, see Cornelis, 2006). IpaB and IpaC trigger initial actin remodelling at sites of entry by recruiting ezrin and activating Src kinase and the Rho GTPase Rac1, respectively (Mounier *et al.*, 2009; Skoudy *et al.*, 1999; Tran Van Nhieu *et al.*, 1999). Ezrin connects the plasma membrane to the actin cytoskeleton, and has been implicated in host cell invasion by several pathogens, including NMEC (Kim, 2006). Another T3SS effector, VirA, also causes Rac1 activation by destabilizing host microtubules (Yoshida *et al.*, 2002). The actin polymerizing effects of these effector molecules are countered by IpaA, which complexes with vinculin and promote actin depolymerization (Mierke, 2009; Park *et al.*, 2011). Together, these and other effectors modulate localized actin and membrane dynamics, facilitating the envelopment and internalization of *Shigella*.

Following release into the host cell cytosol via IpaB- and IpaC-mediated vesicle rupture, *Shigella* can polymerize actin into polarized tails that propel the bacterial cells though the host cytosol (Reis and Horn, 2010). IcsA (VirG), an autotransporter protein that mediates its own translocation across the bacterial outer membrane at one pole, stimulates the formation of actin tails by interacting with the actin regulatory protein N-WASP, which in turn binds and activates the actin nucleating complex Arp2/3 (Goldberg and Theriot, 1995). IscA-mediated motility, aided by VirA-mediated disruption of microtubules, facilitates the cell-to-cell spread of *Shigella* within the colonic epithelium (Bernardini *et al.*, 1989). This process also disrupts intercellular tight junctions, which again provides *Shigella* with greater access to the submucosa (Sakaguchi *et al.*, 2002).

Shigella is able to rapidly multiply while sequestered within the confines of the host cytosol (Parsot, 2005). However, colonic epithelial cells have a rapid turnover rate and may not provide a particularly stable home. To increase the longevity of this sheltered environment, *Shigella* employs several T3SS effectors. IpgD stimulates activation of the pro-survival PI 3-kinase/Akt signalling pathway, while IpaB and OspE work to limit host cell cycle progression and detachment (Iwai *et al.*, 2007, 2009). Other effectors, including OspB, OspG, OspF, and IpaH, act to temper host inflammatory responses that may compromise both host and bacterial viability (Croxen and Finlay, 2010).

The complex interplay among the various T3SS effectors and other virulence factors utilized by *Shigella* to enter and manipulate host cells is remarkable, especially in contrast to UPEC and NMEC that do not typically encode a T3SS (Ren *et al.*, 2004). Results obtained from the analysis of host cell invasion by *Shigella* has not only shed light on an important set of virulence factors, but have also provided new tools and perspectives for deciphering host cell function. Ultimately, this line of research holds much promise for understanding basic cell biology processes as well as the development of novel therapeutics.

AIEC: ileal irritation

Inflammation within the gastrointestinal tract that is broadly categorized as inflammatory bowel disease (IBD) is associated with abdominal pain, rectal bleeding, cramps, vomiting, and bloody diarrhoea (Sasaki *et al.*, 2007). Two major manifestations of IBD are ulcerative colitis (UC) and Crohn's disease (CD) (Baumgart and Carding, 2007). The latter often correlates with the formation of lesions in ileum, which are frequently colonized by a type of invasive bacteria known as adherent invasive *E. coli* (AIEC) (Andersen *et al.*, 2012; Boudeau *et al.*, 1999; Darfeuille-Michaud, 2002; Darfeuille-Michaud *et al.*, 2004). Indeed, AIEC-like bacteria are present in nearly all biopsies taken from CD patients, but are found at much lower frequencies in UC patients or normal controls (Sasaki *et al.*, 2007). These sorts of observations suggest a causative link between the presence of AIEC and CD, but this possibility is far from proven and is likely clouded by numerous other factors, including host genetic background and composition of the intestinal microbiota (Carvalho *et al.*, 2009a). Nonetheless, understanding the pathogenic behaviour of AIEC remains a top priority for deciphering the seemingly convoluted aetiology of CD.

AIEC binds and invades intestinal epithelial cells (enterocytes) within the ileum via an actin- and microtubule-dependent mechanism (Glasser *et al.*, 2001). These events are dependent upon type 1 pili as well as OmpA, reminiscent of NMEC interactions with BMECs (Glasser *et al.*, 2001; Rolhion *et al.*, 2010). The type 1 pilus adhesin FimH binds CEACAM6 present on the apical surface of polarized enterocytes, while OmpA interacts with Gp96, a homologue of the OmpA receptor bound by NMEC within the blood brain barrier (Barnich *et al.*, 2007; Carvalho *et al.*, 2009b). In the case of AIEC, OmpA need not be directly associated with the bacterial surface, and instead is presented in association with outer membrane vesicles that bleb from AIEC (Ellis and Kuehn, 2010; Rolhion *et al.*, 2010). Interestingly, both CEACAM6 and Gp96 are expressed at abnormally high levels by enterocytes in CD patients (Rolhion *et al.*, 2010). Elevated levels of the cytokines TNFα and INFγ, as found in CD patients, can enhance expression of CEACAM6 (Barnich *et al.*, 2007; Pache *et al.*, 2009). Flagella expression can also affect AIEC adherence to enterocytes, possibly via regulatory effects on the biosynthesis of type 1 pili (Barnich *et al.*, 2003).

Although type 1 pili promote AIEC adherence, they are not sufficient to trigger AIEC entry into enterocytes (Boudeau *et al.*, 2001). The specific bacterial factors that mediate AIEC invasion of epithelial cells remain undefined. Once internalized, AIEC can disrupt endosomal membranes and enter the host cytosol where it can multiply to high levels (Boudeau *et al.*, 1999). Autophagic mechanisms within host cells may modulate these events, as also suggested for UPEC (Lapaquette *et al.*, 2010; Wang *et al.*, 2012).

AIEC can also bind M cells within the ileum, employing adhesive organelles known as Long Polar Fimbriae (LPF) (Chassaing *et al.*, 2011). These fibres are induced in the presence of bile salts within the intestinal tract, and mediate bacterial interactions with glycoprotein 2 (GP2) localized on the apical surface of M cells (Chassaing *et al.*, 2013). As noted earlier, GP2 is also a receptor for FimH (Hase *et al.*, 2009). LPF–mediated interactions with GP2 promote the uptake and translocation of AIEC through M cells into the submucosa where the bacteria interact with macrophages (Carvalho *et al.*, 2009a; Chassaing *et al.*, 2011). AIEC enter macrophages via a phagocytic mechanism, independent of type 1 pili (Bringer *et al.*, 2007; Glasser *et al.*, 2001). Once internalized, AIEC can multiply within macrophages without triggering host cell death (Glasser *et al.*, 2001). However, this process does induce macrophages to express TNFα, which in turn can up-regulate CEACAM6 expression in neighbouring enterocytes, enhancing the ability of AIEC to bind the intestinal epithelium (Darfeuille-Michaud, 2002; Glasser *et al.*, 2001; Komatsu *et al.*, 2001).

It is easy to imagine how these pro-inflammatory events and the self-enhancing sustainability of AIEC within the intestinal tract could contribute to the development and persistence of CD. Determining the cause and effect association between AIEC and CD requires that the mechanisms and consequences of AIEC interactions with resident intestinal cells and inflammatory pathways be better defined. Results will help solidify prevention and treatment strategies for CD, and potentially other syndromes classified as IBD.

Conclusion

UPEC, NMEC, EIEC (*Shigella*), and AIEC represent a huge burden in terms of both financial and health costs. Though they target different host niches, these pathogens manipulate many of the same host pathways and, in some cases, use the same virulence factors to bind and invade target host cells. Initial scepticism concerning the physiological relevance of host cell invasion by *E. coli* pathogens like UPEC and AIEC is giving way to evidence showing that intracellular events such as IBC formation and the establishment of intracellular quiescent reservoirs can significantly impact disease progression and outcome. The development of therapies aimed at preventing host cell invasion or reducing the intracellular growth and survival of these oftentimes-recalcitrant pathogens will help shore up our defences against the oncoming onslaught of antibiotic resistant bacteria, while also providing insight into the remarkable interplay taking place between us and invasive pathogens.

References

Allen, P.M., Roberts, I., Boulnois, G.J., Saunders, J.R., and Hart, C.A. (1987). Contribution of capsular polysaccharide and surface properties to virulence of *Escherichia coli* K1. Infect. Immun. 55, 2662–2668.

Anderson, G.G., Palermo, J.J., Schilling, J.D., Roth, R., Heuser, J., and Hultgren, S.J. (2003). Intracellular bacterial biofilm-like pods in urinary tract infections. Science 301, 105–107.

Anderson, G.G., Goller, C.C., Justice, S., Hultgren, S.J., and Seed, P.C. (2010). Polysaccharide capsule and sialic acid-mediated regulation promote biofilm-like intracellular bacterial communities during cystitis. Infect. Immun. *78*, 963–975.

Andersen, V., Olsen, A., Carbonnel, F., Tjønneland, A., and Vogel, U. (2012). Diet and risk of inflammatory bowel disease. Digest. Liver Dis. *44*, 185–194.

Apodaca, G. (2004). The uroepithelium: not just a passive barrier. Traffic *5*, 117–128.

Arnaout, M.A., Mahalingam, B., and Xiong, J.P. (2005). Integrin structure, allostery, and bidirectional signaling. Annu. Rev. Cell Dev. Biol. *21*, 381–410.

Bahrani-Mougeot, F.K., Buckles, E.L., Lockatell, C.V., Hebel, J.R., Johnson, D.E., Tang, C.M., and Donnenberg, M.S. (2002). Type 1 fimbriae and extracellular polysaccharides are preeminent uropathogenic *Escherichia coli* virulence determinants in the murine urinary tract. Mol. Microbiol. *45*, 1079–1093.

Baorto, D.M., Gao, Z., Malaviya, R., Dustin, M.L., van der Merwe, A., Lublin, D.M., and Abraham, S.N. (1997). Survival of FimH-expressing enterobacteria in macrophages relies on glycolipid traffic. Nature *389*, 636–639.

Barnich, N., Boudeau, J., Claret, L., and Darfeuille-Michaud, A. (2003). Regulatory and functional co-operation of fla- gella and type 1 pili in adhesive and invasive abilities of AIEC strain LF82 isolated from a patient with Crohn's disease. Mol. Microbiol. *48*, 781–794.

Barnich, N., Carvalho, F.A., Glasser, A.-L., Darcha, C., Jantscheff, P., Allez, M., Peeters, H., Bommelaer, G., Desreumaux, P., Colombel, J.-F., et al. (2007). CEACAM6 acts as a receptor for adherent-invasive *E. coli*, supporting ileal mucosa colonization in Crohn disease. J. Clin. Invest. *117*, 1566–1574.

Baumgart, D.C., and Carding, S.R. (2007). Inflammatory bowel disease: cause and immunobiology. Lancet *369*, 1627–1640.

van den Beld, M.J.C., and Reubsaet, F.A.G. (2012). Differentiation between *Shigella*, enteroinvasive *Escherichia coli* (EIEC) and noninvasive *Escherichia coli*. Eur. J. Clin. Microbiol. Infect. Dis. *31*, 899–904.

Berger, C.N., Billker, O., Meyer, T.F., Servin, A.L., and Kansau, I. (2004). Differential recognition of members of the carcinoembryonic antigen family by Afa/Dr adhesins of diffusely adhering *Escherichia coli* (Afa/Dr DAEC). Mol. Microbiol. *52*, 963–983.

Bernardini, M.L., Mounier, J., d'Hauteville, H., Coquis-Rondon, M., and Sansonetti, P.J. (1989). Identification of icsA, a plasmid locus of *Shigella flexneri* that governs bacterial intra- and intercellular spread through interaction with F-actin. Proc. Natl. Acad. Sci. *86*, 3867–3871.

Berry, R.E., Klumpp, D.J., and Schaeffer, A.J. (2009). Urothelial cultures support intracellular bacterial community formation by uropathogenic *Escherichia coli*. Infect. Immun. *77*, 2762–2772.

Bishop, B.L., Duncan, M.J., Song, J., Li, G., Zaas, D., and Abraham, S.N. (2007). Cyclic AMP-regulated exocytosis of *Escherichia coli* from infected bladder epithelial cells. Nat. Med. *13*, 625–630.

Björn, E., Mayer-Scholl, A., Walker, T., Tang, C., Weinrauch, Y., and Zychlinksy, A. (2010). Neutrophil antimicrobial proteins enhance *Shigella flexneri* adhesion and invasion. Cell. Microbiol. *12*, 1134–1143.

Blango, M.G., and Mulvey, M.A. (2010). Persistence of uropathogenic *Escherichia coli* in the face of multiple antibiotics. Antimicrob. Agents Chemother. *54*, 1855–1863.

Bokil, N.J., Totsika, M., Carey, A.J., Stacey, K.J., Hancock, V., Saunders, B.M., Ravasi, T., Ulett, G.C., Schembri, M.A., and Sweet, M.J. (2011). Intramacrophage survival of uropathogenic *Escherichia coli*: differences between diverse clinical isolates and between mouse and human macrophages. Immunobiology *216*, 1164–1171.

Boudeau, J., Glasser, A.-L., Masseret, E., Joly, B., and Darfeuille-Michaud, A. (1999). Invasive Ability of an *Escherichia coli* Strain Isolated from the Ileal Mucosa of a Patient with Crohn's Disease. Infect. Immun. *67*, 4499–4509.

Boudeau, J., Barnich, N., and Darfeuille-Michaud, A. (2001). Type 1 pili-mediated adherence of *Escherichia coli* strain LF82 isolated from Crohn's disease is involved in bacterial invasion of intestinal epithelial cells. Mol. Microbiol. *39*, 1272–1284.

Bower, J.M., Eto, D.S., and Mulvey, M.A. (2005). Covert operations of uropathogenic *Escherichia coli* within the urinary tract. Traffic *6*, 18–31.

Boyer, L., Magoc, L., Dejardin, S., Cappillino, M., Paquette, N., Hinault, C., Charriere, G.M., Ip, W.K., Fracchia, S., Hennessy, E., et al. (2011). Pathogen-derived effectors trigger protective immunity via activation of the Rac2 enzyme and the IMD or Rip kinase signaling pathway. Immunity *35*, 536–549.

Bringer, M.-A., Rolhion, N., Glasser, A.-L., and Darfeuille-Michaud, A. (2007). The oxidoreductase DsbA plays a key role in the Ability of the Crohn's disease-associated adherent-invasive *Escherichia coli* strain LF82 to resist macrophage killing. J. Bacteriol. *189*, 4860–4871.

Brown, W.J., Chambers, K., and Doody, A. (2003). Phospholipase A2 (PLA2) Enzymes in membrane trafficking: mediators of membrane shape and function. Traffic *4*, 214–221.

Brumbaugh, A.R., and Mobley, H.L. (2012). Preventing urinary tract infection: progress toward an effective *Escherichia coli* vaccine. Expert Rev. Vaccines *11*, 663–676.

Burns, J.L., Griffith, A., Barry, J.J., Jonas, M., and Chi, E.Y. (2001). Transcytosis of gastrointestinal epithelial cells by *Escherichia coli* K1. Pediatr. Res. *49*, 30–37.

Carvalho, F.A., Barnich, N., Sivignon, A., Darcha, C., Chan, C.H., Stanners, C.P., and Darfeuille-Michaud, A. (2009a). Crohn's disease adherent-invasive *Escherichia coli* colonize and induce strong gut inflammation in transgenic mice expressing human CEACAM. J. Exp. Med. *206*, 2179–2189.

Carvalho, F.A., Barnich, N., Sivignon, A., Darcha, C., Chan, C.H.F., Stanners, C.P., and Darfeuille-Michaud, A. (2009b). Crohn's disease adherent-invasive *Escherichia coli* colonize and induce strong gut inflammation in transgenic mice expressing human CEACAM. J. Exp. Med. *206*, 2179–2189.

Chassaing, B., Rolhion, N., Vallée, A.D., Salim, S.Y., Prorok-Hamon, M., Neut, C., Campbell, B.J., Söderholm, J.D., and Hugot, J. (2011). Crohn disease–associated adherent-invasive *E. coli* bacteria target mouse and human Peyer's patches via long polar fimbriae. J. Clin. Invest. *121*, 966–975.

Chassaing, B., Etienne-Mesmin, L., Bonnet, R., and Darfeuille-Michaud, A. (2013). Bile salts induce long polar fimbriae expression favouring Crohn's disease-associated adherent-invasive *Escherichia coli* interaction with Peyer's patches. Environ. Microbiol. *15*, 355–371.

Chi, F., Bo, T., Wu, C.-H., Jong, A., and Huang, S.-H. (2012). Vimentin and PSF Act in Concert to Regulate IbeA+ *E. coli* K1 Induced Activation and Nuclear Translocation of NF-κB in Human Brain Endothelial Cells. PLoS One 7, e35862.

Choudhury, D. (1999). X-ray Structure of the FimC-FimH Chaperone-Adhesin Complex from Uropathogenic *Escherichia coli*. Science *285*, 1061–1066.

Chung, J.W., Hong, S.J., Kim, K.J., Goti, D., Stins, M.F., Shin, S., Dawson, V.L., Dawson, T.M., and Kim, K.S. (2003). 37 kDa Laminin receptor precursor modulates cytotoxic necrotizing factor 1-mediated RhoA activation and bacterial uptake. J. Biol. Chem. *278*, 16857–16862.

Connell, I., Agace, W., Klemm, P., Schembri, M., Marild, S., and Svanborg, C. (1996). Type 1 fimbrial expression enhances *Escherichia coli* virulence for the urinary tract. Proc. Natl. Acad. Sci. U.S.A. *93*, 9827–9832.

Cornelis, G.R. (2006). The type III secretion injectisome. Nature Rev. Microbiol. *4*, 811–825.

Croxen, M.A., and Finlay, B.B. (2010). Molecular mechanisms of *Escherichia coli* pathogenicity. Nat. Rev. Microbiol. *8*, 26–38.

Darfeuille-Michaud, A. (2002). Adherent-invasive *Escherichia coli*: a putative new *E. coli* pathotype associated with Crohn's disease. Int. J. Med. Microbiol. *292*, 185–193.

Darfeuille-Michaud, A., Boudeau, J., Bulois, P., Neut, C., Glasser, A.-L., Barnich, N., Bringer, M.-A., Swidsinski, A., Beaugerie, L., and Colombel, J.-F. (2004). High Prevalence of Adherent-Invasive *Escherichia coli* Associated With Ileal Mucosa in Crohn's Disease. Gastroenterology *127*, 412–421.

Das, A., Astryan, L., Reddy, M.A., Wass, C.A., Stins, M.F., Joshi, S., Bonventre, J.V., and Kim, K.S. (2001). Differential role of cytosolic phospholipase A2 in the INVASION OF BRAIN MICROVASCULAR ENDOTHELIAL Cells by *Escherichia coli* and *Listeria monocytogenes*. J. Infect. Dis. *184*, 732–737.

Dhakal, B.K., and Mulvey, M.A. (2009). Uropathogenic *Escherichia coli* invades host cells via an HDAC6-modulated microtubule-dependent pathway. J. Biol. Chem. *284*, 446–454.

Dhakal, B.K., and Mulvey, M.A. (2012). The UPEC pore-forming toxin alpha-hemolysin triggers proteolysis of host proteins to disrupt cell adhesion, inflammatory, and survival pathways. Cell Host Microbe *11*, 58–69.

Dhakal, B.K., Kulesus, R.R., and Mulvey, M.A. (2008). Mechanisms and consequences of bladder cell invasion by uropathogenic *Escherichia coli*. Eur. J. Clin. Invest. *38*(Suppl. 2), 2–11.

Dielubanza, E.J., and Schaeffer, A.J. (2011). Urinary tract infections in women. Med. Clin. North Am. *95*, 27–41.

Dompierre, J.P., Godin, J.D., Charrin, B.C., Cordelieres, F.P., King, S.J., Humbert, S., and Saudou, F. (2007). Histone deacetylase 6 inhibition compensates for the transport deficit in Huntington's disease by increasing tubulin acetylation. J. Neurosci. *27*, 3571–3583.

Doye, A., Mettouchi, A., Bossis, G., Clément, R., Buisson-Touati, C., Flatau, G., Gagnoux, L., Piechaczyk, M., Boquet, P., and Lemichez, E. (2002). CNF1 Exploits the Ubiquitin-Proteasome Machinery to Restrict Rho GTPase Activation for Bacterial Host Cell Invasion. Cell *111*, 553–564.

Duncan, M.J., Li, G., Shin, J.S., Carson, J.L., and Abraham, S.N. (2004). Bacterial penetration of bladder epithelium through lipid rafts. J. Biol. Chem. *279*, 18944–18951.

Ellis, T.N., and Kuehn, M.J. (2010). Virulence and Immunomodulatory Roles of Bacterial Outer Membrane Vesicles. Microbiol. Mol. Biol. Rev. *74*, 81–94.

Espina, M., Olive, A.J., Kenjale, R., Moore, D.S., Ausar, S.F., Kaminski, R.W., Oaks, E.V., Middaugh, C.R., Picking, W.D., and Picking, W.L. (2006). IpaD localizes to the tip of the type III secretion system needle of *Shigella flexneri*. Infect. Immun. *74*, 4391–4400.

Eto, D.S., Sundsbak, J.L., and Mulvey, M.A. (2006). Actin-gated intracellular growth and resurgence of uropathogenic *Escherichia coli*. Cell. Microbiol. *8*, 704–717.

Eto, D.S., Jones, T.A., Sundsbak, J.L., and Mulvey, M.A. (2007). Integrin-mediated host cell invasion by type 1-piliated uropathogenic *Escherichia coli*. PLoS Pathog *3*, e100.

Eto, D.S., Gordon, H.B., Dhakal, B.K., Jones, T.A., and Mulvey, M.A. (2008). Clathrin, AP-2, and the NPXY-binding subset of alternate endocytic adaptors facilitate FimH-mediated bacterial invasion of host cells. Cell. Microbiol. *10*, 2553–2567.

Fagan, R.P., and Smith, S.G.J. (2007). The Hek outer membrane protein of *Escherichia coli* is an auto-aggregating adhesin and invasin. FEMS Microbiol. Lett. *269*, 248–255.

Fagan, R.P., Lambert, M.A., and Smith, S.G.J. (2008). The Hek outer membrane protein of *Escherichia coli* strain RS218 binds to proteoglycan and utilizes a single extracellular loop for adherence, invasion, and autoaggregation. Infect. Immun. *76*, 1135–1142.

de Figueiredo, P., Doody, A., Polizotto, R.S., Drecktrah, D., Wood, S., Banta, M., Strang, M.S., and Brown, W.J. (2001). Inhibition of transferrin recycling and endosome tubulation by phospholipase A2 antagonists. J. Biol. Chem. *276*, 47361–47370.

Foxman, B., and Brown, P. (2003). Epidemiology of urinary tract infections: transmission and risk factors, incidence, and costs. Infect. Dis. Clin. North. Am. *17*, 227–241.

Fukushi, Y., Orikasa, S., and Kagayama, M. (1979). An electron microscopic study of the interaction between vesical epithelium and *E. Coli*. Invest. Urol. *17*, 61–68.

Garcia, M.I., Jouve, M., Nataro, J.P., Gounon, P., and Le Bouguenec, C. (2000). Characterization of the AfaD-like family of invasins encoded by pathogenic *Escherichia coli* associated with intestinal and extra-intestinal infections. FEBS Lett. *479*, 111–117.

Garofalo, C.K., Hooton, T.M., Martin, S.M., Stamm, W.E., Palermo, J.J., Gordon, J.I., and Hultgren, S.J. (2007). *Escherichia coli* from urine of female patients with urinary tract infections is competent for intracellular bacterial community formation. Infect. Immun. *75*, 52–60.

Gbarah, A., Gahmberg, C.G., Ofek, I., Jacobi, U., and Sharon, N. (1991). Identification of the leukocyte adhesion molecules CD11 and CD18 as receptors for type 1-fimbriated (mannose-specific) *Escherichia coli*. Infect. Immun. *59*, 4524–4530.

Glasser, A.-L., Boudeau, J., Barnich, N., Perruchot, M.-H., Colombel, J.-F., and Darfeuille-Michaud, A. (2001). Adherent invasive *Escherichia coli* strains from patients with Crohn's disease survive and replicate within macrophages without inducing host Cell death. Infect. Immun. *69*, 5529–5537.

Glenn, H.L., and Jacobson, B.S. (2002). Arachidonic acid signaling to the cytoskeleton: the role of cyclooxygenase and cyclic AMP-dependent protein kinase in actin bundling. Cell Motil. Cytoskel. *53*, 239–250.

Goldberg, M.B., and Theriot, J.A. (1995). *Shigella flexneri* surface protein IcsA is sufficient to direct actin-based motility. Proc. Natl. Acad. Sci. *92*, 6572–6576.

Goller, C.C., and Seed, P.C. (2010). Revisiting the *Escherichia coli* polysaccharide capsule as a virulence factor during urinary tract infection: contribution to intracellular biofilm development. Virulence *1*, 333–337.

Grimmer, S., van Deurs, B., and Sandvig, K. (2002). Membrane ruffling and macropinocytosis in A431 cells require cholesterol. J. Cell. Sci. *115*, 2953–2962.

Guichon, A., Hersh, D., Smith, M.R., and Zychlinksy, A. (2001). Structure–function analysis of the *Shigella* virulence factor IpaB. J. Bacteriol. *183*, 1269–1276.

Guignot, J., Hudault, S., Kansau, I., Chau, I., and Servin, A.L. (2009). Human decay-accelerating factor and CEACAM receptor-mediated internalization and intracellular lifestyle of Afa/Dr diffusely adhering *Escherichia coli* in epithelial cells. Infect. Immun. *77*, 517–531.

Hart, A., Pham, T., Nowicki, S., Whorton, E.B., Jr., Martens, M.G., Anderson, G.D., and Nowicki, B.J. (1996). Gestational pyelonephritis--associated *Escherichia coli* isolates represent a nonrandom, closely related population. Am. J. Obstet. Gynecol. *174*, 983–989.

Hart, A., Nowicki, B.J., Reisner, B., Pawelczyk, E., Goluszko, P., Urvil, P., Anderson, G., and Nowicki, S. (2001). Ampicillin-resistant *Escherichia coli* in gestational pyelonephritis: increased occurrence and association with the colonization factor Dr adhesin. J. Infect. Dis. *183*, 1526–1529.

Hase, K., Kawano, K., Nochi, T., Pontes, G.S., Fukuda, S., Ebisawa, M., Kadokura, K., Tobe, T., Fujimura, Y., Kawano, S., *et al.* (2009). Uptake through glycoprotein 2 of FimH(+) bacteria by M cells initiates mucosal immune response. Nature *462*, 226–230.

Hayward, R.D., Cain, R.J., McGhie, E.J., Philips, N., Garner, M.J., and Koronakis, V. (2005). Cholesterol binding by the bacterial type III translocon is essential for virulence effector delivery into mammalian cells. Mol. Microbiol. *56*, 590–603.

He, L., Chi, F., Bo, T., Wang, L., Wu, C., Jong, A., and Huang, S. (2012). p54nrb, a PSF protein partner, contributes to Meningitic *Escherichia coli* K1-mediated pathogenicities. Open J. Appl. Sci. *2*, 1–10.

Hicks, R.M. (1975). The mammalian urinary bladder: an accommodating organ. Biol. Rev. Camb. Philos. Soc. *50*, 215–246.

High, N., Mounier, J., Prévost, M.C., and Sansonetti, P.J. (1992). IpaB of *Shigella flexneri* causes entry into epithelial cells and escape from the phagocytic vacuole. EMBO J. *11*, 1991–1999.

Hoffman, J.A., Wass, C., Stins, M.F., and Kim, K.S. (1999). The capsule supports survival but not traversal of *Escherichia coli* K1 across the blood-brain barrier. Infect. Immun. *67*, 3566–3570.

Houdouin, V., Bonacorsi, S., Bidet, P., Blanco, J., De La Rocque, F., Cohen, R., Aujard, Y., and Bingen, E. (2008). Association between mortality of *Escherichia coli* meningitis in young infants and non-virulent clonal groups of strains. Clin. Microbiol. Infect. *14*, 685–690.

Iwai, H., Kim, M., Yoshikawa, Y., Ashida, H., Ogawa, M., Fujita, Y., Muller, D., Kirikae, T., Jackson, P.K., Kotani, S., *et al.* (2007). A bacterial effector targets Mad2L2, an APC inhibitor, to modulate host cell cycling. Cell *140*, 611–623.

Iwai, H., Kim, M., Yoshikawa, Y., Ashida, H., Ogawa, M., Fujita, Y., Muller, D., Kirikae, T., Jackson, P.K., Kotani, S., *et al.* (2009). Bacteria hijack integrin-linked kinase to stabilize focal adhesions and block cell detachment. Nature *459*, 578–582.

Jensen, V.B., Harty, J.T., and Jones, B.D. (1998). Interactions of the Invasive Pathogens *Salmonella typhimurium*, *Listeria monocytogenes*, and *Shigella flexneri* with M cells and murine Peyer's patches. Infect. Immun. *66*, 3758–3766.

Jones, C.H., Pinkner, J.S., Roth, R., Heuser, J., Nicholes, A.V., Abraham, S.N., and Hultgren, S.J. (1995). FimH adhesin of type 1 pili is assembled into a fibrillar tip structure in the Enterobacteriaceae. Proc. Natl. Acad. Sci. U.S.A. *92*, 2081–2085.

Jost, S.P. (1989). Cell cycle of normal bladder urothelium in developing and adult mice. Virchows Arch. B Cell. Pathol. Incl. Mol. Pathol. *57*, 27–36.

Jouve, M., Garcia, M.I., Courcoux, P., Labigne, A., Gounon, P., and Le Bouguenec, C. (1997). Adhesion to and invasion of HeLa cells by pathogenic *Escherichia coli* carrying the afa-3 gene cluster are mediated by the AfaE and AfaD proteins, respectively. Infect. Immun. *65*, 4082–4089.

Justice, S.S., Hung, C., Theriot, J.A., Fletcher, D.A., Anderson, G.G., Footer, M.J., and Hultgren, S.J. (2004). Differentiation and developmental pathways of uropathogenic *Escherichia coli* in urinary tract pathogenesis. Proc. Natl. Acad. Sci. U.S.A. *101*, 1333–1338.

Justice, S.S., Hunstad, D.A., Seed, P.C., and Hultgren, S.J. (2006a). Filamentation by *Escherichia coli* subverts innate defenses during urinary tract infection. Proc. Natl. Acad. Sci. U.S.A. *103*, 19884–19889.

Justice, S.S., Lauer, S.R., Hultgren, S.J., and Hunstad, D.A. (2006b). Maturation of intracellular *Escherichia coli* communities requires SurA. Infect. Immun. *74*, 4793–4800.

Justice, S.S., Li, B., Downey, J.S., Dabdoub, S.M., Brockson, M.E., Probst, G.D., Ray, W.C., and Goodman, S.D. (2012). Aberrant Community Architecture and Attenuated Persistence of Uropathogenic *Escherichia coli* in the Absence of Individual IHF Subunits. PLoS One 7, e48349.

Kansau, I., Berger, C., Hospital, M., Amsellem, R., Nicolas, V., Servin, A.L., and Bernet-Camard, M.F. (2004). Zipper-like internalization of Dr-positive *Escherichia coli* by epithelial cells is preceded by an adhesin-induced mobilization of raft-associated molecules in the initial step of adhesion. Infect. Immun. *72*, 3733–3742.

Kaper, J.B., Nataro, J.P., and Mobley, H.L. (2004). Pathogenic *Escherichia coli*. Nat. Rev. Microbiol. *2*, 123–140.

Khan, N.A., Wang, Y., Kim, K.J., Chung, J.W., Wass, C.A., and Kim, K.S. (2002). Cytotoxic necrotizing factor-1 contributes to *Escherichia coli* K1 invasion of the central nervous system. J. Biol. Chem. *277*, 15607–15612.

Khan, N.A., Kim, Y., Shin, S., and Kim, K.S. (2007). FimH-mediated *Escherichia coli* K1 invasion of human brain microvascular endothelial cells. Cell. Microbiol. *9*, 169–178.

Kim, K.J., Elliott, S.J., Di Cello, F., Stins, M.F., and Kim, K.S. (2003). The K1 capsule modulates trafficking of *E. coli*-containing vacuoles and enhances intracellular bacterial survival in human brain microvascular endothelial cells. Cell. Microbiol. *5*, 245–252.

Kim, K.J., Chung, J.W., and Kim, K.S. (2005). 67-kDa laminin receptor promotes internalization of cytotoxic necrotizing factor 1-expressing *Escherichia coli* K1 into human brain microvascular endothelial cells. J. Biol. Chem. *280*, 1360–1368.

Kim, K.S. (2003). Pathogenesis of bacterial meningitis: from bacteremia to neuronal injury. Nature Rev. Neurosci. *4*, 376–385.

Kim, K.S. (2006). Microbial translocation of the blood–brain barrier. Int. J. Parasitol. *36*, 607–614.

Kim, K.S. (2010). Acute bacterial meningitis in infants and children. Lancet Infect. Dis. *10*, 32–42.

Kim, K.S. (2012). Current concepts on the pathogenesis of *Escherichia coli* meningitis: implications for therapy and prevention. Curr. Opin. Infect. Dis. *25*, 273–278.

Komatsu, M., Kobayashi, D., Saito, K., Furuya, D., Yagihashi, A., Araake, H., Tsuji, N., Sakamaki, S., Niitsu, Y., and Watanabe, N. (2001). Tumor necrosis factor-α in serum of patients with inflammatory bowel disease as measured by a highly sensitive immuno-PCR. Clin. Chem. *47*, 1297–1301.

Kraehenbuhl, J.-P. (2000). Epithelial M Cells: Differentiation and Function. Annu. Rev. Cell Dev. Biol. *16*, 301–332.

Kreft, M.E., Jezernik, K., Kreft, M., and Romih, R. (2009). Apical plasma membrane traffic in superficial cells of bladder urothelium. Ann. N.Y. Acad. Sci. *1152*, 18–29.

Krishnan, S., Fernandez, G.E., Sacks, D.B., and Prasadarao, N.V. (2012). IQGAP1 mediates the disruption of adherens junctions to promote *Escherichia coli* K1 invasion of brain endothelial cells. Cell. Microbiol. *14*, 1415–1433.

Krishnan, S., Chen, S., Turcatel, G., Arditi, M., and Prasadarao, N.V. (2013). Regulation of Toll-like receptor 2 interaction with Ecgp96 controls *Escherichia coli* K1 invasion of brain endothelial cells. Cell. Microbiol. *15*, 63–81.

Kuncewicz, T., Balakrishnan, P., Snuggs, M.B., and Kone, B.C. (2001). Specific association of nitric oxide synthase-2 with Rac isoforms in activated murine macrophages. Am. J. Physiol. Renal Physiol. *281*, F326–336.

Kuroda, S., Fukata, M., Kobayashi, K., Nakafuku, M., Nomura, N., Iwamatsu, A., and Kaibuchi, K. (1996). Identification of IQGAP as a putative target for the small GTPases, Cdc42 and Rac1. J. Biol. Chem. *271*, 23363–23367.

Kwik, J., Boyle, S., Fooksman, D., Margolis, L., Sheetz, M.P., and Edidin, M. (2003). Membrane cholesterol, lateral mobility, and the phosphatidylinositol 4,5-bisphosphate-dependent organization of cell actin. Proc. Natl. Acad. Sci. U.S.A. *100*, 13964–13969.

Lafont, F., Tran Van Nhieu, G., Hanada, K., Sansonetti, P., and van der Goot, F.G. (2002). Initial steps of *Shigella* infection depend on the cholesterol/sphingolipid raft-mediated CD44–IpaB interaction. EMBO J. *21*, 4449–4457.

Lan, R., Lumb, B., Ryan, D., and Reeves, P.R. (2001). Molecular Evolution of Large Virulence Plasmid in *Shigella* Clones and Enteroinvasive *Escherichia coli*. Infect. Immun. *69*, 6303–6309.

Lan, R., Alles, M.C., Donohoe, K., Martinez, M.B., and Reeves, P.R. (2004). Molecular Evolutionary Relationships of Enteroinvasive *Escherichia coli* and *Shigella* spp. Infect. Immun. *72*, 5080–5088.

Langermann, S., Palaszynski, S., Barnhart, M., Auguste, G., Pinkner, J.S., Burlein, J., Barren, P., Koenig, S., Leath, S., Jones, C.H., et al. (1997). Prevention of mucosal *Escherichia coli* infection by FimH-adhesin-based systemic vaccination. Science *276*, 607–611.

Lapaquette, P., Glasser, A.-L., Huett, A., Xavier, R.J., and Darfeuille-Michaud, A. (2010). Crohn's disease-associated adherent-invasive *E. coli* are selectively favoured by impaired autophagy to replicate intracellularly. Cell. Microbiol. *12*.

Le Bouguenec, C. (1999). Diarrhea-associated diffusely adherent *Escherichia coli*. Clin. Microbiol. Rev. *12*, 180–181.

Lemonnier, M., Landraud, L., and Lemichez, E. (2007). Rho GTPase-activating bacterial toxins: from bacterial virulence regulation to eukaryotic cell biology. FEMS Microbiol. Rev. *31*, 515–534.

Lerm, M., Pop, M., Fritz, G., Aktories, K., and Schmidt, G. (2002). Proteasomal degradation of cytotoxic necrotizing factor 1-activated rac. Infect. Immun. *70*, 4053–4058.

Le Trong, I., Aprikian, P., Kidd, B.A., Forero-Shelton, M., Tchesnokova, V., Rajagopal, P., Rodriguez, V., Interlandi, G., Klevit, R., Vogel, V., *et al.* (2010). Structural basis for mechanical force regulation of the adhesin FimH via finger trap-like beta sheet twisting. Cell *141*, 645–655.

Lewis, K. (2012). Persister cells: molecular mechanisms related to antibiotic tolerance. Handb. Exp. Pharmacol. 121–133.

Li, B., Smith, P., Horvath, D.J., Jr., Romesberg, F.E., and Justice, S.S. (2010). SOS regulatory elements are essential for UPEC pathogenesis. Microb. Infect. *12*, 662–668.

Li, K., Zhou, W., Hong, Y., Sacks, S.H., and Sheerin, N.S. (2009). Synergy between type 1 fimbriae expression and C3 opsonisation increases internalisation of *E. coli* by human tubular epithelial cells. BMC Microbiol. *9*, 64.

Loughman, J.A., and Hunstad, D.A. (2012). Induction of indoleamine 2,3-dioxygenase by uropathogenic bacteria attenuates innate responses to epithelial infection. J. Infect. Dis. *205*, 1830–1839.

McTaggart, L.A., Rigby, R.C., and Elliott, T.S. (1990). The pathogenesis of urinary tract infections associated with *Escherichia coli, Staphylococcus saprophyticus* and *S. epidermidis.* J. Med. Microbiol. *32*, 135–141.

Malaviya, R., Gao, Z., Thankavel, K., van der Merwe, P.A., and Abraham, S.N. (1999). The mast cell tumor necrosis factor alpha response to FimH-expressing *Escherichia coli* is mediated by the glycosylphosphatidylinositol-anchored molecule CD48. Proc. Natl. Acad. Sci. U.S.A. *96*, 8110–8115.

Martinez, J.J., and Hultgren, S.J. (2002). Requirement of Rho-family GTPases in the invasion of Type 1-piliated uropathogenic *Escherichia coli.* Cell. Microbiol. *4*, 19–28.

Martinez, J.J., Mulvey, M.A., Schilling, J.D., Pinkner, J.S., and Hultgren, S.J. (2000). Type 1 pilus-mediated bacterial invasion of bladder epithelial cells. EMBO J. *19*, 2803–2812.

Maruvada, R., Argon, Y., and Prasadarao, N.V. (2008). *Escherichia coli* interaction with human brain microvascular endothelial cells induces signal transducer and activator of transcription 3 association with the C-terminal domain of Ec-gp96, the outer membrane protein A receptor for invasion. Cell. Microbiol. *10*, 2326–2338.

Mierke, C.T. (2009). The Role of Vinculin in the Regulation of the Mechanical Properties of Cells. Cell Biochem. Biophys. *53*, 115–126.

Mills, M., Meysick, K.C., and O'Brien, A.D. (2000). Cytotoxic necrotizing factor type 1 of uropathogenic *Escherichia coli* kills cultured human uroepithelial 5637 cells by an apoptotic mechanism. Infect. Immun. *68*, 5869–5880.

Moran, M., and Miceli, M.C. (1998). Engagement of GPI-linked CD48 contributes to TCR signals and cytoskeletal reorganization: a role for lipid rafts in T cell activation. Immunity *9*, 787–796.

Mossman, K.L., Mian, M.F., Lauzon, N.M., Gyles, C.L., Lichty, B., Mackenzie, R., Gill, N., and Ashkar, A.A.

(2008). Cutting edge: FimH adhesin of type 1 fimbriae is a novel TLR4 ligand. J. Immunol. *181*, 6702–6706.

Mounier, J., Popoff, M.R., Enninga, J., Frame, M.C., Sansonetti, P.J., and Tran Van Nhieu, G. (2009). The IpaC carboxyterminal effector domain mediates Src-dependent actin polymerization during *Shigella* invasion of epithelial Cells. PLoS Pathog. *5*, e1000271.

Muller, C.M., Aberg, A., Straseviciene, J., Emody, L., Uhlin, B.E., and Balsalobre, C. (2009). Type 1 fimbriae, a colonization factor of uropathogenic *Escherichia coli,* are controlled by the metabolic sensor CRP-cAMP. PLoS Pathog *5*, e1000303.

Mulvey, M.A. (2002). Adhesion and entry of uropathogenic *Escherichia coli.* Cell. Microbiol. *4*, 257–271.

Mulvey, M.A., Lopez-Boado, Y.S., Wilson, C.L., Roth, R., Parks, W.C., Heuser, J., and Hultgren, S.J. (1998). Induction and evasion of host defenses by type 1-piliated uropathogenic *Escherichia coli.* Science *282*, 1494–1497.

Mulvey, M.A., Schilling, J.D., Martinez, J.J., and Hultgren, S.J. (2000). Bad bugs and beleaguered bladders: interplay between uropathogenic *Escherichia coli* and innate host defenses. Proc. Natl. Acad. Sci. U.S.A. *97*, 8829–8835.

Mulvey, M.A., Schilling, J.D., and Hultgren, S.J. (2001). Establishment of a persistent *Escherichia coli* reservoir during the acute phase of a bladder infection. Infect. Immun. *69*, 4572–4579.

Mysorekar, I.U., and Hultgren, S.J. (2006). Mechanisms of uropathogenic *Escherichia coli* persistence and eradication from the urinary tract. Proc. Natl. Acad. Sci. U.S.A. *103*, 14170–14175.

Mysorekar, I.U., Mulvey, M.A., Hultgren, S.J., and Gordon, J.I. (2002). Molecular regulation of urothelial renewal and host defenses during infection with uropathogenic *Escherichia coli.* J. Biol. Chem. *277*, 7412–7419.

Nazareth, H., Genagon, S.A., and Russo, T.A. (2007). Extraintestinal pathogenic *Escherichia coli* survives within neutrophils. Infect. Immun. *75*, 2776–2785.

Nicholson, T.F., Watts, K.M., and Hunstad, D.A. (2009). OmpA of uropathogenic *Escherichia coli* promotes postinvasion pathogenesis of cystitis. Infect. Immun. *77*, 5245–5251.

Niyogi, S.K. (2005). Shigellosis. J. Microbiol. *43*, 133–143.

Nizet, V., and Klein, J.O. (2010). Bacterial Sepsis and Meningitis, 7th edition (Philidelphia, Elsevier Saunders).

Norton, J.P., and Mulvey, M.A. (2012). Toxin-Antitoxin Systems Are Important for Niche-Specific Colonization and Stress Resistance of Uropathogenic *Escherichia coli.* PLoS Pathog. *8*, e1002954.

Nowicki, B., Hart, A., Coyne, K.E., Lublin, D.M., and Nowicki, S. (1993). Short consensus repeat-3 domain of recombinant decay-accelerating factor is recognized by *Escherichia coli* recombinant Dr adhesin in a model of a cell–cell interaction. J. Exp. Med. *178*, 2115–2121.

Nowicki, B., Martens, M., Hart, A., and Nowicki, S. (1994). Gestational age-dependent distribution of *Escherichia coli* fimbriae in pregnant patients with pyelonephritis. Ann. N.Y. Acad. Sci. *730*, 290–291.

Ørskov, F., Ørskov, I., Sutton, A., Schneerson, R., Lin, W., Egan, W., Hoff, G.E., and Robbins, J.B. (1979). Form variation in *Escherichia coli* K1: determined by O-acetylation of the capsular polysaccharide. J. Exp. Med. *149*, 669–685.

Pache, I., Rogler, G., and Felley, C. (2009). TNF-α blockers in inflammatory bowel diseases: Practical consensus recommendations and a user's guide. Swiss Med. Week. *139*, 278–287.

Park, H., Valencia-Gallardo, C., Sharff, A., Tran Van Nhieu, G., and Izard, T. (2011). Novel Vinculin Binding Site of the IpaA Invasin of *Shigella*. J. Biol. Chem. *286*, 23214–23221.

Parsot, C. (2005). *Shigella* spp. and enteroinvasive *Escherichia coli* pathogenicity factors. FEMS Microbiol. Lett. *252*, 11–18.

Parthasarathy, G., Yao, Y., and Kim, K.S. (2007). Flagella Promote *Escherichia coli* K1 Association with and Invasion of Human Brain Microvascular Endothelial Cells. Infect. Immun. *75*, 2937–2945.

Perdomo, J.J., Gounon, P., and Sansonetti, P.J. (1994). Polymorphonuclear leukocyte transmigration promotes invasion of colonic epithelial monolayer by *Shigella flexneri*. J. Clin. Invest. *93*, 633–643.

Prasadarao, N.V., Wass, C.A., Stins, M.F., Shimada, H., and Kim, K.S. (1999). Outer Membrane Protein A-Promoted Actin Condensation of Brain Microvascular Endothelial Cells Is Required for *Escherichia coli* Invasion. Infect. Immun. *67*, 5775–5783.

Pratt, L.A., and Kolter, R. (1998). Genetic analysis of *Escherichia coli* biofilm formation: roles of flagella, motility, chemotaxis and type I pili. Mol. Microbiol. *30*, 285–293.

Pupo, G.M., Karaolis, D.K.R., Lan, R., and Reeves, P.R. (1997). Evolutionary Relationships among Pathogenic and Nonpathogenic *Escherichia coli* Strains Inferred from Multilocus Enzyme Electrophoresis and mdh Sequence Studies. Infect. Immun. *65*, 2685–2692.

Reed, N.A., Cai, D., Blasius, T.L., Jih, G.T., Meyhofer, E., Gaertig, J., and Verhey, K.J. (2006). Microtubule acetylation promotes kinesin-1 binding and transport. Curr. Biol. *16*, 2166–2172.

Reis, R.S.d., and Horn, G. (2010). Enteropathogenic *Escherichia coli*, *Samonella*, *Shigella* and *Yersinia*: cellular aspects of host–bacteria interactions in enteric diseases. Gut Pathogens 2, 8.

Ren, C., Chaudhuri, R.R., Fivian, A., Bailey, C.M., Antonio, M., Barnes, W.M., and Pallen, M.J. (2004). The ETT2 Gene Cluster, Encoding a Second Type III Secretion System from *Escherichia coli*, Is Present in the Majority of Strains but Has Undergone Widespread Mutational Attrition. J. Bacteriol. *186*, 3547–3560.

Rolhion, N., Barnich, N., Bringer, M.-A., Glasser, A.-L., Ranc, J., Hébuterne, X., Hofman, P., and Darfeuille-Michaud, A. (2010). Abnormally expressed ER stress response chaperone Gp96 in CD favours adherent-invasive *Escherichia coli* invasion. Gut *59*, 1355–1362.

Rolland, K., Lambert-Zechovsky, N., Picard, B., and Denamur, E. (1998). *Shigella* and enteroinvasive *Escherichia coli* strains are derived from distinct ancestral strains of *E. coli*. Microbiology *144*, 2667–2672.

Romih, R., Veranic, P., and Jezernik, K. (1999). Actin filaments during terminal differentiation of urothelial cells in the rat urinary bladder. Histochem. Cell Biol. *112*, 375–380.

Rosen, D.A., Hooton, T.M., Stamm, W.E., Humphrey, P.A., and Hultgren, S.J. (2007). Detection of intracellular bacterial communities in human urinary tract infection. PLoS Med. *4*, e329.

Russell, P.W., and Orndorff, P.E. (1992). Lesions in two *Escherichia coli* type 1 pilus genes alter pilus number and length without affecting receptor binding. J. Bacteriol. *174*, 5923–5935.

Sakaguchi, T., Köhler, H., Gu, X., McCormick, B.A., and Reinecker, H.-C. (2002). *Shigella flexneri* regulates tight junction-associated proteins in human intestinal epithelial cells. Cell. Microbiol. *4*, 367–381.

Sansonetti, P.J., and Phalipon, A. (2007). *Shigella*'s ways of manipulating the host intestinal innate and adaptive immune system: a tool box for survival? Immunol. Cell Biol. 1–11.

Sansonetti, P.J., d'Hauteville, H., Ecobichon, C., and Poucel, C. (1983). Molecular comparison of virulence plasmids in *Shigella* and enteroinvasive *Escherichia coli*. Ann. Microbiol. (Paris) *134A*, 295–318.

Sansonetti, P.J., Phalipon, A., Arondel, A., Thirumalai, K., Banerjee, S., Akira, S., Takeda, K., and Zychlinsky, A. (2000). Caspase-1 Activation of IL-1β and IL-18 Are Essential for *Shigella flexneri*–Induced Inflammation. Immunity *12*, 581–590.

Sasaki, M., Sitaraman, S.V., Babbin, B.A., Gerner-Smidt, P., Ribot, E.M., Garrett, N., Alpern, J.A., Akyildiz, A., Theiss, A.L., Nusrat, A., *et al.* (2007). Invasive *Escherichia coli* are a feature of Crohn's disease. Lab. Invest. *87*, 1042–1054.

Sauter, S.L., Rutherfurd, S.M., Wagener, C., Shively, J.E., and Hefta, S.A. (1991). Binding of nonspecific cross-reacting antigen, a granulocyte membrane glycoprotein, to *Escherichia coli* expressing type 1 fimbriae. Infect. Immun. *59*, 2485–2493.

Sauter, S.L., Rutherfurd, S.M., Wagener, C., Shively, J.E., and Hefta, S.A. (1993). Identification of the specific oligosaccharide sites recognized by type 1 fimbriae from *Escherichia coli* on nonspecific cross-reacting antigen, a CD66 cluster granulocyte glycoprotein. J. Biol. Chem. *268*, 15510–15516.

Schembri, M.A., and Klemm, P. (2001). Biofilm formation in a hydrodynamic environment by novel fimh variants and ramifications for virulence. Infect. Immun. *69*, 1322–1328.

Schilling, J.D., Mulvey, M.A., Vincent, C.D., Lorenz, R.G., and Hultgren, S.J. (2001). Bacterial invasion augments epithelial cytokine responses to *Escherichia coli* through a lipopolysaccharide-dependent mechanism. J. Immunol. *166*, 1148–1155.

Schwan, W.R. (2011). Regulation of fim genes in uropathogenic *Escherichia coli*. World J. Clin. Infect. Dis. *1*, 17–25.

Schwan, W.R., Lee, J.L., Lenard, F.A., Matthews, B.T., and Beck, M.T. (2002). Osmolarity and pH growth conditions regulate fim gene transcription and type

1 pilus expression in uropathogenic *Escherichia coli*. Infect. Immun. *70*, 1391–1402.

Schwartz, D.J., Chen, S.L., Hultgren, S.J., and Seed, P.C. (2011). Population dynamics and niche distribution of uropathogenic *Escherichia coli* during acute and chronic urinary tract infection. Infect. Immun. *79*, 4250–4259.

Servin, A.L. (2005). Pathogenesis of Afa/Dr diffusely adhering *Escherichia coli*. Clin. Microbiol. Rev. *18*, 264–292.

Shanmuganathan, M.V., Krishnan, S., Fu, X., and Prasadarao, N.V. (2013). Attenuation of biopterin synthesis prevents *Escherichia coli* K1 invasion of brain endothelial cells and the development of meningitis in newborn mice. J. Infect. Dis. *207*, 61–71.

Shin, J.S., Gao, Z., and Abraham, S.N. (2000). Involvement of cellular caveolae in bacterial entry into mast cells. Science *289*, 785–788.

Silver, R.P., Aaronson, W., and Vann, W.F. (1988). The K1 capsular polysaccharide of *Escherichia coli*. Rev. Infect. Dis. *10*(Suppl. 2), S282–S286.

Singer, M., and Sansonetti, P.J. (2004). IL-8 Is a Key Chemokine Regulating Neutrophil Recruitment in a New Mouse Model of *Shigella*-Induced Colitis. J. Immunol. *173*, 4197–4206.

Skoudy, A., Tran Van Nhieu, G., Mantis, N., Arpin, M., Mounier, J., Gounon, P., and Sansonetti, P. (1999). A functional role for ezrin during *Shigella flexneri* entry into epithelial cells. J. Cell Sci. *112*, 2059–2068.

Smith, Y.C., Grande, K.K., Rasmussen, S.B., and O'Brien, A.D. (2006). Novel three-dimensional organoid model for evaluation of the interaction of uropathogenic *Escherichia coli* with terminally differentiated human urothelial cells. Infect. Immun. *74*, 750–757.

Sokurenko, E.V., Chesnokova, V., Doyle, R.J., and Hasty, D.L. (1997). Diversity of the *Escherichia coli* type 1 fimbrial lectin. Differential binding to mannosides and uroepithelial cells. J. Biol. Chem. *272*, 17880–17886.

Sokurenko, E.V., Chesnokova, V., Dykhuizen, D.E., Ofek, I., Wu, X.R., Krogfelt, K.A., Struve, C., Schembri, M.A., and Hasty, D.L. (1998). Pathogenic adaptation of *Escherichia coli* by natural variation of the FimH adhesin. Proc. Natl. Acad. Sci. U.S.A. *95*, 8922–8926.

Song, J., Bishop, B.L., Li, G., Grady, R., Stapleton, A., and Abraham, S.N. (2009). TLR4-mediated expulsion of bacteria from infected bladder epithelial cells. Proc. Natl. Acad. Sci. U.S.A. *106*, 14966–14971.

Southgate, J., Kennedy, W., Hutton, K.A., and Trejdosiewicz, L.K. (1995). Expression and *in vitro* regulation of integrins by normal human urothelial cells. Cell Adhes. Commun. *3*, 231–242.

Sukumaran, S.K., and Prasadarao, N.V. (2003). *Escherichia coli* K1 Invasion increases human brain microvascular endothelial cell monolayer permeability by disassembling vascular-endothelial cadherins at tight junctions. J. Infect. Dis. *188*, 1295–1309.

Teng, C.-H., Xie, Y., Shin, S., Di Cello, F., Paul-Satyaseela, M., Cai, M., and Kim, K.S. (2006). Effects of ompA deletion on expression of Type 1 fimbriae in *Escherichia coli* K1 strain RS218 and on the association of *E. coli* with human brain microvascular endothelial cells. Infect. Immun. *74*, 5609–5616.

Teng, C.-H., Tseng, Y.-T., Maruvada, R., Pearce, D., Xie, Y., Paul-Satyaseela, M., and Kim, K.S. (2010). NlpI contributes to *Escherichia coli* K1 strain RS218 interaction with human brain microvascular endothelial cells. Infect. Immun. *78*, 3090–3096.

Thanassi, D.G., Bliska, J.B., and Christie, P.J. (2012). Surface organelles assembled by secretion systems of Gram-negative bacteria: diversity in structure and function. FEMS Microbiol. Rev. *36*, 1046–1082.

Thumbikat, P., Berry, R.E., Zhou, G., Billips, B.K., Yaggie, R.E., Zaichuk, T., Sun, T.-T., Schaeffer, A.J., and Klumpp, D.J. (2009). Bacteria-induced uroplakin signaling mediates bladder response to infection. PLoS Pathog. *5*, e1000415.

Torrino, S., Visvikis, O., Doye, A., Boyer, L., Stefani, C., Munro, P., Bertoglio, J., Gacon, G., Mettouchi, A., and Lemichez, E. (2011). The E3 ubiquitin-ligase HACE1 catalyzes the ubiquitylation of active Rac1. Dev. Cell *21*, 959–965.

Tran Van Nhieu, G., Caron, E., Hall, A., and Sansonetti, P.J. (1999). IpaC induces actin polymerization and filopodia formation during *Shigella* entry into epithelial cells. EMBO J. *18*, 3249–3262.

Tseng, Y.-t., Wang, S.-W., Kim, K.S., Wang, Y.-H., Yao, Y., Chen, C.-C., Chiang, C.-W., Hsieh, P.-C., and Teng, C.-H. (2012). NlpI facilitates deposition of C4bp on *Escherichia coli* by blocking classical complement-mediated killing, which results in high-level bacteremia. Infect. Immun. *80*, 3669–3678.

Tunkel, A.R., Hartman, B.J., Kaplan, S.L., Kaufman, B.A., Roos, K.L., Scheld, W.M., and Whitley, R.J. (2004). Practice guidelines for the management of bacterial meningitis. Clin. Infect. Dis. *39*, 1267–1284.

Van Loy, C.P., Sokurenko, E.V., and Moseley, S.L. (2002). The major structural subunits of Dr and F1845 fimbriae are adhesins. Infect. Immun. *70*, 1694–1702.

Visvikis, O., Boyer, L., Torrino, S., Doye, A., Lemonnier, M., Lorès, P., Rolando, M., Flatau, G., Mettouchi, A., Bouvard, D., *et al.* (2011). *Escherichia coli* producing CNF1 toxin hijacks tollip to trigger Rac1-dependent cell invasion. Traffic *12*, 579–590.

Wang, C., Mendonsa, G.R., Symington, J.W., Zhang, Q., and Cadwell, K. (2012). Atg16L1 deficiency confers protection from uropathogenic *Escherichia coli* infection *in vivo*. Proc. Natl. Acad. Sci. U.S.A. *109*, 11008–11013.

Wang, G., Moniri, N.H., Ozawa, K., Stamler, J.S., and Daaka, Y. (2006). Nitric oxide regulates endocytosis by S-nitrosylation of dynamin. Proc. Natl. Acad. Sci. U.S.A. *103*, 1295–1300.

Wang, H., Liang, F.X., and Kong, X.P. (2008). Characteristics of the phagocytic cup induced by uropathogenic *Escherichia coli*. J. Histochem. Cytochem. *56*, 597–604.

Wang, H., Min, G., Glockshuber, R., Sun, T.-T., and Kong, X.-P. (2009). Uropathogenic *E. coli* adhesin-induced host cell receptor conformation changes: implications in transmembrane signaling transduction. J. Mol. Biol. *392*, 352–361.

Wang, Z., Humphrey, C., Frilot, N., Wang, G., Nie, Z., Moniri, N.H., and Daaka, Y. (2011). Dynamin2- and endothelial nitric oxide synthase-regulated invasion of bladder epithelial cells by uropathogenic *Escherichia coli*. J. Cell. Biol. *192*, 101–110.

Watarai, M., Funato, S., and Sasakawa, C. (1996). Interaction of Ipa proteins of *Shigella* flexneri with α5β1 integrin promotes entry of the bacteria into mammalian cells. J. Exp. Med. *183*, 991–999.

Weissman, S.J., Chattopadhyay, S., Aprikian, P., Obata-Yasuoka, M., Yarova-Yarovaya, Y., Stapleton, A., Ba-Thein, W., Dykhuizen, D., Johnson, J.R., and Sokurenko, E.V. (2006). Clonal analysis reveals high rate of structural mutations in fimbrial adhesins of extraintestinal pathogenic *Escherichia coli*. Mol. Microbiol. *59*, 975–988.

White, C.D., Brown, M.D., and Sacks, D.B. (2009). IQGAPs in cancer: a family of scaffold proteins underlying tumorigenesis. FEBS Lett. *583*, 1817–1824.

Wiles, T.J., Dhakal, B.K., Eto, D.S., and Mulvey, M.A. (2008a). Inactivation of host Akt/protein kinase B signaling by bacterial pore-forming toxins. Mol. Biol. Cell *19*, 1427–1438.

Wiles, T.J., Kulesus, R.R., and Mulvey, M.A. (2008b). Origins and virulence mechanisms of uropathogenic *Escherichia coli*. Exp. Mol. Pathol. *85*, 11–19.

Wood, M.W., Breitschwerdt, E.B., Nordone, S.K., Linder, K.E., and Gookin, J.L. (2012). Uropathogenic *E. coli* promote a paracellular urothelial barrier defect characterized by altered tight junction integrity, epithelial cell sloughing and cytokine release. J. Comp. Pathol. *147*, 11–19.

Wooster, D.G., Maruvada, R., Blom, A.M., and Prasadarao, N.V. (2006). Logarithmic phase *Escherichia coli* K1 efficiently avoids serum killing by promoting C4bp-mediated C3b and C4b degradation. Immunology *117*, 482–493.

Wright, K.J., and Hultgren, S.J. (2006). Sticky fibers and uropathogenesis: bacterial adhesins in the urinary tract. Future Microbiol. *1*, 75–87.

Wright, K.J., Seed, P.C., and Hultgren, S.J. (2007). Development of intracellular bacterial communities of uropathogenic *Escherichia coli* depends on type 1 pili. Cell Microbiol. *9*, 2230–2241.

Wu, X.-R., Sun, T.-T., and Medina, J.J. (1996). In vitro binding of type 1-fimbriated *Escherichia coli* to uroplakins Ia and Ib: Relation to urinary tract infections. Proc. Natl. Acad. Sci. *93*, 9630–9635.

Wu, X.R., Kong, X.P., Pellicer, A., Kreibich, G., and Sun, T.T. (2009). Uroplakins in urothelial biology, function, and disease. Kidney Int. *75*, 1153–1165.

Yakovenko, O., Sharma, S., Forero, M., Tchesnokova, V., Aprikian, P., Kidd, B., Mach, A., Vogel, V., Sokurenko, E., and Thomas, W.E. (2008). FimH forms catch bonds that are enhanced by mechanical force due to allosteric regulation. J. Biol. Chem. *283*, 11596–11605.

Yoshida, S., Katayama, E., Kuwae, A., Mimuro, H., Suzuki, T., and Sasakawa, C. (2002). *Shigella* deliver an effector protein to trigger host microtubule destabilization, which promotes Rac1 activity and efficient bacterial internalization. EMBO J. *21*, 2923–2935.

Yu, S., and Lowe, A.W. (2009). The pancreatic zymogen granule membrane protein, GP2, binds *Escherichia coli* Type 1 fimbriae. BMC Gastroenterol. *9*, 58.

Zalewska, B., Piatek, R., Cieslinski, H., Nowicki, B., and Kur, J. (2001). Cloning, expression, and purification of the uropathogenic *Escherichia coli* invasin DraD. Protein Expr. Purif. *23*, 476–482.

Zhou, G., Mo, W.J., Sebbel, P., Min, G., Neubert, T.A., Glockshuber, R., Wu, X.R., Sun, T.T., and Kong, X.P. (2001). Uroplakin Ia is the urothelial receptor for uropathogenic *Escherichia coli*: evidence from *in vitro* FimH binding. J. Cell Sci. *114*, 4095–4103.

Zhu, L., Maruvada, R., Sapirstein, A., Malik, K.U., Peters-Golden, M., and Kim, K.S. (2010). Arachidonic Acid Metabolism Regulates *Escherichia coli* Penetration of the Blood-Brain Barrier. Infect. Immun. *78*, 4302–4310.

Zou, Y., He, L., and Huang, S.-H. (2006). Identification of a surface protein on human brain microvascular endothelial cells as vimentin interacting with *Escherichia coli* invasion protein IbeA. Biochem. Biophys. Res. Comm. *351*, 625–630.

Vaccines Against Enteric *Escherichia coli* Infections in Animals

14

Eric Cox, Vesna Melkebeek, Bert Devriendt, Bruno M. Goddeeris
and Daisy Vanrompay

Abstract

Several *Escherichia coli* animal pathogens exist, which have an important economical impact, such as enterotoxigenic *E. coli*, avian pathogenic *E. coli* and mammary pathogenic *E. coli*. Furthermore, ruminants are carrier of enterohaemorrhagic *E. coli*, which are not pathogenic for them, but can be transmitted to humans via direct contact and contamination of food or drinks, resulting in severe disease. Nevertheless, only few *E. coli* vaccines are available. This chapter summarizes the current status of vaccines for preventing intestinal *E. coli* infections in animals and will deal with vaccines against ETEC infections in pigs and ruminants and vaccines, which prevent EHEC colonization in cattle.

Introduction

Animal-pathogenic *Escherichia coli* mainly belong to the diarrhoeagenic *E. coli* (DEC), uropathogenic *E. coli* (UPEC), septicaemic *E. coli* (SePEC) among which the avian-pathogenic *E. coli* (APEC) and the mammary-pathogenic *E. coli* (MPEC). The diarrhoeagenic *E. coli* can be further divided into the enterotoxigenic *E. coli* (ETEC) producing enterotoxins causing hypersecretion of electrolytes and water by enterocytes in pigs and ruminants and the enteropathogenic *E. coli* (EPEC) producing attaching and effacing (AE) lesions in most mammals. EPEC are major pathogens in rabbits. *E. coli* producing Shigatoxin STx2e (STEC, verotoxigenic *E. coli* or VTEC), are the cause of oedema disease in pigs, whereas cattle that produce STx and AE lesions cause subclinical or non-clinical infections in ruminants. The latter strains are also called enterohaemorrhagic *E. coli* or EHEC and can cause severe disease in humans, mainly elderly and children (reviewed by Mainil, 2013).

Even though several *E. coli* animal pathogens exist, which have an important economical impact, only few *E. coli* vaccines are available. This chapter summarizes the current status of vaccines for preventing intestinal *E. coli* infections in animals and will deal with vaccines against ETEC infections in pigs and ruminants and vaccines, which prevent EHEC colonization in cattle.

Vaccines against enterotoxigenic *E. coli*

Intestinal infections with enterotoxigenic *Escherichia coli* (ETEC) are prevalent in humans, pigs, and sheep. In neonatal and recently weaned piglets, ETEC-associated diarrhoea results in morbidity and mortality (Gyles, 1994) and is considered as one of the economically most important diseases in swine husbandry (Chen *et al.*, 2004; Frydendahl, 2002; Van den Broeck *et al.*, 1999a). Also in calves and lambs ETEC infections in the first four days of life can be responsible for severe diarrhoea with high mortality (reviewed by Nagy and Fekete, 1993). ETEC express long, proteinaceous appendages or fimbriae on their surface, which bind to specific receptors on small intestinal enterocytes and allow the bacteria to colonize the small intestine. This is the initial step in the establishment of the enteric infection (Nagy *et al.*, 1985). Fimbriae associated with ETEC strains involved in diarrhoea in pigs are F4 (K88), F5 (K99), F6 (987P), F7 (F41) and F18, whereas in calves F5, F7 and F17 fimbriae are frequently identified (Wilson and Francis, 1986; Casey et al., 1992; Nagy and Fekete, 1999, Nguyen et al., 2011). In lambs F5 and F41 fimbriae producing ETEC have been identified (Nagy and Fekete, 1999).

In pigs, F4 fimbriae are most frequently found in ETEC causing diarrhoea and mortality in newborn, suckling and newly weaned piglets, whereas F5, F6 and F41 are associated with neonatal diarrhoea and F18 fimbriae are typically associated with post-weaning diarrhoea. F4 and F18 fimbriae occur in different antigenic variants. F4 fimbriae occur as F4ab, F4ac and F4ad (Orskov et al., 1964; Guinée and Jansen, 1979), but the F4ac variant is by far the most common (Westermann et al., 1988; Fairbrother et al., 2005). For F18, 2 antigenic variants were discovered, namely F18ab and F18ac (Rippinger et al., 1995). In general, post-weaning diarrhoea is caused by the F18ac variant whereas F18ab is more related to oedema disease, although there are exceptions (Hide et al., 1995, Rippinger et al., 1995). In addition to the porcine ETEC strains expressing the classical fimbrial antigens (F4, F5, F6, F7, and F18), a subgroup of porcine ETEC expressing an afimbrial adhesin involved in diffuse adherence (AIDA) was described (Ngeleka et al., 2003).

In ruminants F5, F41 and F17 are found in neonatal ETEC outbreaks. Four different antigenic variants of the major subunit of F17 have been identified, namely F17a (FY or Att25), F17b (F17-like, Vir adhesin), F17c (20K, G) and F17d (Att111 or F111). F17a and F17d are found on bovine ETEC strains, F17b on septicaemic E. coli in calves and lambs and F17c on E. coli in calves with diarrhoea or septicaemia, lambs with nephrosis or septicaemia, kids with septicaemia and humans with urinary tract infections (Girardeau et al., 1979; Pohl et al., 1982; Bertels et al., 1989; El Mazouari et al., 1994; Bertin et al., 1996a; Cid et al., 1999; Mainil et al., 2000). The adhesin of F17 fimbriae is the F17G protein. Until now only two variants of the $f17G$ gene have been detected by PCR, $f17GI$, associated mainly with F17a and F17d, and $f17GII$, associated mainly with F17b and F17c fimbriae even though differences in receptor affinity of the adhesin protein, have been observed for all F17 variants (Bertin et al., 1996a,b).

After colonization, ETEC produce enterotoxins inducing a severe watery diarrhoea by disrupting the water and electrolyte balance in the intestine (Gyles, 1994; Blanco et al., 1997). These enterotoxins include heat-labile toxin (LT), porcine and human heat-stable toxin a (StaP or STaH), heat-stable toxin b (STb), and enteroaggregative heat-stable toxin 1 (EAST1). Porcine ETEC strains are positive for one or more of these toxins, nevertheless a role of EAST1 in ETEC-induced diarrhoea remains questionable (Zhang et al., 2006; Ruan et al., 2012). In ruminants ETEC isolates mainly produce StaP, rarely LT or STb are identified (Mainil et al., 1990; Fairbrother et al., 2005; Nguyen et al., 2011). EAST1 is not prevalent on bovine ETEC, but highly associated with the adhesin CS31, which is prevalent on isolates from calves with E. coli septicaemia (Bertin et al., 1998; Veilleux and Dubreuil, 2006).

In general, most neonatal infections can be prevented by passive colostral and lactogenic immunity obtained by vaccination of dams (Rutter and Jones, 1973; Contrepois et al., 1986, Deprez et al., 1986). However, as ETEC infections are non-invasive gastrointestinal infections, mucosal, i.e. lactogenic immunity rather than systemic i.e. colostral immunity will be important to fight the disease. Several maternal vaccines are on the market. These either contain inactivated bacteria with fimbriae or purified fimbriae with or without LT toxoid and are applied mainly parenterally in the pregnant dam (Snodgrass et al., 1982; Nagy and Fekete, 2005). However, this passive protection decreases with ageing and lactogenic immunity suddenly stops by weaning. This does not pose a major problem in ruminants because a high proportion of fatal ETEC infections occur during the first few days after birth. Later on calves are not susceptible to colonization by ETEC. Commercial vaccines for cows contain killed E. coli F5 isolates and/or the F5 adhesin. None of these vaccines seems to contain F17 or CS31 and therefore are not expected to protect calves (or lambs) against infections with F17+ or CS31+ E. coli. The impact of this in the field is not clear. Commercial vaccines for sows contain F4, F5, F6 and/or F41 fimbriae, either purified or as inactivated E. coli expressing these fimbriae with or without the LT toxoid. The ageing pig remains susceptible to infection with F4+ ETEC and also becomes susceptible to F18+ *E. coli* infections at two to three weeks of age (Coddens *et al.*, 2007). Consequently, the newly weaned piglet, which is devoid of lactogenic protection, becomes highly susceptible to enteropathogens, e.g. F4+ ETEC and F18+ ETEC and VTEC (Hampson, 1994). To protect these newly weaned piglets, an active intestinal mucosal immune response is required, in which the production of antigen-specific secretory IgA plays an important role (McGhee *et al.*, 1992). Recently much effort has been put in developing a vaccine against post-weaning *E. coli* infections in pigs. Since *E. coli* infections seem to occur immediately post weaning, the major challenge remains to induce a protective intestinal immunity at weaning.

Oral subunit vaccines

The concept of fimbriae as colonization factors playing a crucial role in the occurrence of infection led to the development of oral fimbriae-containing vaccines. In 1999, Van den Broeck *et al* demonstrated for the first time that purified F4ac fimbriae are powerful oral immunogens. Oral delivery of 1 mg during 2 consecutive days to weaned pigs resulted in F4-specific IgM- and IgA-secreting cells (ASC) in the Peyers' patches, mesenteric lymph nodes, blood and lamina propria as soon as 4, 7, 9 and 11 days, respectively. Moreover, the induced IgA responses were comparable to those obtained upon an oral ETEC challenge infection (Van den Broeck *et al.*, 1999b). Furthermore, when weaned pigs were orally vaccinated during 3 consecutive days and boosted on day 16 post primary vaccination, complete protection was obtained against oral challenge with virulent F4+ ETEC (Van den Broeck *et al.*, 1999c). Presence of the F4 receptor (F4R) on small intestinal enterocytes was pivotal to obtain the protective mucosal response, as it is to obtain disease upon infection. Indeed, no systemic or mucosal responses could be observed after oral immunization and a subsequent challenge infection in pigs that did not express the F4R (Van den Broeck *et al.*, 1999c). As already mentioned, pigs need to have active immunity at weaning, and therefore immunization has to occur during the suckling period often in the presence of colostral and lactogenic

immunity. Since glycoproteins and milk antibodies can interfere with the oral immunization (Shahriar *et al.*, 2006), several delivery systems for oral immunization have been evaluated such as transgenic edible alfalfa plants (Joensuu *et al.*, 2006), enteric coated pellets (Snoeck *et al.*, 2003) and Gantrez®AN nanoparticles (Vandamme *et al.*, 2011). Induction of an F4-specific mucosal immune response by oral immunization of piglets with plant FaeG was restricted to a rather weak response and a limited reduction of F4+ *E. coli* excretion following F4+ ETEC challenge. Co-administration of the mucosal adjuvant CT significantly reduced the F4+ *E. coli* excretion to a level similar to that observed after oral F4 immunization. A major advantage is that large amounts of plant FaeG, which remain stable over prolonged storage, can be produced.

In contrast to the results for F4, oral delivery of purified F18 could not induce a protective response against an F18+ *E. coli* challenge, even if doses of up to 30 mg F18 fimbriae and the mucosal adjuvant LT (R192G) were used (Verdonck *et al.*, 2007). This difference in immunogenicity between F18 and F4 fimbriae might be explained by the different structure of both fimbriae. For F18 fimbriae, the adhesin is the minor subunit FedF (Imberechts *et al.*, 1996; Smeds *et al.*, 2001), whereas for F4 frimbriae it is the major fimbrial subunit FaeG, present in multiple copies on a single fimbria (Bakker *et al.*, 1992). In addition, Tiels *et al.* (2007) suggested that the interaction between the FedF adhesin and the major subunit FedA is not as stable as subunit–subunit interactions in other fimbriae like F4, which may lead to a lower stability in the gastro-intestinal tract and or a lower capacity of the F18 fimbriae to bind to their specific receptor on intestinal enterocytes, the blood group ABH type 1 determinants (Coddens *et al.*, 2009). In a more recent study, a reduced faecal excretion of F18+ *E. coli* could be observed when weaned pigs were orally immunized with recombinant FedF that was covalently coupled to F4 fimbriae (Tiels *et al.*, 2008). Although further research is needed to improve the stability and efficacy of this vaccine, this approach opens new perspectives towards the development of a combined oral vaccine against both F4 and F18 infections.

Live oral vaccines

Another promising category of ETEC vaccines are the live attenuated and the live wild-type avirulent *E. coli* vaccines. The theory behind these vaccines is that after oral delivery, the vaccine colonizes the small intestinal mucosa thereby triggering the immune response of the animal. Fimbriae-specific sIgA antibodies will be induced which will prevent adherence to the enterocytes of pathogenic bacteria expressing similar fimbriae. In 2002, Bozic and co-workers demonstrated that a single oral vaccination of pigs with a live attenuated *E. coli* strain expressing F4ac on day 4 after weaning failed to give protection against challenge infection with a virulent strain 7 days later. However, priming with levamisole, an immunomodulator with anti-anergic properties, via intramuscular injection during three successive days immediately before the oral vaccination, significantly reduced the clinical signs of post-weaning diarrhoea (PWD) (Bozic *et al.*, 2003). In 2003, ARKO released the only US federally licensed modified live avirulent *E. coli* vaccines for swine, Entero Vac and Oedema Vac with claimed efficacy against virulent F4+ *E. coli* and F18+ *E. coli*, respectively. However, no studies on the efficacy of both vaccines could be found. Since 2008, a live vaccine against F4+ ETEC induced PWD, called Coliprotec, has been commercialized in Canada by Prevtec microbia. It was also introduced in Brazil in 2011 and should be available in the USA and Mexico in 2012 and in Europe in 2014 (www.prevtecmicrobia.com/en/section-41-r-d). Coliprotec is an oral live vaccine, containing naturally avirulent *E. coli* bacteria that express F4 fimbriae but do not produce toxins. Clinical studies have demonstrated that Coliprotec significantly reduced colonization of pig intestines after challenge with a virulent F4+ ETEC strain and consequently, that the duration and severity of diarrhoea, as well as the accumulation of fluids in the intestines after infection, were significantly reduced. A field study carried out in pig farms across three Canadian provinces demonstrated the safety of the vaccine. As Entero Vac and Oedema Vac, Coliprotec can be delivered via drinking water and is recommended for the vaccination of healthy weaned pigs at the age of 17/18 days or more. However, since PWD caused by F4+ ETEC occurs shortly after weaning, mostly 3–10 days, a gap might exist the first days after weaning during which the pigs are not protected. To avoid this, vaccination or at least a first priming of the immune system should ideally occur during the suckling period. It seems likely that during this period, the presence of lactogenic antibodies in the gastro-intestinal tract might hinder effective colonization by the vaccine strain and possibly, successful vaccination. Promising results were obtained with a live F18ac-positive *E. coli* vaccine in suckling pigs born to sows with F18-specific colostrum IgA antibodies (Bertschinger *et al.*, 2000). In this experiment, the pigs were vaccinated during 3 consecutive days, starting from 10 days before weaning, and received a challenge infection on 3 consecutive days beginning 9 or 11 days post weaning. Although colonization by the challenge strain was not prevented in the vaccinated group, a significant rise in F18ac-specific serum IgA and a $> 3 \log_{10}$ decrease in faecal shedding of the F18ac-positive challenge strain was observed compared to the unvaccinated group. The role of the F18 fimbriae as protective antigens was further highlighted by the need to immunize with a culture in the fimbriated phase. Curiously, in that study cross-protection against F18ab could not be observed, even though the tip adhesin of F18ab and F18ac, FedF, is highly conserved (Tiels *et al.*, 2005). This suggest that no or a weak antibody response occurred against the adhesin. The authors suggest that an enhanced binding of the vaccine to M cells via the specific maternal IgA antibodies present in the gut lumen, a mechanism that has been demonstrated in mice, rats and rabbits (Weltzin *et al.*, 1989), might contribute to the efficacy of their vaccine in suckling pigs. However, they did not determine the titre or the presence of lactogenic F18-specific IgA at the moment of vaccination, making the role of maternal antibodies in the outcome of the vaccination controversial. In addition, IgA-mediated M cell uptake has not been demonstrated in pigs until now. Furthermore, the use of live fimbriae-producing toxin-free bacteria involves the risk of reversion to virulence. For instance the plasmid-encoded pathogenicity island (PAI2173) has been discovered to carry STa and STb enterotoxin genes on a conjugative plasmid (pTC2173). PAI2173 has proven to be transposable and seems to be characteristic of

F18+ ETEC strains isolated from cases of porcine post-weaning diarrhoea (Fekete *et al.*, 2003). Other examples are the conjugative plasmids EntP307 and pCG86 isolated from porcine ETEC strains and carrying the LT and STb enterotoxin genes. Some of these conjugative plasmids carry antibiotic resistance genes (Mazaitis *et al.*, 1982).

Encapsulated subunit and live vaccines

In order to enhance the efficacy of live oral as well as subunit vaccines, encapsulation strategies have been suggested to protect the vaccine during transport through the gastro-intestinal tract and/or against neutralization by maternal antibodies and other milk factors, like receptor analogues present at fat globule membranes (Atroshi *et al.*, 1983) or in milk serum (Snoeck, 2004), when applied during the suckling period. Felder *et al.* (2001) were the first to test the feasibility of an oral immunization of pigs with microencapsulated live *E. coli*, formalin inactivated *E. coli*, crude or purified F18 fimbriae. They encapsulated the bacteria and fimbriae in poly(lactide-co-glycolide) (PLGA) microspheres by spray-drying and delivered them orally to newborn suckling as well as to weaned pigs. Oral delivery of encapsulated F18 fimbriae to weaned pigs did not result in a significant serum antibody response or in a reduction in colonization of *E. coli* after challenge infection. These results are in line with the study of Verdonck *et al.* (2007) demonstrating that purified F18 fimbriae are not capable of inducing protective immunity upon oral delivery. F18-positive *E. coli* encapsulated in PLGA particles, either live or inactivated, also failed to induce a significant response after delivery to weaned or suckling pigs. These results are in contrast with the successful immunization by Bertschinger *et al.* (2000) with a live F18-positive *E. coli* strain orally delivered to suckling pigs. Although the encapsulation procedure did not reduce the antigenicity of isolated F18 fimbriae as demonstrated by parenteral immunization, a reduced viability of the encapsulated live bacteria cannot be excluded (Felder *et al.*, 2001). Furthermore, microparticles have been suggested to be taken up via M cells before releasing their antigen (des Rieux *et al.*, 2006). This might prevent live bacteria from colonizing the small intestine, which usually is the initial step in the induction of an immune response against live *E. coli* bacteria. Therefore, the use of an encapsulation system that disintegrates within the small intestine might be more appropriate. Snoeck *et al.* (2003) compared the use of F4 fimbriae in enteric-coated pellets to F4 fimbriae in solution in suckling pigs. An advantage of these pellets is that they also can be mixed with creep feed. The pellets released the F4 fimbriae in the beginning of the jejunum, near the jejunal Peyers' patches, which are the major inductive site of the F4-specific immune response, allowing the fimbriae to bind to specific receptors present on enterocytes and/or M cells (Snoeck *et al.*, 2006). Encapsulation of F4 fimbriae in pellets resulted in a one to six log reduction in excretion of F4+ ETEC upon challenge, whereas only a one to two log reduction was observed when suckling pigs were orally vaccinated with F4 fimbriae in solution, demonstrating that protection of the antigen against neutralization by milk factors and against degradation and inactivation by enzymes in the cranial gastrointestinal tract has beneficial effects. Nevertheless, intestinal colonization with F4+ ETEC after challenge infection could not be prevented, indicating the need for further optimization. An oral tablet formulation consisting of carboxymethyl high amylose starch was proposed to deliver live F4+ *E. coli* (Calinescu *et al.*, 2005) or purified F4 fimbriae (Calinescu *et al.*, 2007). Formulation resulted in a better survival rate of the bacteria and a higher stability of the F4 fimbriae in simulated gastric fluid. Release of the antigen is based on the fast swelling of the tablets during passage from gastric acidity to the alkaline enzymatic intestinal environment, triggering their rapid, almost total dissolution. Recently, we tested the adjuvanticity of methylvinylether-co-maleic anhydride (Gantrez®) on oral delivery of F4 fimbriae to weaned pigs. Encapsulation of F4 in Gantrez® nanoparticles raised the serum antibody response against F4, but did not improve protection as compared to soluble F4 fimbriae. Moreover, the best effect was observed when empty nanoparticles were added to soluble F4 fimbriae, suggesting that adjuvant properties rather than protection of the antigen against gastrointestinal degradation were responsible for the enhanced humoral immune response, making this system less suitable for vaccination during the suckling period (Vandamme *et al.*, 2011).

Parenteral vaccines

An alternative strategy to vaccinate suckling pigs in the presence of lactogenic maternal antibodies might be parenteral immunization. However, parenteral vaccines in general stimulate the systemic rather than the mucosal immune system. Intramuscular injection of formalin-killed F4+ *E. coli* during the suckling period even resulted in a state of suppression that was reflected by a strongly reduced intestinal response compared to the control group upon a subsequent challenge with live F4+ *E. coli* (Bianchi *et al.*, 1996). Nevertheless, intramuscular immunization of pigs with 1 mg F4 fimbriae induced a systemic IgA response only slightly lower than that following an ETEC infection (Van den Broeck *et al.*, 1999b). This response could be improved by lowering the vaccine dose to 0.1 mg, suggesting that besides the immunization route, the antigen dose also plays a role in the induction of serum IgA (Van der Stede *et al.*, 2002). Furthermore, in a small scale experiment intramuscular immunization of gnotobiotic pigs with a tripartite FaeG-FedF-LT$_{192}$A2:B fusion protein resulted in the induction of IgA antibodies in serum, faeces and intestinal washes that neutralized CT, inhibited adherence of F4 and F18 and protected pigs against clinical signs after an F4+ ETEC challenge (Ruan *et al.*, 2011). The use of a fusion protein might be an interesting strategy towards the development of a multivalent vaccine against ETEC-induced post-weaning diarrhoea. However, care should be taken when interpreting the results of these parenteral immunisation studies. Systemic induction of a serum IgA response might result in overflow of sIgA to the intestinal tract as long as the systemic response is on going, but this is not necessarily associated with the presence of sIgA antibody secreting cells (ASC) at the GALT or with the induction of a mucosal memory response.

In order to modulate a systemic response towards an intestinal mucosal IgA response, the use of 1α,25 dihydroxyvitamin D$_3$ (vitamin D$_3$) has been proposed (Daynes *et al.*, 1996; Enioutina *et al.*, 1999, 2000; Van der Stede *et al.*, 2004). In mice, co-administration of vitamin D$_3$

via the subcutaneous route resulted in increased numbers of IgA and IgG ASC in the lamina propria (Daynes *et al.*, 1996). This was at least partly due to an altered migration of antigen-pulsed DC from the local draining lymph nodes towards the Peyers' patches, where the immune response was initiated (Enioutina *et al.*, 1999, 2000). In pigs, a slightly increased number of IgA ASC in the Peyers' patches was observed following intramuscular immunization with antigen and vitamin D$_3$ accompanied with a significantly reduced faecal excretion after challenge infection, suggesting that F4-primed memory B cells were present at the GALT at the time the challenge was performed (Van der Stede *et al.*, 2003).

Vaccines to prevent *E. coli* O157:H7 colonization and shedding by ruminants

Enterohaemorrhagic *E. coli* (EHEC), a subgroup of Shiga toxin-producing *E. coli* (STEC), are worldwide recognized as important zoonotic pathogens. EHEC cause diarrhoea, haemorrhagic colitis and the haemolytic uraemic syndrome (HUS) in humans. The bacteria colonize the host gastrointestinal tract via attaching and effacing (A/E) lesions, which are characterized by bacteria firmly attaching to the epithelial cells, leading to localized damage to brush border microvilli and rearrangement of host cytoskeleton proteins beneath the intimately attached bacterial colonies (Knutton *et al.*, 1989). The key virulence factors in EHEC are encoded at the locus of enterocyte effacement (LEE) (McDaniel *et al.*, 1995). Although many STEC serotypes have been associated with human illness, *E. coli* serotype O157:H7 is by far the most frequently implicated in human disease (Nataro and Kaper, 1998).

E. coli O157:H7 is widespread on cattle farms throughout the world. More recently, sheep and goats have emerged as important sources of human infection, particularly with the widespread popularity of petting farms and the increased use of sheep and goat food products, including unpasteurized milk (La Ragione *et al.*, 2009). Nevertheless, cattle still remain the main source of zoonotic *E. coli* O157:H7 infections.

Longitudinal studies have shown that shedding

of *E. coli* O157:H7 is transient (Smith *et al.*, 2009a). Three distinct patterns were reported: (i) non-persistent shedding lasting not more than 7 days, (ii) moderately persistent shedding lasting approximately one month, or (iii) persistent shedding lasting several months to even a year (Rice *et al.*, 2003; Robinson *et al.*, 2004; Baines *et al.*, 2008). Natural fluctuations of shedding within an animal over time are caused by intrinsic factors such as gut flora, gut health, behaviour or other anatomical, physiological or physical stresses (Robinson *et al.*, 2009) or by external factors such as the diet (Jacob *et al.*, 2009). Shedding appears to be related to weaning and age.

Individual animals have been mentioned to be shedding *E. coli* O157:H7 at high levels ($> 10^4$ CFU/g of faeces) for a longer period, the so-called 'super shedders'. This would have a disproportionate effect on cattle hide and subsequent carcass contamination. However, at the moment there is debate going on if super shedders really exist and if most animals are not excreting high numbers of bacteria for some period during the course of infection. In a longitudinal study on 3 EHEC positive farms, 30 1-year-old clinically healthy bulls were sampled each six weeks during 36 to 42 weeks for EHEC excretion. Seventy-seven per cent of the animals were excreting EHEC on one or more occasions and only one animal was shedding $\geq 10^3$ cfu/g faeces at one sampling time point (Joris *et al.*, 2012). Presently it is unknown why certain animals would become super-shedders. However, interventions that would prevent high level shedding in the population most certainly will reduce *E. coli* O157:H7 transmission in production and lairage environments resulting in reduced risk of beef carcass contamination and thus a safer finished product (Arthur *et al.*, 2010; Matthews *et al.*, 2006).

Vaccination of cattle

Several researchers explored vaccination of cattle to prevent and/or clean up intestinal colonization and subsequent *E. coli* O157:H7 shedding. Vaccination to prevent *E. coli* O157:H7 intestinal colonization and faecal excretion should be based on priming of the animal's mucosal immune system against antigens of the bacterium involved in colonization of its intestinal tract.

The importance of LEE-encoded type III secreted proteins in the colonization of the ruminant's intestine has been demonstrated by the inability of *E. coli* O157:H7 mutants with *eae*, *espA*, *espB*, *tir*, or combinations of these genes, disrupted or deleted to colonize the intestine of ruminants (Dean-Nystrom *et al.*, 1998; Bretschneider *et al.*, 2007; Naylor *et al.*, 2007). This was supported by *in vitro* experiments, which showed that *E. coli* O157:H7 attachment to HEp-2 cells could be inhibited by sera from cattle immunized with *E. coli* O157:H7 type III secreted proteins, making these proteins important ruminant vaccination targets (Asper *et al.*, 2007).

Several researchers evaluated type III secreted protein-based vaccines. Some strategies were promising, while others were unsuccessful. For instance, intramuscular vaccination of 14-day-old calves with recombinant EspA (100 μg protein/dose) formulated in Freund's incomplete adjuvant followed by intranasal boosting (300 μg protein/dose) with the same protein mixed with cholera toxin B as a mucosal adjuvant, induced high antigen-specific serum IgG1 and salivary IgA responses. However, this strategy was unable to protect animals against intestinal colonization following oral challenge with 10^{10} of *E. coli* O157:H7 (Dziva *et al.*, 2007). Also unsuccessful was intramuscular (100 μg protein/dose) priming and intranasal (500 μg protein/dose) boosting of 13 day-old calves with recombinant intimin in combination with the putative recombinant adherence factor Efa-1, formulated in an aluminium hydroxide oil-based adjuvant, followed by oral challenge with 10^{10} CFU of *E. coli* O157:H7 (van Diemen *et al.*, 2007). More successful was an oral vaccination of 1- to 2-week-old calves with live attenuated recombinant *Salmonella enterica* Dublin (10^9–10^{10} CFU) expressing *E. coli* O157:H7 intimin (Khare *et al.*, 2010). The live salmonella vaccine strain effectively colonized the intestines of calves. Vaccination resulted in a transient clearance and subsequently reduced colonization and shedding of *E. coli* O157:H7 following challenge. Transient clearance of *E. coli* O157:H7 was not associated with an enhanced IgA-mediated mucosal immune response. Nevertheless, a protective vaccine based on intimin would be challenging, as more than 17 serologically distinct variants have been identified even though a high degree of cross-reactivity is seen in the field (Joris *et al.*, 2012).

So far, the most promising T3SS protein-based vaccinations were those using multiple proteins, either a combination of EspA, EspB and Tir or, one or more of those former proteins combined with intimin (Potter *et al.*, 2004; Peterson *et al.*, 2007a,b; Smith *et al.*, 2009b; McNeilly *et al.*, 2010; Allen *et al.*, 2011; Vilte *et al.*, 2011). Potter *et al.* (2004) vaccinated calves and yearling cattle subcutaneously with a supernatant-based type III secreted protein cocktail containing both Esps and Tir. VSA3 was used as adjuvant, which is the same as Emulsigen D (MVP laboratories, USA), an oil-in-water emulsion containing dimethyl-dioctadecylammoniumbromide (DDA), claimed to be a T cell immunostimulant. The animals were challenged (10^8 CFU of *E. coli* O157:H7) by orogastric intubation, 2 weeks following the last immunization. Three subsequent immunizations with 50 µg proteins significantly reduced the number of bacteria shed in the faeces, the number of animals excreting bacteria as well as the duration of shedding. Anti-Tir and anti-EspA serum antibodies were already detected following primo vaccination. Vaccination of cattle in a clinical trial using the same procedure also significantly reduced the prevalence of *E. coli* O157:H7 (Potter *et al.*, 2004). The efficacy depended on the administration frequency, since decreasing the number of immunizations from 3 to 2 could not significantly reduce the number of faecal *E. coli* O157:H7 shedders in 218 pens of feedlot cattle (Van Donkersgoed *et al.*, 2005). Van Donkersgoed *et al.* (2005) used the adjuvant Emulsigen, which does not contain DDA (oil-in-water emulsion; MVP Technologies, Ohama, NE, USA).

Variations in vaccination outcome using T3SS proteins can be attributed to differences in: (i) adjuvants, (ii) age of experimental animals, (iii) immune status (passive immunity/active immunity), (iv) immunization regime, (number of doses) (v) infection pressure and/or (vi) EHEC serotypes occurring on the farm (Van Donkersgoed *et al.*, 2005; Asper *et al.*, 2007; McNeilly *et al.*, 2010). Indeed, there is antigenic variation for EspA and intimin between *E. coli* serotypes, but these antigens show little variation within *E. coli* O157:H7 strains based on analysis

of currently available sequences (NCBI-blastp) (Adu-Bobie *et al.*, 1998; Crepin *et al.*, 2005). Possible occurrence of multiple EHEC serotypes stresses the advantage of including multiple T3SS antigens as it makes vaccine breakdown less likely than with a monovalent composition. In October 2008, the vaccine Econiche™ (Bioniche Life Sciences Inc., Belleville, Ontario, Canada), based on T3SS proteins, received full licensing approval from the Canadian Food Inspection Agency. Econiche™ is the first commercially available *E. coli* O157:H7 vaccine for use in cattle. Econiche™ is a concentrated *E. coli* 0157:H7 supernatant, containing EspA and Tir, but possibly also additionally secreted proteins such as intimin. However, we could find no information on the exact composition of the vaccine or on standardization of production procedures. Econiche™ contains Emulsigen D (MVP Technologies). The vaccine license is conditional for the US market, as additional potency and efficacy studies are needed. However, the vaccine was not capable of completely clearing the infection in cattle and even after three vaccine doses contamination of the hide during transport to the slaughterhouse still occurred (Moxley *et al.*, 2009; Smith *et al.*, 2009a). Nevertheless, in 2012 the UK Veterinary Medicines Directorate (VMD) has approved the use of the company's cattle vaccine against *E coli* under the conditions of a Special Treatment Certificate (STC) scenario. STCs are issued to veterinary surgeons when an appropriate remedy for an animal disease is not available in the UK, but can be accessed from another country. The company is still pursuing further formal regulatory approvals in Europe for the vaccine.

Meanwhile, alternative vaccine candidate antigens are being tested. Many of the LEE encoded proteins, such as Tir, EspA, EspB, and EspD, are indeed critical for the virulence of EHEC (Misyurina *et al.*, 2010). However, the discovery of non-LEE effectors, such as NleA, TccP, and NleB, whose genes are located in small pathogenicity islands and prophages, and whose functions are also associated with colonization and virulence of EHEC, has caught the eye of vaccine designers (Garmendia *et al.*, 2004; Gruenheid *et al.*, 2004; Roe *et al.*, 2007; Misyurina *et al.*, 2010). Recently, Asper *et al.* (2011) evaluated the immunogenicity of 29 non-LEE proteins using sera of naturally infected cattle. Surprisingly, NleA was the only non-LEE protein that appeared to be highly immunogenic. *In vitro* studies with EPEC strains in human Caco-2 cells revealed that NleA disrupts host intestinal tight junctions that are important for maintaining the intestinal barrier function (Thanabalasuriar *et al.*, 2010). Moreover, NleA localized to the host cell secretory pathway and inhibited vesicle trafficking by interacting with the Sec24 subunit of mammalian coatamer protein complex-II (COPII). Recently, Thanabalasuriar *et al.* (2012) provided evidence that interaction and inhibition of COPII by NleA is an important aspect of EHEC-mediated disease in humans. The role of NleA in the pathogenesis of ruminant EHEC infections is still unknown and the potential of NleA as a novel EHEC vaccine candidate antigen remains to be examined.

H7 flagella may present an additional target for *E. coli* O157:H7 vaccine design, as they are important during initial binding of *E. coli* O157:H7 to bovine primary rectal cells (Mahajan *et al.*, 2009) and to bovine intestinal tissue explants (Erdem *et al.*, 2007), whereafter the T3SS takes over. McNeilly *et al.* (2008) were the first to explore the protective effect of systemic and mucosal immunization with purified H7 flagellin, the main structural component of H7 flagella, on subsequent *E. coli* O157:H7 colonization in cattle. H7 flagellin plus Quil A was administered intramuscularly, while H7 flagellin alone or incorporated into poly(D,L-lactide-co-glycolide) (PLG) microparticles was applied directly onto the rectal mucosa. Calves (9 weeks of age) were immunized on three separate occasions at 2-week intervals. Ten days after the last immunization, calves were orally challenged with 10^{10} *E. coli* O157:H7 Stx deletion mutant strain. Rectal immunization did not induce systemic antibodies and only rectal immunization with free flagellin induced rectal IgA antibodies. However both rectal immunizations had no effect on bacterial colonization or shedding. Reduced colonization rates and delayed peak bacterial shedding were only observed in the intramuscularly immunized calves, but no reduction in total bacterial shedding occurred. Systemic immunization induced high levels of flagellin-specific IgG1, IgG2 and IgA in both serum and nasal

secretions and detectable levels of both antibody isotypes in rectal secretions. They concluded that H7 flagellin might be a useful component in a systemic *E. coli* O157:H7 vaccine (McNeilly *et al.*, 2008). However, two years later they discovered that systemically induced H7-specific IgG1 and IgG2 antibodies, but not IgA, may impair innate immune responses to *E. coli* via neutralization of TLR5 activation and thus might reduce vaccine efficacy (McNeilly *et al.*, 2010).

In 2013, Bioniche Life Sciences has been granted $500,000 by the National Research Council of Canada Industrial Research Assistance Program (IRAP) to develop a second-generation vaccine for *E. coli* in cattle. This new vaccine is aimed to be safer and to be more readily produced and with a higher yields than the recent Econiche vaccine. The new vaccine may also have the potential to cross-protect against other *E. coli* serotypes. *E. coli* O157 is the most common serotype causing human infection in North America. Other serotypes include O26, O111, O103, O121, O45 and O145.

In 2009, other researchers developed a different vaccination strategy, immunizing cattle against the bacterial outer membrane siderophore receptor and porin (SRP) proteins (Fox *et al.*, 2009; Thronton *et al.*, 2009). Antibodies against SRP proteins blocked iron transport into the cell, which rendered *E. coli* O157:H7 into a competitive disadvantage in a mixed microbial environment. Two subcutaneous injections with the *E. coli* O157:H7 SRP vaccine significantly reduced the number of calves that were faecal culture positive in orally challenged calves. Three immunizations were more effective than two. A feedlot study demonstrated that the SRP vaccine reduced the prevalence of *E. coli* O157:H7, the number of days cattle tested positive and the number of days cattle were identified as high-shedders. In 2009, Epitopix (Epitopix LLC, Willmar, MN, USA) received a conditional license for the U.S. market for an SRP®-based *E. coli* O157:H7 vaccine. Three immunizations are needed to create acceptable protection (Thomson *et al.*, 2009). A study was conducted to examine the performance, health, and shedding characteristics of beef calves that were vaccinated with an *E. coli* O157:H7 SRP bacterial extract. Some of the calves had been born to cows vaccinated prepartum with the same vaccine. The conclusion was that the timing of vaccination of calves needs to be examined further to maximize the field efficacy of this vaccine (Wileman *et al.*, 2011).

Recently, a new technical approach for *E. coli* O157:H7 vaccine design was evaluated, namely bacterial ghosts of *E. coli* O157:H7. The bacterial ghosts were subcutaneously injected. Two doses of vaccine composed of 10 mg of *E. coli* O157:H7 bacterial ghosts were injected with an interval of 21 days. Fourteen days after the second immunization, calves were orally challenged with 10^9 CFU of *E. coli* O157:H7 (stx2, eae-γ). Vaccinated animals elicited significant levels of bacterial ghost-specific IgG, but not IgA antibodies in serum and low levels of IgA and IgG antibodies in saliva. Following oral challenge with *E. coli* O157:H7, a significant reduction in both the duration and total bacterial shedding was observed in vaccinated calves compared to the non-immunized group (Vilte *et al.*, 2012). BGs are produced by the controlled expression of the PhiX174 lysis gene E. E-mediated lysis of bacteria results in the formation of empty bacterial cell envelopes, which have the same cell surface composition as their living counterparts. They display all surface components, including colonization factors in a non-denatured form and are able to induce a strong mucosal immune response (Jalava *et al.*, 2003). Further research is needed, as the specificity of the bacterial ghost-induced antibodies as well as the protective antigen(s) in the bacterial ghost composition are presently unknown. Moreover, the lack of antibodies against *E. coli* O157:H7 LPS was a rather unexpected finding, which needs to be clarified.

Recently, Snedeker *et al.* (2011) performed a systematic search of eight databases and land-grant university research reports using an algorithm adapted from a previously conducted systematic review of pre-harvest interventions against *E. coli* O157 to locate all reports of in vivo trials of *E. coli* O157 vaccines in ruminants published between 1990 and 2010. They concluded that type III and SRP protein vaccines significantly reduce the prevalence of *E. coli* O157 in beef cattle faeces. Moreover, according to their statistical meta-analysis further studies are needed, particularly with regards to the SRP protein vaccine, to examine the efficacy of vaccines at feedlot pen densities.

Although the above-mentioned vaccination methods are promising for reducing *E. coli* O157 colonization and shedding, it is at the moment unclear to what extent pre-harvest burden needs to be reduced to effectively prevent contamination of consumable products. The infectious dose to colonize a ruminant has been estimated to be lower than 300 CFU (Besser *et al.*, 2001), indicating that such levels should be drastically reduced to curtail transmission between animals. Furthermore, the dose to infect humans is estimated to be between 1 and 100 CFU (Reida *et al.*, 1994; O'Donnell *et al.*, 2002). Therefore, interventions to reduce infection should be applied on different levels, starting in the animal itself, in the slaughterhouse, in the processing plants, and last but not least, the consumer should take minimal precautions. Only with a multilevel strategy can maximal reductions in bacterial load be achieved, ensuring a safe food chain from farm to fork.

Small ruminants

Atef Yekta *et al.* (2011) immunized sheep intramuscularly with a combination of recombinant EspA, recombinant EspB and the recombinant C-terminal part (380 amino acids) of intimin-γ. The animals received 100 µg of each protein suspended in incomplete Freund's adjuvant. Sheep were injected three times with 2 weeks interval and challenged orally 10 days after the last immunization with 10^{10} CFU of *E coli* O157:H7 (Stx negative strain). Systemic vaccination using T3SS proteins could significantly reduce *E coli* O157:H7 excretion in a sheep infection model. Protection correlated with serum antibody titres against EspA, EspB and intimin-γ.

Recently, Zhang *et al.* (2012) immunized goats intranasal with the recombinant Stx2B-Tir-Stx1B-Zot fusion protein. The zonula occludens toxin (Zot) is a 44.8 kDa single polypeptide chain encoded by the filamentous bacteriophage CTXφ of *Vibrio cholerae*. The zonula occludens toxin binds a receptor on intestinal epithelial cells and increases the permeability of the small intestine by affecting the structure of epithelial tight junctions. It allows the passage of macromolecules through the paracellular route. The zonula occludens toxin is a promising tool for mucosal antigen delivery. Three months old goats were intranasally immunized with 200 µg antigen on day 1 and boost-immunized 3 weeks later. Goats were orally challenged 14 days later with 10^{10} CFU of *E. coli* O157:H7. All immunized goats elicited significant Stx2B-Tir-Stx1B-Zot-specific serum IgG and 50% of the immunized goats contained IgA in the faeces. Vaccination significantly reduced EHEC shedding.

Conclusions and comments

Neonatal ETEC infections can be controlled via vaccination of the dams. Nevertheless, there are ETEC, which produce colonization factors not included in the presently available vaccines such as F17 fimbriae absent in cattle vaccines.

ETEC strains involved in post-weaning diarrhoea mostly express F4 or F18 fimbriae. To prevent colonization, an F4- and/or F18-specific sIgA response is needed in the small intestine at the moment of infection, which occurs mostly at weaning. The oral route is the most logical route to obtain a protective immune response and several oral vaccination strategies have been successfully performed in weaned pigs, including subunit vaccines as well as live oral vaccines. However, post-weaning diarrhoea often occurs within 3–10 days after weaning. Therefore, it would be more ideal to vaccinate during the suckling period. Several strategies have been evaluated with limited success until now. Research that clarifies what exactly happens when suckling pigs having high levels of maternal antibodies are orally vaccinated is still missing. This knowledge could help to improve the current methods.

E. coli O157:H7 is a continuous risk for humans and controlling the food chain is not always sufficient to prevent infection. Vaccination could be interesting to completely eradicate *E. coli* O157:H7 from the food chain. Current vaccination strategies only succeed partially in reducing *E. coli* O157:H7 excretion. Understanding of *E. coli* O157:H7 host cell interactions, and particularly of differences between the pathogenesis in humans and in the ruminant reservoir, can improve current control strategies.

Besides vaccines against ETEC and EHEC, vaccination also occurs against avian *E. coli* infections. Poulvac® *E. coli* contains a live aroA gene deleted *E. coli* (type O78 *E. coli* strain EC34195) containing 5.2×10^6 to 9.1×10^8 CFU that in one dose given as a spray at 1 day-of-age protects from 2 until 8 weeks-of-age for reduction of lesions and until 12 weeks of age for reduction of mortality. Cross-protection was shown for reduction of incidence and severity of airsacculitis caused by *E. coli* serotypes O1, O2 and O18. For these serotypes no onset of immunity or duration of immunity was established. The benefits of Poulvac *E. coli* are its significant reduction of incidence of lesions typical of colibacillosis (pericarditis, perihepatitis, airsacculitis) and a significant reduction of mortality due to *E. coli* O78 infections. Recently, another mutant of *E. coli* serovar O78 has been produced using an allelic exchange procedure (Nagano *et al.*, 2012). The mutant AESN1331, carrying a deletion in the crp gene, lost tryptophan deaminase activity. The mutant strain additionally no longer fermented sugars other than glucose and L-arabinose, did not harbour four known virulence-associated genes (iss, tsh, cvaA, papC), and was susceptible to many antimicrobials, with the exception of nalidixic acid. The lethal dose (LD50 value) of the mutant strain on intravenous challenge in chickens was approximately 10-fold higher than that of the parent strain. Additionally, the mutant strain was rapidly eliminated from chickens, being detected in the respiratory tract only on the first day post-inoculation by fine spray. Administration of the mutant strain via various routes, such as spray and eye drop for chickens as well as *in ovo* inoculation, evoked an effective immune response that protected against a virulent wild-type *E. coli* O78 strain. Other APEC vaccines had to be administered twice.

Other important *E. coli* vaccines are these against coliform mastitis. These are based on the use of *Escherichia coli* J5 bacterins to enhance resistance against intramammary infection caused by Gram-negative bacteria during the periparturient period and early lactation (Gonzales *et al.*, 1983). Field trials and experimental infection trials have shown that the use of *E. coli* J5 bacterins often reduced the severity and duration of clinical mastitis (Hogan *et al.*, 1992a,b; Wilson *et al.*,

2009). The highest risk period for coliform mastitis is during the immediate periparturient period. Therefore, a vaccine may be judged effective if it successfully reduces symptoms of coliform mastitis during this limited 'at-risk' period. But also here there are still possibilities for improvement.

References

Adu-Bobie, J., Frankel, G., Bain, C., Goncalves, A.G., Trabulsi, L.R., Douce, G., Knutton, S., and Dougan, G. (1998). Detection of intimins alpha, beta, gamma, and delta, four intimin derivatives expressed by attaching and effacing microbial pathogens. J. Clin. Microbiol. 36, 662–668.

Allen, K.J., Rogan, D., Finlay, B.B., Potter, A.A., A and Asper, D.J. (2011). Vaccination with type III secreted proteins leads to decreased shedding in calves after experimental infection with Escherichia coli O157. Can. J. Vet. Res. 75, 98–105.

Arthur, T.M., Brichta-Harhay, D.M., Bosilevac, J.M., Kalchayanand, N., Shackelford, S.D., Wheeler, T.L., and Koohmaraie, M. (2010). Super shedding of Escherichia coli O157:H7 by cattle and the impact on beef carcass contamination. Meat Sci. 86, 32–37.

Asper, D.J., Sekirov, I., Finlay, B.B., Rogan, D., and Potter, A.A. (2007). Cross reactivity of enterohemorrhagic Escherichia coli O157:H7-specific sera with non-O157 serotypes. Vaccine 25, 8262–8269.

Asper, D.J., Karmali, M.A., Townsend, H., Rogan, D., and Potter, A.A. (2011). Serological Response of Shiga Toxin-Producing Escherichia coli Type III Secreted Proteins in Sera from Vaccinated Rabbits, Naturally Infected Cattle, and Humans. Clin. Vac. Immunol. 18, 1052–1057.

Atef Yekta, M., Goddeeris, B.M., Vanrompay, D., and Cox, E. (2011). Immunization of sheep with a combination of intiminγ, EspA and EspB decreases Escherichia coli O157:H7 shedding. Vet. Immunol. Immunopathol. 140, 42–46.

Atroshi, F., Alaviuhkola, T., Schildt, R., and Sandholm, M. (1983). Fat globule membrane of sow milk as a target for adhesion of K88-positive Escherichia coli. Comp. Immunol. Microbiol. Infect. Dis. 6, 235–245.

Bakker, D., Willemsen, P.T., Willems, R.H., Huisman, T.T., Mooi, F.R., Oudega, B., Stegehuis, F., and de Graaf, F.K. (1992). Identification of minor fimbrial subunits involved in biosynthesis of K88 fimbriae. J. Bacteriol. 174, 6350–6358.

Baines, D., Lee, B., and McAllister, T. (2008). Heterogeneity in enterohemorrhagic Escherichia coli O157:H7 fecal shedding in cattle is related to Escherichia coli O157:H7 colonization of the small and large intestine. Can. J. Microbiol. 54, 984–995.

Bertels, A., Pohl, P., Schlicker, C., Van Driessche, E., Charlier, G., de Greve, H., and Lintermans, P. (1989). Isolation of the F111 fimbrial antigen on the surface of a bovine Escherichia coli isolated out of calf diarrhoea: characterization and discussion of the need to adapt recent vaccines against neonatal calf diarrhoea. Vlaams Diergeneeskd. Tijdschr. 58, 118–122.

Bertin, A., Girardeau, J.P., Darfeuille-Michaud A., and Contrepois, M. (1996a). Characterization of 20K, a new adhesin of septicemic and diarrhoea-associated Escherichia coli strains, that belongs to a family of adhesins with N-acetyl-D-glucosamine recognition. Infect. Immun. 64, 332–342.

Bertin, Y., Martin, C., Oswald, E., and Girardeau, J.P. (1996b). Rapid and specific detection of F17-related pilin and adhesin genes in diarrheic and septicemic Escherichia coli strains by multiplex PCR. J. Clin. Microbiol. 34, 2921–2928.

Bertin, Y., Martin, C., Girardeau, J.P., Pohl, P., and Contrepois, M. (1998). Association of genes encoding P fimbriae, CS31A antigen and EAST 1 toxin among CNF1-producing Escherichia coli strains from cattle with septicemia and diarrhoea. FEMS Microbiol. Lett. 162, 235–239.

Bertschinger, H.U., Nief, V., and Tschäpe, H. (2000). Active oral immunization of suckling piglets to prevent colonization after weaning by enterotoxigenic Escherichia coli with fimbriae F18. Vet. Microbiol. 71, 255–267.

Besser, T.E., Richards, B.L., Rice, D.H., and Hancock, D.D. (2001). Escherichia coli O157:H7 infection of calves: infectious dose and direct contact transmission. Epidemiol. Infect. 127, 555–560.

Bianchi, A.T., Scholten, J.W., van Zijderveld, A.M., van Zijderveld, F.G., and Bokhout, B.A. (1996). Parenteral vaccination of mice and piglets with F4+ Escherichia coli suppresses the enteric anti-F4 response upon oral infection. Vaccine 14, 199–206.

Blanco, M., Blanco, J.E., Gonzales, E.A., Mora, A., Jansen, W., Gomes, T.A.T., Zerbini, L.F., Yano, T., Pestana de Castro, A.F., and Blanco, J. (1997). Genes coding for enterotoxins and verotoxins in porcine Escherichia coli strains belonging to different O:K:H: serotypes: relationship with toxic phenotypes. J. Clin. Microbiol. 35, 2958–2963.

Bozić, F., Lacković, G., Stokes, C.R., and Valpotić, I. (2002). Recruitment of intestinal CD45RA+ and CD45RC+ cells induced by a candidate oral vaccine against porcine post-weaning colibacillosis. Vet. Immunol. Immunopathol. 86, 137–146.

Bozić, F., Bilić, V., and Valpotić, I., (2003). Levamisole mucosal adjuvant activity for a live attenuated Escherichia coli oral vaccine in weaned pigs. J. Vet. Pharmacol. Ther. 26, 225–231.

Bretschneider, G., Berberov, E.M., and Moxley, R.A. (2007). Isotype-specific antibody responses against Escherichia coli O157:H7 locus of enterocyte effacement proteins in adult beef cattle following experimental infection, Vet. Immunol. Immunopathol. 118, 229–238.

Calinescu, C., Mulhbacher, J., Nadeau, E., Fairbrother, J.M., and Mateescu, M.A. (2005). Carboxymethyl high amylose starch (CM-HAS) as excipient for Escherichia coli oral formulations. Eur. J. Pharm. Biopharm. 60, 53–60.

Calinescu, C., Nadeau, E., Mulhbacher, J., Fairbrother, J.M., and Mateescu, M.A. (2007). Carboxymethyl high amylose starch for F4 fimbriae gastro-resistant oral formulation. Int. J. Pharm. 343, 18–25.

Casey, T.A., Nagy, B., and Moon, H.W., 1992. Pathogenicity of porcine enterotoxigenic *Escherichia coli* that do not express K88, K99, F41, or 987P adhesins. Am. J. Vet. Res. *53*, 1488–1492.

Coddens, A., Verdonck, F., Tiels, P., Rasschaert, K., Goddeeris, B.M., and Cox, E. (2007). The age-dependent expression of the F18+ *E. coli* receptor on porcine gut epithelial cells is positively correlated with the presence of histo-blood group antigens. Vet. Microbiol. *122*, 332–341.

Coddens A., Diswall, M., Angström, J., Breimer, M.E., Goddeeris, B., Cox, E., and Teneberg, S. (2009). Recognition of blood group ABH type 1 determinants by the FedF adhesin of F18-fimbriated *Escherichia coli*. J. Biol. Chem. *284*, 9713–9726.

Chen, X., Gao, S., Jiao, X., and Liu, X.F. (2004). Prevalence of serogroups and virulence factors of *Escherichia coli* strains isolated from pigs with postweaning diarrhoea in eastern China. Vet. Microbiol. *103*, 13–20.

Cid, D., Sanz, R., Marín, I., de Greve, H., Ruiz-Santa-Quiteria, J.A., Amils, R., and de la Fuente, R. (1999). Characterization of nonenterotoxigenic *Escherichia coli* strains producing F17 fimbriae isolated from diarrheic lambs and goat kids. J. Clin. Microbiol. *37*, 1370–1375.

Contrepois, M., Dubourguier, H.C., Parodi, A.L., Girardeau, J.P., and Ollier, J.L. (1986). Septicaemic *Escherichia coli* and experimental infection of calves. Vet. Microbiol. *12*, 109–118.

Crepin, V.F., Shaw, R., Knutton, S., and Frankel, G. (2005). Molecular basis of antigenic polymorphism of EspA filaments: development of a peptide display technology. J. Mol. Biol. *350*, 42–52.

Daynes, R.A., Enioutina, E.Y., Butler, S., Mu, H.H., McGee, Z.A., and Araneo, B.A. (1996). Induction of common mucosal immunity by hormonally immunomodulated peripheral immunization. Infect. Immun. *64*, 1100–1109.

Dean-Nystrom, E.A., Bosworth, B.T., Moon, H.W., and O'Brien, A.D. (1998). *Escherichia coli* O157:H7 requires intimin for enteropathogenicity in calves. Infect. Immun. *66*, 4560–4563.

Deprez, P., Van den Hende, C., Muylle, E., and Oyaert, W. (1986). The influence of the administration of sow's milk on the post-weaning excretion of hemolytic *E. coli* in the pig. Vet. Res. Commun. *10*, 469–478.

van Diemen, M., Dziva, F., Abu-Median, A., Wallis, T.S., van den Bosch, H., Dougan, G., Chanter, N., Frankel, G., and Stevens, M.P. (2007). Subunit vaccines based on intimin and Efa-1 polypeptides induce humoral immunity in cattle but do not protect against intestinal colonisation by enterohaemorrhagic *Escherichia coli* O157:H7 or O26:H–. Vet. Immunol. Immunopathol. *116*, 47–58.

Dziva, F., Mahajan, A., Cameron, P., Currie, C., McKendrick, I.J., Wallis, T.S., Smith, D.G., and Stevens, M.P. (2007). EspP, a Type V-secreted serine protease of enterohaemorrhagic *Escherichia coli* O157:H7, influences intestinal colonization of calves and adherence to bovine primary intestinal epithelial cells. FEMS Microbiol. Lett. *271*, 258–264.

El Mazouari, K., Oswald, E., Hernalsteens, J.P., Lintermans, P., and De Greve, H. (1994). F17-like fimbriae from an invasive *Escherichia coli* strain producing cytotoxic necrotizing factor type 2 toxin. Infect. Immun. *62*, 2633–2638.

Enioutina, E.Y., Visic, D., McGee, Z.A., and Daynes, R.A. (1999). The induction of systemic and mucosal immune responses following the subcutaneous immunization of mature adult mice: characterization of the antibodies in mucosal secretions of animals immunized with antigen formulations containing a vitamin D3 adjuvant. Vaccine *17*, 3050–3064.

Enioutina, E.Y., Visic, D., and Daynes, R.A. (2000). The induction of systemic and mucosal immune responses to antigen-adjuvant compositions administered into the skin: alterations in the migratory properties of dendritic cells appears to be important for stimulating mucosal immunity. Vaccine *18*, 2753–2767.

Erdem, A.L., Avelino, F., Xicohtencatl-Cortes, J., and Girón, J.A. (2007). Host protein binding and adhesive properties of H6 and H7 flagella of attaching and effacing *Escherichia coli*. J. Bacteriol. *189*, 7426–7435.

Fairbrother, J.M., Nadeau, E., and Gyles, C.L. (2005). *Escherichia coli* in postweaning diarrhoea in pigs: an update on bacterial types, pathogenesis, and prevention strategies. Anim. Health. Res. Rev. *6*, 17–39.

Fekete, P.Z., Schneider, G., Olasz, F., Blum-Oehler, G., Hacker, J.H., and Nagy, B. (2003). Detection of a plasmid-encoded pathogenicity island in F18+ enterotoxigenic and verotoxigenic *Escherichia coli* from weaned pigs. Int. J. Med. Microbiol. *293*, 287–298.

Felder, C.B., Vorlaender, N., Gander, B., Merkle, H.P., and Bertschinger, H.U. (2001). Microencapsulated enterotoxigenic *Escherichia coli* and detached fimbriae for peroral vaccination of pigs. Vaccine *19*, 706–715.

Fox, J.T., Thomson, D.U., Drouillard, J.S., Thornton, A.B., Burkhardt, D.T., Emery, D.A., and Nagaraja, T.G. (2009). Efficacy of *Escherichia coli* O157:H7 siderophore receptor/porin proteins-based vaccine in feedlot cattle naturally shedding *E. coli* O157. Foodborne Pathog. Dis. *6*, 893–899.

Frydendahl, K. (2002). Prevalence of serogroups and virulence genes in *Escherichia coli* associated with postweaning diarrhoea and edema disease in pigs and a comparison of diagnostic approaches. Vet. Microbiol. *85*, 169–182.

Garmendia, J., Phillips, A., Chong, Y., Schuller, S., Marches, O., Dahan, S., Oswald, E., Shaw, R.K., Knutton, S., and Frankel, G. (2004). TccP is an enterohaemorrhagic *E. coli* O157:H7 type III effector protein that couples Tir to the actin-cytoskeleton. Cell. Microbiol. *6*, 1167–1183.

Girardeau, J.P., Dubourguier, H.C., and Contrepois, M. (1979). Attachement des *Escherichia coli* entéropathogènes à la muqueuse intestinale. In Gastro-entérites Néonatales du Veau, Espinasse, J., Ed. Proceedings of the Annual Conference of the French Buiatrics Society, Vichy, France, 25–26th October, pp. 53–66.

Gonzales, R.N., Cullor, J.S., Jasper, D.E., Farver, T.B., Bushnell, R.B., and Oliver, M.N. (1983). Prevention of clinical coliform mastitis in dairy cows by a mutant *Escherichia coli* vaccine. Can. J. Vet. Res. *53*, 301–305.

Gruenheid, S., Sekirov, I., Thomas, N.A., Deng, W., O'Donnell, P., Goode, D., Li, Y., Frey, E.A., Brown, N.F., Metalnikov, P., Pawson, T., Ashman, K., and Finlay, B.B. (2004). Identification and characterization of NleA, a non-LEE-encoded type III translocated virulence factor of enterohaemorrhagic *Escherichia coli* O157:H7. Mol. Microbiol. *51*, 1233–1249.

Guinée, P.A., and Jansen, W.H. (1979). Behavior of *Escherichia coli* K antigens K88ab, K88ac, and K88ad in immunoelectrophoresis, double diffusion, and hemagglutination. Infect. Immun. *23*, 700–705.

Gyles C.L. (1994). *Escherichia coli* enterotoxins. In *Escherichia coli* in Domestic Animals and Humans, Gyles, C.L., Ed. (CAB International, Wallingford, Oxon, UK), p. 337.

Hampson, D.J. (1994). Postweaning diarrhoea in pigs. In *Escherichia coli* in domestic animals and humans, Gyles, C.L., Ed. (Wallingford, Oxon, UK: CAB International), p. 171.

Hogan, J.S., Smith, K.L., Todhunter, D.A., and Schoenberger, P.S. (1992a). Field trial to determine efficacy of an *Escherichia coli* J5 mastitis vaccine. J. Dairy Sci. 75, 78–84.

Hogan, J.S., Weiss, W.P., Todhunter, D.A., Smith, K.L., and Schoenberger, P.S. (1992b). Efficacy of an *Escherichia coli* J5 vaccine in an experimental challenge trial. J. Dairy Sci. 75, 415–422.

Hide, E.J., Connaughton, I.D., Driesen, S.J., Hasse, D., Monckton, R.P., and Sammons, N.G. (1995). The prevalence of F107 fimbriae and their association with Shiga-like toxin II in *Escherichia coli* strains from weaned Australian pigs. Vet. Microbiol. 47, 235–243.

Imberechts, H., Wild, P., Charlier, G., De Greve, H., Lintermans, P., and Pohl, P. (1996). Characterization of F18 fimbrial genes fedE and fedF involved in adhesion and length of enterotoxemic *Escherichia coli* strain 107/86. Microb. Pathog. *21*, 183–192.

Jacob, M.E., Callaway, T.R., and Nagaraja, T.G. (2009). Dietary interactions and interventions affecting *Escherichia coli* O157 colonization and shedding in cattle. Foodborne Pathog. Dis. 6, 785–792.

Jalava, K., Eko, F.O., Riedmann, E., and Lubitz, W. (2003). Bacterial ghosts as carrier and targeting systems for mucosal antigen delivery. Expert. Rev. Vaccines. 2, 45–51.

Joensuu, J.J., Verdonck, F., Ehrström, A., Peltola, M., Siljander-Rasi, H., Nuutila, A.M., Oksman-Caldentey, K.M., Teeri, T.H., Cox, E., Goddeeris, B.M., and Niklander-Teeri, V. (2006). F4 (K88) fimbrial adhesin FaeG expressed in alfalfa reduces F4+ enterotoxigenic *Escherichia coli* excretion in weaned piglets. Vaccine 24, 2387–2394.

Joris, A.M. (2012). Prevalence, characterization and long term follow up of enterohemorrhagic *E. coli* and TTSS specific antibodies in cattle. PhD thesis, UGent, Belgium.

Khare, S., Alali, W., Zhang, S., Hunter, D., Pugh, R., Fang, F.C., Libby, S.J., and Adams, L.G. (2010). Vaccination with attenuated *Salmonella enterica* Dublin expressing *E. coli* O157:H7 outer membrane protein Intimin induces transient reduction of fecal shedding of E coli O157:H7 in cattle. BMC Vet. Res. 6, 35.

Knutton, S., Baldwin, T., Williams, P.H., and McNeish, A.S. (1989). Actin accumulation at sites of bacterial adhesion to tissue culture cells: basis of a new diagnostic test for enteropathogenic and enterohemorrhagic *Escherichia coli*. Infect. Immun. 57, 1290–1298.

La Ragione, R.M., Best, A., Woodward, M.J., and Wales, A.D. (2009). *Escherichia coli* O157:H7 colonization in small domestic ruminants. FEMS Microbiol. Rev. 33, 394–410.

McDaniel, T.K., Jarvis, K.G., Donnenberg, M.S., and Kaper, J.B. (1995). A genetic locus of enterocyte effacement conserved among diverse enterobacterial pathogens, Proc. Natl. Acad. Sci. U.S.A. 92, 1664–1668.

McGhee, J.R., Mestecky, J., Dertzbaugh, M.T., Eldridge, J.H., Hirasawa, M., and Kiyono, H. (1992). The mucosal immune system: from fundamental concepts to vaccine development. Vaccine *10*, 75–88.

McNeilly, T.N., Naylor, S.W., Mahajan, A., Mitchell, M.C., McAteer, S., Deane, D., Smith, D.G., Low, J.C., Gally, D.L., and Huntley, J.F. (2008). *Escherichia coli* O157:H7 colonization in cattle following systemic and mucosal immunization with purified H7 flagellin. Infect. Immun. 76, 2594–2602.

McNeilly, T.N., Mitchell, M.C., Rosser, T., McAteer, S., Erridge, C., Low, J.C., Smith, D.G., Huntley, J.F., Mahajan, A., and Gally, D.L. (2010a). Immunization of cattle with a combination of purified intimin-531, EspA and Tir significantly reduces shedding of *Escherichia coli* O157:H7 following oral challenge. Vaccine 28, 1422–1428.

McNeilly, T.N., Mitchell, M.C., Nisbet, A.J., McAteer, S., Erridge, C., Inglis, N.F., Smith, D.G., Low, J.C., Gally, D.L., Huntley, J.F., and Mahajan, A. (2010b). IgA and IgG antibody responses following systemic immunization of cattle with native H7 flagellin differ in epitope recognition and capacity to neutralise TLR5 signalling. Vaccine 28, 1412–1421.

Mainil, J.G., Gérardin, J., and Jacquemin, E. (2000). Identification of the F17 fimbrial subunit- and adhesin-encoding (f17A and f17G) gene variants in necrotoxigenic *Escherichia coli* from cattle, pigs and humans. Vet. Microbiol. 73, 327–335.

Mahajan, A., Currie, C.G., Mackie, S., Tree, J., McAteer, S., McKendrick, I., McNeilly, T.N., Roe, A., La Ragione, R.M., Woodward, M.J., *et al.* (2009). An investigation of the expression and adhesin function of H7 flagella in the interaction of *Escherichia coli* O157:H7 with bovine intestinal epithelium. Cell. Microbiol. *11*, 121–137.

Mainil, J.G., Bex, F., Jacquemin, E., Pohl, P., Couturier, M., and Kaeckenbeeck, A. (1990). Prevalence of four enterotoxin (STaP, STaH, STb, and LT) and four adhesin subunit (K99, K88, 987P, and F41) genes among *Escherichia coli* isolates from cattle. Am. J. Vet. Res. *51*, 187–190.

Matthews, L., Low, J.C., Gally, D.L., Pearce, M.C., Mellor, D.J., Heesterbeek, J.A., Chase-Topping, M., Naylor, S.W., Shaw, D.J., Reid, S.W., *et al.* (2006). Heterogeneous shedding of *Escherichia coli* O157 in cattle and its implications for control. Proc. Natl. Acad. Sci. U.S.A. *103*, 547–552.

Mazaitis, A.J., Maas, R., and Maas, W.K. (1981). Structure of a naturally occurring plasmid with genes for enterotoxin production and drug resistance. J. Bacteriol. *145*, 97–105.

Misyurina, O., Asper, D.J., Deng, W., Finlay, B.B., Rogan, D., and Potter, A.A. (2010). The role of Tir, EspA, and NleB in the colonization of cattle by Shiga toxin producing *Escherichia coli* O26:H11. Can. J. Microbiol. *56*, 739–747.

Moxley, R.A., Smith, D.R., Luebbe, M., Erickson, G.E., Klopfenstein, T.J., and Rogan, D. (2009). *Escherichia coli* O157:H7 vaccine dose-effect in feedlot cattle. Foodborne Pathog. Dis. *6*, 879–884.

Nagano, T., Kitahara, R., and Nagai, S. (2012). An attenuated mutant of avian pathogenic *Escherichia coli* serovar O78: a possible live vaccine strain for prevention of avian colibacillosis. Microbiol. Immunol. *56*, 605–612.

Nagy, B., and Fekete, P.Z. (1999). Enterotoxigenic *Escherichia coli* (ETEC) in farm animals. Vet. Res. *30*, 259–284.

Nagy, B., and Fekete, P.Z. (2005). Enterotoxigenic *Escherichia coli* in veterinary medicine. Int. J. Med. Microbiol. *295*, 443–454.

Nataro, J.P., and Kaper, J.B. (1989). Diarrheagenic *Escherichia coli*. Clin. Microbiol. Rev. *11*, 142–201.

Naylor, S.W., Flockhart, A., Nart, P., Smith, D.G., Huntley, J., Gally, D.L., and Low, J.C. (2007). Shedding of *Escherichia coli* O157:H7 in calves is reduced by prior colonization with the homologous strain. Appl. Environ. Microbiol. *73*, 3765–3767.

Nguyen, T.D., Vo, T.T., and Vu-Khac, H. (2011). Virulence factors in *Escherichia coli* isolated from calves with diarrhoea in Vietnam. J. Vet. Sci. *12*, 159–164.

O'Donnell, J.M., Thornton, L., McNamara, E.B., Prendergast, T., Igoe, D., and Cosgrove, C. (2002). Outbreak of Verocytotoxin-producing *Escherichia coli* O157 in a child day care facility. PHLS *5*, 54–58.

Orskov, I., Orskov, F., Sojka, W.J., and Wittig, W. (1964). K antigens K88ab(L) and K88ac(L) in *E. coli*. A new O antigen: 0147 and a new K antigen: K89(B). Acta Pathol. Microbiol. Scand. *62*, 439–447.

Peterson, R.E., Klopfenstein, T.J., Moxley, R.A., Erickson, G.E., Hinkley, S., Bretschneider, G., Berberov, E.M., Rogan, D., and Smith, D.R. (2007a). Effect of a vaccine product containing type III secreted proteins on the probability of *Escherichia coli* O157:H7 fecal shedding and mucosal colonization in feedlot cattle. J. Food Prot. *70*, 2568–2577.

Peterson, R.E., Klopfenstein, T.J., Moxley, R.A., Erickson, G.E., Hinkley, S., Rogan, D., and Smith, D.R. (2007b). Efficacy of dose regimen and observation of herd immunity from a vaccine against *Escherichia coli* O157:H7 for feedlot cattle. J. Food Prot. *70*, 2561–2567.

Pohl P., Lintermans, P., Van Muylem K., S and chotte, M. (1982). Colibacilles entérotoxinogènes de veau possédant un antigène d'attachement différent de l'antigène K99. Ann. Méd. Vét. *126*, 569–571.

Potter, A.A., Klashinsky, S., Li, Y., Frey, E., Townsend, H., Rogan, D., Erickson, G., Hinkley, S., Klopfenstein, T., Moxley, R.A., Smith, D.R., and Finlay, B.B. (2004).

Decreased shedding of *Escherichia coli* O157:H7 by cattle following vaccination with type III secreted proteins. Vaccine *22*, 362–369.

Reida, P., Wolff, M., Pohls, H.W., Kuhlmann, W., Lehmacher, A., Alecsic', S., Karch, H., and Bockemühl, J. (1994). An outbreak due to enterohaemorrhagic *Escherichia coli* O157:H7 in a children day care centre characterized by person-to-person transmission and environmental contamination. Zbl. Bakt. *281*, 534–543.

Rice, D.H., Sheng, H.Q., Wynia, S.A., and Hovde, C.J. (2003). Rectoanal mucosal swab culture is more sensitive than fecal culture and distinguishes *Escherichia coli* O157:H7-colonized cattle and those transiently shedding the same organism. J. Clin. Microbiol. *41*, 4924–4929.

des Rieux, A., Fievez, V., Garinot, M., Schneider, Y.J., and Préat, V. (2006). Nanoparticles as potential oral delivery systems of proteins and vaccines: a mechanistic approach. J. Control. Release. *116*, 1–27.

Rippinger, P., Bertschinger, H.U., Imberechts, H., Nagy, B., Sorg, I., Stamm, M., Wild, P., and Wittig, W. (1995). Designations F18ab and F18ac for the related fimbrial types F107, 2134P and 8813 of *Escherichia coli* isolated from porcine postweaning diarrhoea and from oedema disease. Vet. Microbiol. *45*, 281–295.

Robinson, S.E., Wright, E.J., Hart, C.A., Bennett, M., and French, N.P. (2004). Intermittent and persistent shedding of *Escherichia coli* O157 in cohorts of naturally infected calves. J. Appl. Microbiol. *97*, 1045–1053.

Robinson, S.E., Brown, P.E., Wright, E.J., Hart, C.A., and French, N.P. (2009). Quantifying within- and between-animal variation and uncertainty associated with counts of *Escherichia coli* O157 occurring in naturally infected cattle faeces, J. R. Soc. Interface. *6*, 169–177.

Roe, A.J., Tysall, L., Dransfield, T., Wang, D., Fraser-Pitt, D., Mahajan, A., Constandinou, C., Inglis, N., Downing, A., Talbot, R., *et al.* (2007). Analysis of the expression, regulation and export of NleA-E in *Escherichia coli* O157:H7. Microbiology. *153*, 1350–1360.

Ruan, X., Liu, M., Casey, T.A., and Zhang, W. (2011). A tripartite fusion, FaeG-FedF-LT(192)A2:B, of enterotoxigenic *Escherichia coli* (ETEC) elicits antibodies that neutralize cholera toxin, inhibit adherence of K88 (F4) and F18 fimbriae, and protect pigs against K88ac/heat-labile toxin infection. Clin. Vaccine Immunol. *18*, 1593–1599.

Ruan, X., Crupper, S.S., Schultz, B.D., Robertson, D.C., and Zhang, W. (2012). *Escherichia coli* expressing EAST1 toxin did not cause an increase of cAMP or cGMP levels in cells, and no diarrhoea in 5-day old gnotobiotic pigs. PLoS ONE 7(8), e43203.

Rutter, J.M., and Jones, G.W. (1973). Protection against enteric disease caused by *Escherichia coli*-a model for vaccination with a virulence determinant? Nature *242*, 531–532.

Shahriar, F., Ngeleka, M., Gordon, J.R., and Simko, E. (2006). Identification by mass spectroscopy of F4ac-fimbrial-binding proteins in porcine milk and characterization of lactadherin as an inhibitor of

F4ac-positive *Escherichia coli* attachment to intestinal villi *in vitro*. Dev. Comp. Immunol. *30*, 723–734.

Smeds, A., Hemmann, K., Jakava-Viljanen, M., Pelkonen, S., Imberechts, H., and Palva, A. (2001). Characterization of the adhesin of *Escherichia coli* F18 fimbriae. Infect. Immun. *69*, 7941–7945.

Smith, D.R., Moxley, R.A., Klopfenstein, T.J., and Erickson, G.E. (2009a). A randomized longitudinal trial to test the effect of regional vaccination within a cattle feedyard on *Escherichia coli* O157:H7 rectal colonization, fecal shedding, and hide contamination. Foodborne Pathog. Dis. *6*, 885–892.

Smith, D.R., Moxley, R.A., Peterson, R.E., Klopfenstein, T.J., Erickson, G.E., Bretschneider, G., Berberov, E.M., and Clowser, S. (2009b). A two-dose regimen of a vaccine against type III secreted proteins reduced *Escherichia coli* O157:H7 colonization of the terminal rectum in beef cattle in commercial feedlots. Foodborne Pathog. Dis. *6*, 155–161.

Snedeker, K.G., Campbell, M., and Sargeant, J.M. (2011). A systematic review of vaccinations to reduce the shedding of *Escherichia coli* O157 in the faeces of domestic ruminants. Zoo. Public Health *59*, 126–138.

Snodgrass, D.R., Nagy, L.K., Sherwood, D., and Campbell, I. (1982). Passive immunity in calf diarrhoea: vaccination with K99 antigen of enterotoxigenic *Escherichia coli* and rotavirus. Infect. Immun. *37*, 586–591.

Snoeck, V. (2004). Targeting intestinal induction sites for oral immunisation of piglets against F4+ *Escherichia coli* infection. PhD Thesis, UGent. Belgium.

Snoeck, V., Huyghebaert, N., Cox, E., Vermeire, A., Vancaeneghem, S., Remon, J.P., and Goddeeris, B.M. (2003). Enteric-coated pellets of F4 fimbriae for oral vaccination of suckling piglets against enterotoxigenic *Escherichia coli* infections. Vet. Immunol. Immunopathol. *96*, 219–227.

Snoeck, V., Verfaillie, T., Verdonck, F., Goddeeris, B.M., and Cox, E. (2006). The jejunal Peyer's patches are the major inductive sites of the F4-specific immune response following intestinal immunisation of pigs with F4 (K88) fimbriae. Vaccine *24*, 3812–3820.

Thanabalasuriar, A., Koutsouris, A., Weflen, A., Mimee, M., Hecht, G., and Gruenheid, S. (2010). The bacterial virulence factor NleA is required for the disruption of intestinal tight junctions by enteropathogenic *Escherichia coli*. Cell. Microbiol. *12*, 31–41.

Thanabalasuriar, A., Bergeron, J., Gillingham, A., Mimee, M., Thomassin, J.L., Strynadka, N., Kim, J., and Gruenheid, S. (2012). Sec24 interaction is essential for localization and virulence-associated function of the bacterial effector protein NleA. Cell. Microbiol. *14*, 1206–1218.

Thomson, D.U., Loneragan, G.H., Thornton, A.B., Lechtenberg, K.F., Emery, D.A., Burkhardt, D.T., and Nagaraja, T.G. (2009). Use of a siderophore receptor and porin proteins-based vaccine to control the burden of *Escherichia coli* O157:H7 in feedlot cattle. Foodborne Pathog. Dis. *6*, 871–877.

Thornton, A.B., Thomson, D.U., Loneragan, G.H., Fox, J.T., Burkhardt, D.T., Emery, D.A., and Nagaraja, T.G. (2009). Effects of a siderophore receptor and porin proteins-based vaccination on fecal shedding of *Escherichia coli* O157:H7 in experimentally inoculated cattle. J. Food Prot. *72*, 866–869.

Tiels, P., Verdonck, F., Smet, A., Goddeeris, B., and Cox, E. (2005). The F18 fimbrial adhesin FedF is highly conserved among F18+*Escherichia coli* isolates. Vet Microbiol. *110*, 277–283.

Tiels, P., Verdonck, F., Coddens, A., Ameloot, P., Goddeeris, B.M., and Cox, E. (2007). Monoclonal antibodies reveal a weak interaction between the F18 fimbrial adhesin FedF and the major subunit FedA. Vet. Microbiol. *119*, 115–120.

Tiels, P., Verdonck, F., Coddens, A., Goddeeris, B., and Cox, E. (2008). The excretion of F18+ *E. coli* is reduced after oral immunisation of pigs with a FedF and F4 fimbriae conjugate. Vaccine *16*, 2154–2163.

Van den Broeck, W., Cox, E., and Goddeeris, B.M. (1999a). Seroprevalence of F4+ enterotoxigenic *Escherichia coli* in regions with different pig farm densities. Vet. Microbiol. *69*, 207–216.

Van den Broeck, W., Cox, E., and Goddeeris, B.M. (1999b). Induction of immune responses in pigs following oral administration of purified F4 fimbriae. Vaccine *17*, 2020–2029.

Van den Broeck, W., Cox, E., and Goddeeris, B.M. (1999c). Receptor-dependent immune responses in pigs after oral immunization with F4 fimbriae. Infect. Immun. *67*, 520–526.

Vandamme, K., Melkebeek, V., Cox, E., Remon, J.P., and Vervaet, C. (2011). Adjuvant effect of Gantrez®AN nanoparticles during oral vaccination of piglets against F4+enterotoxigenic *Escherichia coli*. Vet. Immunol. Immunopathol. *139*, 148–155.

Van Der Stede, Y., Verfaillie, T., Cox, E., Verdonck, F., and Goddeeris, B.M. (2004). 1alpha,25-dihydroxyvitamin D3 increases IgA serum antibody responses and IgA antibody-secreting cell numbers in the Peyer's patches of pigs after intramuscular immunization. Clin. Exp. Immunol. *135*, 380–390.

Van Donkersgoed, J., Hancock, D., Rogan, D., and Potter, A.A. (2005). *Escherichia coli* O157:H7 vaccine field trial in 9 feedlots in Alberta and Saskatchewan. Can. Vet. J. *46*, 724–728.

Van der Stede, Y., Cox, E., and Goddeeris, B.M. (2002). Antigen dose modulates the immunoglobulin isotype responses of pigs against intramuscularly administered F4-fimbriae. Vet. Immunol. Immunopathol. *88*, 209–216.

Van der Stede, Y., Cox, E., Verdonck, F., Vancaeneghem, S., and Goddeeris, B.M. (2003). Reduced faecal excretion of F4+*E coli* by the intramuscular immunisation of suckling piglets by the addition of 1alpha,25-dihydroxyvitamin D3 or CpG-oligodeoxynucleotides. Vaccine *21*, 1023–1032.

Veilleux, S., and Dubreuil, J.D. (2006). Presence of *Escherichia coli* carrying the EAST1 toxin gene in farm animals. Vet. Res. *37*, 3–13.

Verdonck, F., Tiels, P., van Gog, K., Goddeeris, B.M., Lycke, N., Clements, J., and Cox, E. (2007). Mucosal immunization of piglets with purified F18 fimbriae does not protect against F18+ *Escherichia coli* infection. Vet. Immunol. Immunopathol. *120*, 69–79.

Vilte, D.A., Larzábal, M., Garbaccio, S., Gammella, M., Rabinovitz, B.C., Elizondoa, A.M., Cantet, R.J., Delgado, F., Meikle, V., Cataldi, A., and Mercado, E.C. (2011). Reduced faecal shedding of *Escherichia coli* O157:H7 in cattle following systemic vaccination with γ-intimin C_{280} and EspB proteins. Vaccine *29*, 3962–3968.

Vilte, D.A., Larzábal, M., Mayr, U.B., Garbaccio, S., Gammella, M., Rabinovitz, B.C., Delgado, F., Meikle, V., Cantet, R.J., Lubitz, P., *et al.* (2012). A systemic vaccine based on *Escherichia coli* O157:H7 bacterial ghosts (BGs) reduces the excretion of *E. coli* O157:H7 in calves. Vet. Immunol. Immunopathol. *146*, 169–176.

Weltzin, R., Lucia-Jandris, P., Michetti, P., Fields, B.N., Kraehenbuhl, J.P., and Neutra, M.R. (1989). Binding and transepithelial transport of immunoglobulins by intestinal M cells: demonstration using monoclonal IgA antibodies against enteric viral proteins. J. Cell. Biol. *108*, 1673–1685.

Westerman, R.B., Mills, K.W., Phillips, R.M., Fortner, G.W., and Greenwood, J.M. (1988). Predominance of the ac variant in K88-positive *Escherichia coli* isolates from swine. J. Clin. Microbiol. *26*, 149–150.

Wileman, B.W., Thomson, D.U., Olson, K.C., Jaeger, J.R., Pacheco, L.A., Bolte, J., Burkhardt, D.T., Emery, D.A., and Straub, D. (2011). *Escherichia coli* O157:H7 shedding in vaccinated beef calves born to cows vaccinated prepartum with *Escherichia coli* O157:H7 SRP vaccine. J. Food Prot. *74*, 1599–1604.

Wilson, R.A., and Francis, D.H. (1986). Fimbriae and enterotoxins associated with *Escherichia coli* serogroups isolated from pigs with colibacillosis. Am. J. Vet. Res. *47*, 213–217.

Wilson, D.J., Mallard, B.A., Burton, J.L., Schukken, Y.H., and Grohn, Y.T. (2009). Association of *Escherichia coli* J5-specific serum antibody responses with clinical mastitis outcome for J5 vaccinate and control dairy cattle. Clin. Vaccine Immunol. *16*, 209–217.

Zhang, W., Berberov, E.M., Freeling, J., He, D., Moxley, R.A., and Francis, D.H. (2006). Significance of heat-stable and heat-labile enterotoxins in porcine colibacillosis in an additive model for pathogenicity studies. Infect Immun. *74*, 3107–3114.

Zhang, X.H., He, K.W., Zhao, P.D., Ye, Q., Luan, X.T., Yu, Z.Y., Wen, L.B., Ni, Y.X., Li, B., Wang, X.M., et al. (2012). Intranasal immunisation with Stx2B-Tir-Stx1B-Zot protein leads to decreased shedding in goats after challenge with *Escherichia coli* O157:H7. Vet. Rec. *170*, 178.

Perspective of use of Vaccines for Preventing Enterotoxigenic *Escherichia coli* Diarrhoea in Humans

15

Weiping Zhang

Abstract

Enterotoxigenic *Escherichia coli* (ETEC) strains are the leading bacteria that cause diarrhoea to young children living in the developing countries and children and adults travelling to these areas. ETEC strains produce adhesins that mediate bacteria initial attachment to host epithelial cells and subsequent colonization at host small intestines, and enterotoxins including heat-labile toxin (LT) and heat-stable toxin (STa) that disrupt fluid homeostasis in host epithelial cells to cause electrolyte-rich fluid hyper-secretion and diarrhoea. As country-wide implementation of clean drinking water and effective sanitation systems, which can effectively limit ETEC infections, are still an out of reach goal in many developing countries, vaccination is the most practical prevention approach. Vaccines inducing host anti-adhesin immunity to block ETEC attachment and colonization and also antitoxin immunity to neutralize enterotoxicity are considered optimal against ETEC diarrhoea. However, although a cholera vaccine (Dukoral®) that stimulates anti-CT immunity provides short–term cross protection against ETEC diarrhoea for travellers, vaccines effectively protecting against ETEC are currently still lacking. Vaccines under development are whole-cell oral vaccines and intend to stimulate intestinal mucosal immunity, and newer experimental ETEC vaccine candidates are aimed to provide long-lasting and more broad-based protection.

Introduction

Diarrhoeal disease is the second leading cause of death to children younger than 5 years, and continues to be a major problem to global health (Black *et al.*, 2010). Enterotoxigenic *Escherichia coli* (ETEC), *E. coli* strains producing enterotoxins, are the leading bacteria that cause diarrhoeal disease in young children who live in the developing countries and children and adults who travel from developed countries to ETEC endemic countries or regions. Despite treatment with oral rehydration solution and gradual improvement of drinking water supply have decreased diarrhoeal disease incidences and associated morbidity and mortality, ETEC strains are still responsible for 280 to 400 million diarrhoea cases occurred annually in children younger than 5 and 100 million more cases in children older than 5 years, that results in annual deaths of 300,000–500,000 (WHO, 2006). ETEC are also the most common cause of diarrhoea to adult travellers and military personnel deployed at endemic regions (Jiang *et al.*, 2002; Sack *et al.*, 2007; Sanders *et al.*, 2005; Steffen *et al.*, 2005). It is estimated that ETEC cause about 400 million adult traveller diarrhoeal cases annually (Qadri *et al.*, 2005; Wenneras and Erling, 2004; WHO, 2006). The virulence determinants of ETEC in diarrhoea are identified as bacterial adhesins and enterotoxins. ETEC adhesins include colonization factor antigens (CFAs), putative colonization factors (PFCs) and coli surface antigens (CSs). These adhesins mediate ETEC bacteria initial attachment to host epithelial cells and promote bacteria colonization at host small intestines. Enterotoxins are mainly heat-labile toxin (LT) and heat-stable type Ib toxin (STa), and they enzymatically disrupt fluid homeostasis in host epithelial cells to cause electrolyte-rich fluid hyper-secretion that leads to diarrhoea.

ETEC adhesins

Bacterial adhesins facilitate an initial adherence between bacteria and host cells (for additional information, see Chapter 6). Initial adherence between bacteria and host epithelial cells is the first step of ETEC infection, and without the initial attachment subsequent infection steps cannot occur. Moreover, adhesins not only promote ETEC bacterial attachment and colonization at human small intestines, but also bring bacteria to close proximity to small intestinal epithelial cells so that enterotoxins produced by ETEC can be effectively delivered to host cells. Adhesins thereby play a critical role in ETEC-associated diarrhoea. Adhesins produced by ETEC strains are mainly fimbrial or fibrillar biopolymeric filaments, but can be a helix or non-fimbrial structure, displayed at bacteria surface (Gaastra and Svennerholm, 1996). ETEC adhesins are identified based on structural morphology variations, as well as heterogeneities in genetic structure and more importantly antigenic characteristics. Since recognized first in 1975 (Evans *et al.*, 1975), ETEC adhesins have been classified with different nomenclature systems that include colonization factor antigen (CFA), colonization factor (CF) or putative colonization factor (PCF), and coli surface antigen (CS). Currently, the CS system becomes more commonly used, particularly after evidences revealed that up to three immunologically heterogeneous adhesins were mixed under a single CFA name. Those CFA adhesins were re-designated as CSs; but CFA/I, the most studied ETEC adhesin, is retained for its nomenclature. So far, there are up to 28 CSs have been assigned (Nada *et al.*, 2011), but 23 ETEC adhesins are commonly acknowledged (Gaastra and Svennerholm, 1996; Turner *et al.*, 2006b) (Table 15.1). Among them, seven adhesins (CFA/I, CS1, CS2, CS3, CS4, CS5, CS6) are produced by ETEC strains that cause about 80% of ETEC diarrhoea cases and also the moderate to severe diarrhoea cases (Isidean *et al.*, 2011; Sack *et al.*, 2007; Svennerholm and Tobias, 2008; Wolf, 1997).

CFA/I group: ETEC adhesins cluster into groups based on morphological and genetic similarity, and particularly the amino acid sequences of the N-terminus of major structural subunits and the minor structural tip proteins (Gaastra and Svennerholm, 1996; Li *et al.*, 2009). CFA/I group is the largest ETEC adhesin genetic cluster. It includes CFA/I, CS1, CS2, CS4, CS14, CS17, CS19 and PCFO71 adhesins (Anantha *et al.*, 2004; Chattopadhyay *et al.*, 2012; Gaastra and Svennerholm, 1996). These adhesins are of the class 5 fimbriae and generally organized as a four-gene operon that consists of a structural major subunit gene, a minor subunit (tip) gene, and two genes encode a chaperone and an usher protein. Within the group, the most closely related (genetically but not antigenically) adhesins to CFA/I are CS4 and CS14, whereas CS1, CS17, CS19 and PCFO71 form another subgroup, and CS2 alone falls in a third subgroup (Chattopadhyay *et al.*, 2012; Gaastra and Svennerholm, 1996).

CS5 group: The CS5 cluster includes CS5, CS7, CS13, CS18, and CS20 adhesins (Gaastra and Svennerholm, 1996; Nuccio and Baumler, 2007; Qadri *et al.*, 2005; Valvatne *et al.*, 2004). The CS5 operon contains 6 genes to code 1 major subunit, 2 minor subunits, 2 chaperons and 1 usher protein (Duthy *et al.*, 2002). Adhesin CS23 which contains 9 open reading frames (ORFs) may belong to this group since its major structural subunit AlaE is closely related to CS13 pilin protein CshE (Del Canto *et al.*, 2012). Adhesins in this group are also genetically related to the 987P and K88 fimbriae expressed by ETEC associated with pig diarrhoea (de Graaf, 1994; Del Canto *et al.*, 2012); thereby the CS5 group is referred sometimes as the 987P group.

T4P group: The third cluster includes CS8 (CFA/III) and CS21 (longus). Adhesins in this group are genetically related to the type IV family fimbriae or pili (T4P) expressed by non-ETEC strains such as *Vibrio cholerae* or enteropathogenic *E. coli* (Gaastra and Svennerholm, 1996; Giron *et al.*, 1997). Adhesin CS21 is expressed by ETEC strains expressing LT and STa (Gomez-Duarte and Kaper, 1995), and is found more prevalent in ETEC strains associated with travel diarrhoea (22%) versus non-travel diarrhoea (11%) (Isidean *et al.*, 2011).

Table 15.1 ETEC adhesins, derived from Gaastra and Svennerholm (1996) and Turner *et al.* (2006)

CS adhesins	CFA, PCF, CS designation	Morphology (diameter)	Structural subunit(s) and sizes (kDa)	References
	CFA/I	Fimbrial (7 nm)	CfaB, 15.0	Evans and Evans (1978), Jordi *et al.* (1992)
CS1	CFA/II	Fimbrial (7 nm)	CooA(CsoA), 16.5	Perez-Casal *et al.* (1990), Marron and Smyth (1995)
CS2		Fimbrial (7 nm)	CotA, 15.3	Froehlich *et al.* (1995)
CS3		Fibrillar (2–3 nm)	CstH, 15.1	Jalajakumari *et al.* (1989)
CS4	CFA/IV, PCF8775	Fimbrial (6 nm)	CsaB, 17.0	Thomas *et al.* (1985)
CS5		Helical (5 nm)	CsfA, 21.0	Clark *et al.* (1992)
CS6		Non-fimbrial	CssA/CssB, 14.5/16.0	Thomas *et al.* (1985)
CS7	CS7	Helical (3–6 nm)	CsvA, 21.5	Hibberd *et al.* (1991)
CS8	CFA/III	Fimbrial (7 nm)	CofA, 18.0	Taniguchi *et al.* (1995)
CS10	2230	Nonfimbrial	16.0	Darfeuille-Michaud *et al.* (1986)
CS11	PCFO148	Fibrillae (3 nm)		Knutton *et al.* (1987)
CS12	PCFO159	Fimbrial (7 nm)	CswA, 19.0	Tacket *et al.* (1987)
CS13	PCFO9	Fibrillae	CshE, 27.0	Heuzenroeder *et al.* (1990)
CS14	PCFO166	Fimbrial (7 nm)	CsuA1/CsuA2, 15.5/17.0	McConnell *et al.* (1989)
CS15	8786	Nonfimbrial	NfaA, 16.3	Aubel *et al.* (1991)
CS17	CS17	Fimbrial (7 nm)	CsbA, 17.5	McConnell *et al.* (1990)
CS18	PCFO20	Fimbrial (7 nm)	FotA, 25.0	Viboud *et al.* (1993, 1996)
CS19	CS19	Fimbrial (7 nm)	CsdA, 16.0	Grewal *et al.* (1997)
CS20	CS20	Fimbrial (7 nm)	CsnA, 19.8	Valvatne *et al.* (1996)
CS21	Longus	Fimbrial (7 nm)	LngA, 22.0	Giron *et al.* (1994)
CS22	CS22	Nonfimbrial	CseA, 17.1	Pichel *et al.* (2000)
CS23	CS23	Fimbrial	AalE, 28.0	Del Canto *et al.* (2012)
	PCFO71	Fimbrial	CosA, 17.4	Anantha *et al.* (2004)

CS15 group: Two non-fimbrial adhesins, CS15 and CS22, are similar genetically, and they are homologous to the SEF14 fimbriae of *Salmonella enterica* serovar *enteritidis* (Aubel *et al.*, 1991). These two adhesins are clustered together as the CS15 group (Elsinghorst, 2002; Gomez-Duarte *et al.*, 1999; Ogunniyi *et al.*, 1997), or as a distinctive sister group to the CFA/I group based on a phylogenetic analysis of multilocus sequence typing data (Nada *et al.*, 2011).

Non-homologous group: CS3, CS6, CS10, CS11 and CS12 are shown no clear homology with any known fimbriae (Gaastra and Svennerholm, 1996). A recently phylogenetic analysis, however, suggests that CS12 is closely related to the 987P fimbriae, and CS3 may represent a distinctive basal lineage to the T4P group (Nada *et al.*, 2011).

ETEC enterotoxins

Initial attachment and colonization at host small intestines is the first step of ETEC infection, but it is the enterotoxins produced by ETEC strains that enzymatically disrupt fluid homeostasis in host epithelial cells to cause electrolyte-rich fluid hyper-secretion that leads to diarrhoea (Nataro and Kaper, 1998). Enterotoxins produced by ETEC strains associated with human diarrhoea are mainly heat-labile toxin (LT) and heat-stable toxin type Ib (STa) (additional information can be found later in this chapter). Overall, ETEC strains expressing LT alone, STa alone, and LT together with STa are 25–27%, 40–46% and 29–33%, respectively (Isidean *et al.*, 2011; Wolf, 1997). Other enterotoxins, such as enteroaggregative heat-stable toxin 1 (EAST1) and occasionally heat-stable toxin type II (STb) are also detected in ETEC strains isolated from diarrhoeal patients.

LT

LT, a typical A:B type holotoxin, is closely related to cholera toxin (CT) produced by *Vibrio cholerae*. LT and CT share many structural and functional characteristics including a holotoxin structure, 80% homology in amino acid sequences of the A and B subunits, GM_1 as the primary binding receptor, and enzymatic activities (Nataro and Kaper, 1998; Sixma *et al.*, 1993). LT toxin produced by ETEC strains is composed of one 28 kDa subunit A (LT_A) and five identical 11.6 kDa B (LT_B) subunits. The LT_A and LT_B subunits (encoded by *eltAB* cistron genes) are biosynthesized independently (Hirst *et al.*, 1984a), translocated across the cytoplasmic membrane (Hirst *et al.*, 1984b), involved in formation of outer membrane vesicles (Chutkan and Kuehn, 2011; Horstman and Kuehn, 2000; McBroom *et al.*, 2006; Wai *et al.*, 1995), and then are assembled as a holotoxin. The LT_A subunit is a polypeptide consisted of a large A1 peptide and a small A2 peptide, joined by a disulfide bond (Sixma *et al.*, 1991, 1993). The five LT_B subunits form a ring-like pentamer that binds mainly to host GM1 receptors located at small intestinal epithelial cells (Sixma *et al.*, 1992). The LT A1 peptide is responsible for the enzymatic activity of the toxin, and the LT A2 peptide covalently associates the LT_A subunit to the centre of the LT_B pentamer for the formation of a holotoxin.

Once bound to host GM_1 receptors, holotoxin structured LT enters host small intestinal epithelial cells through endocytosis (Lencer *et al.*, 1995; Spangler, 1992) or possibly internalization of the entire outer membrane vesicle (Kesty *et al.*, 2004). Subsequently, LT A1 peptide is proteolytically cleaved to activate its enzymatic activity and translocated through the cell in a process involving trans-Golgi vesicular transport apparatus (Donta *et al.*, 1993; Lencer *et al.*, 1995). But precisely where the A1 peptide cleavage takes place, either inside the outer membrane vesicle upon its binding to host cells or after holotoxin entering host cells, is less clear. Once inside the host cell, the A1 peptide acts as an ADP-ribosyltransferase to stimulate adenylate cyclase activity via activation of the G protein (G*sa*). That leads to increased

levels of intracellular cyclic AMP, and results in activation of protein kinase A. The activated kinase A causes the supernormal phosphorylation of chloride channels located in the apical epithelial cell membrane, like cystic fibrosis transmembrane conductance regulator (CFTR), that promotes secretion of Cl^- but inhibits absorption of Na^+ (Spangler, 1992). That directly disrupts osmotic balance in host small intestinal epithelial cells and causes net water secretion to the gut lumen that results in diarrhoea (Cieplak *et al.*, 1995; Nataro and Kaper, 1998).

In addition to stimulation of adenylate cyclase activity to increase intracellular cAMP levels in host small intestinal epithelial cells that leads to net water secretion, LT enhances ETEC infection through facilitation of bacterial colonization (Berberov *et al.*, 2004; Glenn *et al.*, 2009; Zhang *et al.*, 2006). Young gnotobiotic pigs, when challenged with an LT+ ETEC strain, had significantly more bacteria colonized at small intestines and developed severer diarrhoea and dehydration, compared to piglets challenged with an ETEC strain expressing STb toxin or an LT-deleted mutant strain (Berberov *et al.*, 2004; Zhang *et al.*, 2006). In addition, piglets challenged with an isogenic *E. coli* strain having its LT genes mutated to express less toxic LT_{R192G} toxoid showed no diarrhoea and a significant reduction in bacterial colonization at small intestines compared to those challenged with the LT+ strain (Santiago-Mateo *et al.*, 2012).

There is a different type of heat-labile toxin, LTII, found in *E. coli* strains (Green *et al.*, 1983; Guth *et al.*, 1986a; Guth *et al.*, 1986b; Holmes *et al.*, 1986). LTII is also an A:B type holotoxin-structured protein, but is antigenically distinctive from LT (LTI) as they share only 57% genetic similarity at their A subunits (Pickett *et al.*, 1987, 1989). The role of LTII played in ETEC is not determined. LTII is often detected among *E. coli* strains isolated from animals, but rarely humans, with no diseases (Nataro and Kaper, 1998). But LTII is also found in *E. coli* or ETEC strains isolated from calves (Rigobelo *et al.*, 2006) including derived beef products (Cerqueira *et al.*, 1994; Franco *et al.*, 1991), pigs (Celemin *et al.*, 1994, 1995), ostriches (Nardi *et al.*, 2005) and humans (Seriwatana *et al.*, 1988) with diarrhoea. Additionally, a recent study demonstrated that purified LTII (IIa and IIb) stimulates fluid accumulation in ligated intestinal loops in calves but not in pigs or rabbits (Casey et al., 2012). However, neonatal piglets develop severe water diarrhoea when inoculated with an E. coli strain expressing LTII as the only known toxin (Casey et al., 2012).

STa

STa (heat-stable toxin type Ib), or hSTa, produced by ETEC associated with human diarrhoea is a 2 kDa peptide with 19 amino acids (an analogous STa in ETEC strains associated with animal diarrhoea has 18 amino acids and is called as heat-stable type Ia, or porcine-type STa – pSTa) (Dreyfus et al., 1983; Thompson and Giannella, 1985). STa, coded by gene estA, is initially produced as a pre-pro peptide of 72 amino acids. During its translocation across the inner membrane, the pre-pro peptide is proteolytically cleaved to release the 19 amino acid peptide (Rasheed et al., 1990). This 19 amino acid peptide possesses six cysteine residues that form three disulfide bonds catalysed by enzyme DsbA (disulfide bond isomerase) (Yamanaka et al., 1994). The mature STa peptide is exported across the outer membrane through outer membrane protein transporter TolC (Yamanaka et al., 1998).

STa binds to a transmembrane guanylate cyclase C (GC-C) receptor at apical membrane of host intestinal epithelial cells, and activates guanylate cyclase pathway leading to an increase of intracellular cyclic GMP levels (Crane et al., 1992; Mezoff et al., 1992; Nair and Takeda, 1998). Similar to the consequence of increased cAMP levels by LT, an increase of cGMP levels leads to activation of CFTR via cGMP-dependent protein kinase II phosphorylation. Phosphorylation of protein kinase II promotes secretion of water and salts but inhibits absorption of sodium chloride in host intestinal epithelial cells, that results in net water secretion and diarrhoea (Almenoff et al., 1993; Cohen, 1992; Cohen et al., 1993; Sears and Kaper, 1996).

In addition to binding to GC-C and stimulating guanylate cyclase activity to increase intracellular cGMP levels in host small intestinal epithelial cells, STa may elevate intracellular cGMP levels via a second pathway. It is found that STa can trick

a rapid rise of inositol triphosphate, and activation of inositol triphosphate leads to rapid release of cytosolic calcium that promotes translocation of protein kinase C from cytosol to membrane and production of cGMP in cultural human colonic cells (Bhattacharya and Chakrabarti, 1998; Ganguly et al., 2001; Gupta et al., 2005). It is also reported that STa can cause deterioration of epithelial barrier integrity in human colonic cells (T-84 cells) (Nakashima et al., 2012). But whether STa also enhances ETEC colonization at host small intestines is unclear.

STb

STb (heat-stable toxin type II) is genetically and antigenically distinctive from STa, and is found mostly in ETEC strains isolated form young pigs with diarrhoea and oedema disease (Burgess et al., 1978; Handl and Flock, 1992; Moon et al., 1980; Moon et al., 1986; Nagy and Fekete, 1999; Zhang et al., 2007). STb is occasionally detected in ETEC strains isolated from human diarrhoeal patients (Lortie et al., 1991; Okamoto et al., 1993; Weikel et al., 1986), but is believed to be not directly associated with human diarrhoea (Chapman et al., 2006; Nataro and Kaper, 1998). Coded by estB gene, STb is synthesized initially as a peptide of 71 amino acids, but is proteolytically cleaved to a 48 amino acid peptide (5.1 kDa) during translocation across the inner membrane (Dreyfus et al., 1992; Fujii et al., 1991; Kupersztoch et al., 1990; Lee et al., 1983; Picken et al., 1983). This 48 amino acid peptide has four cysteine residues that form two disulfide bonds catalysed by DsbA inside the periplasm, and the mature STb peptide is translocated across the outer membrane through TolC protein (Foreman et al., 1995).

STb binds to receptor sulfatide at host small intestinal epithelial cells (Dubreuil, 2008; Goncalves et al., 2008; Rousset and Dubreuil, 1999; Rousset et al., 1998). Once internalized in host cells (Albert et al., 2011), unlike LT or STa, STb does not stimulate an increase of intracellular cAMP or cGMP concentrations (Hitotsubashi et al., 1992b), but rather stimulates a pertussis toxin-sensitive GTP-binding regulatory protein $G_{ai}3$ that leads to release of calcium (Dreyfus et al., 1993; Harville and Dreyfus, 1996). Release of calcium activates calmodulin-dependent protein kinase II to open an intestinal ion channel, may also protein kinase C, and then CFTR (Dreyfus et al., 1993; Fujii et al., 1997). The increased cytosolic calcium levels stimulate production of prostaglandin E_2 (PGE_2) and serotonin or 5-hydroxytryptamine (5-HT) that promotes water and electrolytes transported out of host intestinal cells (Harville and Dreyfus, 1995; Hitotsubashi et al., 1992a,b; Peterson and Whipp, 1995).

STb was found to shorten the villous length in pig jejunal loops and to cause morphological deterioration (Whipp et al., 1985, 1986, 1987). STb also induces intrinsic apoptosis in animal and human epithelial cells (Syed and Dubreuil, 2012). Unlike LT that facilitates ETEC colonization, STb does not enhance ETEC colonization at host small intestines. In contrast, 5–8-week old pigs challenged with ETEC stains expressing STb have fewest colonized bacteria detected, compared to those challenged with a ΔSTb knockout mutant strain, a ΔLTΔSTb knockout strain, or a LT⁺ ETEC strain (Erume et al., 2013).

EAST1

EAST1 coded by astA gene is a peptide of 38 amino acids (4.1 kDa). Interactions between EAST1 and host cells are less well studied. It is believed that the EAST1 peptide is translocated across the inner membrane via a Sec-dependent pathway. The peptide likely forms two disulfide bonds in the periplasm catalysed by DsbA enzyme. But mechanism involved in outer membrane translocation remains unknown (Turner et al., 2006b). EAST1 is reported shown enterotoxicity in the Ussing chambers mounted with rabbit intestinal ileal strips (Savarino et al., 1991) or stimulation of intracellular cGMP levels in rabbit ileal epithelial cells (Savarino et al., 1993). A recent study, however, demonstrates that EAST1 shows no enterotoxicity in human T-84 or porcine IPEC-1 and IPEC-J2 cell lines in Ussing chamber assays and does not stimulate an increase of intracellular cAMP or cGMP levels in these cells (Ruan et al., 2012).

EAST1 toxin gene is initially detected from enteroaggregative E. coli (EAEC) strain 17-2 isolated from diarrhoeal patients in Chile (Nataro et al., 1987), and following case–control studies suggest EAST1 is a putative agent of EAEC-associated diarrhoeal disease (Howell et al., 1987; Vial et al., 1988). Noticeably, EAST1 is commonly detected in ETEC strains isolated from children and travellers with diarrhoea (Bhan et al., 1989; Gascon et al., 1998, 2000; Itoh et al., 1997), and is highly prevalent among enterohaemorrhagic E. coli (EHEC) and enteropathogenic E. coli (EPEC) strains isolated from humans with diarrhoea (Paiva de Sousa and Dubreuil, 2001; Savarino et al., 1993, 1996; Veilleux et al., 2008; Yamamoto and Taneike, 2000). In addition, EAST1 is commonly detected in E. coli or ETEC strains isolated from animals with diarrhoea (Bertin et al., 1998; Choi et al., 2001; Frydendahl, 2002; Han et al., 2002; Ngeleka et al., 2003; Noamani et al., 2003; Osek, 2003; Veilleux and Dubreuil, 2006; Yamamoto et al., 1997). It is found that 35% of E. coli strains isolated from young pigs with post-weaning diarrhoea carry the EAST1 genes alone or combined with LT and/or ST (Zhang et al., 2007).

However, this EAST1 gene is also commonly detected in E. coli strains isolated from healthy or asymptomatic children or adults (Dougan et al., 1986; Fujihara et al., 2009; Gascon et al., 1998, 2000; Howell et al., 1987; Savarino et al., 1996; Vial et al., 1988; Yamamoto and Echeverria, 1996; Yamamoto and Nakazawa, 1997), and healthy animals as well (Ngeleka et al., 2003; Zajacova et al., 2012b). Furthermore, data from recent challenge studies using the pig model indicate that EAST1 alone is not sufficiently virulent to cause diarrhoea, and suggest EAST1 is not a virulence determinant in ETEC diarrhoea (Ruan et al., 2012; Zajacova et al., 2012a).

Other ETEC virulence factors

Whereas type I fimbria commonly produced by E. coli strains show no clear involvement in ETEC virulence (Levine et al., 1982), there are evidences indicated that non-fimbrial adhesins Tia, TibA, and EtpA promote ETEC adherence to small intestinal epithelial cells (Elsinghorst and Kopecko, 1992; Roy et al., 2009b). Tia, coded by tia gene as a 25 kDa outer membrane protein (Fleckenstein et al., 2000), interacts with heparin sulfate proteoglycans at host epithelial cell surface (Fleckenstein et al., 2002), and facilitates adherence and invasion of ETEC O157:H7 to host epithelial cells (Fleckenstein *et al.*, 1996; Mammarappallil and Elsinghorst, 2000).

TibA, coded by the *tibA* gene form *tibDBCA* operon, is a 100 kDa glycosylated outer membrane protein (Elsinghorst and Weitz, 1994; Lindenthal and Elsinghorst, 1999). TibA directs ETEC bacteria to bind and to invade into host epithelial cells (Lindenthal and Elsinghorst, 2001; Sherlock *et al.*, 2005). In addition, TibA is suggested to facilitate bacterial aggregation to form biofilm and to play a role in bacterial pathogenesis (Sherlock *et al.*, 2005).

EtpA is a recently identified outer membrane glycoprotein functioned as an adhesin (Fleckenstein *et al.*, 2006). EtpA is not involved in direct binding of ETEC to host epithelial cells, rather as a molecular bridge binding to receptors at host cells and to the tips of ETEC flagella (Roy *et al.*, 2008b, 2009b).

In addition to the nonfimbrial Tia, TibA and EtpA, two putative toxins, EatA and ClyA are thought to be involved in ETEC virulence (Patel *et al.*, 2004). EatA is highly homologous to the serine protease autotransporters, which are implicated as toxins and extracellular proteases (Henderson *et al.*, 2004), and is suggested as an ETEC virulence factor by damaging host epithelial cell surface (Patel *et al.*, 2004). ClyA (HlyE or SheA) is a 34 kDa pore-forming cytotoxin identified in *E. coli* K12 and *E. coli* clinical isolates including ETEC (Ludwig *et al.*, 1999). Expression of ClyA can be negatively but also positively regulated, and switch of the ClyA expression likely affects infectious outcomes (Ludwig *et al.*, 2004; Oscarsson *et al.*, 1996; Westermark *et al.*, 2000). However, the role of ClyA played precisely in ETEC diarrhoea has not been well defined.

Challenges in ETEC vaccine development

Despite the association between *E. coli* and diarrhoea was discovered over a century ago (reviewed by Zhang and Sack, 2012), the role of ETEC as a cause of severe diarrhoea in humans was described in 1970s (Gorbach *et al.*, 1971; Sack, 1980; Sack *et al.*, 1971), and despite ETEC are recognized as the most common bacteria that cause diarrhoea, effective prevention against ETEC diarrhoea has

yet to be achieved. ETEC-associated diarrhoea is a faecal–oral transmitted disease, thereby it should be preventable. But as country-wide safe drinking water is simply not available in the developing countries endemic for ETEC and medical facilities to treat diarrhoeal patients (oral rehydration) are largely not existing in vast rural areas, ETEC diarrhoea continues to be a major public health threat. Eventually, improved water supply and sanitation systems would provide the long term solution for preventing ETEC diarrhoea, but until this is achieved, vaccination is the most practical approach to reduce the public health burden from ETEC (Zhang and Sack, 2012).

Unfortunately, there is no effective vaccine currently available against ETEC diarrhoea. Major challenges in developing protective ETEC vaccines include that (1) different ETEC strains express immunologically heterogeneous adhesins and enterotoxins, therefore only vaccines inducing broadly protective anti-adhesin immunity and antitoxin immunity could effectively protect against ETEC strains; (2) enterotoxins produced by ETEC strains are the virulence determinants and thus must be included as antigens, but both LT and STa are potent toxic and cannot be used as safe antigens; (3) the 19-amino-acid peptide STa enterotoxin is poorly immunogenic, itself is unable to induce anti-STa immunity; (4) children in endemic areas are frequently infected and re-infected with ETEC strains in their early years, an effective vaccine has to induce long-lasting small intestinal mucosal immunity for prolonged protection against ETEC; (5) suitable animal models are required to effectively develop vaccines, but we lack an optimal animal challenge model with which to unambiguously assess ETEC vaccine candidacy; and finally (6) ETEC diarrhoea is endemic mostly to the developing countries, only practical and inexpensive vaccines will be suitable (Zhang and Sack, 2012).

Heterogeneity of ETEC strains

ETEC strains are identified and characterized by the enterotoxin(s) and adhesins, and also the O (lipopolysaccharide) and H (flagella) serotypes (Wolf, 1997). We now know that there are at least 23 adhesins and two toxins expressed by ETEC strains isolated from diarrhoea patients. In addition, more than 100 different O serogroups (Tobias and Svennerholm, 2012; Wolf, 1997) and 34 H flagellar antigens (Wolf, 1997) are typed from ETEC isolates. Moreover, mosaic associations between O and H serotypes among ETEC strains make it even more complicated. Due to extreme diversity in lipopolysaccharide and flagellar serogroups and complex O:H associations among individual strains, the O and H antigens have to be precluded in ETEC vaccine development (Qadri et al., 2005).

Although some ETEC adhesins fall into clusters based on genetic similarity, the 23 adhesins expressed by ETEC strains associated with human diarrhoea largely remain antigenically distinctive. In addition, early ETEC prevalence survey (Wolf, 1997) and recent systematic reviews of published data found that a significant proportion (20–34%) of ETEC isolates have no adhesins detected (Isidean et al., 2011; Qadri et al., 2005). That suggests that additional adhesins likely will be identified. Since ETEC adhesins are immunologically different, induced host anti-adhesin immunity will be antigen specific. Experimental anti-adhesin vaccines carrying one or two CFA adhesin antigens only protect against adherence of ETEC strains expressing homologous CFA adhesins, but not against ETEC strains with different CFA adhesins (Boedeker, 2005). To effectively protect against ETEC adherence and colonization, an effective vaccine must induce broad-spectrum anti-adhesin immunity against a majority (if not all) of the heterogeneous ETEC adhesins.

The two ETEC enterotoxins, LT and STa, are distinctive from each other as they share no homology genetically or antigenically. ETEC strains express LT, STa, or both (Wolf, 1997), and occasionally also express porcine-type STa and STb toxins (Chapman et al., 2006). Since these ETEC toxins are antigenically distinct, induced antitoxin immunity against an individual toxin will not provide cross protection. Indeed, early studies showed that induced anti-LT immunity protects only against ETEC strains expressing LT, but is not protective against STa-positive ETEC strains (Frantz and Robertson, 1981; Frantz and Mellencamp, 1983). In addition, in contrast to the LT from animal ETEC strains (pLT) that shows no heterogeneity (Zhang et al., 2009), LT expressed by human ETEC strains are polymorphic in structure and toxicity (Lasaro et al., 2008), and whether a given anti-LT immunity protects against other polymorphic LT remains to be verified.

Moreover, ETEC strains show geological variations in profiles of adhesins and toxins. It is found that ETEC phenotypes vary considerably in different regions and countries (Isidean et al., 2011). Therefore, an ETEC vaccine product shown protective in one country or region could be less or even not effective in other countries and regions. Differences of genetic background among human subjects and populations complicate host–pathogen interaction and add challenges in developing effective ETEC vaccines. Veterinary clinical studies have long noticed that pigs of different genetic background express different receptors specifically recognized by ETEC fimbrial adhesins (Rutter et al., 1975; Sellwood et al., 1975). Only pigs that express receptors recognized by adhesins of the challenge ETEC strains become colonized and develop diarrhoea, whereas other pigs that do not express the same receptors remain healthy even challenged with the same ETEC strain (Erickson et al., 1994; Francis et al., 1998). Moreover, experimental vaccine studies indicate that only pigs expressing specific receptors recognizing adhesins of a live attenuated vaccine candidate can be colonized and develop strong mucosal immunity (Ruan and Zhang, 2013; Santiago-Mateo et al., 2012). Humans likely have genetically determined receptors specific for different adhesins, which may explain why only 70% of immunologically naïve volunteers developed diarrhoea when challenged by a high dose of the most virulent ETEC strain H10407 (Harro et al., 2011a). On the other hand, individuals who do not express specific receptors may also respond less well to a given live attenuated vaccine candidate, and that confounds the evaluation of a correlation between immune response and clinical protection (Zhang and Sack, 2012).

Potent toxicity of LT and STa, and poor immunogenicity of STa

Potent toxicity of ETEC enterotoxins was initially demonstrated in 1956 by Dr S.N. De, who used rabbit ileal loops to study E. coli isolated from stools of adult patients with cholera-like symptoms but with no *V. cholerae* detected (De *et al.*, 1956). But this finding was hidden in literature for over a decade, until, in 1968, Dr R.B. Sack rediscovered that the live culture and culture filtrates of *E. coli* isolated from diarrhoea patients cause cholera-like fluid accumulation in ligated rabbit loops (Sack, 1968). Since then, LT and STa enterotoxicity was demonstrated in suckling mice (Hughes *et al.*, 1978), in ligated ileal loops of rabbits, mice and dogs, and skins of rabbits and guinea pigs (Evans *et al.*, 1973; Hitotsubashi *et al.*, 1992b; Sack *et al.*, 1976; Taylor *et al.*, 1961). It is shown that as little as 6 ng STa and 200 ng CT are sufficient to cause positive secretory responses in mouse ligated intestinal loops (Hitotsubashi *et al.*, 1992b). The potent toxicity of LT and STa is confirmed in human volunteers. Human adults (7 out 10) developed cholera-like illness when infected with ETEC strains isolated from two diarrhoeal American soldiers in Vietnam (DuPont *et al.*, 1971).

The potent toxicity precludes native LT and STa from being used as antigens in safe vaccine development. Furthermore, the 19-amino-acid STa is poorly immunogenic. Thus, STa itself does not induce anti-STa immunity and is not useful in vaccine development. But as over two thirds of ETEC strains isolated from diarrhoea patients express STa alone or together with LT (Wolf, 1997), anti-STa immunity must be also induced for effective protection against ETEC diarrhoea (Zhang and Sack, 2012).

Induction of long-last mucosal immunity and effective protection

Prolonged protection against ETEC diarrhoea is needed because young children living in developing countries are frequently infected with ETEC (Qadri *et al.*, 2000a; Rao *et al.*, 2003; Steinsland *et al.*, 2003b). On average, a child living in endemic areas experiences 3 to 6 or more diarrhoea infections per year (Black *et al.*, 2010). To provide prolonged protection against ETEC diarrhoea, a vaccine product needs to induce durable immune response in hosts or to be administrated repeatedly. The challenge is that even if vaccine products, especially oral vaccines that have induced great immune responses and

shown good protection when given to children at the developed countries, such as oral polio vaccine, rotavirus and CVD103Hgr cholera vaccines, fail to deliver similar results to children who live in the developing countries, likely due to immune hypo-responsiveness (Paul, 2009; Sack, 2008). With limited is known regarding causes of immune hypo-responsiveness in children from the developing countries to otherwise good vaccine candidates, although interference from maternal antibodies, alteration of intestinal microbiota development, nutrient deficiency and abnormality of intestinal mucosa and microvilli are among the possible factors (Hallander *et al.*, 2002; Levine, 2010; Sack, 2008; Steinsland *et al.*, 2003a), we will continue to have challenges in developing ETEC vaccines to induce strong immune responses among young children living in the developing countries.

Just as induction of potent immune responses to children in the poor countries remains a difficult task, elevation of host local mucosal immune responses becomes even more challenging. Evidences from experimental studies and clinical trials suggest the mucosal immunity mediated by secretory IgA antibodies (directed against adhesins and enterotoxins for ETEC) play a critical role in protecting against enteric pathogens including ETEC (Svennerholm and Holmgren, 1995). Oral vaccine products to deliver ETEC adhesin and toxin antigens directly at host small intestinal areas are preferred. However, oral vaccination will have to first overcome immune hyporesponsiveness (or environmental enteropathy) to be effective. Increases in delivery doses could help antigen uptake to boost host immune responses (Daley *et al.*, 2007). However, administration of a high oral dose is already shown to cause adverse effects, such as vomiting, to young children especially (Qadri *et al.*, 2006).

Moreover, antibody (IgG or IgA) response and clinical protection, especially against ETEC, are not always correlated. The protective immune response may be a complex relationship in which different components of the immune response are needed to protect against ETEC, while a single response may not be sufficient to provide protection (Rao *et al.*, 2005; Tobias *et al.*, 2008a). Evaluation of systemic and mucosal immunity continues to be important to determine optimal immunization doses and strategies, but no single measure of antibody response will be able to predict clinical protection (Zhang and Sack, 2012).

Animal challenge models and clinical trial design

Preclinical evaluation of ETEC vaccine candidates requires an animal model to assess safety and immunogenicity, and ideally also potential efficacy, of vaccine candidates prior to human volunteer studies and field trials for protective efficacy against ETEC infection. A lack of a suitable animal model to unambiguously evaluate vaccine candidates has severely hampered efforts in ETEC vaccine development (Boedeker, 2005; Zhang and Sack, 2012). An ideal animal model should use animals having a normal and regular intestine, with a functional and uncompromised immune system, and would develop diarrhoea similar to the human diarrhoea in intensity and duration after infection with ETEC as a similar dose (Spira et al., 1981). Since mice do not develop diarrhoea after inoculation of ETEC, this most commonly used murine model is not helpful for vaccine development against ETEC diarrhoea. Of other animal models, it appears that the rabbit RITARD (removable intestinal tie-adult rabbit diarrhoea model) and the pig models are best suited. However, rabbits do not always react to STa toxin, and pigs do not express receptors recognized by adhesins expressed by most ETEC strains associated with human diarrhoea, unless combined, neither satisfies all the needs for ETEC vaccine development. The non-human primate model, the *Aotus* monkey, could also be used to study ETEC vaccine candidacy. However, the cost and more importantly the lack of characterization for ETEC pathogenesis in monkeys may prevent this non-human primate model from being practically used in ETEC vaccine development.

Carefully conducted adult volunteer studies are needed to provide valuable evidence for safety and preparation for field efficacy trials. The vaccine candidate must be evaluated in age-descending dose-finding studies prior to a field trial(s). Importantly, the vaccine candidate must be evaluated in large-scale and controlled field trials. Given that the most important outcome is

the prevention of moderate and severe diarrhoea in infants and children, field trials will need to include a sufficiently large sample size so that this outcome can be determined using methods similar to those used for testing rotavirus vaccines (Madhi et al., 2010; Zaman et al., 2010).

Brief history in ETEC vaccine development

ETEC diarrhoea occurs most commonly in infants and children under 3 years living in the endemic areas, but rates tend to decline with age (Merson et al., 1980; Qadri et al., 2000a). This age-associated trend does not occur to people who travel from non-endemic countries to the endemic areas unless for a prolonged stay (Levine, 2008; Qadri et al., 2005; Wenneras and Erling, 2004). This suggests that people develop immunity against ETEC after natural exposure and that acquired host immunity provides protection against subsequent ETEC infections. Moreover, trials of a whole cell/B subunit cholera vaccine (WC/BS) indicate that this cholera vaccine induced short-term protection against LT-producing ETEC strains, apparently due to the cross protection of LT and CT (Clemens et al., 1988; Peltola et al., 1991; Scerpella et al., 1995). These clinical observations and cholera vaccine studies provide evidences that vaccines protecting against ETEC should be possible.

Whole-cell ETEC vaccine candidates

Killed whole-cell vaccine candidates

The whole-cell/recombinant CT_B subunit (WC/rBS) cholera vaccine (Dukoral®) was attempted as a candidate against ETEC. Field trials showed this cholera vaccine protects against *Vibrio cholerae* but also ETEC, due to cross protection from the induced anti-CT_B immunity against LT produced by ETEC strains (Clemens et al., 1988). Following volunteer studies indicated that 52% of immunized adults travelling from Finland to Morocco and 50% of US young adults staying in Mexico for a summer were protected against ETEC diarrhoea (Peltola et al., 1991; Scerpella et al., 1995). Although Dukoral is licensed for prevention of travellers' diarrhoea in some countries, it is as a cholera vaccine, not as a ETEC vaccine due to its short lived and also lower than satisfactory protection against ETEC diarrhoea. However, protection observed from WC/rBS cholera vaccine against ETEC diarrhoea has encouraged efforts for ETEC vaccine development.

The first killed oral whole-cell ETEC vaccine candidate is ETEC prototype strain H10407 (O78:H11, LT$^+$STa$^+$CFA/I$^+$) inactivated with colicin E2 (Evans et al., 1988b). Adults inoculated with killed H10407 cells (2 doses of 3×10^{10} CFUs) developed anti-CFA/I and anti-LT IgA antibodies and were protected against homologous challenge (Evans et al., 1988a,b). However, immunity induced by the killed H10407 is not expected to protect against ETEC strains expressing immunologically heterologous adhesins and STa toxin. Nevertheless, these studies demonstrated that killed oral whole-cell ETEC bacteria can induce immunity protecting against ETEC strains expressing homologous adhesins and LT toxin, and leaded to concept of killed or live attenuated oral ETEC vaccines. It was concluded that such an ETEC vaccine should contain the B subunit of CT (or LT) along with several ETEC strains expressing the most prevalent adhesins (Svennerholm et al., 1989).

The SBL (rCTB-CF) ETEC vaccine candidate, which carries five formalin-killed E. coli strains expressing CFA/I, CS1, CS2, CS3, CS4 and CS5 adhesins and STa toxin, plus recombinant CT$_B$ subunit, shown first time the feasibility in developing broad protective vaccines against heterogeneous ETEC strains. Even though at least 23 adhesins are found among ETEC strains, CFA/I, CFA/II (CS1, CS2, CS3) and CFA/IV (CS4, CS5, CS6) are expressed by ETEC strains associated with about 80% diarrhoea cases (Qadri et al., 2005) and the moderate to severe diarrhoea cases (Sack et al., 2007). Therefore, these adhesins become primarily targeted in ETEC vaccine development. This SBL product was found safe and immunogenic in Swedish adults (Ahren et al., 1993, 1998; Jertborn et al., 1998; Wenneras et al., 1992), and also in adult volunteers from endemic areas such as Egypt and Israeli (Cohen et al., 2000; Savarino et al., 1998). Additionally, Egyptian and Bangladeshi children, aged 2–12 years, tolerated

SBL oral administration well, and over 90% of the immunized children developed detectable antibody response to CTB, CFA/I, CS2 and CS4 (Qadri et al., 2000b, 2003; Savarino et al., 1999). However, Egyptian children less than 2 years did not respond well nor did they tolerate the same product as well, and experienced no significant protection against ETEC diarrhoea (Savarino, 2003; Savarino et al., 2002; Svennerholm, 2004). Protective efficacy of SBL against ETEC diarrhoea was also assessed among US adults travelling to Mexico and Guatemala (Sack et al., 2007). This study concluded that the killed oral SBL vaccine candidate protected against moderate to severe travellers' diarrhoea caused by the homologous ETEC strains, but that it did not reduce the overall rates of travellers' diarrhoea (Sack et al., 2007).

Live attenuated ETEC vaccine candidates

The first live attenuated ETEC vaccine candidate is a spontaneous toxin gene-deletion mutant derived from ETEC strain E1392 (O6:H16, LT$^+$STa$^+$CS1/CS3) (Levine, 1990; Sack et al., 1988). With losses of the LT and STa toxin genes, the mutant strain became less virulent (Turner et al., 2001). Rabbits orally immunized with this live strain were protected against homologous ETEC strains (Sack et al., 1988). When human volunteers were given this strain at a dose of 2×10^{10} CFUs, 75% were protected against challenge of an LT$^+$/STa$^+$ ETEC strain expressing the same CFAs. However, 15% of the volunteers developed diarrhoea following receipt of the vaccine strain, suggesting that this product needed to be further attenuated (Tacket, 1997).

Further deletion of the aro gene, a heat shock protein gene (HtrA), or one or two omp genes coding outer membrane proteins in E1392/75-2A resulted in safer 'PTL strains' (Turner et al., 2001). Safety and immunogenicity of the PTL strains were verified in mice and human volunteers (McKenzie et al., 2006; Turner et al., 2001). Unfortunately, a following double-blind placebo-controlled study indicated that PTL was unable to protect against challenge with ETEC strain E24377A (O139:H28, LT$^+$STa$^+$CS1/CS3), an ETEC strain has the same CFA antigens but different O serotype (McKenzie et al., 2008).

Live attenuated *Salmonella* and *Shigella* strains have been used to express ETEC adhesin and/or toxin antigens as ETEC vaccine candidates. Live *Salmonella* vaccine strains expressing CFA/I or LT_B (or with LT toxoid LT_{R192G}, as an adjuvant) elicited anti-CFA/I and anti-LT antibodies (Guillobel *et al.*, 2000; Khan *et al.*, 2007; Lasaro *et al.*, 2004; Yang *et al.*, 2011). Unfortunately, all these products have only one or two ETEC antigens expressed in live *Salmonella* vaccine strains and are unlikely to be broadly effective against the heterogeneous ETEC strains (Zhang and Sack, 2012). Live attenuated *Shigella flexneri*, *S. sonnei*, and *S. dysenteriae* were also used to express ETEC CFA adhesin or LT toxoid antigens for protecting against both *Shigella* and ETEC diarrhoea (Altboum *et al.*, 2001, 2003; Barry *et al.*, 2003, 2006; Koprowski *et al.*, 2000). Guinea pigs tolerated these products and developed immune response to CFA or LT antigens. Particularly interested are the products containing two, three or five strains expressing multiple ETEC CFA adhesins and LT toxoid antigens that stimulate antibodies against each included CFA adhesin, and LT as well (Barry *et al.*, 2003, 2006). But published data from clinical efficacy studies with these vaccine candidates so far are still lacking.

In addition, live *Vibrio cholerae* vaccine strains including Peru-15 were used to express ETEC antigens for protection against ETEC diarrhoea (Walker *et al.*, 2007). Recent studies demonstrate that CT_B or CFA/I antigen can be expressed at high levels in *Vibrio cholerae* strains to elicit anti-CFA/I IgG and IgA and neutralizing anti-CTB antibodies (Roland *et al.*, 2007; Tobias *et al.*, 2008b). Additional studies including human volunteer studies and clinical trials will be needed to assess their eventual usefulness as ETEC vaccines (Zhang and Sack, 2012).

ETEC subunit vaccine candidates

Antitoxin subunit vaccine candidates
Antitoxin vaccine development has been largely focused on anti-LT immunity, as the small STa molecule naturally does not stimulate anti-STa immunity. The non-toxic LT_B (or the homologous CT_B) subunit, in forms of recombinant proteins (Clemens *et al.*, 1988; Svennerholm, 1991), plant expressed editable proteins (Haq *et al.*, 1995; Karaman *et al.*, 2012; Kim *et al.*, 2010; Mason *et al.*, 2002; Rosales-Mendoza *et al.*, 2008, 2009; Tacket *et al.*, 2004) and protein conjugates (Szu *et al.*, 1989, 1994), was tried to induce anti-LT immunity against ETEC. In addition, native LT was expressed in potatoes as edible vaccine candidate (Mason *et al.*, 1998) or directly applied on a patch for transcutaneous immunization (TCI) (Glenn *et al.*, 1998, 2000, 2007; Guerena-Burgueno *et al.*, 2002; McKenzie *et al.*, 2007; Scharton-Kersten *et al.*, 1999). Among these products, the LT patch was shown promising initially, as early studies indicated the LT patch stimulated strong host immune responses to LT (90–100% in serum and mucosal IgA and IgG) (Guerena-Burgueno *et al.*, 2002; McKenzie *et al.*, 2007). Moreover, a followed small field trial showed immunized adults travelling to Mexico or Guatemala had 75% protection against moderated-to-severe diarrhoea from any cause, as well as moderate-to-severe ETEC diarrhoea (Frech *et al.*, 2008). However, a subsequent field study at a larger scale indicated that LT patch had about 60% protection against LT^+ ETEC diarrhoea (Ellingsworth, 2011). Like all other LT or LT_B derived products, LT patch had no protection against STa^+ ETEC or any other cause diarrhoea (Ellingsworth, 2011).

Anti-adhesin subunit vaccine candidates:
Although an early study indicated purified adhesins induced anti-adhesin antibodies that protect against ETEC (Tacket *et al.*, 1988), the immunological heterogeneity among ETEC adhesins and the high cost of adhesin preparation basically make anti-adhesin subunit vaccines unpractical (Svennerholm and Tobias, 2008; Walker *et al.*, 2007). In addition, ETEC adhesins are sensitive to proteolytic degradation, although application of biodegradable poly ($_{DL}$-lactide-co-glycolide, PLG) microspheres to capsulate purified ETEC adhesins can decrease degradation by stomach acids (Edelman *et al.*, 1993; Reid *et al.*, 1993). PLG microencapsulation was suitable and proven safe, and were supposed to improve delivery and uptake at mucosal inductive sites, but immune

responses induced by encapsulated antigen were generally very low (Katz *et al.*, 2003; Tacket *et al.*, 1994). Co-administration with LT_{R192G} (mLT) improved stimulation of immune responses specific to the encapsulated adhesin (CS3 or CS6) in mice, but the mice became noticeably distressed after being co-immunized with mLT (Byrd and Cassels, 2006a,b), and the increase of immune responses was at unproportionally low magnitudes (Lapa *et al.*, 2008). Purified CS6 adhesin was once included in the LT patch and induced moderate antibody responses in adult volunteers (Guerena-Burgueno *et al.*, 2002). Unfortunately, vaccine candidates including one or two adhesin antigens are unable to stimulate broad immunity against ETEC diarrhoea.

Whereas major structure subunits of ETEC adhesins are immunologically heterogeneous, the minor subunits, the adhesin fimbrial tips, are found to have greater genetic conservation (Anantha *et al.*, 2004), thus become a favour target for developing broadly protective anti-adhesin vaccines (Savarino, 2005). The CFA/I tip, CfaE, was demonstrated to induce immunity protecting against CFA/I ETEC in non-human primate challenge studies (Savarino, 2005). When purified CfaE was given transcutaneously, together with LT, LT_{R192G} or $LT_{R192G/A211A}$ (dmLT), mice and rabbits developed strong anti-CFA/I antibodies, and the elicited antibodies inhibited agglutination of CFA/I, CS2, CS4, CS14 and CS17 ETEC strains (Savarino, 2011).

Non-fimbrial adhesin EtpA is reported to induce immunity against ETEC adherence. It is noted that mice repeatedly exposed to ETEC developed anti-EtpA antibodies and had significant reduction in colonization by subsequent challenged ETEC strain H10407 (Roy *et al.*, 2008a). Moreover, mice immunized with recombinant EtpA glycoprotein developed anti-EtpA IgG and IgA antibodies in serum and faeces, and the elicited antibodies shown adherence inhibition against challenge with ETEC expressing CFA/I or CS1 and CS3 (Roy *et al.*, 2009a). However, whether anti-EtpA immunity could provide truly broad protection, especially against ETEC strains of different serotypes need to be further characterized.

Recent progress in ETEC vaccine development

Significant progress has been made in recent years towards development of a broadly protective ETEC vaccine. That include: (1) introduction of a pig study model and refinement of human challenge model, that result in better assessment of immunogenicity of target antigens and protective efficacy of induced immunity; (2) improvement on current vaccine candidates by expressing additional antigens to induce broader immunity, or by expressing antigens at a higher level to enhance specific mucosal immunity and to have fewer bacterial organisms included in an administration dose so that potential adverse effects can be reduced or eliminated; (3) construction of LT and STa toxoid fusions that serve as safe antigens and induce neutralizing anti-STa and anti-LT antibodies; and (4) application of dmLT as a mucosal adjuvant to enhance anti-adhesin immunogenicity and also to stimulate anti-LT immunity.

Progress in ETEC challenge models

A suitable animal study model is required to assess ETEC vaccine candidacy (Boedeker, 2005). The rabbit model, especially the RITARD model developed in 1980s to study cholera and ETEC pathogenesis and anti-adhesin immunity protection against diarrhoea (Cray *et al.*, 1983; Spira and Sack, 1982; Spira *et al.*, 1981), is found effective to assess induced anti-adhesin immunity against adherence from homologous adhesins and anti-LT immunity against LT enterotoxicity (Ahren and Svennerholm, 1985; Sack *et al.*, 1988; Svennerholm *et al.*, 1990). However, it was noticed that although rabbits consistently react to CT or LT toxin, they do not always react to STa toxin (Sack *et al.*, 1988), thus become less effective in examining protection against STa-producing ETEC strains.

Pigs, especially young pigs, always react to STa and LT toxins, and are naturally susceptible to ETEC diarrhoea. Moreover, pigs develop clinically identical disease as human diarrhoeal patients when inoculated with ETEC of a similar dose (Berberov *et al.*, 2004; Smith and Linggood, 1971; Zhang *et al.*, 2006, 2008). Additionally, pigs and humans are similar in physiology, organ development and more importantly, immune systems. Indeed, the system of piglets expressing the K88 receptors and porcine ETEC strains expressing K88 fimbrial adhesins has served as a model to characterize ETEC diarrhoea since the 1970s (Smith and Linggood, 1971), and the biological relevance between diarrhoea in K88ac receptor-positive pigs and the inoculation of K88ac fimbrial ETEC strains has been well characterized (Erickson *et al.*, 1992, 1994; Francis *et al.*, 1998). This established biological correlation eliminates any ambiguity caused by natural resistance to ETEC infection that likely occurs to other experimental animals and human volunteers. Moreover, this pig model allows us to evaluate local mucosal immunity for protection against ETEC diarrhoea (Zhang and Francis, 2010; Zhang *et al.*, 2010). In addition, LT and STa toxins expressed in porcine and human ETEC strains are highly homologous, and pigs develop diarrhoea after inoculation of ETEC strains expressing human-type LT or STa toxin (Zhang *et al.*, 2008). Together, that makes the pig model perhaps the best animal model to assess protective efficacy of antitoxin vaccine candidates. Unfortunately, since adhesins expressed by porcine-type ETEC strains are different than human-type ETEC strains (except adhesins of the 987P cluster), receptors at human and pig small intestinal epithelial cells are apparently host-specific. Receptors expressed by pigs typically do not recognize CFA adhesins expressed by human ETEC strains. Therefore, pigs are unlikely colonized by human-type ETEC strains. That makes the pig model not useful in assessing protective efficacy from an anti-adhesin vaccine candidate. However, a dual-model application, with the RITARD model to examine anti-adhesin immunity for adherence inhibition and the pig model to evaluate antitoxin immunity for enterotoxicity neutralization, can unambiguously assess protective efficacy of any ETEC vaccine candidates (Zhang and Sack, 2012).

Recently refinement in human challenge model leads to more accurate assessment of ETEC vaccine candidacy. Human volunteer challenge has long been used in clinic trials of vaccines against enteric diseases, and is the ultimate model to evaluate ETEC vaccine candidates (Walker *et al.*, 2007). Though volunteer studies have been extremely helpful, we have learned that (1) different ETEC strains have variable virulence, and the most commonly used challenge strain H10407 is the most virulent challenge strain (Porter *et al.*, 2011), and (2) a too high challenge dose may mislead to unsatisfied protection from an otherwise very promising vaccine candidate, due to that a high inoculum can overwhelm host intestinal immunity. We should keep in mind that an inoculum experienced by subjects from natural exposure in a field trial is likely to be much less and with a much lower attack rate. Indeed, when a reduced challenge dose of strain H10407 was used in a recent study, 'false negative' result was eliminated (Harro *et al.*, 2011a).

Improved rCTB-CF killed vaccine

Due to a lack of significant protection in field trials (Sack *et al.*, 2007; Savarino *et al.*, 2002; Svennerholm, 2004), the SBL ETEC vaccine product was modified with over expression of CFA antigens and replacement of STa+ strains with non-toxic *E. coli* strains, as well as of recombinant CTB with a LT-like LTB/CTB hybrid (LCTBA) protein. Two products were generated subsequently: one consists of a formalin-killed *E. coli* recombinant strain overexpressing CFA/I plus 1 mg of recombinant LCTBA (OEV-120), and the other includes recombinant non-toxigenic *E. coli* strains expressing much higher yield of CFA/I, CS3, CS5 and CS6 (to replace original STa+ strains), and the recombinant LT-like LTB/CTB hybrid (LCTBA) protein to substitute the CT_B recombinant protein. Both products also include dmLT as an adjuvant (and an additional antigen) that may further enhance the immune response to CFA and toxin (Holmgren, 2011; Norton *et al.*, 2011). In addition, high expression of CFA adhesin antigens results in not only fewer *E. coli* cells being included in a delivery dose that reduces adverse effects, especially vomiting among infants, and also an increase of anti-adhesin mucosal immune responses. Replacing rCTB with this LT-like LCTBA protein is to induce antitoxin immunity more specifically against LT produced by ETEC strains.

Recent animal studies show that these modified killed ETEC vaccines, with and without dmLT, are well tolerated even in large amounts, and the immunized mice develop strong serum and mucosal anti-CFA and anti-LT immune responses (Holmgren, 2011). It is noted that the induced immune responses far exceed levels induced by the original SBL products. Moreover, human volunteers showed no significant adverse effect when orally given high doses of OEV-120 (2.5×10^{10} CFUs, 1×10^{11} CFUs), but developed much greater anti-CFA/I and anti-CT IgA immune response (Lundgren, 2011). Compared to the original SBL product, the modified products are not only safer but also induce much high levels of immune responses to ETEC adhesins and LT. Undergoing safety and immunogenicity studies and potential field trials will provide additional evidence to provide better assessment of these 'generation two' killed whole-cell ETEC vaccine candidates.

Live attenuated ACE527 vaccine

Currently the most promising live attenuated ETEC vaccine candidate is ACE527. ACE527 is derived from the live attenuated PTL products by including additional CFA adhesin antigens and toxin antigens to stimulate broad based protection against heterogeneous ETEC strains. ACE527 consists of three live strains, ACAM 2025 (CFA/I, LTB) (Turner et al., 2006a), ACAM 2022 (CS5/CS6, LTB) (Turner et al., 2011), and ACAM 2027 (CS1/CS2/CS3, LTB) (Turner et al., 2011), and is expected to induce immunity against CFA/I, CFA/II, CFA/VI adhesins and LT (Harro et al., 2011b; Turner et al., 2011). Safety and immunogenicity studies show that ACE527 is safe and stimulate vigorous immune responses against LTB and CFAs (CFA/I, CS3, CS6) in adult volunteers (Darsley et al., 2012; Harro C, 2011a). A recent double blind, randomized, placebo-controlled study demonstrate that ACE527 has a significantly impact on intestinal colonization by the challenge strain H10407 (LT/STa, CFA/I, O78:H11), and show reduced severity of ETEC diarrhoea (Darsley et al., 2012; Harro C, 2011c). These data indicate that ACE527 has a dual mode of action by targeting both colonization factors and the LT enterotoxicity, and suggest ACE527 should be evaluated in more advanced trials for protective efficacy against ETEC diarrhoea (Darsley et al., 2012).

Other progress in ETEC vaccine development

In addition to introduction of a pig study model and refinement of human challenge model, improvement in killed whole-cell rCTB-CF and live attenuated ACE527 products, other noticeable progress made recently in ETEC vaccine development includes application of dmLT adjuvants and induction of protective anti-STa immunity. This dmLT ($LT_{R192g/L211A}$) shows reduced activity in stimulating cAMP level in Caco-2 cells and no toxicity in a patent mouse assay even when 250 μg dmLT is used, but maintains LT antigenicity (Norton et al., 2011). Thus dmLT can serve as a safer antigen for ETEC vaccine development. More importantly, dmLT maintains the mucosal adjuvanticity of native LT. Data from a phase I human volunteer study indicate dmLT stimulates host anti-LT immunity to protect against LT-producing ETEC but also exhibits ability to act as a mucosal adjuvant to enhance protective efficacy of the existing oral vaccine candidates (Holmgren, 2011; Sjokvist Ottsjo et al., 2013).

While protective anti-LT immunity can be induced by LT_B or dmLT, induction of protective anti-STa immunity becomes the top priority in antitoxin vaccine development. Since anti-LT immunity cannot cross protect against STa (Frantz and Robertson, 1981; Frantz and Mellencamp, 1983), and over two thirds of ETEC strains associated with diarrhoea express STa (alone or together with LT) (Isidean et al., 2011; Wolf, 1997), only vaccines also inducing anti-STa immunity can provide effective protection against ETEC. However, to be included as a vaccine antigen, STa must have its potent toxicity remarkably reduced and immunogenicity enhanced (Taxt et al., 2010; Zhang et al., 2010). STa was shown to stimulate anti-STa immunity when was coupled with a strongly immunogenic carrier protein and presented as a fusion or a conjugate antigen (Batisson and Der Vartanian, 2000; Batisson et al., 2000; Clements, 1990; Frantz and Robertson, 1981; Sanchez et al., 1986, 1988a,b; Svennerholm, 1992; Zeng et al., 2012). But native STa maintains its toxicity, thus these compounds are not suitable for safe vaccine development. Earlier studies indicated that STa can have toxicity eliminated or reduced with its disulfide bonds disrupted or shorter peptide with single amino acids mutated (Svennerholm, 1988; Yamasaki, 1988). Disruption of the disulfide bonds, however, results in significant change of STa antigenic topology; thus even if these STa derivatives became immunogenic (when carried by a carrier protein), the induced immunity would not be able to neutralize native STa toxin. Recent studies demonstrate a single substitution at some amino acid residues of a full-length STa results in elimination or significantly reduction of its toxicity but retention of much of its native structure. These STa toxoids become immunogenic when are fused to a LT toxoid, and resultant toxoid fusion antigens elicit antibodies neutralizing both native STa and LT toxins (Liu et al., 2011a,b; Zhang and Francis, 2010; Zhang et al., 2010). With the LT/CT hybrid replaced by the LT-STa toxoid fusion, or with additional expression of the LT-STa toxoid fusion, the modified killed whole-cell rCTB-CF or the ACE527 are expected to induce anti-adhesin immunity to block adherence of ETEC strains expressing CFA/I, CS1-CS6 and also antitoxin immunity to neutralize both LT and STa toxins; thus broadly protective vaccines against ETEC will finally be developed.

Conclusion remarks

Diarrhoea is the second leading cause (only behind pneumonia) of deaths to children under 5 years, and ETEC strains continue to be the leading bacteria causing diarrhoea to young children living in developing countries and also children and adults travelling to these areas. Development of broadly protective vaccines against ETEC diarrhoea should remain as a top priority for global and regional health programmes. OWING to the nature of the disease, challenges continue to exist. But with significant progress made recently, a broadly protective ETEC vaccine protecting against the most prevalent ETEC strains or strains associated with moderate to severe diarrhoea should be developed in a very near future.

References

Ahren, C.M., and Svennerholm, A.M. (1985). Experimental enterotoxin-induced Escherichia coli diarrhoea and protection induced by previous infection with bacteria of the same adhesin or enterotoxin type. Infect. Immun. 50, 255–261.

Ahren, C., Wenneras, C., Holmgren, J., and Svennerholm, A.M. (1993). Intestinal antibody response after oral immunization with a prototype cholera B subunit-colonization factor antigen enterotoxigenic Escherichia coli vaccine. Vaccine 11, 929–934.

Ahren, C., Jertborn, M., and Svennerholm, A.M. (1998). Intestinal immune responses to an inactivated oral enterotoxigenic Escherichia coli vaccine and associated immunoglobulin A responses in blood. Infect. Immun. 66, 3311–3316.

Albert, M.A., Kojic, L.D., Nabi, I.R., and Dubreuil, J.D. (2011). Cell type-dependent internalization of the Escherichia coli STb enterotoxin. FEMS Immunol. Med. Microbiol. 61, 205–217.

Almenoff, J.S., Williams, S.I., Scheving, L.A., Judd, A.K., and Schoolnik, G.K. (1993). Ligand-based histochemical localization and capture of cells expressing heat-stable enterotoxin receptors. Mol. Microbiol. 8, 865–873.

Altboum, Z., Barry, E.M., Losonsky, G., Galen, J.E., and Levine, M.M. (2001). Attenuated Shigella flexneri 2a Delta guaBA strain CVD 1204 expressing enterotoxigenic Escherichia coli (ETEC) CS2 and CS3 fimbriae as a live mucosal vaccine against Shigella and ETEC infection. Infect. Immun. 69, 3150–3158.

Altboum, Z., Levine, M.M., Galen, J.E., and Barry, E.M. (2003). Genetic characterization and immunogenicity of coli surface antigen 4 from enterotoxigenic Escherichia coli when it is expressed in a Shigella live-vector strain. Infect. Immun. 71, 1352–1360.

Anantha, R.P., McVeigh, A.L., Lee, L.H., Agnew, M.K., Cassels, F.J., Scott, D.A., Whittam, T.S., and Savarino, S.J. (2004). Evolutionary and functional relationships of colonization factor antigen i and other class 5 adhesive fimbriae of enterotoxigenic Escherichia coli. Infect. Immun. 72, 7190–7201.

Aubel, D., Darfeuille-Michaud, A., and Joly, B. (1991). New adhesive factor (antigen 8786) on a human enterotoxigenic Escherichia coli O117:H4 strain isolated in Africa. Infect. Immun. 59, 1290–1299.

Barry, E.M., Altboum, Z., Losonsky, G., and Levine, M.M. (2003). Immune responses elicited against multiple enterotoxigenic Escherichia coli fimbriae and mutant LT expressed in attenuated Shigella vaccine strains. Vaccine 21, 333–340.

Barry, E.M., Wang, J., Wu, T., Davis, T., and Levine, M.M. (2006). Immunogenicity of multivalent Shigella-ETEC candidate vaccine strains in a guinea pig model. Vaccine 24, 3727–3734.

Batisson, I., and Der Vartanian, M. (2000). Contribution of defined amino acid residues to the immunogenicity of recombinant Escherichia coli heat-stable enterotoxin fusion proteins. FEMS Microbiol. Lett. 192, 223–229.

Batisson, I., Der Vartanian, M., Gaillard-Martinie, B., and Contrepois, M. (2000). Full capacity of recombinant Escherichia coli heat-stable enterotoxin fusion proteins for extracellular secretion, antigenicity, disulfide bond formation, and activity. Infect. Immun. 68, 4064–4074.

Berberov, E.M., Zhou, Y., Francis, D.H., Scott, M.A., Kachman, S.D., and Moxley, R.A. (2004). Relative importance of heat-labile enterotoxin in the causation of severe diarrhoeal disease in the gnotobiotic piglet model by a strain of enterotoxigenic *Escherichia coli* that produces multiple enterotoxins. Infect. Immun. *72*, 3914–3924.

Bertin, Y., Martin, C., Girardeau, J.P., Pohl, P., and Contrepois, M. (1998). Association of genes encoding P fimbriae, CS31A antigen and EAST 1 toxin among CNF1-producing *Escherichia coli* strains from cattle with septicemia and diarrhoea. FEMS Microbiol. Lett. *162*, 235–239.

Bhan, M.K., Raj, P., Levine, M.M., Kaper, J.B., Bhandari, N., Srivastava, R., Kumar, R., and Sazawal, S. (1989). Enteroaggregative *Escherichia coli* associated with persistent diarrhoea in a cohort of rural children in India. J. Infect. Dis. *159*, 1061–1064.

Bhattacharya, J., and Chakrabarti, M.K. (1998). Rise of intracellular free calcium levels with activation of inositol triphosphate in a human colonic carcinoma cell line (COLO 205) by heat-stable enterotoxin of *Escherichia coli*. Biochim. Biophys. Acta *1403*, 1–4.

Black, R.E., Cousens, S., Johnson, H.L., Lawn, J.E., Rudan, I., Bassani, D.G., Jha, P., Campbell, H., Walker, C.F., Cibulskis, R., *et al.* (2010). Global, regional, and national causes of child mortality in 2008: a systematic analysis. Lancet *375*, 1969–1987.

Boedeker, E.C. (2005). Vaccines for enterotoxigenic *Escherichia coli*: current status. Curr. Opin. Gastroenterol. *21*, 15–19.

Burgess, M.N., Bywater, R.J., Cowley, C.M., Mullan, N.A., and Newsome, P.M. (1978). Biological evaluation of a methanol-soluble, heat-stable *Escherichia coli* enterotoxin in infant mice, pigs, rabbits, and calves. Infect. Immun. *21*, 526–531.

Byrd, W., and Cassels, F.J. (2006a). The encapsulation of enterotoxigenic *Escherichia coli* colonization factor CS3 in biodegradable microspheres enhances the murine antibody response following intranasal administration. Microbiology *152*, 779–786.

Byrd, W., and Cassels, F.J. (2006b). Intranasal immunization of BALB/c mice with enterotoxigenic *Escherichia coli* colonization factor CS6 encapsulated in biodegradable poly(DL-lactide-co-glycolide) microspheres. Vaccine *24*, 1359–1366.

Casey, T.A., Connell, T.D., Holmes, R.K., and Whipp, S.C. (2012). Evaluation of heat-labile enterotoxins type IIa and type IIb in the pathogenicity of enterotoxigenic *Escherichia coli* for neonatal pigs. Vet. Microbiol. *159*, 83–89.

Celemin, C., Anguita, J., Naharro, G., and Suarez, S. (1994). Evidence that *Escherichia coli* isolated from the intestine of healthy pigs hybridize with LT-II, ST-Ib and SLT-II DNA probes. Microb. Pathog. *16*, 77–81.

Celemin, C., Rubio, P., Echeverria, P., and Suarez, S. (1995). Gene toxin patterns of *Escherichia coli* isolated from diseased and healthy piglets. Vet. Microbiol. *45*, 121–127.

Cerqueira, A.M.F., Tibana, A., Gomes, T.A.T., and Guth, B.E.C. (1994). Search for Lt-Ii and Stb DNA-Sequences among *Escherichia-Coli* Isolated from Bovine Meat-Products by Colony Hybridization. J. Food Prot. *57*, 734–736.

Chapman, T.A., Wu, X.Y., Barchia, I., Bettelheim, K.A., Driesen, S., Trott, D., Wilson, M., and Chin, J.J. (2006). Comparison of virulence gene profiles of *Escherichia coli* strains isolated from healthy and diarrheic swine. Appl. Environ. Microbiol. *72*, 4782–4795.

Chattopadhyay, S., Tchesnokova, V., McVeigh, A., Kisiela, D.I., Dori, K., Navarro, A., Sokurenko, E.V., and Savarino, S.J. (2012). Adaptive evolution of class 5 fimbrial genes in enterotoxigenic *Escherichia coli* and its functional consequences. J. Biol. Chem. *287*, 6150–6158.

Choi, C., Cho, W., Chung, H., Jung, T., Kim, J., and Chae, C. (2001). Prevalence of the enteroaggregative *Escherichia coli* heat-stable enterotoxin 1 (EAST1) gene in isolates in weaned pigs with diarrhoea and/or edema disease. Vet. Microbiol. *81*, 65–71.

Chutkan, H., and Kuehn, M.J. (2011). Context-dependent activation kinetics elicited by soluble versus outer membrane vesicle-associated heat-labile enterotoxin. Infect. Immun. *79*, 3760–3769.

Cieplak, W., Jr., Mead, D.J., Messer, R.J., and Grant, C.C. (1995). Site-directed mutagenic alteration of potential active-site residues of the A subunit of *Escherichia coli* heat-labile enterotoxin. Evidence for a catalytic role for glutamic acid 112. J. Biol. Chem. *270*, 30545–30550.

Clark, C.A., Heuzenroeder, M.W., and Manning, P.A. (1992). Colonization factor antigen CFA/IV (PCF8775) of human enterotoxigenic *Escherichia coli*: nucleotide sequence of the CS5 determinant. Infect. Immun. *60*, 1254–1257.

Clemens, J.D., Sack, D.A., Harris, J.R., Chakraborty, J., Neogy, P.K., Stanton, B., Huda, N., Khan, M.U., Kay, B.A., Khan, M.R., *et al.* (1988). Cross-protection by B subunit-whole cell cholera vaccine against diarrhoea associated with heat-labile toxin-producing enterotoxigenic *Escherichia coli*: results of a large-scale field trial. J. Infect. Dis. *158*, 372–377.

Clements, J.D. (1990). Construction of a nontoxic fusion peptide for immunization against *Escherichia coli* strains that produce heat-labile and heat-stable enterotoxins. Infect. Immun. *58*, 1159–1166.

Cohen, D., Orr, N., Haim, M., Ashkenazi, S., Robin, G., Green, M.S., Ephros, M., Sela, T., Slepon, R., Ashkenazi, I., *et al.* (2000). Safety and immunogenicity of two different lots of the oral, killed enterotoxigenic *escherichia coli*-cholera toxin B subunit vaccine in Israeli young adults. Infect. Immun. *68*, 4492–4497.

Cohen, M.B. (1992). The heat-stable enterotoxin receptor: a probe for ligand hunting. J. Pediatr. Gastroenterol. Nutr. *15*, 337–338.

Cohen, M.B., Jensen, N.J., Hawkins, J.A., Mann, E.A., Thompson, M.R., Lentze, M.J., and Giannella, R.A. (1993). Receptors for *Escherichia coli* heat stable enterotoxin in human intestine and in a human intestinal cell line (Caco-2). J. Cell. Physiol. *156*, 138–144.

Crane, J.K., Wehner, M.S., Bolen, E.J., Sando, J.J., Linden, J., Guerrant, R.L., and Sears, C.L. (1992). Regulation of intestinal guanylate cyclase by the heat-stable enterotoxin of *Escherichia coli* (STa) and protein kinase C. Infect. Immun. *60*, 5004–5012.

Cray, W.C., Jr., Tokunaga, E., and Pierce, N.F. (1983). Successful colonization and immunization of adult rabbits by oral inoculation with *Vibrio cholerae* O1. Infect. Immun. *41*, 735–741.

Daley, A., Randall, R., Darsley, M., Choudhry, N., Thomas, N., Sanderson, I.R., Croft, N.M., and Kelly, P. (2007). Genetically modified enterotoxigenic *Escherichia coli* vaccines induce mucosal immune responses without inflammation. Gut *56*, 1550–1556.

Darfeuille-Michaud, A., Forestier, C., Joly, B., and Cluzel, R. (1986). Identification of a nonfimbrial adhesive factor of an enterotoxigenic *Escherichia coli* strain. Infect. Immun. *52*, 468–475.

Darsley, M.J., Chakraborty, S., Denearing, B., Sack, D.A., Feller, A., Buchwaldt, C., Bourgeois, A.L., Walker, R., and Harro, C.D. (2012). The Oral, Live Attenuated Enterotoxigenic *Escherichia coli* Vaccine ACE527 Reduces the Incidence and Severity of Diarrhea in a Human Challenge Model of Diarrheal Disease. Clin. Vaccine Immunol. *19*, 1921–1931.

De, S.N., Bhattacharya, K., and Sarkar, J.K. (1956). A study of the pathogenicity of strains of *Bacterium coli* from acute and chronic enteritis. J. Pathol. Bacteriol. *71*, 201–209.

Del Canto, F., Botkin, D.J., Valenzuela, P., Popov, V., Ruiz-Perez, F., Nataro, J.P., Levine, M.M., Stine, O.C., Pop, M., Torres, A.G., et al. (2012). Identification of Coli Surface Antigen 23, a novel adhesin of enterotoxigenic *Escherichia coli*. Infect. Immun. *80*, 2791–2801.

Donta, S.T., Beristain, S., and Tomicic, T.K. (1993). Inhibition of heat-labile cholera and *Escherichia coli* enterotoxins by brefeldin A. Infect. Immun. *61*, 3282–3286.

Dougan, G., Sellwood, R., Maskell, D., Sweeney, K., Liew, F.Y., Beesley, J., and Hormaeche, C. (1986). *In vivo* properties of a cloned K88 adherence antigen determinant. Infect. Immun. *52*, 344–347.

Dreyfus, L.A., Frantz, J.C., and Robertson, D.C. (1983). Chemical properties of heat-stable enterotoxins produced by enterotoxigenic *Escherichia coli* of different host origins. Infect. Immun. *42*, 539–548.

Dreyfus, L.A., Urban, R.G., Whipp, S.C., Slaughter, C., Tachias, K., and Kupersztoch, Y.M. (1992). Purification of the STB enterotoxin of *Escherichia coli* and the role of selected amino acids on its secretion, stability and toxicity. Mol. Microbiol. *6*, 2397–2406.

Dreyfus, L.A., Harville, B., Howard, D.E., Shaban, R., Beatty, D.M., and Morris, S.J. (1993). Calcium influx mediated by the *Escherichia coli* heat-stable enterotoxin B (STB). Proc. Natl. Acad. Sci. U.S.A. *90*, 3202–3206.

Dubreuil, J.D. (2008). *Escherichia coli* STb toxin and colibacillosis: knowing is half the battle. FEMS Microbiol. Lett. *278*, 137–145.

DuPont, H.L., Formal, S.B., Hornick, R.B., Snyder, M.J., Libonati, J.P., Sheahan, D.G., LaBrec, E.H., and Kalas, J.P. (1971). Pathogenesis of *Escherichia coli* diarrhoea. N. Engl. J. Med. *285*, 1–9.

Duthy, T.G., Manning, P.A., and Heuzenroeder, M.W. (2002). Identification and characterization of assembly proteins of CS5 pili from enterotoxigenic *Escherichia coli*. J. Bacteriol. *184*, 1065–1077.

Edelman, R., Russell, R.G., Losonsky, G., Tall, B.D., Tacket, C.O., Levine, M.M., and Lewis, D.H. (1993). Immunization of rabbits with enterotoxigenic *E. coli*

colonization factor antigen (CFA/I) encapsulated in biodegradable microspheres of poly (lactide-co-glycolide). Vaccine *11*, 155–158.

Ellingsworth, L.R. (2011). Transcutaneous immunization and the travelers' diarrhoea vaccine system: A phase III pivotal efficacy study. In The 6th International Conference on Vaccines for Enteric Diseases, VED 2011 (Cannes, France).

Elsinghorst, E.A. (2002). Enterotoxigenic *Escherichia coli*. In *Escherichia coli*, virulence mechanisms of a versatile pathogen, Donnenberg, M.S., ed. (Academic Press, New York, NY), pp. 155–187.

Elsinghorst, E.A., and Kopecko, D.J. (1992). Molecular cloning of epithelial cell invasion determinants from enterotoxigenic *Escherichia coli*. Infect. Immun. *60*, 2409–2417.

Elsinghorst, E.A., and Weitz, J.A. (1994). Epithelial cell invasion and adherence directed by the enterotoxigenic *Escherichia coli* tib locus is associated with a 104 kilodalton outer membrane protein. Infect. Immun. *62*, 3463–3471.

Erickson, A.K., Willgohs, J.A., McFarland, S.Y., Benfield, D.A., and Francis, D.H. (1992). Identification of two porcine brush border glycoproteins that bind the K88ac adhesin of *Escherichia coli* and correlation of these glycoproteins with the adhesive phenotype. Infect. Immun. *60*, 983–988.

Erickson, A.K., Baker, D.R., Bosworth, B.T., Casey, T.A., Benfield, D.A., and Francis, D.H. (1994). Characterization of porcine intestinal receptors for the K88ac fimbrial adhesin of *Escherichia coli* as mucin-type sialoglycoproteins. Infect. Immun. *62*, 5404–5410.

Erume, J., Wijemanne, P., Berberov, E.M., Kachman, S.D., Oestmann, D.J., Francis, D.H., and Moxley, R.A. (2013). Inverse relationship between heat stable enterotoxin-b induced fluid accumulation and adherence of F4ac-positive enterotoxigenic *Escherichia coli* in ligated jejunal loops of F4ab/ac fimbria receptor-positive swine. Vet. Microbiol. *25*, 315–324.

Evans, D.G., and Evans, D.J., Jr. (1978). New surface-associated heat-labile colonization factor antigen (CFA/II) produced by enterotoxigenic *Escherichia coli* of serogroups O6 and O8. Infect. Immun. *21*, 638–647.

Evans, D.G., Evans, D.J., Jr., and Pierce, N.F. (1973). Differences in the response of rabbit small intestine to heat-labile and heat-stable enterotoxins of *Escherichia coli*. Infect. Immun. *7*, 873–880.

Evans, D.G., Silver, R.P., Evans, D.J., Jr., Chase, D.G., and Gorbach, S.L. (1975). Plasmid-controlled colonization factor associated with virulence in *Esherichia coli* enterotoxigenic for humans. Infect. Immun. *12*, 656–667.

Evans, D.G., Evans, D.J., Jr., Opekun, A.R., and Graham, D.Y. (1988a). Non-replicating oral whole cell vaccine protective against enterotoxigenic *Escherichia coli* (ETEC) diarrhoea: stimulation of anti-CFA (CFA/I) and anti-enterotoxin (anti-LT) intestinal IgA and protection against challenge with ETEC belonging to heterologous serotypes. FEMS Microbiol. Immunol. *1*, 117–125.

Evans, D.J., Jr., Evans, D.G., Opekun, A.R., and Graham, D.Y. (1988b). Immunoprotective oral whole cell vaccine for enterotoxigenic *Escherichia coli* diarrhoea prepared by in situ destruction of chromosomal and plasmid DNA with colicin E2. FEMS Microbiol. Immunol. *1*, 9–18.

Fleckenstein, J.M., Kopecko, D.J., Warren, R.L., and Elsinghorst, E.A. (1996). Molecular characterization of the tia invasion locus from enterotoxigenic *Escherichia coli*. Infect. Immun. *64*, 2256–2265.

Fleckenstein, J.M., Lindler, L.E., Elsinghorst, E.A., and Dale, J.B. (2000). Identification of a gene within a pathogenicity island of enterotoxigenic *Escherichia coli* H10407 required for maximal secretion of the heat-labile enterotoxin. Infect. Immun. *68*, 2766–2774.

Fleckenstein, J.M., Holland, J.T., and Hasty, D.L. (2002). Interaction of an outer membrane protein of enterotoxigenic *Escherichia coli* with cell surface heparan sulfate proteoglycans. Infect. Immun. *70*, 1530–1537.

Fleckenstein, J.M., Roy, K., Fischer, J.F., and Burkitt, M. (2006). Identification of a two-partner secretion locus of enterotoxigenic *Escherichia coli*. Infect. Immun. *74*, 2245–2258.

Foreman, D.T., Martinez, Y., Coombs, G., Torres, A., and Kupersztoch, Y.M. (1995). TolC and DsbA are needed for the secretion of STB, a heat-stable enterotoxin of *Escherichia coli*. Mol. Microbiol. *18*, 237–245.

Francis, D.H., Grange, P.A., Zeman, D.H., Baker, D.R., Sun, R., and Erickson, A.K. (1998). Expression of mucin-type glycoprotein K88 receptors strongly correlates with piglet susceptibility to K88(+) enterotoxigenic *Escherichia coli*, but adhesion of this bacterium to brush borders does not. Infect. Immun. *66*, 4050–4055.

Franco, B.D., Gomes, T.A., Jakabi, M., and Marques, L.R. (1991). Use of probes to detect virulence factor DNA sequences in *Escherichia coli* strains isolated from foods. Int. J. Food Microbiol. *12*, 333–338.

Frantz, J.C., and Robertson, D.C. (1981). Immunological properties of *Escherichia coli* heat-stable enterotoxins: development of a radioimmunoassay specific for heat-stable enterotoxins with suckling mouse activity. Infect. Immun. *33*, 193–198.

Frech, S.A., Dupont, H.L., Bourgeois, A.L., McKenzie, R., Belkind-Gerson, J., Figueroa, J.F., Okhuysen, P.C., Guerrero, N.H., Martinez-Sandoval, F.G., Melendez-Romero, J.H., *et al.* (2008). Use of a patch containing heat-labile toxin from *Escherichia coli* against travellers' diarrhoea: a phase II, randomised, double-blind, placebo-controlled field trial. Lancet *371*, 2019–2025.

Froehlich, B.J., Karakashian, A., Sakellaris, H., and Scott, J.R. (1995). Genes for CS2 pili of enterotoxigenic *Escherichia coli* and their interchangeability with those for CS1 pili. Infect. Immun. *63*, 4849–4856.

Frydendahl, K. (2002). Prevalence of serogroups and virulence genes in *Escherichia coli* associated with postweaning diarrhoea and edema disease in pigs and a comparison of diagnostic approaches. Vet. Microbiol. *85*, 169–182.

Fujihara, S., Arikawa, K., Aota, T., Tanaka, H., Nakamura, H., Wada, T., Hase, A., and Nishikawa, Y. (2009). Prevalence and properties of diarrhoeagenic *Escherichia coli* among healthy individuals in Osaka City, Japan. Jpn. J. Infect. Dis. *62*, 318–323.

Fujii, Y., Hayashi, M., Hitotsubashi, S., Fuke, Y., Yamanaka, H., and Okamoto, K. (1991). Purification and characterization of *Escherichia coli* heat-stable enterotoxin II. J. Bacteriol. *173*, 5516–5522.

Fujii, Y., Nomura, T., Yamanaka, H., and Okamoto, K. (1997). Involvement of Ca(2+)-calmodulin-dependent protein kinase II in the intestinal secretory action of *Escherichia coli* heat-stable enterotoxin II. Microbiol. Immunol. *41*, 633–636.

Gaastra, W., and Svennerholm, A.M. (1996). Colonization factors of human enterotoxigenic *Escherichia coli* (ETEC). Trends Microbiol. *4*, 444–452.

Ganguly, U., Chaudhury, A.G., Basu, A., and Sen, P.C. (2001). STa-induced translocation of protein kinase C from cytosol to membrane in rat enterocytes. FEMS Microbiol. Lett. *204*, 65–69.

Gascon, J., Vargas, M., Quinto, L., Corachan, M., Jimenez de Anta, M.T., and Vila, J. (1998). Enteroaggregative *Escherichia coli* strains as a cause of traveler's diarrhoea: a case–control study. J. Infect. Dis. *177*, 1409–1412.

Gascon, J., Vargas, M., Schellenberg, D., Urassa, H., Casals, C., Kahigwa, E., Aponte, J.J., Mshinda, H., and Vila, J. (2000). Diarrhea in children under 5 years of age from Ifakara, Tanzania: a case–control study. J. Clin. Microbiol. *38*, 4459–4462.

Giron, J.A., Levine, M.M., and Kaper, J.B. (1994). Longus: a long pilus ultrastructure produced by human enterotoxigenic *Escherichia coli*. Mol. Microbiol. *12*, 71–82.

Giron, J.A., Gomez-Duarte, O.G., Jarvis, K.G., and Kaper, J.B. (1997). Longus pilus of enterotoxigenic *Escherichia coli* and its relatedness to other type-4 pili--a minireview. Gene *192*, 39–43.

Glenn, G.M., Scharton-Kersten, T., Vassell, R., Mallett, C.P., Hale, T.L., and Alving, C.R. (1998). Transcutaneous immunization with cholera toxin protects mice against lethal mucosal toxin challenge. J. Immunol. *161*, 3211–3214.

Glenn, G.M., Taylor, D.N., Li, X., Frankel, S., Montemarano, A., and Alving, C.R. (2000). Transcutaneous immunization: a human vaccine delivery strategy using a patch. Nature Med. *6*, 1403–1406.

Glenn, G.M., Villar, C.P., Flyer, D.C., Bourgeois, A.L., McKenzie, R., Lavker, R.M., and Frech, S.A. (2007). Safety and immunogenicity of an enterotoxigenic *Escherichia coli* vaccine patch containing heat-labile toxin: use of skin pretreatment to disrupt the stratum corneum. Infect. Immun. *75*, 2163–2170.

Gomez-Duarte, O.G., and Kaper, J.B. (1995). A plasmid-encoded regulatory region activates chromosomal eaeA expression in enteropathogenic *Escherichia coli*. Infect. Immun. *63*, 1767–1776.

Gomez-Duarte, O.G., Ruiz-Tagle, A., Gomez, D.C., Viboud, G.I., Jarvis, K.G., Kaper, J.B., and Giron, J.A. (1999). Identification of lngA, the structural gene of longus type IV pilus of enterotoxigenic *Escherichia coli*. Microbiology *145(Pt 7)*, 1809–1816.

Goncalves, C., Berthiaume, F., Mourez, M., and Dubreuil, J.D. (2008). *Escherichia coli* STb toxin binding to sulfatide and its inhibition by carrageenan. FEMS Microbiol. Lett. *281*, 30–35.

Gorbach, S.L., Banwell, J.G., Chatterjee, B.D., Jacobs, B., and Sack, R.B. (1971). Acute undifferentiated human diarrhoea in the tropics. I. Alterations in intestinal micrflora. J. Clin. Invest. 50, 881–889.

de Graaf, F.K., and Gaastra, W. (1994). Fimbriae of enterotoxigneic Escherichia coli. In Fimbriae: adhesion, genetics, biogenesis, and vaccines, Klemm, P., ed. (Baco Raton, FL: CRC press), pp. 53–83.

Green, B.A., Neill, R.J., Ruyechan, W.T., and Holmes, R.K. (1983). Evidence that a new enterotoxin of Escherichia coli which activates adenylate cyclase in eucaryotic target cells is not plasmid mediated. Infect. Immun. 41, 383–390.

Grewal, H.M., Valvatne, H., Bhan, M.K., van Dijk, L., Gaastra, W., and Sommerfelt, H. (1997). A new putative fimbrial colonization factor, CS19, of human enterotoxigenic Escherichia coli. Infect. Immun. 65, 507–513.

Guerena-Burgueno, F., Hall, E.R., Taylor, D.N., Cassels, F.J., Scott, D.A., Wolf, M.K., Roberts, Z.J., Nesterova, G.V., Alving, C.R., and Glenn, G.M. (2002). Safety and immunogenicity of a prototype enterotoxigenic Escherichia coli vaccine administered transcutaneously. Infect. Immun. 70, 1874–1880.

Guillobel, H.C., Carinhanha, J.I., Cardenas, L., Clements, J.D., de Almeida, D.F., and Ferreira, L.C. (2000). Adjuvant activity of a nontoxic mutant of Escherichia coli heat-labile enterotoxin on systemic and mucosal immune responses elicited against a heterologous antigen carried by a live Salmonella enterica serovar Typhimurium vaccine strain. Infect. Immun. 68, 4349–4353.

Gupta, D.D., Saha, S., and Chakrabarti, M.K. (2005). Involvement of protein kinase C in the mechanism of action of Escherichia coli heat-stable enterotoxin (STa) in a human colonic carcinoma cell line, COLO-205. Toxicol. Appl. Pharmacol. 206, 9–16.

Guth, B.E., Pickett, C.L., Twiddy, E.M., Holmes, R.K., Gomes, T.A., Lima, A.A., Guerrant, R.L., Franco, B.D., and Trabulsi, L.R. (1986a). Production of type II heat-labile enterotoxin by Escherichia coli isolated from food and human feces. Infect. Immun. 54, 587–589.

Guth, B.E., Twiddy, E.M., Trabulsi, L.R., and Holmes, R.K. (1986b). Variation in chemical properties and antigenic determinants among type II heat-labile enterotoxins of Escherichia coli. Infect. Immun. 54, 529–536.

Hallander, H.O., Paniagua, M., Espinoza, F., Askelof, P., Corrales, E., Ringman, M., and Storsaeter, J. (2002). Calibrated serological techniques demonstrate significant different serum response rates to an oral killed cholera vaccine between Swedish and Nicaraguan children. Vaccine 21, 138–145.

Han, D.U., Choi, C., Kim, J., Cho, W.S., Chung, H.K., Ha, S.K., Jung, K., and Chae, C. (2002). Anti-microbial susceptibility for east1 + Escherichia coli isolated from diarrheic pigs in Korea. J. Vet. Med. B Infect. Dis. Vet. Public Health 49, 346–348.

Handl, C.E., and Flock, J.I. (1992). STb producing Escherichia coli are rarely associated with infantile diarrhoea. J. Diarrhoeal Dis. Res. 10, 37–38.

Haq, T.A., Mason, H.S., Clements, J.D., and Arntzen, C.J. (1995). Oral immunization with a recombinant bacterial antigen produced in transgenic plants. Science 268, 714–716.

Harro, C., Chakraborty, S., Feller, A., DeNearing, B., Cage, A., Ram, M., Lundgren, A., Svennerholm, A.M., Bourgeois, A.L., Walker, R.I., et al. (2011a). Refinement of a human challenge model for evaluation of enterotoxigenic Escherichia coli vaccines. Clin. Vaccine Immunol. 18, 1719–1727.

Harro, C., Sack, D., Bourgeois, A.L., Walker, R., Denearing, B., Feller, A., Chakraborty, S., Buchwaldt, C., and Darsley, M.J. (2011b). A Combination Vaccine Consisting of Three Live Attenuated Enterotoxigenic Escherichia coli Strains Expressing a Range of Colonization Factors and Heat-Labile Toxin Subunit B Is Well Tolerated and Immunogenic in a Placebo-Controlled Double-Blind Phase I Trial in Healthy Adults. Clin. Vaccine Immunol. 18, 2118–2127.

Harro, C., Sack, D., Darsley, M., Bourgeois, A.L., DeNearing, B., Feller, A., Chakraborty, S., Marcum, A., Comendador, R., Buchwaldt, C., and Walker, R. (2011c). Volunteers receiving live attenuated ETEC vaccine (ACE527) have reduced severity of illness following H10407 challenge. In The 6th International Conference on Vaccines for Enteric Diseases, (VED) 2011 (Cannes, France).

Harville, B.A., and Dreyfus, L.A. (1995). Involvement of 5-hydroxytryptamine and prostaglandin E2 in the intestinal secretory action of Escherichia coli heat-stable enterotoxin B. Infect. Immun. 63, 745–750.

Harville, B.A., and Dreyfus, L.A. (1996). Release of serotonin from RBL-2H3 cells by the Escherichia coli peptide toxin STb. Peptides 17, 363–366.

Henderson, I.R., Navarro-Garcia, F., Desvaux, M., Fernandez, R.C., and Ala'Aldeen, D. (2004). Type V protein secretion pathway: the autotransporter story. MMBR 68, 692–744.

Heuzenroeder, M.W., Elliot, T.R., Thomas, C.J., Halter, R., and Manning, P.A. (1990). A new fimbrial type (PCFO9) on enterotoxigenic Escherichia coli 09:H– LT+ isolated from a case of infant diarrhoea in central Australia. FEMS Microbiol. Lett. 54, 55–60.

Hibberd, M.L., McConnell, M.M., Willshaw, G.A., Smith, H.R., and Rowe, B. (1991). Positive regulation of colonization factor antigen I (CFA/I) production by enterotoxigenic Escherichia coli producing the colonization factors CS5, CS6, CS7, CS17, PCFO9, PCFO159:H4 and PCFO166. J. Gen. Microbiol. 137, 1963–1970.

Hirst, T.R., Randall, L.L., and Hardy, S.J. (1984a). Cellular location of enterotoxin in Escherichia coli. Biochem. Soc. Trans. 12, 189–191.

Hirst, T.R., Sanchez, J., Kaper, J.B., Hardy, S.J., and Holmgren, J. (1984b). Mechanism of toxin secretion by Vibrio cholerae investigated in strains harboring plasmids that encode heat-labile enterotoxins of Escherichia coli. Proc. Natl. Acad. Sci. U.S.A. 81, 7752–7756.

Hitotsubashi, S., Akagi, M., Saitou, A., Yamanaka, H., Fujii, Y., and Okamoto, K. (1992a). Action of Escherichia *coli* heat-stable enterotoxin II on isolated sections of mouse ileum. FEMS Microbiol. Lett. *69*, 249–252.

Hitotsubashi, S., Fujii, Y., Yamanaka, H., and Okamoto, K. (1992b). Some properties of purified *Escherichia coli* heat-stable enterotoxin II. Infect. Immun. *60*, 4468–4474.

Holmes, R.K., Twiddy, E.M., and Pickett, C.L. (1986). Purification and characterization of type II heat-labile enterotoxin of *Escherichia coli*. Infect. Immun. *53*, 464–473.

Holmgren, J., Blomquist, M., Bourgeois, L., Carlin, N., Clements, J., Ekman, A., Gustafsson, B., Hellman, M., Lundgren, A., Lofstrand, M., Nygren, E., *et al.* (2011). Preclinical evaluation of an oral inactivated ETEC vaccine, based on *E. coli* overexpressing CFA/I, CS3, CS5 and CS6 together with LTB/CTB subunit, and of the adjuvant effect of co-administration with dmLT. In The 6th International Conference on Vaccines for Enteric Diseases, VED 2011 (Cannes, France).

Horstman, A.L., and Kuehn, M.J. (2000). Enterotoxigenic *Escherichia coli* secretes active heat-labile enterotoxin via outer membrane vesicles. J. Biol. Chem. *275*, 12489–12496.

Howell, A., Harland, R.N., Barnes, D.M., Hayward, E., Redford, J., Swindell, R., and Sellwood, R.A. (1987). Endocrine therapy for advanced carcinoma of the breast: effect of tumor heterogeneity and site of biopsy on the predictive value of progesterone receptor estimations. Cancer Res. *47*, 296–299.

Hughes, J.M., Murad, F., Chang, B., and Guerrant, R.L. (1978). Role of cyclic GMP in the action of heat-stable enterotoxin of *Escherichia coli*. Nature *271*, 755–756.

Isidean, S.D., Riddle, M.S., Savarino, S.J., and Porter, C.K. (2011). A systematic review of ETEC epidemiology focusing on colonization factor and toxin expression. Vaccine *29*, 6167–6178.

Itoh, Y., Nagano, I., Kunishima, M., and Ezaki, T. (1997). Laboratory investigation of enteroaggregative *Escherichia coli* O untypeable:H10 associated with a massive outbreak of gastrointestinal illness. J. Clin. Microbiol. *35*, 2546–2550.

Jalajakumari, M.B., Thomas, C.J., Halter, R., and Manning, P.A. (1989). Genes for biosynthesis and assembly of CS3 pili of CFA/II enterotoxigenic *Escherichia coli*: novel regulation of pilus production by bypassing an amber codon. Mol. Microbiol. *3*, 1685–1695.

Jertborn, M., Ahren, C., Holmgren, J., and Svennerholm, A.M. (1998). Safety and immunogenicity of an oral inactivated enterotoxigenic *Escherichia coli* vaccine. Vaccine *16*, 255–260.

Jiang, Z.D., Lowe, B., Verenkar, M.P., Ashley, D., Steffen, R., Tornieporth, N., von Sonnenburg, F., Waiyaki, P., and DuPont, H.L. (2002). Prevalence of enteric pathogens among international travelers with diarrhoea acquired in Kenya (Mombasa), India (Goa), or Jamaica (Montego Bay). J. Infect. Dis. *185*, 497–502.

Jordi, B.J., Willshaw, G.A., van der Zeijst, B.A., and Gaastra, W. (1992). The complete nucleotide sequence of region 1 of the CFA/I fimbrial operon of human enterotoxigenic *Escherichia coli*. DNA sequence: the journal of DNA sequencing and mapping *2*, 257–263.

Karaman, S., Cunnick, J., and Wang, K. (2012). Expression of the cholera toxin B subunit (CT-B) in maize seeds and a combined mucosal treatment against cholera and traveler's diarrhoea. Plant Cell Rep. *31*, 527–537.

Katz, D.E., DeLorimier, A.J., Wolf, M.K., Hall, E.R., Cassels, F.J., van Hamont, J.E., Newcomer, R.L., Davachi, M.A., Taylor, D.N., and McQueen, C.E. (2003). Oral immunization of adult volunteers with microencapsulated enterotoxigenic *Escherichia coli* (ETEC) CS6 antigen. Vaccine *21*, 341–346.

Kesty, N.C., Mason, K.M., Reedy, M., Miller, S.E., and Kuehn, M.J. (2004). Enterotoxigenic *Escherichia coli* vesicles target toxin delivery into mammalian cells. EMBO J. *23*, 4538–4549.

Khan, S., Chatfield, S., Stratford, R., Bedwell, J., Bentley, M., Sulsh, S., Giemza, R., Smith, S., Bongard, E., Cosgrove, C.A., *et al.* (2007). Ability of SPI2 mutant of S. typhi to effectively induce antibody responses to the mucosal antigen enterotoxigenic *E. coli* heat labile toxin B subunit after oral delivery to humans. Vaccine *25*, 4175–4182.

Kim, T.G., Kim, B.G., Kim, M.Y., Choi, J.K., Jung, E.S., and Yang, M.S. (2010). Expression and immunogenicity of enterotoxigenic *Escherichia coli* heat-labile toxin B subunit in transgenic rice callus. Mol. Biotechnol. *44*, 14–21.

Knutton, S., Lloyd, D.R., and McNeish, A.S. (1987). Identification of a new fimbrial structure in enterotoxigenic *Escherichia coli* (ETEC) serotype O148:H28 which adheres to human intestinal mucosa: a potentially new human ETEC colonization factor. Infect. Immun. *55*, 86–92.

Koprowski, H., 2nd, Levine, M.M., Anderson, R.J., Losonsky, G., Pizza, M., and Barry, E.M. (2000). Attenuated *Shigella* flexneri 2a vaccine strain CVD 1204 expressing colonization factor antigen I and mutant heat-labile enterotoxin of enterotoxigenic *Escherichia coli*. Infect. Immun. *68*, 4884–4892.

Kuperzstoch, Y.M., Tachias, K., Moomaw, C.R., Dreyfus, L.A., Urban, R., Slaughter, C., and Whipp, S. (1990). Secretion of methanol-insoluble heat-stable enterotoxin (STB): energy- and secA-dependent conversion of pre-STB to an intermediate indistinguishable from the extracellular toxin. J. Bacteriol. *172*, 2427–2432.

Lapa, J.A., Sincock, S.A., Ananthakrishnan, M., Porter, C.K., Cassels, F.J., Brinkley, C., Hall, E.R., van Hamont, J., Gramling, J.D., Carpenter, C.M., *et al.* (2008). Randomized clinical trial assessing the safety and immunogenicity of oral microencapsulated enterotoxigenic *Escherichia coli* surface antigen 6 with or without heat-labile enterotoxin with mutation R192G. Clin. Vaccine Immunol. *15*, 1222–1228.

Lasaro, M.O., Luiz, W.B., Sbrogio-Almeida, M.E., Nishimura, L.S., Guth, B.E., and Ferreira, L.C. (2004). Combined vaccine regimen based on parenteral priming with a DNA vaccine and administration of an oral booster consisting of a recombinant *Salmonella enterica* serovar Typhimurium vaccine strain for immunization against infection with human-derived enterotoxigenic *Escherichia coli* strains. Infect. Immun. *72*, 6480–6491.

Lasaro, M.A., Rodrigues, J.F., Mathias-Santos, C., Guth, B.E., Balan, A., Sbrogio-Almeida, M.E., and Ferreira, L.C. (2008). Genetic diversity of heat-labile toxin expressed by enterotoxigenic Escherichia coli strains isolated from humans. J. Bacteriol. 190, 2400–2410.

Lee, C.H., Moseley, S.L., Moon, H.W., Whipp, S.C., Gyles, C.L., and So, M. (1983). Characterization of the gene encoding heat-stable toxin II and preliminary molecular epidemiological studies of enterotoxigenic Escherichia coli heat-stable toxin II producers. Infect. Immun. 42, 264–268.

Lencer, W.I., Constable, C., Moe, S., Jobling, M.G., Webb, H.M., Ruston, S., Madara, J.L., Hirst, T.R., and Holmes, R.K. (1995). Targeting of cholera toxin and Escherichia coli heat labile toxin in polarized epithelia: role of COOH-terminal KDEL. J. Cell Biol. 131, 951–962.

Levine, M.M. (1990). Vaccines against enterotoxigenic Escherichia coli infections. In New Generation Vaccines, Woodrow, G.C., Levine, M.M., ed. (Marcel Dekker, New York, NY), pp. 649–660.

Levine, M.M. (2010). Immunogenicity and efficacy of oral vaccines in developing countries: lessons from a live cholera vaccine. BMC Biol. 8, 129.

Levine, M.M., and Svennerholm, A.M. (2008). Immunoprophylaxis and immunologic control. In Travelers' diarrhoea, 2nd edition, Ericsson, C., DuPont, H., Steffen, R., eds. (Hamilton, Canada: B. C Decker Inc.), pp. 215–232.

Levine, M.M., Black, R.E., Brinton, C.C., Jr., Clements, M.L., Fusco, P., Hughes, T.P., O'Donnell, S., Robins-Browne, R., Wood, S., and Young, C.R. (1982). Reactogenicity, immunogenicity and efficacy studies of Escherichia coli type 1 somatic pili parenteral vaccine in man. Scand. J. Infect. Dis. Suppl. 33, 83–95.

Li, Y.F., Poole, S., Nishio, K., Jang, K., Rasulova, F., McVeigh, A., Savarino, S.J., Xia, D., and Bullitt, E. (2009). Structure of CFA/I fimbriae from enterotoxigenic Escherichia coli. Proc. Natl. Acad. Sci. U.S.A. 106, 10793–10798.

Lindenthal, C., and Elsinghorst, E.A. (1999). Identification of a glycoprotein produced by enterotoxigenic Escherichia coli. Infect. Immun. 67, 4084–4091.

Lindenthal, C., and Elsinghorst, E.A. (2001). Enterotoxigenic Escherichia coli TibA glycoprotein adheres to human intestine epithelial cells. Infect. Immun. 69, 52–57.

Liu, M., Ruan, X., Zhang, C., Lawson, S.R., Knudsen, D.E., Nataro, J.P., Robertson, D.C., and Zhang, W. (2011a). Heat-labile- and heat–stable-toxoid fusions (LTRG-STaPF) of human enterotoxigenic Escherichia coli elicit neutralizing antitoxin antibodies. Infect. Immun. 79, 4002–4009.

Liu, M., Zhang, C., Mateo, K., Nataro, J.P., Robertson, D.C., and Zhang, W. (2011b). Modified Heat-Stable Toxins (hSTa) of Enterotoxigenic Escherichia coli Lose Toxicity but Display Antigenicity after Being Genetically Fused to Heat-Labile Toxoid LT(R192G). Toxins (Basel) 3, 1146–1162.

Lortie, L.A., Dubreuil, J.D., and Harel, J. (1991). Characterization of Escherichia coli strains producing heat-stable enterotoxin b (STb) isolated from humans with diarrhoea. J. Clin. Microbiol. 29, 656–659.

Ludwig, A., Bauer, S., Benz, R., Bergmann, B., and Goebel, W. (1999). Analysis of the SlyA-controlled expression, subcellular localization and pore-forming activity of a 34 kDa haemolysin (ClyA) from Escherichia coli K-12. Mol. Microbiol. 31, 557–567.

Ludwig, A., von Rhein, C., Bauer, S., Huttinger, C., and Goebel, W. (2004). Molecular analysis of cytolysin A (ClyA) in pathogenic Escherichia coli strains. J. Bacteriol. 186, 5311–5320.

Lundgren, A., Leach, S., Jertborn, M., Kaim, J., Wiklund, G., Adamsson, J., Eklund, L., Bourgeois, L., Walker, R., Tobias, J., et al. (2011). Clinical trial of an inactivated whole cell ETEC prototype vaccine. In The 6th international conference on vaccines for enteric diseases, VED 2011 (Cannes, France).

McBroom, A.J., Johnson, A.P., Vemulapalli, S., and Kuehn, M.J. (2006). Outer membrane vesicle production by Escherichia coli is independent of membrane instability. J. Bacteriol. 188, 5385–5392.

McConnell, M.M., Chart, H., Field, A.M., Hibberd, M., and Rowe, B. (1989). Characterization of a putative colonization factor (PCFO166) of enterotoxigenic Escherichia coli of serogroup O166. J. Gen. Microbiol. 135, 1135–1144.

McConnell, M.M., Hibberd, M., Field, A.M., Chart, H., and Rowe, B. (1990). Characterization of a new putative colonization factor (CS17) from a human enterotoxigenic Escherichia coli of serotype O114:H21 which produces only heat-labile enterotoxin. J. Infect. Dis. 161, 343–347.

McKenzie, R., Bourgeois, A.L., Engstrom, F., Hall, E., Chang, H.S., Gomes, J.G., Kyle, J.L., Cassels, F., Turner, A.K., Randall, R., et al. (2006). Comparative safety and immunogenicity of two attenuated enterotoxigenic Escherichia coli vaccine strains in healthy adults. Infect. Immun. 74, 994–1000.

McKenzie, R., Bourgeois, A.L., Frech, S.A., Flyer, D.C., Bloom, A., Kazempour, K., and Glenn, G.M. (2007). Transcutaneous immunization with the heat-labile toxin (LT) of enterotoxigenic Escherichia coli (ETEC): protective efficacy in a double-blind, placebo-controlled challenge study. Vaccine 25, 3684–3691.

McKenzie, R., Darsley, M., Thomas, N., Randall, R., Carpenter, C., Forbes, E., Finucane, M., Sack, R.B., Hall, E., and Bourgeois, A.L. (2008). A double-blind, placebo-controlled trial to evaluate the efficacy of PTL-003, an attenuated enterotoxigenic E. coli (ETEC) vaccine strain, in protecting against challenge with virulent ETEC. Vaccine 26, 4731–4739.

Madhi, S.A., Cunliffe, N.A., Steele, D., Witte, D., Kirsten, M., Louw, C., Ngwira, B., Victor, J.C., Gillard, P.H., Cheuvart, B.B., et al. (2010). Effect of human rotavirus vaccine on severe diarrhoea in African infants. N. Engl. J. Med. 362, 289–298.

Mammarappallil, J.G., and Elsinghorst, E.A. (2000). Epithelial cell adherence mediated by the enterotoxigenic Escherichia coli tia protein. Infect. Immun. 68, 6595–6601.

Marron, M.B., and Smyth, C.J. (1995). Molecular analysis of the cso operon of enterotoxigenic Escherichia coli reveals that CsoA is the adhesin of CS1 fimbriae and that the accessory genes are interchangeable with those of the cfa operon. Microbiology *141(Pt 11)*, 2849–2859.

Mason, H.S., Haq, T.A., Clements, J.D., and Arntzen, C.J. (1998). Edible vaccine protects mice against *Escherichia coli* heat-labile enterotoxin (LT): potatoes expressing a synthetic LT-B gene. Vaccine *16*, 1336–1343.

Mason, H.S., Warzecha, H., Mor, T., and Arntzen, C.J. (2002). Edible plant vaccines: applications for prophylactic and therapeutic molecular medicine. Trends Mol. Med. *8*, 324–329.

Merson, M.H., Sack, R.B., Islam, S., Saklayen, G., Huda, N., Huq, I., Zulich, A.W., Yolken, R.H., and Kapikian, A.Z. (1980). Disease due to enterotoxigenic *Escherichia coli* in Bangladeshi adults: clinical aspects and a controlled trial of tetracycline. J. Infect. Dis. *141*, 702–711.

Mezoff, A.G., Giannella, R.A., Eade, M.N., and Cohen, M.B. (1992). *Escherichia coli* enterotoxin (STa) binds to receptors, stimulates guanyl cyclase, and impairs absorption in rat colon. Gastroenterology *102*, 816–822.

Moon, H.W., Kohler, E.M., Schneider, R.A., and Whipp, S.C. (1980). Prevalence of pilus antigens, enterotoxin types, and enteropathogenicity among K88-negative enterotoxigenic *Escherichia coli* from neonatal pigs. Infect. Immun. *27*, 222–230.

Moon, H.W., Schneider, R.A., and Moseley, S.L. (1986). Comparative prevalence of four enterotoxin genes among *Escherichia coli* isolated from swine. Am. J. Vet. Res. *47*, 210–212.

Nada, R.A., Shaheen, H.I., Khalil, S.B., Mansour, A., El-Sayed, N., Touni, I., Weiner, M., Armstrong, A.W., and Klena, J.D. (2011). Discovery and phylogenetic analysis of novel members of class b enterotoxigenic *Escherichia coli* adhesive fimbriae. J. Clin. Microbiol. *49*, 1403–1410.

Nagy, B., and Fekete, P.Z. (1999). Enterotoxigenic *Escherichia coli* (ETEC) in farm animals. Vet. Res *30*, 259–284.

Nair, G.B., and Takeda, Y. (1998). The heat-stable enterotoxins. Microb. Pathog. *24*, 123–131.

Nakashima, H., Yukawa, Y., Imagama, S., Kanemura, T., Kamiya, M., Yanase, M., Ito, K., Machino, M., Yoshida, G., Ishikawa, Y., *et al.* (2012). Complications of cervical pedicle screw fixation for nontraumatic lesions: a multicenter study of 84 patients. J. Neurosurg. Spine *16*, 238–247.

Nardi, A.R., Salvadori, M.R., Coswig, L.T., Gatti, M.S., Leite, D.S., Valadares, G.F., Neto, M.G., Shocken-Iturrino, R.P., Blanco, J.E., and Yano, T. (2005). Type 2 heat-labile enterotoxin (LT-II)-producing *Escherichia coli* isolated from ostriches with diarrhoea. Vet Microbiol. *105*, 245–249.

Nataro, J.P., and Kaper, J.B. (1998). Diarrheagenic *Escherichia coli*. Clin. Microbiol. Rev. *11*, 142–201.

Nataro, J.P., Kaper, J.B., Robins-Browne, R., Prado, V., Vial, P., and Levine, M.M. (1987). Patterns of adherence of diarrhoeagenic *Escherichia coli* to HEp-2 cells. Pediatr. Infect. Dis. J. *6*, 829–831.

Ngeleka, M., Pritchard, J., Appleyard, G., Middleton, D.M., and Fairbrother, J.M. (2003). Isolation and association of *Escherichia coli* AIDA-I/STb, rather than EAST1 pathotype, with diarrhoea in piglets and antibiotic sensitivity of isolates. J. Vet. Diagn. Invest. *15*, 242–252.

Noamani, B.N., Fairbrother, J.M., and Gyles, C.L. (2003). Virulence genes of O149 enterotoxigenic *Escherichia coli* from outbreaks of postweaning diarrhoea in pigs. Vet. Microbiol. *97*, 87–101.

Norton, E.B., Lawson, L.B., Freytag, L.C., and Clements, J.D. (2011). Characterization of a mutant *Escherichia coli* heat-labile toxin, LT(R192G/L211A), as a safe and effective oral adjuvant. Clin. Vaccine Immunol. *18*, 546–551.

Nuccio, S.P., and Baumler, A.J. (2007). Evolution of the chaperone/usher assembly pathway: fimbrial classification goes Greek. MMBR *71*, 551–575.

Ogunniyi, A.D., Kotlarski, I., Morona, R., and Manning, P.A. (1997). Role of SefA subunit protein of SEF14 fimbriae in the pathogenesis of *Salmonella enterica* serovar Enteritidis. Infect. Immun. *65*, 708–717.

Okamoto, K., Fujii, Y., Akashi, N., Hitotsubashi, S., Kurazono, H., Karasawa, T., and Takeda, Y. (1993). Identification and characterization of heat-stable enterotoxin II-producing *Escherichia coli* from patients with diarrhoea. Microbiol. Immunol. *37*, 411–414.

Oscarsson, J., Mizunoe, Y., Uhlin, B.E., and Haydon, D.J. (1996). Induction of haemolytic activity in *Escherichia coli* by the slyA gene product. Mol. Microbiol. *20*, 191–199.

Osek, J. (2003). Detection of the enteroaggregative *Escherichia coli* heat-stable enterotoxin 1 (EAST1) gene and its relationship with fimbrial and enterotoxin markers in *E. coli* isolates from pigs with diarrhoea. Vet. Microbiol. *91*, 65–72.

Paiva de Sousa, C., and Dubreuil, J.D. (2001). Distribution and expression of the astA gene (EAST1 toxin) in *Escherichia coli* and *Salmonella*. Int. J. Med. Microbiol. *291*, 15–20.

Patel, S.K., Dotson, J., Allen, K.P., and Fleckenstein, J.M. (2004). Identification and molecular characterization of EatA, an autotransporter protein of enterotoxigenic *Escherichia coli*. Infect. Immun. *72*, 1786–1794.

Paul, Y. (2009). Why polio has not been eradicated in India despite many remedial interventions? Vaccine *27*, 3700–3703.

Peltola, H., Siitonen, A., Kyronseppa, H., Simula, I., Mattila, L., Oksanen, P., Kataja, M.J., and Cadoz, M. (1991). Prevention of travellers' diarrhoea by oral B-subunit/whole-cell cholera vaccine. Lancet *338*, 1285–1289.

Perez-Casal, J., Swartley, J.S., and Scott, J.R. (1990). Gene encoding the major subunit of CS1 pili of human enterotoxigenic *Escherichia coli*. Infect. Immun. *58*, 3594–3600.

Peterson, J.W., and Whipp, S.C. (1995). Comparison of the mechanisms of action of cholera toxin and the heat-stable enterotoxins of *Escherichia coli*. Infect. Immun. *63*, 1452–1461.

Pichel, M., Binsztein, N., and Viboud, G. (2000). CS22, a novel human enterotoxigenic *Escherichia coli* adhesin, is related to CS15. Infect. Immun. *68*, 3280–3285.

Picken, R.N., Mazaitis, A.J., Maas, W.K., Rey, M., and Heyneker, H. (1983). Nucleotide sequence of the gene for heat-stable enterotoxin II of *Escherichia coli*. Infect. Immun. 42, 269–275.

Pickett, C.L., Weinstein, D.L., and Holmes, R.K. (1987). Genetics of type IIa heat-labile enterotoxin of *Escherichia coli*: operon fusions, nucleotide sequence, and hybridization studies. J. Bacteriol. 169, 5180–5187.

Pickett, C.L., Twiddy, E.M., Coker, C., and Holmes, R.K. (1989). Cloning, nucleotide sequence, and hybridization studies of the type IIb heat-labile enterotoxin gene of *Escherichia coli*. J. Bacteriol. 171, 4945–4952.

Porter, C.K., Riddle, M.S., Tribble, D.R., Louis Bougeois, A., McKenzie, R., Isidean, S.D., Sebeny, P., and Savarino, S.J. (2011). A systematic review of experimental infections with enterotoxigenic *Escherichia coli* (ETEC). Vaccine 29, 5869–5885.

Qadri, F., Das, S.K., Faruque, A.S., Fuchs, G.J., Albert, M.J., Sack, R.B., and Svennerholm, A.M. (2000a). Prevalence of toxin types and colonization factors in enterotoxigenic *Escherichia coli* isolated during a 2-year period from diarrhoeal patients in Bangladesh. J. Clin. Microbiol. 38, 27–31.

Qadri, F., Wenneras, C., Ahmed, F., Asaduzzaman, M., Saha, D., Albert, M.J., Sack, R.B., and Svennerholm, A. (2000b). Safety and immunogenicity of an oral, inactivated enterotoxigenic *Escherichia coli* plus cholera toxin B subunit vaccine in Bangladeshi adults and children. Vaccine 18, 2704–2712.

Qadri, F., Ahmed, T., Ahmed, F., Bradley Sack, R., Sack, D.A., and Svennerholm, A.M. (2003). Safety and immunogenicity of an oral, inactivated enterotoxigenic *Escherichia coli* plus cholera toxin B subunit vaccine in Bangladeshi children 18–36 months of age. Vaccine 21, 2394–2403.

Qadri, F., Svennerholm, A.M., Faruque, A.S., and Sack, R.B. (2005). Enterotoxigenic *Escherichia coli* in developing countries: epidemiology, microbiology, clinical features, treatment, and prevention. Clin. Microbiol. Rev. 18, 465–483.

Qadri, F., Ahmed, T., Ahmed, F., Begum, Y.A., Sack, D.A., and Svennerholm, A.M. (2006). Reduced doses of oral killed enterotoxigenic *Escherichia coli* plus cholera toxin B subunit vaccine is safe and immunogenic in Bangladeshi infants 6–17 months of age: dosing studies in different age groups. Vaccine 24, 1726–1733.

Rao, M.R., Abu-Elyazeed, R., Savarino, S.J., Naficy, A.B., Wierzba, T.F., Abdel-Messih, I., Shaheen, H., Frenck, R.W., Jr., Svennerholm, A.M., and Clemens, J.D. (2003). High disease burden of diarrhoea due to enterotoxigenic *Escherichia coli* among rural Egyptian infants and young children. J. Clin. Microbiol. 41, 4862–4864.

Rao, M.R., Wierzba, T.F., Savarino, S.J., Abu-Elyazeed, R., El-Ghoreb, N., Hall, E.R., Naficy, A., Abdel-Messih, I., Frenck, R.W., Jr., Svennerholm, A.M., et al. (2005). Serologic correlates of protection against enterotoxigenic *Escherichia coli* diarrhoea. J. Infect. Dis. 191, 562–570.

Rasheed, J.K., Guzman-Verduzco, L.M., and Kupersztoch, Y.M. (1990). Two precursors of the heat-stable enterotoxin of *Escherichia coli*: evidence of extracellular processing. Mol. Microbiol. 4, 265–273.

Reid, R.H., Boedeker, E.C., McQueen, C.E., Davis, D., Tseng, L.Y., Kodak, J., Sau, K., Wilhelmsen, C.L., Nellore, R., Dalal, P., et al. (1993). Preclinical evaluation of microencapsulated CFA/II oral vaccine against enterotoxigenic *E. coli*. Vaccine 11, 159–167.

Rigobelo, E.C., Stella, A.E., Avila, F.A., Macedo, C., and Marin, J.M. (2006). Characterization of *Escherichia coli* isolated from carcasses of beef cattle during their processing at an abattoir in Brazil. Int. J. Food Microbiol. 110, 194–198.

Roland, K.L., Cloninger, C., Kochi, S.K., Thomas, L.J., Tinge, S.A., Rouskey, C., and Killeen, K.P. (2007). Construction and preclinical evaluation of recombinant Peru-15 expressing high levels of the cholera toxin B subunit as a vaccine against enterotoxigenic *Escherichia coli*. Vaccine 25, 8574–8584.

Rosales-Mendoza, S., Soria-Guerra, R.E., Lopez-Revilla, R., Moreno-Fierros, L., and Alpuche-Solis, A.G. (2008). Ingestion of transgenic carrots expressing the *Escherichia coli* heat-labile enterotoxin B subunit protects mice against cholera toxin challenge. Plant Cell Rep. 27, 79–84.

Rosales-Mendoza, S., Alpuche-Solis, A.G., Soria-Guerra, R.E., Moreno-Fierros, L., Martinez-Gonzalez, L., Herrera-Diaz, A., and Korban, S.S. (2009). Expression of an *Escherichia coli* antigenic fusion protein comprising the heat labile toxin B subunit and the heat stable toxin, and its assembly as a functional oligomer in transplastomic tobacco plants. Plant J. 57, 45–54.

Rousset, E., and Dubreuil, J.D. (1999). Evidence that *Escherichia coli* STb enterotoxin binds to lipidic components extracted from the pig jejunal mucosa. Toxicon 37, 1529–1537.

Rousset, E., Harel, J., and Dubreuil, J.D. (1998). Sulfatide from the pig jejunum brush border epithelial cell surface is involved in binding of *Escherichia coli* enterotoxin b. Infect. Immun. 66, 5650–5658.

Roy, K., Hamilton, D., Allen, K.P., Randolph, M.P., and Fleckenstein, J.M. (2008a). The EtpA exoprotein of enterotoxigenic *Escherichia coli* promotes intestinal colonization and is a protective antigen in an experimental model of murine infection. Infect. Immun. 76, 2106–2112.

Roy, K., Hamilton, D., Allen, K.P., Randolph, M.P., and Fleckenstein, J.M. (2008b). The EtpA exoprotein of enterotoxigenic *Escherichia coli* promotes intestinal colonization and is a protective antigen in an experimental model of murine infection. Infect. Immun. 76, 2106–2112.

Roy, K., Hamilton, D., Ostmann, M.M., and Fleckenstein, J.M. (2009a). Vaccination with EtpA glycoprotein or flagellin protects against colonization with enterotoxigenic *Escherichia coli* in a murine model. Vaccine 27, 4601–4608.

Roy, K., Hilliard, G.M., Hamilton, D.J., Luo, J., Ostmann, M.M., and Fleckenstein, J.M. (2009b). Enterotoxigenic *Escherichia coli* EtpA mediates adhesion between flagella and host cells. Nature 457, 594–598.

Ruan, X., Crupper, S.S., Schultz, B.D., Robertson, D.C., and Zhang, W. (2012). *Escherichia coli* expressing EAST1 toxin did not cause an increase of cAMP or cGMP levels in cells, and no diarrhoea in 5-day old gnotobiotic pigs. PLoS One 7, e43203.

Ruan, X., and Zhang, W. (2013). Oral immunization of a live attenuated *Escherichia coli* strain expressing a holotoxin-structured adhesin–toxoid fusion (1FaeG-FedF-LT(A2):5LT(B)) protected young pigs against enterotoxigenic *E. coli* (ETEC) infection. Vaccine *31*, 1458–1463.

Rutter, J.M., Burrows, M.R., Sellwood, R., and Gibbons, R.A. (1975). A genetic basis for resistance to enteric disease caused by *E. coli*. Nature *257*, 135–136.

Sack, D.A., Shimko, J., Torres, O., Bourgeois, A.L., Francia, D.S., Gustafsson, B., Karnell, A., Nyquist, I., and Svennerholm, A.M. (2007). Randomised, double-blind, safety and efficacy of a killed oral vaccine for enterotoxigenic *E. Coli* diarrhoea of travellers to Guatemala and Mexico. Vaccine *25*, 4392–4400.

Sack, D.A.Q., and Svennerholm, A.M. (2008). Determinants of responses to oral vaccines in developing countries. Ann. Nestle. [Engl] *66*, 71–79.

Sack, R.B. (1968). Diarrhea procuing factors in cultures of *Escherichia coli*. Paper presented at: Proceeding of the 4th Joint Conference, Japen-US Cooperative Medical Science Program (Unzen, Japan).

Sack, R.B. (1980). Enterotoxigenic *Escherichia coli*: identification and characterization. J. Infect. Dis. *142*, 279–286.

Sack, R.B., Gorbach, S.L., Banwell, J.G., Jacobs, B., Chatterjee, B.D., and Mitra, R.C. (1971). Enterotoxigenic *Escherichia coli* isolated from patients with severe cholera-like disease. J. Infect. Dis. *123*, 378–385.

Sack, R.B., Johnson, J., Pierce, N.F., Keren, D.F., and Yardley, J.H. (1976). Challenge of dogs with live enterotoxigenic *Escherichia coli* and effects of repeated challenges on fluid secretion in jejunal Thiry-Vella loops. J. Infect. Dis. *134*, 15–24.

Sack, R.B., Kline, R.L., and Spira, W.M. (1988). Oral immunization of rabbits with enterotoxigenic *Escherichia coli* protects against intraintestinal challenge. Infect. Immun. *56*, 387–394.

Sanchez, J., Uhlin, B.E., Grundstrom, T., Holmgren, J., and Hirst, T.R. (1986). Immunoactive chimeric ST-LT enterotoxins of *Escherichia coli* generated by in vitro gene fusion. FEBS Lett. *208*, 194–198.

Sanchez, J., Hirst, T.R., and Uhlin, B.E. (1988a). Hybrid enterotoxin LTA::STa proteins and their protection from degradation by *in vivo* association with B-subunits of *Escherichia coli* heat-labile enterotoxin. Gene *64*, 265–275.

Sanchez, J., Svennerholm, A.M., and Holmgren, J. (1988b). Genetic fusion of a non-toxic heat-stable enterotoxin-related decapeptide antigen to cholera toxin B-subunit. FEBS Lett. *241*, 110–114.

Sanders, J.W., Putnam, S.D., Riddle, M.S., and Tribble, D.R. (2005). Military importance of diarrhoea: lessons from the Middle East. Curr. Opin. Gastroenterol. *21*, 9–14.

Santiago-Mateo, K., Zhao, M., Lin, J., Zhang, W., and Francis, D.H. (2012). Avirulent K88 (F4)+ *Escherichia coli* strains constructed to express modified enterotoxins protect young piglets from challenge with a virulent enterotoxigenic *Escherichia coli* strain that expresses the same adhesion and enterotoxins. Vet. Microbiol. *159*, 337–342.

Savarino, S.J., Fasano, A., Robertson, D.C., and Levine, M.M. (1991). Enteroaggregative *Escherichia coli* elaborate a heat-stable enterotoxin demonstrable in an *in vitro* rabbit intestinal model. J. Clin. Invest. *87*, 1450–1455.

Savarino, S.J., Fasano, A., Watson, J., Martin, B.M., Levine, M.M., Guandalini, S., and Guerry, P. (1993). Enteroaggregative *Escherichia coli* heat-stable enterotoxin 1 represents another subfamily of *E. coli* heat-stable toxin. Proc. Natl. Acad. Sci. U.S.A. *90*, 3093–3097.

Savarino, S.J., McVeigh, A., Watson, J., Cravioto, A., Molina, J., Echeverria, P., Bhan, M.K., Levine, M.M., and Fasano, A. (1996). Enteroaggregative *Escherichia coli* heat-stable enterotoxin is not restricted to enteroaggregative *E. coli*. J. Infect. Dis. *173*, 1019–1022.

Savarino, S.J., Brown, F.M., Hall, E., Bassily, S., Youssef, F., Wierzba, T., Peruski, L., El-Masry, N.A., Safwat, M., Rao, M., *et al.* (1998). Safety and immunogenicity of an oral, killed enterotoxigenic *Escherichia coli*-cholera toxin B subunit vaccine in Egyptian adults. J. Infect. Dis. *177*, 796–799.

Savarino, S.J., Hall, E.R., Bassily, S., Brown, F.M., Youssef, F., Wierzba, T.F., Peruski, L., El-Masry, N.A., Safwat, M., Rao, M., *et al.* (1999). Oral, inactivated, whole cell enterotoxigenic *Escherichia coli* plus cholera toxin B subunit vaccine: results of the initial evaluation in children. PRIDE Study Group. J. Infect. Dis. *179*, 107–114.

Savarino, S.J., Hall, E.R., Bassily, S., Wierzba, T.F., Youssef, F.G., Peruski, L.F., Jr., Abu-Elyazeed, R., Rao, M., Francis, W.M., El Mohamady, H., *et al.* (2002). Introductory evaluation of an oral, killed whole cell enterotoxigenic *Escherichia coli* plus cholera toxin B subunit vaccine in Egyptian infants. Pediatr. Infect. Dis. J. *21*, 322–330.

Savarino, S.J., Rao, M., Frenck, R., Abdel-Messih, I., Hall, E., Putnam, S., El-Mohamady, H., Wierzba, T., Pittner, B., Kamal, K., *et al.* (2003). Efficacy of an Oral, Inactivated Whole-Cell Enterotoxigenic *E. coli*/Cholera Toxin B Subunit Vaccine in Egyptian Infants. Paper presented at: 6th Annual Conference on Vaccine Research, National Foundation for Infectious Diseases (Arlington, VA).

Savarino, S.J., Poole, S., Sincock, S.A., McVeigh, A., Lee, L.H., Akay, Y. *et al.* (2005). One step beyond: A new approach for vaccines against enterotoxigenic *Escherichia coli*. In 40th Joint Conference on Cholera and Other Bacteria Enteric Infection and Bill and Melinda Gates Foundation Symposium on Vaccine Development.

Savarino, S.J., O'Dowd, A., Poole, S., Maciel, M., Rollenhagen, J., McVeigh, A., Porter, C., Ellingsworth, L., Glenn, G.M., Gourgeois, A.L., and Riddle, M. (2011). Advancement of an adhesin-based ETEC vaccine into clinical evaluation. 6th International Conference on Vaccines for Enteric Diseases. In The 6th International Conference on Vaccines for Enteric Diseases, VED 2011 (Cannes, France).

Scerpella, E.G., Sanchez, J.L., Mathewson, I.J., Torres-Cordero, J.V., Sadoff, J.C., Svennerholm, A.M., DuPont, H.L., Taylor, D.N., and Ericsson, C.D. (1995). Safety, Immunogenicity, and Protective Efficacy of the Whole-Cell/Recombinant B Subunit (WC/rBS) Oral Cholera Vaccine Against Travelers' Diarrhea. J. Travel Med. 2, 22–27.

Scharton-Kersten, T., Glenn, G.M., Vassell, R., Yu, J., Walwender, D., and Alving, C.R. (1999). Principles of transcutaneous immunization using cholera toxin as an adjuvant. Vaccine 17 (Suppl. 2), S37–43.

Sears, C.L., and Kaper, J.B. (1996). Enteric bacterial toxins: mechanisms of action and linkage to intestinal secretion. Microbiol. Rev. 60, 167–215.

Sellwood, R., Gibbons, R.A., Jones, G.W., and Rutter, J.M. (1975). Adhesion of enteropathogenic Escherichia coli to pig intestinal brush borders: the existence of two pig phenotypes. J. Med. Microbiol. 8, 405–411.

Seriwatana, J., Echeverria, P., Taylor, D.N., Rasrinaul, L., Brown, J.E., Peiris, J.S., and Clayton, C.L. (1988). Type II heat-labile enterotoxin-producing Escherichia coli isolated from animals and humans. Infect. Immun. 56, 1158–1161.

Sherlock, O., Vejborg, R.M., and Klemm, P. (2005). The TibA adhesin/invasin from enterotoxigenic Escherichia coli is self recognizing and induces bacterial aggregation and biofilm formation. Infect. Immun. 73, 1954–1963.

Sixma, T.K., Pronk, S.E., Kalk, K.H., Wartna, E.S., van Zanten, B.A., Witholt, B., and Hol, W.G. (1991). Crystal structure of a cholera toxin-related heat-labile enterotoxin from E. coli. Nature 351, 371–377.

Sixma, T.K., Terwissscha van Scheltinga, A.C., Kalk, K.H., Zhou, K., Wartna, E.S., and Hol, W.G. (1992). X-ray studies reveal lanthanide binding sites at the A/B5 interface of E. coli heat labile enterotoxin. FEBS Lett. 297, 179–182.

Sixma, T.K., Kalk, K.H., van Zanten, B.A., Dauter, Z., Kingma, J., Witholt, B., and Hol, W.G. (1993). Refined structure of Escherichia coli heat-labile enterotoxin, a close relative of cholera toxin. J. Mol. Biol. 230, 890–918.

Sjokvist Ottsjo, L., Flach, C.F., Clements, J., Holmgren, J., and Raghavan, S. (2013). The double mutant heat-labile toxin from Escherichia coli, LT (R192G/L211A) is an effective mucosal adjuvant for vaccination against Helicobacter pylori infection. Infect. Immun. 81, 1532–1540.

Smith, H.W., and Linggood, M.A. (1971). Observations on the pathogenic properties of the K88, Hly and Ent plasmids of Escherichia coli with particular reference to porcine diarrhoea. J. Med. Microbiol. 4, 467–485.

Spangler, B.D. (1992). Structure and function of cholera toxin and the related Escherichia coli heat-labile enterotoxin. Microbiol. Rev. 56, 622–647.

Spira, W.M., and Sack, R.B. (1982). Kinetics of early cholera infection in the removable intestinal tie-adult rabbit diarrhoea model. Infect. Immun. 35, 952–957.

Spira, W.M., Sack, R.B., and Froehlich, J.L. (1981). Simple adult rabbit model for Vibrio cholerae and enterotoxigenic Escherichia coli diarrhoea. Infect. Immun. 32, 739–747.

Steffen, R., Castelli, F., Dieter Nothdurft, H., Rombo, L., and Jane Zuckerman, N. (2005). Vaccination against enterotoxigenic Escherichia coli, a cause of travelers' diarrhoea. J. Travel Med. 12, 102–107.

Steinsland, H., Valentiner-Branth, P., Gjessing, H.K., Aaby, P., Molbak, K., and Sommerfelt, H. (2003a). Protection from natural infections with enterotoxigenic Escherichia coli: longitudinal study. Lancet 362, 286–291.

Steinsland, H., Valentiner-Branth, P., Grewal, H.M., Gaastra, W., Molbak, K.K., and Sommerfelt, H. (2003b). Development and evaluation of genotypic assays for the detection and characterization of enterotoxigenic Escherichia coli. Diagn. Microbiol. Infect. Dis. 45, 97–105.

Svennerholm, A.M., and Holmgren, J. (1992). Immunity to enterotoxin-producing bacteria. In Immunology of gastrointestinal diseases, MacDonald, T.T., ed. (Dordrecht, The Netherlands Kluwer Academic Publishers), pp. 227–246.

Svennerholm, A.M., and Holmgren, J. (1995). Oral vaccines against cholera and enterotoxigenic Escherichia coli diarrhoea. Adv. Exp. Med. Biol. 371B, 1623–1628.

Svennerholm, A.M., and Savarino, S.J. (2004). Oral inactivated whole cell B subunit combination vaccine against enterotoxigenic Escherichia coli. In New Generation Vaccines, 3rd edition, Levinem, K., Rappuoli, R., Liu, M.A., and Good, M.F., eds. (New York: Marcel Dekker).

Svennerholm, A.M., and Tobias, J. (2008). Vaccines against enterotoxigenic Escherichia coli. Expert Rev. Vaccines 7, 795–804.

Svennerholm, A.M., Lindvlad, M., Svennerholm, B., and Holmgren, J. (1988). Synthesis of nontoxic, antibody-binding Escherichia coli heat-stable enterotoxin (STa) peptides. FEMS Microbiol. Lett. 55, 23–28.

Svennerholm, A.M., Holmgren, J., and Sack, D.A. (1989). Development of oral vaccines against enterotoxinogenic Escherichia coli diarrhoea. Vaccine 7, 196–198.

Svennerholm, A.M., Wenneras, C., Holmgren, J., McConnell, M.M., and Rowe, B. (1990). Roles of different coli surface antigens of colonization factor antigen II in colonization by and protective immunogenicity of enterotoxigenic Escherichia coli in rabbits. Infect. Immun. 58, 341–346.

Svennerholm, A.M., Ahren, C., Wenneras, C., and Holmgren, J. (1991). Development of an oral vaccine against enterotoxigenic Escherichia coli diarrhoea. In Molecular pathogenesis of gastrointestinal infections, Wadstrom, P.H.M.T., Svennerholm, A.M., Wolf-Watz, H., eds. (Plenum Press, London, UK), pp. 287–294.

Syed, H.C., and Dubreuil, J.D. (2012). Escherichia coli STb toxin induces apoptosis in intestinal epithelial cell lines. Microb. Pathog. 53, 147–153.

Szu, S.C., Li, X.R., Schneerson, R., Vickers, J.H., Bryla, D., and Robbins, J.B. (1989). Comparative immunogenicities of Vi polysaccharide–protein conjugates composed of cholera toxin or its B subunit as a carrier bound to high- or lower-molecular-weight Vi. Infect. Immun. 57, 3823–3827.

Szu, S.C., Taylor, D.N., Trofa, A.C., Clements, J.D., Shiloach, J., Sadoff, J.C., Bryla, D.A., and Robbins, J.B. (1994). Laboratory and preliminary clinical characterization of Vi capsular polysaccharide–protein conjugate vaccines. Infect. Immun. 62, 4440–4444.

Tacket, C.O., and Levine, M.M. (1997). Vaccines against enerotoxigenic Escherichia coli infections. In New Generation vaccines, Levine, M.M., Woodrow, G.C., Kaper, J.B., and Cobon, G.S., ed. (Marcel Dekker, New York, NY), pp. 875–883.

Tacket, C.O., Maneval, D.R., and Levine, M.M. (1987). Purification, morphology, and genetics of a new fimbrial putative colonization factor of enterotoxigenic Escherichia coli O159:H4. Infect. Immun. 55, 1063–1069.

Tacket, C.O., Losonsky, G., Link, H., Hoang, Y., Guesry, P., Hilpert, H., and Levine, M.M. (1988). Protection by milk immunoglobulin concentrate against oral challenge with enterotoxigenic Escherichia coli. N. Engl. J. Med. 318, 1240–1243.

Tacket, C.O., Reid, R.H., Boedeker, E.C., Losonsky, G., Nataro, J.P., Bhagat, H., and Edelman, R. (1994). Enteral immunization and challenge of volunteers given enterotoxigenic E. coli CFA/II encapsulated in biodegradable microspheres. Vaccine 12, 1270–1274.

Tacket, C.O., Pasetti, M.F., Edelman, R., Howard, J.A., and Streatfield, S. (2004). Immunogenicity of recombinant LT-B delivered orally to humans in transgenic corn. Vaccine 22, 4385–4389.

Taniguchi, T., Fujino, Y., Yamamoto, K., Miwatani, T., and Honda, T. (1995). Sequencing of the gene encoding the major pilin of pilus colonization factor antigen III (CFA/III) of human enterotoxigenic Escherichia coli and evidence that CFA/III is related to type IV pili. Infect. Immun. 63, 724–728.

Taxt, A., Aasland, R., Sommerfelt, H., Nataro, J., and Puntervoll, P. (2010). Heat-stable enterotoxin of enterotoxigenic Escherichia coli as a vaccine target. Infect. Immun. 78, 1824–1831.

Taylor, J., Wilkins, M.P., and Payne, J.M. (1961). Relation of rabbit gut reaction to enteropathogenic Escherichia coli. Br. J. Exp. Pathol. 42, 43–52.

Thomas, L.V., McConnell, M.M., Rowe, B., and Field, A.M. (1985). The possession of three novel coli surface antigens by enterotoxigenic Escherichia coli strains positive for the putative colonization factor PCF8775. J. Gen. Microbiol. 131, 2319–2326.

Thompson, M.R., and Giannella, R.A. (1985). Revised amino acid sequence for a heat-stable enterotoxin produced by an Escherichia coli strain (18D) that is pathogenic for humans. Infect. Immun. 47, 834–836.

Tobias, J., and Svennerholm, A.M. (2012). Strategies to overexpress enterotoxigenic Escherichia coli (ETEC) colonization factors for the construction of oral whole-cell inactivated ETEC vaccine candidates. Appl. Microbiol. Biotechnol. 93, 2291–2300.

Tobias, J., Andersson, K., Bialik, A., and Cohen, D. (2008a). Preexisting antibodies to homologous colonization factors and heat-labile toxin in serum, and the risk to develop enterotoxigenic Escherichia coli-associated diarrhoea. Diagn. Microbiol. Infect. Dis. 60, 229–231.

Tobias, J., Lebens, M., Bolin, I., Wiklund, G., and Svennerholm, A.M. (2008b). Construction of non-toxic Escherichia coli and Vibrio cholerae strains expressing high and immunogenic levels of enterotoxigenic E. coli colonization factor I fimbriae. Vaccine 26, 743–752.

Turner, A.K., Terry, T.D., Sack, D.A., Londono-Arcila, P., and Darsley, M.J. (2001). Construction and characterization of genetically defined aro omp mutants of enterotoxigenic Escherichia coli and preliminary studies of safety and immunogenicity in humans. Infect. Immun. 69, 4969–4979.

Turner, A.K., Beavis, J.C., Stephens, J.C., Greenwood, J., Gewert, C., Thomas, N., Deary, A., Casula, G., Daley, A., Kelly, P., et al. (2006a). Construction and phase I clinical evaluation of the safety and immunogenicity of a candidate enterotoxigenic Escherichia coli vaccine strain expressing colonization factor antigen CFA/I. Infect. Immun. 74, 1062–1071.

Turner, S.M., Scott-Tucker, A., Cooper, L.M., and Henderson, I.R. (2006b). Weapons of mass destruction: virulence factors of the global killer enterotoxigenic Escherichia coli. FEMS Microbiol. Lett. 263, 10–20.

Turner, A.K., Stephens, J.C., Beavis, J.C., Greenwood, J., Gewert, C., Randall, R., Freeman, D., and Darsley, M.J. (2011). Generation and Characterization of a Live Attenuated Enterotoxigenic Escherichia coli Combination Vaccine Expressing Six Colonization Factors and Heat-Labile Toxin Subunit B. Clin. Vaccine Immunol. 18, 2128–2135.

Valvatne, H., Sommerfelt, H., Gaastra, W., Bhan, M.K., and Grewal, H.M. (1996). Identification and characterization of CS20, a new putative colonization factor of enterotoxigenic Escherichia coli. Infect. Immun. 64, 2635–2642.

Valvatne, H., Steinsland, H., Grewal, H.M., Molbak, K., Vuust, J., and Sommerfelt, H. (2004). Identification and molecular characterization of the gene encoding coli surface antigen 20 of enterotoxigenic Escherichia coli. FEMS Microbiol. Lett. 239, 131–138.

Veilleux, S., and Dubreuil, J.D. (2006). Presence of Escherichia coli carrying the EAST1 toxin gene in farm animals. Vet. Res. 37, 3–13.

Veilleux, S., Holt, N., Schultz, B.D., and Dubreuil, J.D. (2008). Escherichia coli EAST1 toxin toxicity of variants 17–2 and O 42. Comp. Immunol. Microbiol. Infect. Dis. 31, 567–578.

Vial, P.A., Robins-Browne, R., Lior, H., Prado, V., Kaper, J.B., Nataro, J.P., Maneval, D., Elsayed, A., and Levine, M.M. (1988). Characterization of enteroadherent-aggregative Escherichia coli, a putative agent of diarrhoeal disease. J. Infect. Dis. 158, 70–79.

Viboud, G.I., Binsztein, N., and Svennerholm, A.M. (1993). A new fimbrial putative colonization factor, PCFO20, in human enterotoxigenic Escherichia coli. Infect. Immun. 61, 5190–5197.

Viboud, G.I., Jonson, G., Dean-Nystrom, E., and Svennerholm, A.M. (1996). The structural gene encoding human enterotoxigenic Escherichia coli PCFO20 is homologous to that for porcine 987P. Infect. Immun. 64, 1233–1239.

Wai, S.N., Takade, A., and Amako, K. (1995). The release of outer membrane vesicles from the strains of enterotoxigenic *Escherichia coli*. Microbiol. Immunol. *39*, 451–456.

Walker, R.I., Steele, D., and Aguado, T. (2007). Analysis of strategies to successfully vaccinate infants in developing countries against enterotoxigenic *E. coli* (ETEC) disease. Vaccine *25*, 2545–2566.

Weikel, C.S., Tiemens, K.M., Moseley, S.L., Huq, I.M., and Guerrant, R.L. (1986). Species specificity and lack of production of STb enterotoxin by *Escherichia coli* strains isolated from humans with diarrhoeal illness. Infect. Immun. *52*, 323–325.

Wenneras, C., and Erling, V. (2004). Prevalence of enterotoxigenic *Escherichia coli*-associated diarrhoea and carrier state in the developing world. J. Health Popul. Nutr. *22*, 370–382.

Wenneras, C., Svennerholm, A.M., Ahren, C., and Czerkinsky, C. (1992). Antibody-secreting cells in human peripheral blood after oral immunization with an inactivated enterotoxigenic *Escherichia coli* vaccine. Infect. Immun. *60*, 2605–2611.

Westermark, M., Oscarsson, J., Mizunoe, Y., Urbonaviciene, J., and Uhlin, B.E. (2000). Silencing and activation of ClyA cytotoxin expression in *Escherichia coli*. J. Bacteriol. *182*, 6347–6357.

Whipp, S.C., Moon, H.W., Kemeny, L.J., and Argenzio, R.A. (1985). Effect of virus-induced destruction of villous epithelium on intestinal secretion induced by heat-stable *Escherichia coli* enterotoxins and prostaglandin E1 in swine. Am. J. Vet. Res. *46*, 637–642.

Whipp, S.C., Moseley, S.L., and Moon, H.W. (1986). Microscopic alterations in jejunal epithelium of 3-week-old pigs induced by pig-specific, mouse-negative, heat-stable *Escherichia coli* enterotoxin. Am. J. Vet. Res. *47*, 615–618.

Whipp, S.C., Kokue, E., Morgan, R.W., Rose, R., and Moon, H.W. (1987). Functional significance of histologic alterations induced by *Escherichia coli* pig-specific, mouse-negative, heat-stable enterotoxin (STb). Vet. Res. Commun. *11*, 41–55.

WHO (2006). Future directions for research on enterotoxigenic *Escherichia coli* vaccines for developing countries. Wkly. Epidemiol. Rec. *81*, 97–107.

Wolf, M.K. (1997). Occurrence, distribution, and associations of O and H serogroups, colonization factor antigens, and toxins of enterotoxigenic *Escherichia coli*. Clin. Microbiol. Rev. *10*, 569–584.

Yamamoto, T., and Echeverria, P. (1996). Detection of the enteroaggregative *Escherichia coli* heat-stable enterotoxin 1 gene sequences in enterotoxigenic *E. coli* strains pathogenic for humans. Infect. Immun. *64*, 1441–1445.

Yamamoto, T., and Nakazawa, M. (1997). Detection and sequences of the enteroaggregative *Escherichia coli* heat-stable enterotoxin 1 gene in enterotoxigenic *E. coli* strains isolated from piglets and calves with diarrhoea. J. Clin. Microbiol. *35*, 223–227.

Yamamoto, T., and Taneike, I. (2000). The sequences of enterohemorrhagic *Escherichia coli* and *Yersinia pestis* that are homologous to the enteroaggregative *E. coli* heat-stable enterotoxin gene: cross-species transfer in evolution. FEBS Lett. *472*, 22–26.

Yamamoto, T., Wakisaka, N., Sato, F., and Kato, A. (1997). Comparison of the nucleotide sequence of enteroaggregative *Escherichia coli* heat-stable enterotoxin 1 genes among diarrhoea-associated *Escherichia coli*. FEMS Microbiol. Lett. *147*, 89–95.

Yamanaka, H., Kameyama, M., Baba, T., Fujii, Y., and Okamoto, K. (1994). Maturation pathway of *Escherichia coli* heat-stable enterotoxin I: requirement of DsbA for disulfide bond formation. J. Bacteriol. *176*, 2906–2913.

Yamanaka, H., Nomura, T., Fujii, Y., and Okamoto, K. (1998). Need for TolC, an *Escherichia coli* outer membrane protein, in the secretion of heat-stable enterotoxin I across the outer membrane. Microb. Pathog. *25*, 111–120.

Yamasaki S., I.H., Hirayama, T., Takeda, Y., and Shimonishi, Y. (1988). Effects on the activity of amino acids replacement at positions 12, 13, and 14 heat-stable enterotoxin (STh) by chemical synthesis. In 24th Joint Conf US-Japan Cooperative Med Sci Program on Cholera and Related Diarrheal Disease Panel (Tokyo, Japan), p. 42.

Yang, X., Thornburg, T., Holderness, K., Suo, Z., Cao, L., Lim, T., Avci, R., and Pascual, D.W. (2011). Serum antibodies protect against intraperitoneal challenge with enterotoxigenic *Escherichia coli*. J. Biomed. Biotechnol. *2011*, 632396.

Zajacova, Z.S., Faldyna, M., Kulich, P., Kummer, V., Maskova, J., and Alexa, P. (2012a). Experimental infection of gnotobiotic piglets with *Escherichia coli* strains positive for EAST1 and AIDA. Vet. Immunol. Immunopathol. *15*, 176–182.

Zajacova, Z.S., Konstantinova, L., and Alexa, P. (2012b). Detection of virulence factors of *Escherichia coli* focused on prevalence of EAST1 toxin in stool of diarrheic and non-diarrheic piglets and presence of adhesion involving virulence factors in astA positive strains. Vet. Microbiol. *154*, 369–375.

Zaman, K., Dang, D.A., Victor, J.C., Shin, S., Yunus, M., Dallas, M.J., Podder, G., Vu, D.T., Le, T.P., Luby, S.P., et al. (2010). Efficacy of pentavalent rotavirus vaccine against severe rotavirus gastroenteritis in infants in developing countries in Asia: a randomised, double-blind, placebo-controlled trial. Lancet *376*, 615–623.

Zeng, W., Azzopardi, K., Hocking, D., Wong, C.Y., Robevska, G., Tauschek, M., Robins-Browne, R.M., and Jackson, D.C. (2012). A totally synthetic lipopeptide-based self-adjuvanting vaccine induces neutralizing antibodies against heat-stable enterotoxin from enterotoxigenic *Escherichia coli*. Vaccine *30*, 4800–4806.

Zhang, C., Rausch, D., and Zhang, W. (2009). Little heterogeneity among genes encoding heat-labile and heat-stable toxins of enterotoxigenic *Escherichia coli* strains isolated from diarrheal pigs. Appl. Environ. Microbiol. *75*, 6402–6405.

Zhang, W., and Francis, D.H. (2010). Genetic fusions of heat-labile toxoid (LT) and heat-stable toxin b (STb) of porcine enterotoxigenic *Escherichia coli* elicit protective anti-LT and anti-STb antibodies. Clin. Vaccine Immunol. *17*, 1223–1231.

Zhang, W., and Sack, D.A. (2012). Progress and hurdles in the development of vaccines against enterotoxigenic *Escherichia coli* in humans. Expert Rev. Vaccines *11*, 677–694.

Zhang, W., Berberov, E.M., Freeling, J., He, D., Moxley, R.A., and Francis, D.H. (2006). Significance of heat-stable and heat-labile enterotoxins in porcine colibacillosis in an additive model for pathogenicity studies. Infect. Immun. *74*, 3107–3114.

Zhang, W., Zhao, M., Ruesch, L., Omot, A., and Francis, D. (2007). Prevalence of virulence genes in *Escherichia coli* strains recently isolated from young pigs with diarrhoea in the USA. Vet. Microbiol. *123*, 145–152.

Zhang, W., Robertson, D.C., Zhang, C., Bai, W., Zhao, M., and Francis, D.H. (2008). *Escherichia coli* constructs expressing human or porcine enterotoxins induce identical diarrhoeal diseases in a piglet infection model. Appl. Environ. Microbiol. *74*, 5832–5837.

Zhang, W., Zhang, C., Francis, D.H., Fang, Y., Knudsen, D., Nataro, J.P., and Robertson, D.C. (2010). Genetic fusions of heat-labile (LT) and heat-stable (ST) toxoids of porcine enterotoxigenic *Escherichia coli* elicit neutralizing anti-LT and anti-STa antibodies. Infect. Immun. *78*, 316–325.

Current books of interest

Full details at www.caister.com

Index

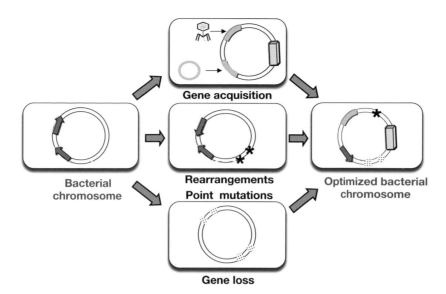

Gene acquisition

Rearrangements
Point mutations

Bacterial
chromosome

Optimized bacterial
chromosome

Gene loss

Figure 3.1 Mechanisms contributing to bacterial genome plasticity and genome optimization. The acquisition of mobile and accessory genetic elements such as bacteriophages, plasmids and genomic islands contributes to the evolution of pathogenic *E. coli* from commensal variants. DNA rearrangements, deletions and point mutations may lead to a reduced expression of virulence genes, inactivation or loss.

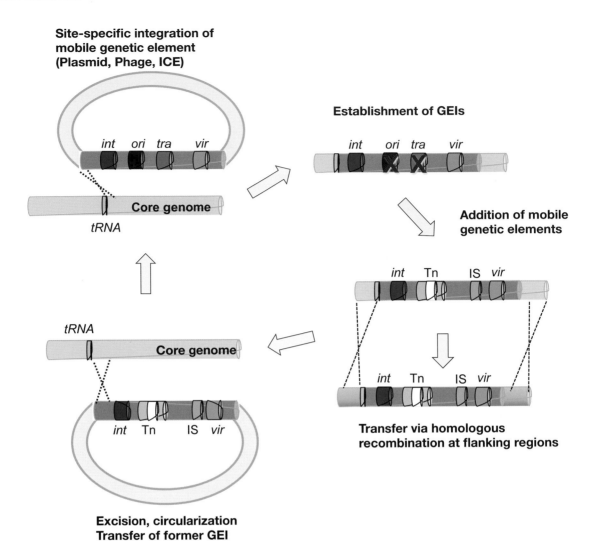

Figure 3.2 Contribution of mobile genetic elements to the evolution of new pathogenic *E. coli* variants. Acquisition of mobile genetic elements (e.g. plasmids, bacteriophages, genomic islands, integrative and conjugative elements, integrons) can result in the chromosomal integration by site-specific recombination. Genomic islands are frequently inserted at the 5′-end of tRNA loci. Genome plasticity due to further gene acquisition, gene loss, or DNA rearrangements results in modification and further evolution of the genomic island. Upon conjugation and homologous recombination between flanking regions, large chromosomal regions including genomic islands can be transferred to suitable recipients. Alternatively, islands can be excised by site-specific recombination and the circularized islands can then be transferred to another recipient.

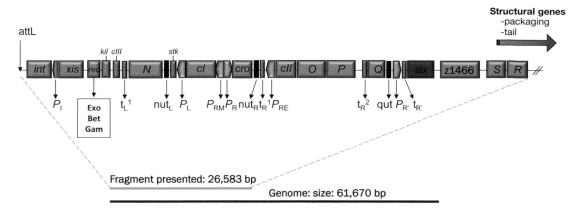

Figure 4.2 Genetic structure of the Stx-phage 933W showing the regulatory region and other relevant genes including *stx*. Not drawn to scale. The scheme was constructed using data from Tyler *et al.*, 2004.

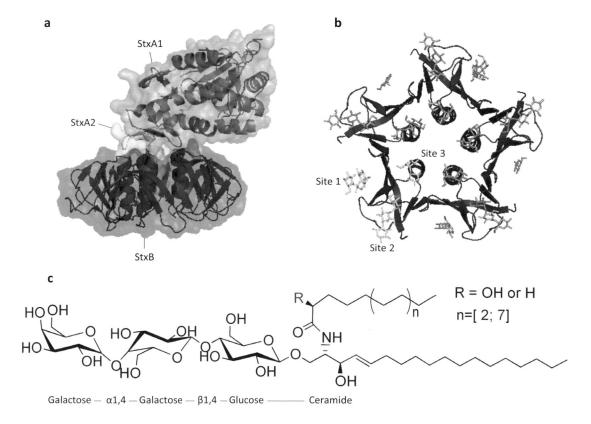

Figure 5.1 The structure of Shiga toxin and its cellular receptor globotriaosylceramide (Gb3). (a) A ribbon diagram of Shiga toxin, consisting of one A-subunit (StxA), which can be cleaved into the fragments A1 (orange) and A2 (yellow), and the homopentameric B-subunit (StxB) that consists of five B-fragments (red). (b) A ribbon diagram of the StxB subunit from the membrane-oriented surface, highlighting one B-fragment (blue) with three Gb3-binding sites. Gb3 is shown as a stick representation. Note the central pore that is lined by α–helices. (c) Chemical structure of the glycosphingolipid Gb3 composed of Galα1–4Galβ1–4GlcCer.

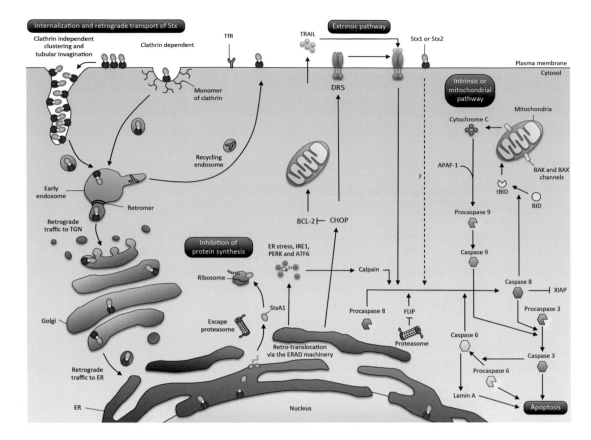

Figure 5.2 (Left part) Intracellular trafficking of Shiga toxins. After binding to its receptor, Stx induces clathrin-dependent and –independent endocytosis. The toxin preferentially localizes to retrograde tubules in early endosomes that are formed in a clathrin-dependent manner and then processed by scission in a retromer-dependent manner. Contrary to the transferrin receptor (TfR) that is recycled back to the plasma membrane, Stx bypasses the late endocytic pathway and is transferred to the *trans*–Golgi network (TGN) and then to the endoplasmic reticulum (ER). There, the A1-fragment (StxA1) utilizes the ER-associated degradation (ERAD) machinery to facilitate its retro-translocation into the host cell cytosol where it inhibits the protein synthesis. (Right part) Apoptosis pathways induced by Shiga toxins. This sketch sums up findings from different cell types, where the apoptosis initiator caspase 8 is activated, resulting in the activation of the executioner caspase 3. In most cases, the enzymatic activity of Shiga holotoxins is required for the induction of apoptosis. APAF-1, apoptotic protease-activating factor 1; ATF6, cyclic AMP-dependent transcription factor 6; BAK, BCL-2-homologous agonist/killer; BCL-2, B cell lymphoma 2; BID, BH3-interacting domain; CHOP, C/EBP-homologous protein; DR5, death receptor 5; ER, endoplasmic reticulum; FLIP, FLICE-like inhibitory protein; Gb3, globotriaosylceramide; PERK, Protein kinase-like ER kinase; tBID, truncated BID; TRAIL, TNF-related apoptosis-inducing ligand; XIAP, X-linked inhibitor of apoptosis protein.

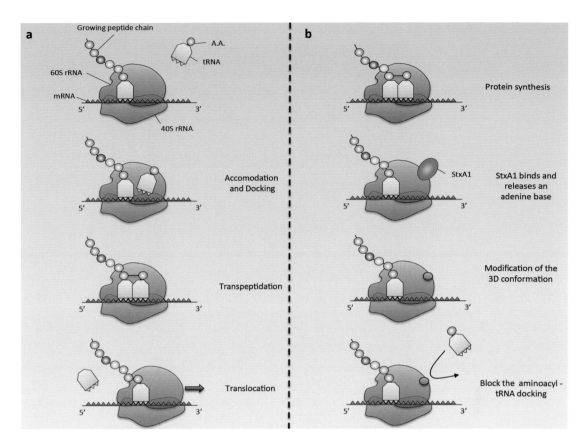

Figure 5.3 Shiga toxin-induced inhibition of protein synthesis. (a) A simplified view on protein synthesis. The 60S- and the 40S-ribosomal subunits assemble around the mRNA and recognize an amino-acyl tRNA specific of the codon. After the accommodation and the docking, the amino-acyl moiety on the tRNA will bind to the growing peptide chain triggered by GTP and elongation factors. Finally, the ribosome moves forward on the mRNA from 5′ to 3′, releasing a free tRNA. (b) StxA1 inhibits the elongation during protein synthesis by binding to the 28S rRNA on the large 60S-subunit of the ribosome and cleaving an adenine base at the position 4324. Thus, the folding of the ribosome changes, modifying the docking site of the amino-acyl tRNA. As consequence, no amino-acyl tRNA could be added to the growing peptide chain resulting in the inhibition of protein synthesis.

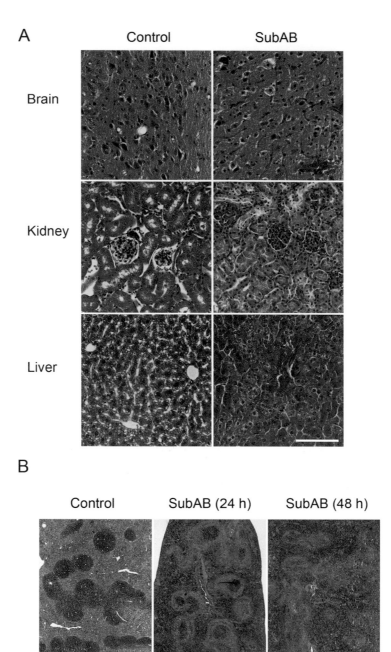

Figure 6.1 SubAB-induced pathology. (A) Haematoxylin–eosin (HE)-stained brain (medulla), kidney and liver sections showing microthrombi and haemorrhage after SubAB injection. Mice were injected intra-peritoneally with 5 μg SubAB or PBS (control) and examined 72 h after treatment (scale bar, 0.1 mm). (B) HE-stained mouse spleen sections showing leucocyte depletion from the white pulp at 24 and 48 h after SubAB injection (scale bar, 0.1 mm).

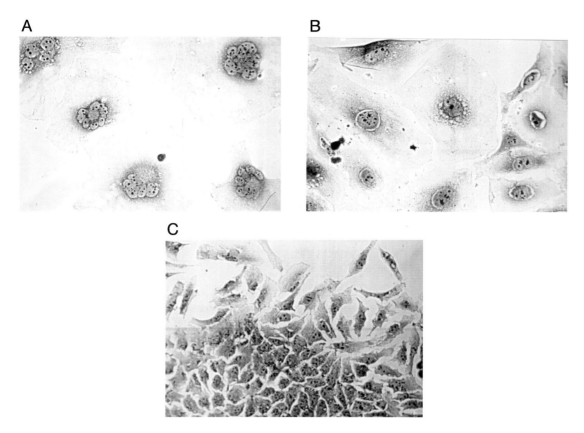

Figure 7.1 CNF- and CDT-specific cytopathic effects on HeLa cells. HeLa cell monolayers were infected with (A) CNF-1 and (B) CDT-IV-producing *E. coli* strains. (C) Untreated HeLa cells served as control. After 3 days' incubation, infecting material was removed, HeLa cells were washed and stained with Giemsa. Morphological changes were investigated by light microscope. The original magnification was 40. Photographs were taken by I. Tóth.

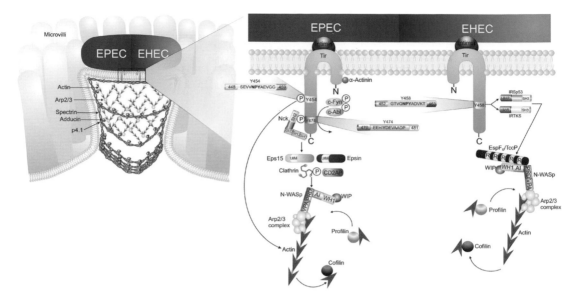

Figure 10.2 EPEC and EHEC recruitment of host proteins at pedestals. An illustration of the organization of crucial components found at EPEC/EHEC pedestals. At sites of A/E lesion formation EPEC and EHEC both generate pedestals and disassemble the localized microvilli. A major structural component of A/E lesions is the highly branched actin network. Here, the Arp2/3 complex is shown at the branch points where new actin filaments are formed; oriented at 70° angles from the pre-existing actin filaments. The pointed (slow-growing) ends of actin filaments are oriented downwards while the barbed (fast-growing) ends points towards the apex of the pedestal. These filaments extend from the sub-apical region of the pedestals to ~2/3 down the pedestal stalk. Spectrin makes-up the basal 2/3 of the pedestal; overlapping with actin in the middle 1/3 of the structures. Spectrin and the associated cytoskeletal elements (adducin and p4.1) also cage the actin along the pedestal membrane. However, adducin does not localize directly beneath the plasma membrane where EPEC/EHEC attach. A closer look at the apical region of the pedestals shows that EPEC and EHEC use the bacterial outer membrane protein intimin and the translocated intimin receptor, Tir, (located in the host plasma membrane) to firmly anchor the bacteria to the surface of host cells. α-actinin binds to the N-terminal arm of EPEC Tir, but it is the C-terminus cytoplasmic domain that is critical for the majority of pedestal formation. Phosphorylation of specific tyrosine (Y) residues by the host tyrosine kinase family members, c-Fyn and c-Abl, on Tir is needed for protein recruitment to ultimately control N-WASp and the Arp2/3 complex for actin polymerization beneath the attached bacteria. EPEC recruits Nck directly to Tir, which is dependent Y474 phosphorylation. Alternatively, phosphorylation of Y454 on Tir can signal actin nucleation in a Nck-independent manner. Downstream of Nck the endocytic proteins Eps15, epsin, CD2AP and clathrin are targeted to the apical region of EPEC pedestals. The ubiquitin interacting motifs (UIM) of Eps15 and epsin are sufficient for localization to pedestals. Unlike EPEC, EHEC does not use Nck or endocytic proteins for pedestal formation. Instead EHEC recruits IRSp53 or IRTKS to Y458 of EHEC Tir. This triggers the docking of a second bacterial protein, EspF_u/TccP to the forming apical complex prior to N-WASp and Arp2/3. Actin filaments are polymerized through Arp2/3-dependent nucleation. Additional abbreviations used in this figure to indicate specific regions used for protein binding include: Src homology 2 (SH2) and 3 (SH3) domain, verprolin, cofilin, acidic (VPA) domain, proline-rich domain (PRD), autoinhibitory (AI) domain, WASP homology 1 (WH1) domain, proline-rich repeats (R), IRSp53-MIM homology (I-BAR) domain.

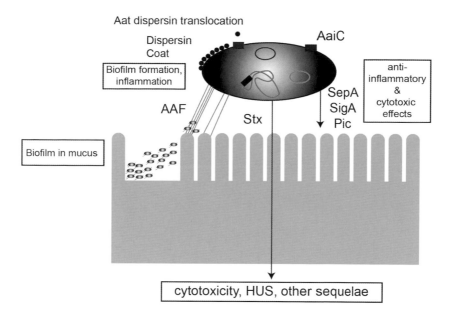

Figure 12.1 Proposed pathogenesis model for the increased virulence associated with the EAEC O104:H4 German outbreak strain. The O104:H4 outbreak strain harbours a Stx2-phage (black box) and a plasmid containing the CTX-M beta lactamase gene (blue), as well as EAEC virulence plasmid pAA (red). The master regulator AggR controls the expression of AAF/I, a dispersin surface coat protein, a dispersin translocator (encoded by *aatPABCD*) as well as a type VI secretion system within a chromosomal island called AAI (encoded by *aaiA-P*). The first stage of infection includes adherence to the intestinal mucosa by virtue of aggregative adherence fimbriae (AAF) followed by stimulation of mucus production, forming a biofilm on the surface of the mucosa; and finally, toxicity (SepA, SigA and Pic) to the mucosa, manifested by cytokine release, cell exfoliation, intestinal secretion and induction of mucosal inflammation. Collectively these factors possibly enable the outbreak strain to bind to and remain closely associated with the intestinal epithelium, which may promote increased uptake of Stx2 into the bloodstream. HUS, haemolytic–uraemic syndrome.

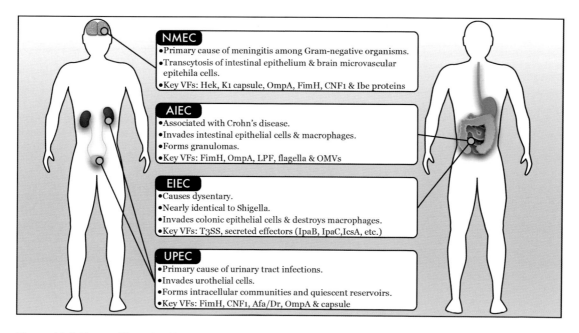

Figure 13.1 Types of invasive *E. coli* strains with associated virulence factors (VFs) used to enter target host cells.

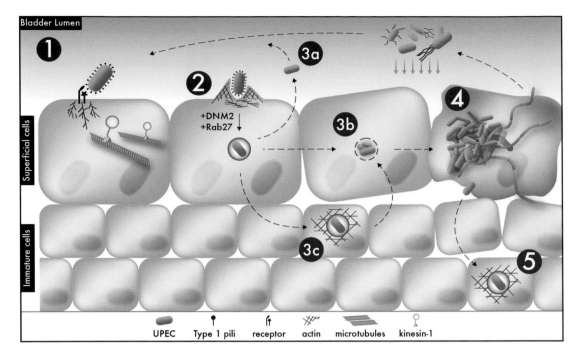

Figure 13.2 UPEC invasion of the urothelium. (**1**) UPEC bind receptors on bladder cells via type 1 pili, triggering actin cytoskeletal rearrangements that are modulated by microtubules, kinesin-1 and many other host factors. (**2**) The host plasma membrane zippers around adherent bacteria, eventually delivering UPEC into a late endosome-like compartment via a DNM2- and Rab27-dependent pathway. Internalized UPEC can be (**3a**) returned back to the extracellular environment via a cAMP-dependent mechanism, (**3b**) exit the endosomal compartment and enter the host cytosol, or (**3c**) remain within actin-bound endosomes in a quiescent state. These reservoirs can persist within either immature urothelial cells or within actin-rich regions of the superficial cells. (**4**) Upon entering the host cytosol UPEC can replicate profusely, forming IBCs. Bacteria are released from IBCs, often exuded as long filamentous forms, and can disseminate further within the urinary tract. Exfoliation of urothelial cells limits the lifespan of IBCs and (**5**) provides UPEC with access to deeper tissue layers.

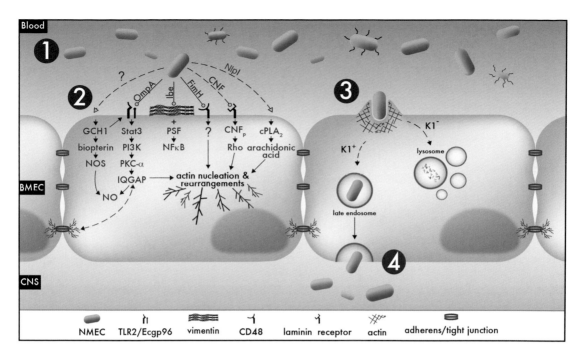

Figure 13.3 Transcytosis of NMEC across the blood brain barrier. NMEC enters the bloodstream via the gastrointestinal tract and (**1**) causes high levels of bacteria. (**2**) Multiple factors expressed by NMEC interact with host receptors and regulatory proteins to promote bacterial attachment and internalization by BMECs, triggering in the process NFκB activation and (**3**) actin cytoskeletal rearrangements. Internalized NMEC are trafficked into late endosomal compartments and (**4**) eventually delivered into the CNS. Bacteria lacking K1 capsule are degraded.

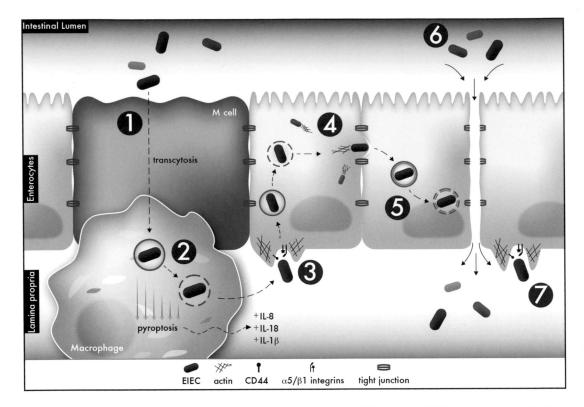

Figure 13.4 EIEC/*Shigella* invasion of the colonic epithelium. (**1**) Bacteria initially cross the epithelium via M cells. (**2**) Within the submucosa, *Shigella* is internalized by macrophages. The secretion of T3SS effector molecules enables *Shigella* to rupture endosomal membranes and kill macrophages via pyroptosis. Inflammatory cytokines released by infected macrophages recruit neutrophils that damage the epithelial barrier, allowing more bacteria to enter the submucosa where they can (**3**) invade enterocytes and (**4**) enter the host cell cytosol via secretion of T3SS effectors. Within enterocytes, *Shigella* can multiply and disseminate, propelling itself within and (**5**) between host cells by polymerizing actin tails. This can disrupt tight junctions, further compromising barrier function and (**6**) providing lumenal bacteria with greater access to the submucosa where they (**7**) can perpetuate the invasion process.